SAS® ESSENTIALS

SAS® ESSENTIALS

Mastering SAS for Data Analytics

Third Edition

ALAN C. ELLIOTT and WAYNE A. WOODWARD

This edition first published 2023
© 2023 John Wiley and Sons, Inc.

Edition History
John Wiley & Sons (2e, 2015)

All rights reserved. No part of this publication may be reproduced, stored in a retrieval system, or transmitted, in any form or by any means, electronic, mechanical, photocopying, recording or otherwise, except as permitted by law. Advice on how to obtain permission to reuse material from this title is available at http://www.wiley.com/go/permissions.

The right of Alan C. Elliott and Wayne A. Woodward to be identified as the authors of this work has been asserted in accordance with law.

Registered Office
John Wiley & Sons, Inc., 111 River Street, Hoboken, NJ 07030, USA

Editorial Office
111 River Street, Hoboken, NJ 07030, USA

For details of our global editorial offices, customer services, and more information about Wiley products visit us at www.wiley.com.

Wiley also publishes its books in a variety of electronic formats and by print-on-demand. Some content that appears in standard print versions of this book may not be available in other formats.

Trademarks: Wiley and the Wiley logo are trademarks or registered trademarks of John Wiley & Sons, Inc. and/or its affiliates in the United States and other countries and may not be used without written permission. All other trademarks are the property of their respective owners. John Wiley & Sons, Inc. is not associated with any product or vendor mentioned in this book.

Limit of Liability/Disclaimer of Warranty
In view of ongoing research, equipment modifications, changes in governmental regulations, and the constant flow of information relating to the use of experimental reagents, equipment, and devices, the reader is urged to review and evaluate the information provided in the package insert or instructions for each chemical, piece of equipment, reagent, or device for, among other things, any changes in the instructions or indication of usage and for added warnings and precautions. While the publisher and authors have used their best efforts in preparing this work, they make no representations or warranties with respect to the accuracy or completeness of the contents of this work and specifically disclaim all warranties, including without limitation any implied warranties of merchantability or fitness for a particular purpose. No warranty may be created or extended by sales representatives, written sales materials or promotional statements for this work. The fact that an organization, website, or product is referred to in this work as a citation and/or potential source of further information does not mean that the publisher and authors endorse the information or services the organization, website, or product may provide or recommendations it may make. This work is sold with the understanding that the publisher is not engaged in rendering professional services. The advice and strategies contained herein may not be suitable for your situation. You should consult with a specialist where appropriate. Further, readers should be aware that websites listed in this work may have changed or disappeared between when this work was written and when it is read. Neither the publisher nor authors shall be liable for any loss of profit or any other commercial damages, including but not limited to special, incidental, consequential, or other damages.

Library of Congress Cataloging-in-Publication Data Applied for:
Paperback ISBN: 9781119901617

Cover Design: Wiley
Cover Image: © piranka/Getty Images

Typeset in 10/12pt Times by Straive, Pondicherry, India.

CONTENTS

About the Authors' Website — vii

PART I   DATA MANIPULATION AND THE SAS® PROGRAMMING LANGUAGE — 1

1   Getting Started — 3
2   Getting Data into SAS® — 20
3   Reading, Writing, and Importing Data — 53
4   Preparing Data for Analysis — 79
5   Preparing to Use SAS® Procedures — 117
6   Preparing Data for Analysis — 141
7   SAS® Advanced Programming Topics Part 2 — 171
8   Controlling Output Using ODS — 205
9   Introduction to PROC SQL — 225

PART II	**STATISTICAL ANALYSIS USING SAS PROCEDURES**	**251**
10	Evaluating Quantitative Data	253
11	Analyzing Counts and Tables	280
12	Comparing Means Using *T*-Tests	306
13	Correlation and Regression	319
14	Analysis of Variance	347
15	Analysis of Variance, Part II	363
16	Nonparametric Analysis	384
17	Logistic Regression	397
18	Factor Analysis	414
19	Creating Custom Graphs	429
20	Creating Custom Reports	467
Appendix A	Options Reference	498
Appendix B	SAS® Function Reference	507
Appendix C	Choosing a SAS® Procedure	528
Appendix D	SAS Code Examples	531
Appendix E	Using SAS® Ondemand for Academics with SAS Essentials	539
Appendix F	SAS® Example BASE Exam Questions	545
Index		557

ABOUT THE AUTHORS' WEBSITE

This book is accompanied by the authors' website:

https://www.alanelliott.com/SASED3/

This site will contain links to:

- The data and program files referenced in the book
- PowerPoint files for instructors
- Teaching videos that relate to the book's content
- Quick tip pages referenced in the book
- Future help or exercises related to the book's content

PART I

DATA MANIPULATION AND THE SAS® PROGRAMMING LANGUAGE

1

GETTING STARTED

LEARNING OBJECTIVES

- To be able to use the SAS® software program in a Windows environment.
- To understand the basic information about getting data into SAS and running a SAS program.
- To be able to run a simple SAS program.

Data analytics is the process of gathering and using data to understand some process, predict a future outcome, or make a decision. The SAS system is a powerful software program designed to give data analysts a wide variety of both data management and data analysis capabilities. Although SAS has millions of users worldwide, it is not the simplest program to learn. With that in mind, we've created this book to provide you with a straightforward approach to learning SAS that can help you surmount the learning curve and successfully use SAS for data analysis.

Two main concepts are involved in learning SAS: (1) how to get data into SAS and ready for analysis using the SAS programming language and (2) how to perform the desired data analysis.

- Part 1 (Chapters 2–9) shows how to get data into SAS and prepare it for analysis.
- Part 2 (Chapters 10–20) shows how to use data to perform statistical analyses, create graphs, and produce reports.

SAS® Essentials: Mastering SAS for Data Analytics, Third Edition. Alan C. Elliott and Wayne A. Woodward.
© 2023 John Wiley & Sons, Inc. Published 2023 by John Wiley & Sons, Inc.
Companion website: https://www.alanelliott.com/SASED3/

This chapter introduces you to the SAS system's use in the Microsoft Windows environment and provides numerous Hands-On examples of how to use SAS to analyze data.

1.1 USING SAS IN A WINDOWS ENVIRONMENT

SAS runs on a number of computer platforms (operating systems) including mainframes and personal computers whose operating systems are UNIX, Linux, or Windows. This book is based on using SAS in a Windows environment where you have the software installed on your local computer. The vast majority of the content in this book will apply to any SAS computer environment. However, we will include occasional references to differences that are present in other operating systems. Most of the differences between versions have to do with file references. Moreover, there are multiple ways to use SAS, notably the programming or Enterprise Guide approaches. This book teaches the programming approach that offers the user the most flexibility. Before discussing the SAS program, we'll review some of the basic file-handling features of Windows. There is a free version of SAS called OnDemand for Academics: SAS Studio for Learning Purposes. Appendix E discusses how to install this version and how it can be used to run many of the examples in this book.

1.1.1 Creating a Folder for Storing Your SAS Files

As there are several versions of Windows currently in use, we present general guidelines that should work in any one of them. To follow our examples, we recommend that you store the data and SAS files provided in this book in a folder named SASDATA. In fact, we recommend that you do this NOW!. The examples in this text will assume that all example files used in this book are in a folder whose file path is C:\SASDATA. You can adapt the examples by using an appropriate file path for other operating systems. Table 1.1 shows several ways to reference a file.

In your own data analysis, you may choose to store your SAS files in a folder with a name that makes sense to you, such as C:\RESEARCH. You can also create subfolders for each analysis; for example, C:\RESEARCH\SMITH or C:\RESEARCH\JONES.

To copy the example files of this book to the C:\SASDATA folder on your own computer, perform the following:

1. Download the example files from the following web site: http://www.alanelliott.com/sas
2. Follow the installation instructions provided on the web site for the 3rd edition of this book. Doing so creates the C:\SASDATA folder on your computer and copies the example files to that folder. (The web site also includes updates concerning the information in this book and other helpful resources.)

TABLE 1.1 Accessing Files in Various Operating Systems

Operating System	Example File Reference
Windows	C:\SASATA\myfile
Windows/Cytrix	\\CLIENT\C$\SASDATA\myfile
UNIX	/home/sasdata/myfile
Open VMS	[username.sasdata]myfile
z/OS	SASDATA.myfile

You can also use SAS on an Apple Mac. To do so, we recommend that you install a virtual Windows environment using commercially available programs such as Bootcamp or Parallels and install SAS in the Windows environment. You can also run SAS OnDemand for Academics on Windows, Mac, or Linux environment. (Refer to Appendix E in this book or search the web for "SAS OnDemand for Academics" for more information.

> The examples in the book are designed to use data and sample programs in the folder C:\SASDATA – in other words, in a folder on your computer's hard drive. You can also put the example files on any rewritable medium such as a flash drive or on a network drive – just remember to adjust the file names and file paths given in the examples in this book accordingly if you store your files in a location other than C:\SASDATA.

1.1.2 Beginning the SAS Program

Since there are multiple ways to use SAS, launching SAS may differ according to your installation and operating system. For a typical Windows environment, where there is a SAS icon (or tile) on your desktop, simply double-click it to launch SAS. If the SAS icon is not on your desktop, go to the Start button and select Start → Programs → SAS (English) to launch SAS. Henceforth, we will refer to this simply as the SAS icon.

(We don't recommend that you click on a .SAS file to launch SAS because it may not open SAS in the way you want. Doing so may open SAS in the Enhanced Editor, Enterprise Guide, or Universal Viewer. (It depends on how SAS was installed on your computer.) This book uses the SAS Windows Enhanced Editor as the primary interface for using SAS. Refer to SAS documentation for information about Enterprise Guide.

1.1.3 Understanding the SAS Windows Interface

Once you begin the SAS program, you will see a screen similar to that shown in Figure 1.1. (The SAS program appearance may be slightly different depending on which versions of Windows and SAS you're using.)

The normal opening SAS screen is divided into three visible windows. The top right window is the **Log Window**, and at the bottom right is the **Editor**. The third window, which appears as a vertical element on the left, is called the **SAS Explorer/Results** window. There are other SAS windows that are not visible on the normal opening screen. These include the **Graph** and **Results Viewer** windows. To open a SAS window that is not currently visible, click its tab at the bottom of the screen. (The Output tab relates to a window that is not often used and we will not discuss it here.) The following is a brief description of the windows that are used in this text.

> **Editor:** Also called the **Enhanced Editor** or **Windows Programming Editor** (WPGM), this is the area where you write SAS code. It is like a simple word processor. When you open a previously saved SAS program, its contents will appear in this window. SAS code is stored in plain ASCII text, so files saved in the ASCII format from any other editor or word processor may be easily opened in this editor. You can also copy (or cut) text from another editor or word processor and paste it into the **Editor Window**.
>
> **Log:** When you run a SAS program, a report detailing how (and if) the program ran appears in the **Log Window**. Typically, when you run a SAS program, you first look at the contents of the **Log Window** to see if any errors in the program were reported. The **Log Window** highlights errors in red. You should also look for warnings and

Figure 1.1 Initial SAS screen. (Source: Created with SAS® software, Version 9.4. Copyright 2022, SAS Institute Inc., Cary, NC, USA. All Rights Reserved. Reproduced with permission of SAS Institute Inc., Cary, NC)

other notes in the **Log Window**, which tell you that some part of your program may not have run correctly.

SAS Explorer/Results: This window appears at the left of the screen and contains two tabs (shown at the bottom of the window): The **Results** tab displays a tree-like listing of your output, making it easy to scroll quickly to any place in your output listing. The **Explorer Window** (which you can display by clicking the **Explorer tab**) displays the available SAS libraries where SAS data files are stored. A SAS library is a nickname for an actual physical location on disk, such as `C:\SASDATA`. This will be described in detail in Chapter 3.

Graph: If your SAS program creates graphics output, SAS will display a **Graph Window tab**. Click that tab to view graphics results.

Results Viewer: Beginning with SAS 9.3, the results of analysis appear in this viewer. It will appear the first time you run an analysis that creates output.

Do not close any of the windows that make up the SAS interface. Move from window to window by clicking on the tabs. If you close one of the windows, its tab at the bottom of the SAS screen will disappear and you will need to go to the **View** pull-down menu and select the appropriate window name to redisplay the element that you closed. (Or restart the program.)

1.2 YOUR FIRST SAS ANALYSIS

Now that you've installed SAS and copied the SAS example data files to your computer, it's time to jump in and perform a quick analysis to see how the program works. (You have downloaded the example files, haven't you? If not, go back to Section 1.1.1 and do so.) Once you have downloaded the example files to your computer, continue with this chapter.

The following steps show you how to open a SAS program file, perform a simple analysis using the data in the file, and create statistical output.

Our first example is a quick overview of how SAS works. You should not be concerned if you don't know much about the information in the SAS program file at this point. The remainder of the book teaches you how to create and run SAS programs.

> **HANDS-ON EXAMPLE 1.1**
>
> In this example, you'll run your first SAS analysis.
>
> 1. Launch SAS. SAS initially appears as shown in Figure 1.1. (If a tutorial or some other initial dialog box appears, dismiss it.)
> 2. Open a program file. This example uses a file named `FIRST.SAS`. To open this file, first make sure that your active window is the **Editor Window**. (Click anywhere in the **Editor Window** to make it the active window.) On the menu bar, select **File → Open Program** to open the file `C:\SASDATA\FIRST.SAS`. You may need to navigate to the `C:\SASDATA` folder to open this file.
> 3. Examine the opened file, the contents of which appear in the **Editor Window**. Maximize the **Editor Window** so that you can see more of the program code if necessary. Figure 1.2 shows the **Editor Window** maximized. We'll learn more in the latter chapters about what these lines of code mean.
>
>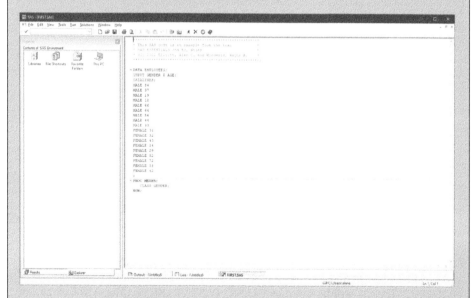
>
> **Figure 1.2** The FIRST.SAS file opened
>
> 4. Run the SAS job. There are three ways to run (sometimes called submit) your SAS code. (1) On the menu bar, select **Run → Submit**, (2) click the "running man" icon on the toolbar, or (3) press F8. Perform any of these to run this code. The SAS
>
> *(Continued)*

8 CHAPTER 1: GETTING STARTED

(*Continued*)

instructions in the file are carried out (or, in computer terminology, the commands are *executed*). The SAS program creates an analysis output that automatically appears in the **Results Viewer**.

5. View the results shown in Figure 1.3.

   a. The analysis requested in this SAS program is a task to calculate basic statistics (PROC MEANS) for each group (GENDER) in the data set.

   b. Note the **Results Window** (as opposed to the **Results Viewer**) on the left of the screen illustrated in Figure 1.3 – it contains a single item called **Means: The SAS System** (the full name is truncated in the image). As you perform more analyses in SAS, this list will grow. You can quickly display the results of any previous analysis by clicking the item in the list in this **Results Window**.

Figure 1.3 Results for FIRST.SAS

6. Print the output. Make sure the **Results Viewer** is active by clicking anywhere in the **Results Viewer**. Print the contents of the viewer by clicking the printer icon on the toolbar or by selecting **File → Print** on the menu bar.

7. Save the output. To save the results shown in the **Results Viewer**, select **File → Save As** on the menu bar and enter an appropriate name for the output file, such as FIRST OUTPUT. The file will be saved as FIRST OUTPUT.MHT. (The .MHT extension refers to a single file web page file and is automatically added to the file name. If you subsequently open this file, it will open into your default web browser.)

8. Save the SAS program code (instruction file) that created the output by first going to the **Editor Window** that contains the FIRST.SAS program code. With the program on the screen, select **File → Save As** and enter a name such as MYFIRST. Make sure

you are saving the code in the `C:\SASDATA` folder. This saves the file as `MYFIRST.SAS`. (The .SAS extension is automatically added to the file name.)

9. Examine the SAS **Log Window**. Click on the **Log Window** tab at the bottom of your screen and note that it contains a report on the SAS program code you just ran. It will include statements such as the following:

```
NOTE: Writing HTML Body file: sashtml.htm
NOTE: There were 18 observations read from the data set
WORK.EMPLOYEES.
```

This indicates that the output is in an html format (web file format) and that the data set contained 18 records.

10. Clear the contents of the **Log Window** by right-clicking and selecting **Edit → Clear All** from the pop-up menu. In the **Editor Window**, right-click and select **Clear All**. (We'll learn a shortcut to this process later in this chapter.)

The Save As command saves the contents of the window that is currently active. Thus, if the **Results Viewer** is active, it saves the output as an `*.MHT` file; if the **Log Window** is active, it saves the contents of that window as an `*.LOG` file; and if the **Editor Window** is the active window, it saves the code as an `*.SAS` (SAS program code) file.

That's it! You've run your first SAS program. But wait, there's more. Now that you have one SAS job under your belt, it's time to try another one. This time, you'll make a small change to the program before you run it.

Before running each Hands-On Example, clear the **Log** and **Editor Windows** so that you will not mix up information from a previous example. To clear each window, use the right-click technique previously described, or go to the window and on the menu bar select **Edit → Clear All.** In Section 1.4.2, we will show a way to shorten this process.

HANDS-ON EXAMPLE 1.2

In this example, you make a small change in the program and see how that change alters the output.

1. In the SAS **Editor Window**, open the program file `SECOND.SAS`. (It should be in your `C:\SASDATA` folder.) The SAS code looks like this:

```
* PUT YOUR TITLE ON THE NEXT LINE;
DATA EXAMPLE;
INPUT AGE @@;
DATALINES;
12 11 12 12 9 11 8 8 7 11 12 14 9 10 7 13
6 11 12 4 11 9 13 6 9 7 13 9 13 12 10 13
11 8 11 15 12 14 10 10 13 13 10 8 12 7 13
11 9 12
```

(*Continued*)

> (*Continued*)
>
> ```
> ;
> PROC MEANS DATA=EXAMPLE;
>     VAR AGE;
> RUN;
> ```
>
> Run the SAS job using any of the methods described in the previous example and observe the output. Note that the title at the top of the output page is "The SAS System."
>
> 2. Under the comment line that reads
>
> ```
> * PUT YOUR TITLE ON THE NEXT LINE;
> ```
>
> type this new line:
>
> ```
> TITLE "HELLO WORLD, MY NAME IS your name.";
> ```
>
> Enter your own name instead of *your name*. Make sure you include the quotation marks and end the statement with a semicolon (;).
>
> 3. Run the SAS program, examine the output, and note how the title has changed in the output.

1.3 HOW SAS WORKS

Look at the SECOND.SAS program file to see how SAS works. It is really very simple. The lines in the file are like the items on a grocery list or a "to do" list. You create a list of things you want SAS to do and when you submit the job, SAS does its best to carry out your instructions. The basic steps are as follows:

1. Define a data set in SAS using the DATA step (which begins with the keyword DATA.) In this case, the data values are a part of the code (although it is not always the case.) The data values to be used in this analysis follow the keyword DATALINES.
2. Once you have a data set of values, you can tell SAS what analysis to perform using a procedure (PROC) statement. In this case, the keywords PROC MEANS initiate the "MEANS" procedure.
3. Run the program and observe the output (in the **Results Viewer**).

Figure 1.4 illustrates this process.

This book teaches you how to create and use SAS by illustrating SAS techniques of data handling and analysis through a number of sample SAS programs.

To exit the SAS program, select **File** → **Exit** on the menu bar. Make sure you've saved all files you wish to save before ending the program. (The program will prompt you to save files if you have not previously done so.)

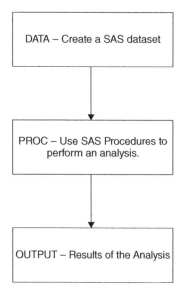

Figure 1.4 How SAS works

1.4 TIPS AND TRICKS FOR RUNNING SAS PROGRAMS

This section contains tips and tricks that can help you if something in your program goes wrong. Now that you have seen two sample SAS jobs, you should have an idea of how SAS programs are constructed. This section provides general rules for writing SAS programs.

Within a SAS program, each statement begins with an identifying keyword (DATA, PROC, INPUT, DATALINES, RUN, etc.) and ends with a semicolon (;). For example:

```
DATA TEMP;
PROC PRINT DATA=TEMP;
```

SAS statements are free-format – they do not have to begin in any particular column. Thus,

- **Statements can begin anywhere and end anywhere**. One statement can continue over several lines with a semicolon at the end of the last line signaling the end of the statement.
- **Several statements may be on the same line**, with each statement terminated with a semicolon.
- **Blanks**, as many as you want but at least one, may be used to separate the components (words) in a SAS program statement.
- **Case (lower- and upper-)** *doesn't* **matter in most SAS statements.**
- **Case** *does matter* **in data and quoted information.** When you enter character data values (such as M or m for "male"), case will matter when you are sorting or selecting the data. Moreover, the contents of a title or footnote are placed in quotation marks and are printed in the output using the uppercase and lowercase characters you specify in the program code.

- **The most common error in SAS programming is a misplaced (or missing) semicolon.** Pay careful attention to the semicolons at the end of the statements; they tell SAS that the statement is completed.
- **A second common error is a missing RUN statement.** You *must* end every program with a RUN statement (in the Windows version of SAS). If you do not end the program with a RUN statement, your program may not terminate normally, and your last statement may not be completed. (For some procedures, you should also end with the QUIT; statement. Those will be discussed when they are mentioned in the text.)
- **A third common error in a SAS program is the presence of unmatched quotation marks.** If you enter a title without matching quotation marks, you will see an error message in your **Log File** such as "The TITLE statement is ambiguous due to . . . unquoted text." This error can cause subsequent statements in your code to be misinterpreted by SAS.
- **A search for errors in a program Log Window should be from the top down.** It is inevitable that you will write programs that contain errors in syntax. When you get lots of errors in a SAS run, always fix the errors from the top down in your program code, because often the early errors lead to confusion that results in the latter errors.
- **If program errors cause problems that result in SAS's "freezing up" or not completing the steps in your program,** you can abort SAS's processes by pressing Ctrl-Break and selecting the "Cancel Submitted Statements" option from the dialog box that appears.
- **If you cannot resolve a problem within SAS**, *save your files*, restart SAS, and reopen the files. Try to find the code that caused the problem, and then re-run your program.
- **The structure of your SAS programs should be easy to read.** Note how the example programs in this book are structured – how some statements are indented for easy reading and how the code contains comments. (Lines that begin with an asterisk (*) are comments that are ignored when the program is run. Comments are discussed more in Chapter 5.)

Now that you have experience running simple programs in SAS, you will want to know more about the syntax and conventions for entering and running a SAS program. The following is a brief introduction to some of the SAS program requirements you need to know.

1.4.1 Using the SAS Enhanced Editor

You may have noted that different parts of the coding in FIRST.SAS or SECOND.SAS appear in various colors and in bold font. The color coding is designed to help you clearly see step boundaries (major SAS commands), keywords, column and table names, and constants. For example, in the SAS Enhanced Editor:

Green is used for comments.
Bold Dark Blue is used for the keyword in major SAS commands.
Blue is used for keywords that have special meaning as SAS commands.
A yellow background is used to highlight data.
A boundary line separates each step.

If you make a mistake in coding your SAS job, the appropriate colors will not appear. In fact, statements that SAS does not understand typically appear in red, which helps you locate potential problems with your code before you attempt to run the program.

TABLE 1.2 SAS Function Keys

Function Key	SAS Command	Result
F2	RESHOW	Reshows window interrupted by system command
F3	END; /*GSUBMIT BUFFER=DEFAULT*/	Submits SAS statements in clipboard
F4	RECALL	Recalls current SAS code to editor
F5	PROGRAM (PGM)	Displays Editor Window
F6	LOG	Displays Log Window
F7	OUTPUT	Displays Output Window
F8	ZOOM OFF;SUBMIT	Submits (runs) the current SAS program
F9	KEYS	Displays this Keys Window
F10	Not defined	
F11	COMMAND BAR	Moves cursor to command bar
F12	Not defined	

A boundary line separates steps (parts) of a SAS program. In the SECOND.SAS example, note that a minus sign in a small box appears on the screen next to the keyword DATA (and another one next to PROC MEANS.) Clicking on this box compresses the step to its first line, and is a way to temporarily hide sections of code. When you run the code, the program still "sees and acts upon" what is in the compressed section.

1.4.2 Using SAS Function Keys

You can use keyboard function keys to move from window to window or to execute certain tasks in SAS. Some people prefer the use of function keys (rather than mouse clicks) as a quicker way of selecting program options. Table 1.2 lists some of the SAS function keys that you can use throughout this book.

You can also customize these keys. To display the full set of function key commands, on the menu bar select **Tools → Options → Keys**.

HANDS-ON EXAMPLE 1.3

In earlier examples, you cleared the information from the **Log Window** manually. In this example, you will create a new F12 function key command that will clear the **Log Window** and the **Results Viewer** and return you to the **Editor Window**. The F12 key you're about to define uses the SAS commands:

- CLEAR LOG – clears the contents of the **Log Window**.
- ODSRESULTS;CLEAR; – clears the contents of the **Results Viewer** (and closes that window). Note that there is no blank between ODS and RESULTS.
- WPGM; returns you to the WPGM (i.e., to the active window most recently edited).

Following these instructions shows how to create a new F12 function key definition.

1. Press the F9 function key to display the **Keys Window**. Table 1.2 shows some of the key definitions that appear on your screen.

(Continued)

> (*Continued*)
>
> 2. Next to the blank F12 option, enter
>
>   CLEAR LOG; ODSRESULTS; CLEAR; WPGM;
>
>   Press the **Enter** key to lock in the new command.
>
> 3. Exit the **Keys Window** (by clicking on the small x at the top right of the window) and try out this function key by re-running one of the previous example programs. With the output displayed on the screen, press F12. The **Results Viewer** will be cleared and closed, the **Log Window** information will be cleared, and the **Editor Window** will be displayed, still containing the current program code. Thus, this command allows you to quickly clear the **Log Window** and **Results Viewer** without deleting the program code. This new command will be used in some of our future examples. (If you press F12 and your **Editor Window** does not reappear, press it again, or press F5 to display the **Editor Window**.)

Use the F12 SAS function key you created to clear the contents of the **Results Viewer** and **Log Window** between examples.

1.4.3 Using the SAS Menus

The SAS pull-down menu items differ depending on which window is currently active (**Editor, Output, Log**, etc.). To see the change in the menu options, click each window to make it active and note the menu choices at the top of the main SAS window, which are particular to the current window.

The SAS window you will use most is the **Enhanced Editor** because that is where you will write and edit SAS code. Therefore, we'll spend a little time describing the menus for the **Editor Window**.

- **File:** Used for opening and saving files, importing and exporting files, and for printing.
- **Edit:** Used to copy, cut, and paste text, as well as to find and replace text.
- **View:** Allows you to go back and forth between viewing the **Editor Window**, the **Log Window**, and the **Output Window**.
- **Tools:** Allows you to open programs for graphics and text editing, along with other options available to customize the program to your preferences.
- **Run:** Allows you to run (submit) a SAS program.
- **Solutions:** Contains options to advanced procedures not discussed in this text.
- **Window:** As in most Windows programs, allows you to choose display options for opened windows such as tile and cascade; also allows you to select (make active) a particular window, such as the **Log Window or Output Viewer**.
- **Help:** Contains options for the SAS help system as well as online documentation. (See the Section 1.4.5.)

1.4.4 Understanding Common File Extensions

Another piece of information you need to be aware of when learning and using SAS concerns the type of files used and created by SAS or mentioned in this book. Like most Windows files, they have specific file name extensions that indicate what application is associated with them (e.g., the way a *.doc or *.docx extension indicates a Microsoft Word file).

- **SAS Code File** (`filename.sas`): This is an ACSII text file and may be edited using the SAS Editor, Notepad, or any text editor that can read an ASCII file.
- **SAS Log File** (`filename.log`): This ASCII text file contains information such as errors, warnings, and data set details for a submitted SAS program.
- **SAS Results File** (`filename.mht` or `filename.html`): This file contains the web-formatted output such as that displayed in the **Results Viewer**. HTML stands for Hyper-Text Markup Language and is the common language of Internet web files. MHT is short for MHTML and stands for Microsoft (or MIME) Hypertext Archive file. It is a type of HTML file that contains the entire html-coded information in a single file (whereas HTML files may access external files for some components such as graphs.)
- **SAS Data File** (`filename.sas7bdat`): This file contains a SAS data set that includes variable names, labels, and the results of calculations you may have performed in a `DATA` step. You cannot open or read this file except in SAS or in a few other programs that can read SAS data files.
- **Raw Data Files** (`filename.dat` or `filename.txt` or `filename.csv`): These ASCII (text) files contain raw data that can be imported into SAS or edited in an editor such as Notepad.
- **Excel File** (`filename.xls` or `filename.xlsx`): The data in a Microsoft Excel file (when properly formatted into a table of columnar data) can be imported into SAS. (We'll discuss data file types that can be imported into SAS in Chapter 3.)

1.4.5 Getting SAS Help

SAS help is available from the SAS **Help** pull-down menu (**Help** → SAS Help and Documentation). When you select this option, SAS displays a help file as shown in Figure 1.5 (with the SAS Products option expanded). As with other software help systems, you can choose to search for keywords (using the Search tab), or you can scroll through the tree-like list to select topics of interest.

In Figure 1.5, note that the SAS Products section has been expanded to display links to help with a number of SAS program components. The options called Base SAS and SAS/STAT are references to the information discussed in this text. Specifically, details about the statistical procedures (`PROC`s) discussed in this book are documented in these two sections of the Help file.

SAS help is also available on the web in a number of forms. SAS Institute provides many of their reference manuals on the web (as pdf files), there are a number of papers presented at SAS conferences on the web, and a number of independent web sites that include SAS examples and help. A search for a particular SAS topic (i.e., "SAS ANOVA") will typically bring up a large number of links.

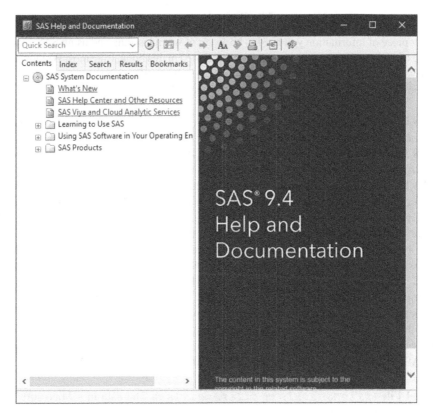

Figure 1.5 Help file display

1.5 SUMMARY

This chapter provided an overview of SAS and examples of how to run an existing SAS program. In the following chapters, we will discuss the components of a SAS program, including how to enter data, how to request analyses, and how to format and read the output.

EXERCISES

1.1 Enter code and run a SAS program.
Make sure you're in the **Editor Window**. **Choose File → New Program**. Enter the following SAS code into the **Editor Window** and run the program. Make sure the **Editor Window** is clear of all other information before you begin.

```
DATA TEMP;
INPUT ID SBP DBP SEX $ AGE WT;
DATALINES;
```

```
1 120 80  M 15 115
2 130 70  F 25 180
3 140 100 M 89 170
4 120 80  F 30 150
5 125 80  F 20 110
;
RUN;
PROC PRINT DATA=TEMP;
TITLE 'Exercise 1.1 - Your Name';
RUN;
PROC MEANS;
RUN;
```

Include your name in the TITLE line. Observe the results in the **Results Viewer**. To save the output to a file, make sure the **Results Viewer** is active and select **File → Save As**, or you can select and copy the contents of the **Results Viewer** and paste the output into an e-mail or word processor document. Hints for running this program are as follows:

a. Begin SAS.
b. Enter the SAS code into the **Editor Window**. Pay attention to the color coding to make sure all of your code has been entered correctly.
c. Before running the program, save the SAS code under the name C:\SASDATA\MYEXERCISE1.1.SAS.
d. Run the program. (Click the "running man" icon or select **Run → Submit** on the menu bar.)
e. Check the **Log Window** for errors.
f. If there are errors in the **Log file**, observe where the error occurred. SAS will usually point out the code that produced the error. If there were errors, proceed to the next step; if not, skip to step k.
g. After determining what changes should be made to your code, press F12 to clear the **Log Window** and **Results Viewer** if you have previously defined the F12 key. Or go to the **Log Window** and right-click, and choose **Edit → Clear All**. (This clears the error messages out of the **Log Window** so that you won't get confused by seeing an old error.)
h. Return to the **Editor Window**.
i. Make corrections in the **Editor Window** to fix the error you discovered in step f. Re-save the code.
j. Go back to step d.
k. Examine the output in the **Results Viewer**.
l. You may optionally print, save, or e-mail the output as directed by your instructor.
m. Optionally end SAS.

1.2 **Change the contents of a SAS program.**
Make sure you're in the **Editor Window**. Choose **File → New Program**. Enter the following SAS code into the **Editor Window** and run the program. Make sure the **Editor Window**, **Log Window**, and **Results Viewer** are clear of all other information

18 CHAPTER 1: GETTING STARTED

before you begin. (You can clear the **Log Window** and **Results Viewer** by pressing F12 if you have previously defined that key.)

```
DATA RECOVERY;
INPUT LNAME $ RECTIME;
DATALINES;
JONES 3.1
SMITH 3.6
HARRIS 4.2
MCCULLEY 2.1
BROWN 2.8
CURTIS 3.8
JOHNSTON 1.8
;
RUN;
PROC PRINT DATA=RECOVERY;
Title 'Exercise 1.2 - Your Name';
RUN;
PROC MEANS DATA=RECOVERY;
RUN;
```

Include your name in the `TITLE` line. Observe the results in the **Results Viewer**. (You may optionally print, save, or e-mail the output as directed by your instructor. These instructions will be assumed from this point on and will not be included in future exercises.)

1.3 (Optional) If you are using SAS Studio, here is how you could access the `FIRST.SAS` example. We'll not provide this detail for all exercises, but this is given here for those who are using SAS Studio.

  a. Begin SAS Studio (See Appendix E for information on installing and beginning SAS Studio.)

  b. Locate the upload icon (↥) at the top left of the SAS Studio screen.

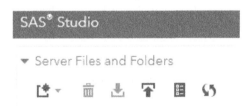

  c. In the **Upload Files** dialog box, click **Choose Files**. From the **Windows File Selector,** go to `C:\SASDATA` and select `FIRST.SAS`. Click Open.

  d. You will return to the **Upload Files** dialog box. This time, choose **Upload**. This puts the `FIRST.SAS` files into the designated SAS Studio library (usually \home\ number). The `FIRST.SAS` file should now appear in the Files (Home) list (as shown here).

e. Double click on FIRST.SAS and the code will appear in the **SAS Studio Code Window**.

f. Click the running man icon to run the FIRST.SAS program
g. Compare the results to what you saw in Figure 1.3

2

GETTING DATA INTO SAS®

LEARNING OBJECTIVES*

- To enter data using freeform list input
- To enter data using the compact method
- To enter data using column input
- To enter data using formatted input
- To enter data using the INFILE technique
- To enter multiple-line data

Before you can perform any analysis, you must have data. SAS contains powerful tools you can use to enter, import, and manage your data. This chapter introduces the most commonly used methods for getting data into SAS. Additional techniques are covered in Chapter 3.

The main way to enter and manipulate data in SAS is by using the SAS DATA step. This DATA step creates a specially formatted type of file referred to as a "SAS dataset" or "SAS data file." A SAS dataset contains the raw data plus other information. It contains variable names, how data values are read and interpreted, how values are displayed in the output, labels associated with the variable names, and other information.

SAS® Essentials: Mastering SAS for Data Analytics, Third Edition. Alan C. Elliott and Wayne A. Woodward.
© 2023 John Wiley & Sons, Inc. Published 2023 by John Wiley & Sons, Inc.
Companion website: https://www.alanelliott.com/SASED3/

TABLE 2.1 Overview of the SAS DATA Step

Task	How Do You Do It?
Getting data: creating a new SAS dataset, either by reading a list of data or by using (importing) data that are in another file	Within the DATA step, use a SAS DATA statement as discussed in this chapter
Saving data: storing data from SAS to a permanent location on your hard drive	Within the DATA step, instruct SAS to write the created dataset to a location on disk or save the dataset by using a SAS library reference as discussed in Chapter 3
Preparing data for analysis: datasets must sometimes be manipulated to make the data ready for analysis – including setting missing values, rearranging data, and revising incorrect values	Within the DATA step, create variable labels, assign missing values, create new variables, and assign labels to data categories as discussed in Chapter 4

Entering and manipulating data are the primary topics of the following several chapters. In the initial SAS examples, you may have noted that the programs begin with the keyword DATA, followed by lines of code defining the data for analysis. As noted in Chapter 1, this is called the DATA step. Within the DATA step, data are manipulated, cleaned, and otherwise prepared for analysis. Depending on how carefully the data have been collected and entered, this process can be quick and simple or long and detailed. Either way, the DATA step is the primary way to get your data into SAS and be prepared for analysis.

Table 2.1 is an overview of how the SAS DATA step works and what will be discussed in the following few chapters.

2.1 USING SAS DATASETS

The DATA step is the primary way in which SAS datasets are created. Therefore, it is important for you to learn how to read, manipulate, and manage SAS datasets using the DATA step. The following sections describe how to create a SAS dataset, beginning with simple examples and later describing more complex dataset creation techniques.

> In this book, we'll refer to a text file containing numbers and other information as a "raw data file." A "SAS data file" is a specially formatted file created and read by SAS programs containing data as well as other information about the structure of the file and the characteristics of the data values.

In the SAS language, the DATA statement signals the creation of a new SAS dataset. For example, a DATA statement in SAS code may look like this:

```
DATA MYDATA;
```

The keyword DATA tells SAS to create a new SAS dataset. The word MYDATA in the DATA statement is the name assigned to the new dataset. (You can choose any name for a SAS dataset that conforms to SAS naming conventions, as described in Section 2.3.) The SAS DATA statement has two major functions, which are as follows:

- It signals the beginning of the DATA step.

- It assigns a name (of your choice) to the dataset created in the DATA step. The general form of the DATA statement is as follows:

```
DATA datasetname;
```

The SAS *datasetname* in the DATA statement can have several forms.

- The SAS dataset name can be a single name (used for **temporary datasets**, those kept only during the current SAS session). The following examples create SAS datasets that are temporary:

```
DATA EXAMPLE;
DATA OCT2023;
```

- The SAS dataset name can be a two-part name. The two-part name tells SAS that this **permanent dataset** will be stored on disk after the current SAS session in a SAS library indicated by the prefix name (the name preceding the dot). The following examples create permanent datasets:

```
DATA RESEARCH.JAN2023;
DATA CLASS.EX1;
DATA JONES.EXP1;
```

Thus, in the first example, the prefix RESEARCH is the name of the SAS library and JAN2023 is the name of the SAS dataset. RESEARCH is a nickname for a hard disk folder. For example, RESEARCH may refer to (be a nickname for) the folder C:\SASDATA. In this case, JAN2023 is the SAS dataset name, and that dataset is located in the C:\SASDATA folder. Details about creating nickname references (i.e., library names) are provided in Chapter 3. Also, if you are using a Citrix Receiver version of SAS, then refer to the section titled "CITRIX Implementation of SAS" in Chapter 3 for information on how to use Windows path names in that platform. Similarly, if you are using the University Edition of SAS, refer to Appendix E for more information on how to use Windows path references for that version of SAS.

- You don't have to use a nickname to refer to a SAS dataset. Another way to refer to a SAS data file that is more familiar to Windows users is to use its Windows filename. For example:

```
DATA "C:\SASDATA\CLASS";
DATA "C:\MYFILES\DECEMBER\MEASLES";
```

As in the previous example, CLASS is the name of a SAS dataset, and that dataset will be located in the C:\SASDATA folder. (Technically, the name of the SAS dataset on disk is C:\SASDATA\CLASS.SAS7BDAT, but you do not have to include the extension .SAS7BDAT in the filename reference in the DATA statement.) This is simply a different way of referring to the same file using a different naming convention. Details of this technique are given in Chapter 3.

- The SAS dataset name can be omitted entirely. This technique is used for temporary datasets, which are kept only during the current SAS session. For example:

```
DATA;
```

In this case, SAS will name the dataset for you. On the first occurrence of such a DATA step, SAS will name the dataset DATA1, then DATA2, and so on.

The DATA statement is the usual beginning for most SAS programs. In the following sections and chapters, you will see some of the power of the DATA statement in reading and preparing data for analysis.

The general syntax for the SAS DATA statement is as follows:

```
DATA datasetname;
   <code that defines the variables in the dataset>;
   <code to enter data>;
   <code to create new variables>;
   <code to assign missing values>;
   <code to output data>;
      <code to assign labels to variables>;
      <and other data tasks>;
```

Each of these components of the DATA statement will be discussed in some detail, although covering the full extent of the DATA statement's capabilities would take a book several times the length of this volume.

Technically, even datasets created with single names as in DATA EXAMPLE are also referred to in SAS as WORK.EXAMPLE. The designation WORK refers to a built-in temporary SAS library (a place that contains data files) that is deleted when you end a SAS session. In your use of temporary SAS datasets such as this, you can refer to them as either EXAMPLE or WORK.EXAMPLE.

2.2 UNDERSTANDING SAS DATASET STRUCTURE

Once you read data into SAS, the program stores the data values in a dataset (also called a data table), which is a collection of data values arranged in a rectangular table. Table 2.2 is a visual illustration of how SAS internally stores a typical dataset – columns are variables (each named) and rows are subjects, observations, or records.

Although the structure of this table may look similar to the way data are stored in a Microsoft Excel spreadsheet, there are some important differences.

- Each **column** represents a variable and is designated with a variable name (ID, SBP, etc.). Every SAS variable (column) must have a name, and the names must

TABLE 2.2 SAS Dataset (Table) Structure

	ID	SBP	DBP	GENDER	AGE	WT
OBS1	001	120	80	M	15	115
OBS2	002	130	70	F	25	180
⋮	⋮	⋮	⋮	⋮	⋮	⋮
OBS100	100	120	80	F	20	110

follow certain naming rules. The data in each column conform to a particular data type (numbers, text, dates, etc.).
- Each **row**, marked here as `OBS1`, `OBS2`, and so on, indicates observations or records. Each row consists of data observed from one subject or entity.

In Section 3.4, you will see how to display a SAS data table in spreadsheet mode.

2.3 RULES FOR SAS VARIABLE NAMES

Each variable (column) in the SAS dataset (table) must have a name, and the names must follow certain rules. This differs from Excel, which imposes no restriction on how (or if) to name columns of data. However, many of the rules for SAS are similar to those for a data table created in Microsoft Access. Each SAS variable name

- must be 1–32 characters long but must not include any blanks;
- must start with the letters A through Z (not case sensitive) or an underscore (_). A name may not include a blank;
- may include numbers (but not as the first character);
- may include upper- and lower-case characters (but variable names are case insensitive);
- should be descriptive of the variable (optional but recommended).

Here are some examples of correct SAS variable names:

```
GENDER
AGE_IN_2023 (notice underscores)
AGEin2023
_OUTCOME (notice the leading underscore)
HEIGHT_IN_CM (notice underscores)
WT_IN_LBS (notice underscores)
```

And here are some examples of incorrect SAS variable names:

```
AGE IN 2023 (name contains blanks)
2023MeaslesCount (name begins with a number)
S S Number (name contains blanks)
Question#5 (name contains special character other than
underscore)
WEIGHT.KG (name contains special character other than
underscore)
AGE-In-2023 (name contains special characters other than the
underscore)
```

These variable names are incorrect because some include blanks (`S S Number`, `AGE IN 2023`, `Question 5`, `WEIGHT in KG`), start with something other than a letter or an underscore (`2023MeaslesCount`), or contain special characters other than the underscore (`AGE-In-2023`).

2.4 UNDERSTANDING SAS VARIABLE TYPES

SAS is designed to differentiate among three basic types of variables: numeric, text, and date.

- **Numeric Variables** (Default): A numeric variable is used to designate values that could be used in arithmetic calculations or are grouping codes for categorical variables. For example, the variables SBP (systolic blood pressure), AGE, and WEIGHT are numeric. However, an ID number, phone number, or Social Security number should not be designated as a numeric variable. For one thing, you typically would not want to use them in a calculation. Moreover, for ID numbers, if ID = 00012 were stored as a number, it would lose the zeros and become 12.
- **Character (Text, String) Variables**: Character variables are used for values that are not used in arithmetic calculations. For example, a variable that uses M and F as codes for gender would be a character variable. For character variables, case matters, because to the computer, a lowercase f is a different character from an uppercase F. It is important to note that a character variable may contain numerical digits. As mentioned previously, a Social Security number (e.g., 450-67-7823) or an ID number (e.g., 043212) should be designated as a character variable because their values should never be used in mathematical calculations. When designating a character variable in SAS, you must indicate to SAS that it is of character type. This is shown in upcoming examples.
- **Date Variables**: A date value may be entered into SAS using a variety of formats, such as 10/15/09, 01/05/2010, JAN052010, and so on. As you will see in upcoming examples, dates are handled in SAS using format specifications that tell SAS how to read or display the date values. For more information about dates in SAS, see Appendix B.

2.5 METHODS OF READING DATA INTO SAS

The following few sections illustrate several techniques for entering data (creating a dataset) in SAS. These are not the only ways to enter data into SAS, but they are the most common. The methods discussed are as follows:

- Reading data using freeform list input
- Reading data using the compact method
- Reading data using column input
- Reading data using formatted input.

Methods of importing data from other file formats (such as Excel) are discussed in the following chapter. Moreover, the initial examples of these data entry methods illustrate how to read data that are within the SAS code; but these methods can also be used to read in data from an external file, as will be illustrated later.

2.5.1 Reading Data Using Freeform List Input

A simple way to enter data into SAS is as freeform input. In this method, variable names in the DATA step are indicated by a list following the INPUT statement keyword. For example:

```
DATA MYDATA;
INPUT ID $ SBP DBP GENDER $ AGE WT;
```

The `DATA` statement gives the SAS dataset the name `MYDATA` (as discussed earlier), and the `INPUT` statement tells SAS the names of the variables in the dataset. Note that in the `INPUT` statement, there is a blank and a dollar sign (`$`) *after* the variable names `ID` and `GENDER`. In freeform input, this notation indicates to SAS that the variables `ID` and `GENDER` are of character type. For `ID`, we want to preserve leading zeros, and for `GENDER`, the values are non-numeric M and F.

When using freeform input, the information for each subject or entity must be listed one row at a time. Each individual row of data contains all the data for that subject, and each data value is separated by at least one blank. (We intentionally put several blanks between some values in the data lines below to illustrate this point.) A `DATALINES` statement signals the beginning of the data values when the data values are listed within the SAS code file. For example:

```
DATALINES;
001 120 80  M     15     115
002 130 70  F 25 180
003 140 100 M 89 170
004 120 80  F 30 150
005 125 80  F 20 110
```

The first data line contains the data for a single subject. The first data value (`001`) in that line corresponds to the first variable (`ID`) in the `INPUT` statement. The second value (`120`) corresponds to `SBP`, and so on. Each data value on each line is separated by at least one blank.

In a SAS program, a special keyword, `DATALINES`, tells SAS that the upcoming lines contain data. The following code (`DFREEFORM.SAS`) is an example of freeform input:

```
DATA MYDATA;
INPUT ID $ SBP DBP GENDER $ AGE WT;
DATALINES;
001 120 80  M     15     115
002 130 70  F 25 180
003 140 100 M 89 170
004 120 80  F 30 150
005 125 80  F 20 110
;
PROC PRINT DATA=MYDATA;
RUN;
```

This program reads in five lines of data and creates a listing of the data (`PROC PRINT`) that is displayed in the **Results Viewer**. Here are the components of this SAS program in the order in which they should occur:

1. The `DATA` statement tells SAS to create a dataset. In this case, the dataset is named `MYDATA`.
2. The `INPUT` statement indicates the variable names (and types for character variables) for the dataset.

TABLE 2.3 A Sample SAS Dataset

ID	SBP	DBP	GENDER	AGE	WT
001	120	80	M	15	115
002	130	70	F	25	180
003	140	100	M	89	170
004	120	80	F	30	150
005	125	80	F	20	110

3. The DATALINES statement tells SAS that the following lines contain data. (You may see older versions of SAS code that use the keyword CARDS instead of DATALINES. The statements are interchangeable.)
4. Each line of data contains the values for one subject, and the values in the data line correspond to the variables named in the INPUT statement; and those data values must be separated on the line by at least one blank.
5. The semicolon following the data indicates the end of the data values. When SAS encounters the semicolon, it stops reading data values and creates an internal SAS dataset table such as the one given in Table 2.3.

    Note that the columns of the dataset table match the variable names in the INPUT statement.
6. The PROC PRINT statement tells SAS to create a listing of the data. We'll learn more about PROC PRINT in Section 5.3.
7. The RUN statement tells SAS to finish this section of the program. Without the RUN statement, SAS may expect that more instructions are pending, and it will not finish the procedure. Therefore, SAS programs *must* conclude with a RUN statement.

One of the most common errors in writing a SAS program is to omit the RUN; statement. Don't let this happen to you!

Advantages of Freeform List Input: Some of the advantages of freeform list input are the following:

- It's easy, with very little to specify.
- Rigid column positions are not required, making data entry easy.
- If you have a dataset in which the data are separated by blanks, this is the quickest way to get your data into SAS.

Rules and Restrictions for Freeform List Input: The following rules and restrictions govern freeform list input:

- Every variable on each data line must be in the order specified by the INPUT statement.
- Data values must be separated by at least one blank. (We'll learn about data with other delimiters in Going Deeper Section 2.6.2.)
- Blank spaces representing missing variables are not allowed. Using blank spaces as missing values in the freeform data entry technique causes values to be out of sync. If there are missing values in the data, a dot (.) should be placed in the position of that

variable in the data line. (We'll learn about other missing value techniques in Section 4.3.1.) For example, a data line with AGE missing might read:

```
4 120 80 F . 150
```

- Data values for character variables have some restrictions. In the freeform technique:
- No embedded blanks (e.g., MR ED) are allowed within the data value for a character field.
- A character field can have a maximum length of eight characters. (A later example will show how to overcome this restriction.)

HANDS-ON EXAMPLE 2.1

In this example, you will enter data using the freeform data entry technique.

1. Open the program file DFREEFORM.SAS. (The code was shown above.) Run the program.
2. Observe that the output listing (shown here) illustrates the same information as in Table 2.3.

TABLE 2.4 Output from PROC PRINT for the Freeform Entry Example

Obs	ID	SBP	DBP	GENDER	AGE	WT
1	001	120	80	M	15	115
2	002	130	70	F	25	180
3	003	140	100	M	89	170
4	004	120	80	F	30	150
5	005	125	80	F	20	110

One reason to display your data using the PROC PRINT statement is to verify that SAS has properly read in your data and created a SAS dataset that can now be used for further analysis.

Note that some "freeform" datasets may use a delimiter other than blank. In that case, additional coding is required to properly read the dataset.

2.5.2 Reading Data Using the Compact Method

A second version of freeform input allows you to have several subjects' data on a single line. This technique is often used in textbooks to save space. For example, here is SAS code using the typical way of reading freeform.

```
DATA WEIGHT;
INPUT TREATMENT LOSS;
DATALINES;
```

```
1  1.0
1  3.0
1 -1.0
1  1.5
1  0.5
1  3.5
2  4.5
2  6.0
2  3.5
2  7.5
2  7.0
2  6.0
2  5.5
3  1.5
3 -2.5
3 -0.5
3  1.0
3   .5
;
PROC PRINT;
RUN;
```

To save space, the data can be compacted using an `@@` option in the `INPUT` statement to tell SAS to continue reading each line until it runs out of data. For example, the saved data using the compact method could look like this:

```
DATA WEIGHT;
INPUT TREATMENT LOSS @@;
DATALINES;
1 1.0 1 3.0 1 -1.0 1 1.5 1 0.5 1 3.5
2 4.5 2 6.0 2 3.5 2 7.5 2 7.0 2 6.0 2 5.5
3 1.5 3 -2.5 3 -0.5 3 1.0 3 .5
;
PROC PRINT;
RUN;
```

This second version of the input procedure takes up less space. Other than the addition of the `@@` indicator, this data entry technique has the same benefits and restrictions as the previous freeform input method.

HANDS-ON EXAMPLE 2.2

In this example, you will edit a freeform data entry program to use the compact entry method.

1. Open the program file `DFREEFORM.SAS`, which was used in the previous example. Run the program and examine the data listing in the Results Viewer. It is the same output given in Table 2.4.

   To return to the editor, and clear the **Results Viewer** and **Log Window**, press the F12 function key. (If you have not defined the function key F12, refer to Section 1.4.2.)

(Continued)

> (*Continued*)
>
> 2. On the INPUT line, place a space and @@ after the WT and before the semicolon so it reads
>
>    ```
>    INPUT ID $ SBP DBP GENDER $ AGE WT @@;
>    ```
>
> 3. Modify the data so that there are two subjects per line. For example, the first two lines become:
>
>    ```
>    1 120 80 M 15 115 2 130 70 F 25 180
>    ```
>
> 4. Run the program and verify that the data are listed correctly in the Results Viewer. (The output should be the same as before.)

2.5.3 Reading Data Using Column Input

Another technique for reading data into SAS is called column input. This data entry technique should be used when your data consist of fixed columns of values that are not necessarily separated by blanks. For example, consider the following data:

```
001120 80M15115
002130 70F25180
003140100M89170
004120 80F30150
005125 80F20110
```

Because the freeform technique will not work with these data, SAS allows you to specify which columns in the raw dataset contain the values for each variable. You must know the starting and ending columns for each variable's values. The INPUT statement uses the following format:

```
INPUT variable startcol-endcol ... ;
```

An example of a column data entry INPUT statement that reads data from fixed columns in the MYDATA dataset is:

```
DATA MYDATA;
INPUT ID $ 1-3 SBP 4-6 DBP 7-9 GENDER $ 10 AGE 11-12 WT 13-15;
```

Note that the primary difference between this column input statement and the freeform list input statement is the inclusion of column ranges telling SAS where to find the information for each variable in the dataset. (As previously, the $ *after* ID and GENDER (and *before* the column designations) specifies that they are character-type [text] variables.)

Here is an example program (DCOLUMN.SAS) using the column input format. This program reads in data and calculates descriptive statistics for the numeric variables using PROC MEANS (which is discussed in detail in Chapter 9).

```
DATA MYDATA;
INPUT ID $ 1-3 SBP 4-6 DBP 7-9 GENDER $ 10 AGE 11-12 WT 13-15;
```

```
DATALINES;
001120 80M15115
002130 70F25180
003140100M89170
004120 80F30150
005125 80F20110
;
RUN;
PROC PRINT DATA=MYDATA;
RUN
```

Note in the `INPUT` statement that each variable name is followed by a number or a range of numbers that tells SAS where to find the data values in the dataset. The values for ID are found in columns 1–3. The numbers for SBP are found in the column range from 4 to 6, and so on.

Advantages of Column input: Some of the advantages of column input are as follows:

- You can input only the variables you need and skip the rest. This is handy when your dataset (perhaps downloaded from a large database) contains variables you're not interested in using. Read only the variables you need.
- Data fields can be defined and read in any order in the `INPUT` statement.
- Blanks are not needed to separate fields.
- Character values can range from 1 to 200 characters. For example:

`INPUT DIAGNOSE $ 1-200;`

- For character data values, embedded blanks are no problem (e.g., John Smith).

Rules and Restrictions for Column Input: The following rules and restrictions govern column input:

- Data values must be in fixed column positions.
- Blank fields are read as missing.
- Character fields are read right justified in the field.
- Column input requires more specifications than freeform input. You must specify the column ranges for each variable.

You should also be aware of how SAS reads in and interprets column data. For example, the `INPUT` statement shown in the following code indicates that the character value data for GENDER appear in columns 1–3. If the value is really only one character wide, SAS reads in the value in the column and right justifies the character data. (The numbers 1234 ... at the top of each example represent column positions.)

```
INPUT GENDER $ 1-3;
 1 2 3 4 5 6 7                    1 2 3 4
     M                                M
   M              -> All read as      M
 M                                    M
```

Note that all Ms are stored right justified at the third column. In a similar way, numeric values can appear anywhere in the specified column range. Consider the use of the following INPUT statement to read values for the numeric variable X:

```
INPUT X 1-6;
1 2 3 4 5 6 7
    2 3 0
  2 3 . 0
  2 . 3 E 1
2 3
- 2 3
+ 2 3
```

SAS reads the information in columns 1–6 and interprets the information as a number. The above numbers are read as 230, 23, 23 (2.3E1 is a number written in scientific notation: 2.3 times 10 to the first power = 23), 23, −23, and 23.

HANDS-ON EXAMPLE 2.3

This example illustrates the column data input method.

1. Open the program file DCOLUMN.SAS. Run the program. The following output is given in Table 2.5.
   Note that ID and GENDER are *left* justified in the cell because they are character variables, and numbers are *right* justified.

   TABLE 2.5 Output from the Column Entry Example

Obs	ID	SBP	DBP	GENDER	AGE	WT
1	001	120	80	M	15	115
2	002	130	70	F	25	180
3	003	140	100	M	89	170
4	004	120	80	F	30	150
5	005	125	80	F	20	110

2. Suppose you are interested only in the variables ID, SBP, and AGE in this dataset. Modify the INPUT statement to include only ID, SBP, and AGE:

   ```
   INPUT ID   $1-3 SBP 4-6 AGE 11-12;
   ```

3. Run the modified program. The data in the columns not specified are ignored during data input, and the listing displays only columns for ID, SBP, and AGE.

One way to tell if a variable in a SAS dataset listing (PROC PRINT) is a character or numeric is this: Character data are listed *left* justified in a cell and numeric data are listed *right* justified in a cell.

2.5.4 Reading Data Using Formatted Input

Another technique for reading data from specific columns is called formatted input. This technique allows you to specify information about how SAS will interpret data as they are read into the program. SAS uses the term *informat* to specify an input format specification that tells it how to read data. (A FORMAT specification tells SAS how to output data into a report. This is discussed in Section 2.5.5.) Using an informat is helpful when you are reading data that might be in an otherwise hard-to-read format, such as date and dollar values. The syntax for formatted input is as follows:

```
DATA MYDATA;
INPUT @col variable1 format. @col variable2 format. ... ;
```

where @col is a pointer that tells SAS from which column to *start* reading the information about a particular variable. Here is a specific example:

```
DATA MYDATA
INPUT @1 SBP 3. @4 DBP 3. @7 GENDER $1. @8 WT 3. @12 OWE COMMA9.;
```

In this example, note that the information for each variable contains three components. Thus, for the first variable, the information is the following:

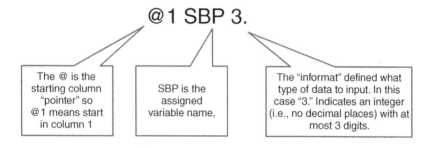

1. **The pointer** (the @ sign) tells SAS to read data for this variable beginning in the indicated column (in this case column 1).
2. **The variable name** (SBP) tells SAS that the name of this first variable is SBP.
3. **The informat** (3.) tells SAS that data at this location in each row are numbers containing three digits and containing no decimal places.

Here is a brief description of how typical informat (input format) specifications are named:

- The *formatname* designation in the Syntax column indicates that formats will have various names
- The $ in the name always indicates a character informat.
- The w indicates the maximum data width (in number of columns) for the variable.
- The d specifies the number of digits to the right of the decimal place. (Only valid for numeric data.)
- All informats contain a decimal point (.) This allows SAS to differentiate an informat from a SAS variable.

Informat Type	Syntax
Character	$*formatname*w.
Numeric	*formatname*w.d
Date and Time	*formatname*w.

Informats not only specify the type and width of the input data, they may "help" SAS interpret data as it is read. For example, note the informat used for the last variable (OWE) in the above example (COMMA9.). The reason this informat is used is that COMMA9. is able to interpret numbers that have dollar signs and commas and convert them into standard numbers. (Try reading data with dollar signs and commas into Excel – it's very difficult. Here is an example program that uses the example INPUT statement: (DINFORMAT1.SAS).

```
DATA MYDATA;
INPUT @1 SBP 3. @4 DBP 3. @7 GENDER $1. @8 WT 3. @12 OWE COMMA9.;
DATALINES;
120 80M115 $5,431.00
130 70F180 $12,122
140100M170 7550
120 80F150 4,523.2
125 80F110 $1000.99
;
PROC PRINT DATA=MYDATA;
RUN;
```

Note that the OWE variable contains some particularly nasty-looking pieces of information that many programs would have trouble reading. With SAS it's a breeze. As you use SAS more, you will find that using informats can be a valuable time saver in preparing data for analysis. Some common informats are listed in Table 2.6. (Additional formats are listed in Appendix A.)

TABLE 2.6 Example SAS Informats (Input Formats)

Informat	Meaning
5.	Five columns of data as numeric data
$5.	Character variable with width 5, removing leading blanks
$CHAR5.	Character variable with width 5, preserving leading blanks
COMMA7.	Seven columns of numeric data and strips out any commas or dollar signs (i.e., $40,000 is read as 40000)
COMMA10.2	Reads 10 columns of numeric data with 2 decimal places (strips commas and dollar signs). $19,020.22 is read as 19020.22
MMDDYY8.	Date as 01/12/23 (watch out for Y2K issue)
MMDDYY10.	Date as 04/07/2023
DATE7.	Date as 20JUL23
DATE9.	Date as 12JAN2023 (No Y2K issue)
ANYDTDTE10.	Read any text that looks like a date into a date variable.

If you read in data using a format such as 5. where the actual number has a decimal place in it, as in 3.4, SAS preserves the value of the number as 3.4 and does not round or truncate it into an integer.

Advantages and restrictions of formatted input: Some of the advantages and restrictions of formatted input are as follows:

- Advantages and restrictions are similar to those for column input.
- The primary difference between the column input and formatted input techniques is the ability to read in data using informat specifications.
- Formatted input is particularly handy for reading date and dollar values.

HANDS-ON EXAMPLE 2.4

In this example, you use the informat method of reading data into a SAS dataset.

1. Open the program file `DINFORMAT1.SAS` (program file shown above) and run the program. The output is given in Table 2.7.
2. Modify the `INPUT` statement by replacing the informat `comma9.` with a `$9.` informat specification. This tells SAS to read in the data for the variable `OWE` as character data rather than as numeric data.
3. Run the modified program and observe the output. How has it changed? Note that the new `OWE` variable read in with the `$9.` informat cannot be used in any calculation because it is not numeric. (Left justification tells you that the column contains a character-type variable.)

TABLE 2.7 Output from Formatted Data Entry Example

Obs	SBP	DBP	GENDER	WT	OWE
1	120	80	M	115	5431.00
2	130	70	F	180	12122.00
3	140	100	M	170	7550.00
4	120	80	F	150	4523.20
5	125	80	F	110	1000.99

2.5.5 Using FORMATs for Output

As this chapter discusses how to use informats for reading in data, it makes sense to look at their counterpart, output formats. Output formats, or simply formats, are used to tell SAS how to display data when it is listed in a procedure such as `PROC PRINT`. In this case, we use a SAS statement named `FORMAT`. For example, the statement (in a `DATA` step)

`FORMAT OWE DOLLAR10.2`

tells SAS to use the format `DOLLAR10.2` when displaying the variable `OWE` in `PROC PRINT` (or any other procedure). The general syntax for the `FORMAT` statement is as follows:

TABLE 2.8 Example SAS (Output) FORMATs

Format	Meaning
5.	Write data using five columns with no decimal points. For example, 12345
5.1	Write data using five columns with one decimal point. For example, 123.4
DOLLAR10.2	Write data with a dollar sign, commas at thousands, and with two decimal places. For example, $1,234.56
COMMA10.1	Write date with commas at thousands and with one decimal place. For example, 1,234.5
MMDDYY10.	Displays dates in common American usage as in 01/09/2023
DATE9.	Displays time using a military style format. For example, 09JAN2023
WORDDATE12.	Displays dates with abbreviated month names such as in January 9, 2023
WORDDATE18.	Displays dates with full month names such as in January 9, 2023
WEEKDATE29.	Displays dates with day of week such as in Saturday, January 9, 2023
BESTw.	Prints numbers with maximum precision according to the designated width w

```
FORMAT VAR1 format1. VAR2 VAR3 VAR4 format2. etc....;
```

FORMAT is the SAS keyword that begins the statement. VAR1 is a variable name to which a format is to be applied and FORMAT1. is the name of the FORMAT to apply to VAR1 in this case. For example, format1. could be 3., COMMA6., DATE9., etc. FORMAT2. applies to VAR2, VAR3, and VAR4. You may define any number of FORMATs to variables within this statement. Some commonly used output formats are listed in Table 2.7. Note that a specification used as an informat can usually also be used as an output format, so there are some that are the same as in Table 2.8.

> If a data value has a decimal such as 4.9 and you output the data value using a format such as 4. that does not have a decimal place, SAS will output a rounded integer (i.e., in this case, 4.9 gets rounded to 5).

The following example illustrates how you could use both informats and formats in a SAS program.

```
DATA REPORT;
INPUT @1 NAME $10. @11 SCORE 5.2 @18 BDATE DATE9.;
FORMAT BDATE WORDDATE12.;
```

This code uses the DATE9. input format to read in a data value called BDATE. The FORMAT statement tells SAS to output the same value, this time using the (output) format WORDDATE12. specification. For the SCORE value, the number is read in using a 5.2 input format. No output format is defined for this value, so it is output using the SAS default format, which in this case is called the BEST12. format, which writes out a number to the most possible decimals available. In this case, there are two decimal places read in, so two are written out.

HANDS-ON EXAMPLE 2.5

In this example, you will use informats to read in date values and formats to write out date values.

1. Open the program file `DINFORMAT2.SAS`. Run the program and observe the output given in Table 2.9.

   ```
   DATA REPORT;
   INPUT @1 NAME $10. @11 SCORE 5.2 @17 BDATE DATE9.;
   FORMAT BDATE WORDDATE12.;
   DATALINES;
   Bill       22.12 09JAN2016
   Jane       33.01 02FEB2000
   Clyde      15.45 23MAR1999
   ;
   PROC PRINT DATA=REPORT;
   RUN;
   ```

2. In the `FORMAT` statement, change the format for `BDATE` to `WEEKDATE29`. (Be sure to include the period at the end of the `FORMAT` specification. Also, *do not* change the format for `BDATE` in the `INPUT` statement.) Rerun the program the program. How does that change the output? Add `SCORE 5.1` to the `FORMAT` statement to specify an output format for the `SCORE` variable. Rerun the program. How does that change the `SCORE` column?

3. Enter a new line of data containing your first name and your birthdate. Make sure your new data line up (i.e., are in the same columns) with the current data. Rerun the program. On what day of the week were you born?

4. Move the entire `FORMAT` statement (including the ;) to a line after `PROC PRINT` but before the last `RUN` statement. Rerun the program. The output should be the same in step 3. The `FORMAT` statement can appear either in a `DATA` step or in a `PROC` step. However, when it appears in the `DATA` step, it *permanently* applies that format to the designated variable. When the `FORMAT` statement appears in a `PROC` statement, it only affects the output for that `PROC`.

5. Put an asterisk before the `FORMAT` line. An asterisk changes a SAS statement into a comment that is ignored when a program is run. Rerun the edited SAS program – how does that change the output? Why? (The answer is that SAS stores dates as integers, and when there is no format assigned to a date variable, the internal representation of the date is output instead of a date value. See Appendix C for more information about dates.)

TABLE 2.9 Output from Formatted Data Entry Example Including Dates

Obs	NAME	SCORE	BDATE
1	Bill	22.12	Jan 9, 2016
2	Jane	33.01	Feb 2, 2000
3	Clyde	15.45	Mar 23, 1999

38 CHAPTER 2: GETTING DATA INTO SAS®

2.5.6 Using the SAS INFORMAT Statement

There is a SAS statement named `INFORMAT` that you could use in the freeform data entry case. For example,

```
DATA PEOPLE;
   INFORMAT LASTNAME FIRSTNAME $12. AGE 3. SCORE 4.2;
   INPUT LASTNAME FIRSTNAME AGE SCORE;
   DATALINES;
Lincoln    George     35 3.45
Ryan       Lacy       33 5.5
;
PROC PRINT DATA=PEOPLE;
RUN;
```

In this case, the `INFORMAT` statement can specify that a freeform input text value is longer than the default eight characters. Note that you still cannot have any blanks in the string variables (`LASTNAME` and `FIRSTNAME`). The code for this example is found within the program file `DINFORMAT3.SAS`.

> We will not use the `INFORMAT` statement in any future examples. We will prefer instead to specify informats in the `INPUT` statement where needed.

2.5.7 Reading External Data Using INFILE

In the examples used so far, the `DATALINES` statement has been used to tell SAS that data records follow the program statements in the `DATA` step. That is, the data are in the same file as your SAS code. If your data are already stored in certain types of separate computer files, you can instruct SAS to read the records from the file using the `INFILE` statement.

For example, in this world of data, data everywhere, datasets are available by download from a number of sources. Data are often in a text (raw data) file where the values are in fixed columns. (Typical file extensions for such files are `DAT`, `TXT`, or `ASC`.) In these situations, you can use either the column input or formatted input technique to read the data into SAS. In practice, because these datasets may contain hundreds or thousands of rows (thus making it difficult to embed them into your SAS code), it is usually wise to read the data from these files using the `INFILE` technique.

The `INFILE` statement is used to identify the filename of an external ASCII (text) file. When you use the `INFILE` statement, you do not also use a `DATALINES` statement. The `INFILE` statement takes the place of the `DATALINES` statement. The general form for the `INFILE` statement is as follows:

```
INFILE filespecification options;
```

For example, the file `EXAMPLE.TXT` contains 50 lines of data in ACSII format. (You can look at this data file using Notepad or any text editor or word processor.) The first few lines are as follows:

```
101 A    12 22.3 25.3 28.2 30.6 5 0
102 A    11 22.8 27.5 33.3 35.8 5 0
104 B    12 22.8 30.0 32.8 31.0 4 0
110 A    12 18.5 26.0 29.0 27.9 5 1
```

TABLE 2.10 Using the INFILE Statement

The MEANS Procedure

Variable	N	Mean	Std Dev	Minimum	Maximum
AGE	50	10.4600000	2.4261332	4.0000000	15.0000000
TIME1	50	21.2680000	1.7169551	17.0000000	24.2000000
TIME2	50	27.4400000	2.6590623	21.3000000	32.3000000
TIME3	50	30.4920000	3.0255942	22.7000000	35.9000000

Example SAS code that will read in the first six data values (columns of data each separated by blanks) of this data file is as follows:

```
DATA MYDATA;
INFILE 'C:\SASDATA\EXAMPLE.TXT';
INPUT  ID $ 1-3 GP $ 5 AGE 6-9
       TIME1 10-14 TIME2 15-19 TIME3 20-24;
```

Note the difference between the `INFILE` and the `INPUT` statements. The `INPUT` statement is nothing new. It uses the column entry designation discussed earlier. However, unlike before, there is no `DATALINES` statement because the `INFILE` statement takes the place of `DATALINES`. It tells SAS that the data are to be found in the file specified by the `INFILE` statement. (Note that the `INFILE` statement must appear in the code *before* the `INPUT` statement so, when the `INPUT` statement is read by SAS, it knows where to go to find the data.)

The output from this program uses a procedure named `PROC MEANS` that displays descriptive statistics for the dataset. When run, this code produces the output given in Table 2.10. To summarize:

- If data are in the program code: Use a `DATALINES` statement.
- If data are to be read from external source: Use an `INFILE` statement.

A more thorough look at the `INFILE` statement is given later in this chapter.

HANDS-ON EXAMPLE 2.6

In this example, you will read in data from an external file and limit the number of variables read into the SAS dataset.

1. Open the program file `DINFILE1.SAS`.

```
DATA MYDATA;
INFILE 'C:\SASDATA\EXAMPLE.TXT';
INPUT ID $ 1-3 GP $ 5 AGE 6-9
      TIME1 10-14 TIME2 15-19 TIME3 20-24;
PROC MEANS DATA=MYDATA;
RUN;
```

Run the program and observe the output given in Table 2.10.

(*Continued*)

(Continued)
2. Modify the `INPUT` statement to read in only the variables `ID`, `GP`, `AGE`, and `TIME3`.
3. Rerun the program and observe the difference in output. Note that the new output includes statistics for `AGE` and `TIME3`. Why are statistics for `ID` and `GP` not included in the output?

2.6 GOING DEEPER: MORE TECHNIQUES FOR ENTERING DATA

SAS includes myriad ways to read in data. Techniques covered thus far are the most common. The following additional techniques can also be helpful when your data are listed in multiple records per subject and when your data are in a comma-delimited format.

2.6.1 Reading Multiple Records per Observation

Occasionally, data will be stored in such a way that multiple lines (or records) are used to hold one subject's information. SAS is able to associate observations that span more than one line to one subject. Suppose input records are designed so that each subject's information is contained on three lines:

```
10011 M 15 115
      120 80 254
      15 65 102
10012 F 25 180
      130 70 240
      34 120 132
10013 M 89 170
      140 100 279
      19 89 111
etc.
```

One method for reading these data into a SAS dataset is to have three consecutive `INPUT` statements. Each `INPUT` statement advances to the next record (but stays with the same subject). Thus, the three records in the data file combine to become one "subject" in the SAS dataset:

```
INPUT ID $ SEX $ AGE WT;
INPUT SBP DBP BLDCHL;
INPUT OBS1 OBS2 OBS3;
```

Another method for reading in these data uses the / indicator in the `INPUT` statement to advance to the next line. Each time a / is seen in the `INPUT` statement, it tells SAS to go to the next physical record in the data file to find additional data for the same subject:

```
INPUT ID $ SEX $ AGE WT/ SBP DBP BLDCHL/ OBS1 OBS2 OBS3;
```

A third way to read these data is by using the #*n* indicator to advance to the first column of the *n*th record in the group. In this case, #2 instructs SAS to begin reading the second line in the data file. The highest #*n* used in the `INPUT` statement tells SAS how many lines are used for each subject:

```
INPUT ID $ SEX $ AGE WT #2 SBP DBP BLDCHL #3 OBS1 OBS2 OBS3;
```

or in another order:

```
INPUT #2 SBP DBP BLDCHL #1 ID $ SEX $ AGE WT #3 OBS1 OBS2 OBS3
```

An advantage of this technique is that you can read the data lines into SAS in any order, as specified by the #n. Note that all of these methods require that there are *the same number of records for each subject*. If there are differing numbers of records per subject, SAS can still read in the data, but that technique is beyond the scope of this book.

An example of a program (DMULTLINE.SAS) to read multiple-line data into SAS is the following:

```
DATA MYDATA;
INPUT ID $ SEX $ AGE WT/ SBP DBP BLDCHL/ OBS1 OBS2 OBS3;
DATALINES;
10011 M 15 115
   120 80 254
    15 65 102
10012 F 25 180
   130 70 240
    34 120 132
10013 M 89 170
   140 100 279
    19 89 111
;
PROC PRINT DATA=MYDATA;
RUN;
```

In this case, we used the '/' advance indicator technique to read the three data lines per person. The output for this program is given in Table 2.11. Note that even though there are nine lines of data, the DATA step reads in the data as three records.

TABLE 2.11 Using Multiline Entry

Obs	ID	SEX	AGE	WT	SBP	DBP	BLDCHL	OBS1	OBS2	OBS3
1	10011	M	15	115	120	80	254	15	65	102
2	10012	F	25	180	130	70	240	34	120	132
3	10013	M	89	170	140	100	279	19	89	111

> **HANDS-ON EXAMPLE 2.7**
>
> 1. Open the program file DMULTILINE.SAS. Change the INPUT line to read in data using each of the three different INPUT statement techniques.
> 2. Change in the INPUT line to read in data using line number technique, but read in first line one (#1), then line three (#3), and finally line two (#2). Run the program and observe the output.

Some of the previous examples used `INPUT` column pointer control such as @5 to move the pointer to column 5. Table 2.12 lists commonly used `INPUT` pointer controls and arguments.

2.6.2 Using Advanced INFILE Statement Options

When you acquire a dataset from another source that you intend to read into SAS using an `INFILE` statement, the dataset might not be in the exact format you need. To help you read in external data, SAS includes a number of options you can use to customize your `INFILE` statement. Some of these `INFILE` options are listed in Table 2.13.

The purpose of the options in Table 2.12 is to give you more flexibility in reading data files into SAS using the `INFILE` statement. The following examples illustrate some of these options.

TABLE 2.12 INPUT Pointer Control Commands

INPUT Pointer Control or Arguments	Meaning
@N	Moves the pointer to the Nth column (N can be a numeric variable or a numeric expression)
+N	Moves the pointer ahead N columns (N can be a numeric variable or a numeric expression)
@	Holds an input record for the execution of the next `INPUT` item and allows you to specify some conditional input criteria
@@	Must be the last item in an `INPUT` line. Reads data for multiple records within a single line of data
#N	Moves the pointer to record N (N can be a numeric variable or a numeric expression)
/	Advances to the next record

TABLE 2.13 INFILE Options

INFILE Option	Meaning
DLM='char'	Allows you to define a delimiter to be something other than a blank. For example, if data are separated by commas, include the option DLM=',' in the `INFILE` statement
DLMSTR='string'	Allows you to define a string as a delimiter in the `INFILE` statement. For example, DLMSTR="~/~"
DSD	Instructs SAS to recognize two consecutive delimiters as a missing value. For example, the information M,15, 115 would be read as M, 15, Missing, 115. Moreover, this option permits the use of a delimiter within quoted strings. For example, the data value "Robert Downey, Jr." would be read properly, meaning that SAS wouldn't interpret the comma in the name as signaling a new value
MISSOVER	Indicates to SAS that, when reading a data line, if it encounters the end of a data line without finding enough data to match all of the variables in the `INPUT` statement, it should continue reading data from the next input data line

GOING DEEPER: MORE TECHNIQUES FOR ENTERING DATA   43

TABLE 2.13 (*Continued*)

INFILE Option	Meaning
TRUNCOVER	Forces the INPUT statement to stop reading data when it gets to the end of a short line
PAD and LRECL=n	This option pads short lines with extra blanks. It is used in conjunction with the LRECL= option that specifies the length of the input lines
FIRSTOBS=n	Tells SAS on what line you want it to start reading your raw data file. This is handy if your data file contains one or more header lines or if you want to skip the first portion of the data lines
OBS=n	Indicates which line in your raw data file should be treated as the last record to be read by SAS
FLOWOVER, STOPOVER, and SCANOVER	These commands are not discussed but may also be used to tell SAS how to read unusual input lines. See SAS documentation for more information

---

HANDS-ON EXAMPLE 2.8

1. Open the program file DINFILE2.SAS.

```
DATA MYDATA;
INFILE 'C:\SASDATA\EXAMPLE.CSV' DLM=',' FIRSTOBS=2 OBS=11;
INPUT GROUP $ AGE TIME1 TIME2 TIME3 Time4 SOCIO;
PROC PRINT DATA=MYDATA;
RUN;
```

This code is used to import the first 10 lines of data in the raw data (comma separated) file EXAMPLE.CSV. The first few lines of this data file are as follows:

```
GROUP,AGE,TIME1,TIME2,TIME3,TIME4,SOCIO
A,12,22.3,25.3,28.2,30.6,5
A,11,22.8,27.5,33.3,35.8,5
B,12,22.8,30.0,32.8,31.0,4
A,12,18.5,26.0,29.0,27.9,5
B,9,19.5,25.0,25.3,26.6,5
B,11,23.5,28.8,34.2,35.6,5
```

The first line in this file is a header line containing variable names. The DLM=',' option tells SAS that the data are comma delimited. The FIRSTOBS=2 option instructs SAS to start reading the data with the second line, and OBS=11 tells SAS to end with line 11, which means that it will read in the first 10 lines of data.

2. Run this program. The PROC PRINT produces the output shown in Table 2.14.
3. Change the OBS=11 statement to read in the first 25 records of the dataset and rerun the program. (Recall that the first data record is on line 2.)
4. Suppose you only want to read in records 6 through 31. Change the FIRSTOBS and OBS statements to read in only those records and rerun the program. (Recall that record 6 begins on line 7.)

(*Continued*)

(*Continued*)

TABLE 2.14 Output from Reading CSV File

Obs	GROUP	AGE	TIME1	TIME2	TIME3	Time4	SOCIO
1	A	12	22.3	25.3	28.2	30.6	5
2	A	11	22.8	27.5	33.3	35.8	5
3	B	12	22.8	30.0	32.8	31.0	4
4	A	12	18.5	26.0	29.0	27.9	5
5	B	9	19.5	25.0	25.3	26.6	5
6	B	11	23.5	28.8	34.2	35.6	5
7	C	8	22.6	26.7	28.0	33.4	3
8	B	8	21.0	26.7	27.5	29.5	5
9	B	7	20.9	28.9	29.7	25.9	2
10	A	11	22.5	29.3	32.6	33.7	2

The DLM option allows you to define the delimiter in the data. The following example illustrates how to use this option when there is an unusual delimiter.

HANDS-ON EXAMPLE 2.9

1. Open the program file DINFILE3.SAS.

```
DATA PLACES;
INFILE DATALINES DLMSTR='!~!';
INPUT CITY $ STATE $ ZIP;
DATALINES;
DALLAS!~!TEXAS!~!75208
LIHUE!~!HI!~!96766
MALIBU!~!CA!~!90265
;
PROC PRINT;
RUN;
```

The data are within this code so you can easily see the delimiter, which is !~!. The DLMSTR= options are placed in the INFILE statement, and in this case the INFILE DATALINES statement tells SAS how to read the data from the DATALINES statement.

2. Run this program and note that the data are read correctly into the defined variables. The output is given in Table 2.15.
3. Add a new record using the city of your choice, making sure to include the correct delimiter. Rerun the program to make sure it works.

TABLE 2.15 Output from Reading Data with an Unusual Delimiter

Obs	CITY	STATE	ZIP
1	DALLAS	TEXAS	75208
2	LIHUE	HI	96766
3	MALIBU	CA	90265

A more challenging problem is reading in a dataset using INPUT where there are blank values in the data. Consider the following dataset. The first three lines represent a ruler to help you see the columns where the data values are located. (It is NOT part of the dataset.) That is, the last name is in column 01 through 21 (20 characters long). The maximum length of records is 45 characters.

```
0         1         2         3         4
123456789012345678901234567890123456789012345
---------------------------------------------
ABRAMS               JJ        C0001  Producer
BRISTOW              SYDNEY    C0015  Teacher
VAUGHN               MICHAEL   C0033  Agent
FLINKMAN             MARSHALL  C0123  Analyst
SLOAN                ARVIN     C0666
BRISTOW              JACK
DIXON                MARCUS    C0233  Chief
```

HANDS-ON EXAMPLE 2.10

1. Open the program file DINFILE4.SAS.

```
DATA TEST;
  INFILE "C:\SASDATA\DINFILEDAT.TXT";
  INPUT LAST $1-21 FIRST $ 22-30 ID $ 31-36 ROLE $ 37-44;
RUN;
PROC PRINT DATA=TEST;RUN;
```

2. Run this program and observe the output. Note that the data are not correctly read. The output is shown in Table 2.16.

   A check of the SAS **Log Window** shows this statement:

```
The minimum record length was 25.
The maximum record length was 44.
NOTE: SAS went to a new line when INPUT statement reached past
the end of a line.
```

TABLE 2.16 Output from Reading CSV File with Problems

Obs	LAST	FIRST	ID	ROLE
1	ABRAMS	JJ	C0001	Producer
2	BRISTOW	SYDNEY	C0015	VAUGHN
3	FLINKMAN	MARSHALL	C0123	SLOAN

This statement tells you that there are different record lengths. To solve this issue, do the following:

3. Add the option TRUNCOVER to the end of the INFILE statement so it reads:

```
INFILE "C:\SASDATA\DINFILEDAT.txt" TRUNCOVER;
```

(*Continued*)

46 CHAPTER 2: GETTING DATA INTO SAS®

(*Continued*)

4. Rerun the program and observe that the data are read in correctly. The TRUNCOVER option forces the INPUT statement to stop reading when it gets to the end of a short line. The results are shown in Table 2.17.
5. Change the input statement to

    INFILE "C:\SASDATA\TESTDATA.TXT" LRECL=44 PAD;

    Run the program and observe the results. This also reads in the data correctly. The PAD option pads short lines with blanks to the length of the LRECL= option. Try running this code without PAD and observe the results.
6. Change the input statement to

    INFILE "C:\SASDATA\TESTDATA.TXT" MISSOVER;

    Run the program and observe the results. The data are not read in correctly. MISSOVER prevents SAS from going to a new input line if it does not find values for all of the variables in the current line of data. Because the data value TEACHER didn't fill out the entire space (i.e., had only 7 characters and ended on column 43), it was skipped.

TABLE 2.17 Output from Reading Data from a .CSV File Using TRUNCOVER

Obs	LAST	FIRST	ID	ROLE
1	ABRAMS	JJ	C0001	Producer
2	BRISTOW	SYDNEY	C0015	Teacher
3	VAUGHN	MICHAEL	C0033	Agent
4	FLINKMAN	MARSHALL	C0123	Analyst
5	SLOAN	ARVIN	C0666	
6	BRISTOW	JACK		
7	DIXON	MARCUS	C0233	Chief

This example illustrates several of the options available in the INFILE statement that allow you to read in data that are not easily readable using the default settings.

2.7 SUMMARY

One of the most powerful features of SAS, and a reason it is used in many research and corporate environments, is that it is very adaptable for reading in data from a number of sources. This chapter showed only the tip of the iceberg. More information about getting data into SAS is provided in the following chapter. For more advanced techniques, see the SAS documentation.

The following Figures 2.1 and 2.2 provide "Basic Tips" of some of the main points from Chapter 1 and in this chapter. These sheets are also available to print out from www.alanelliott.com. Click on the Quick Tips link.

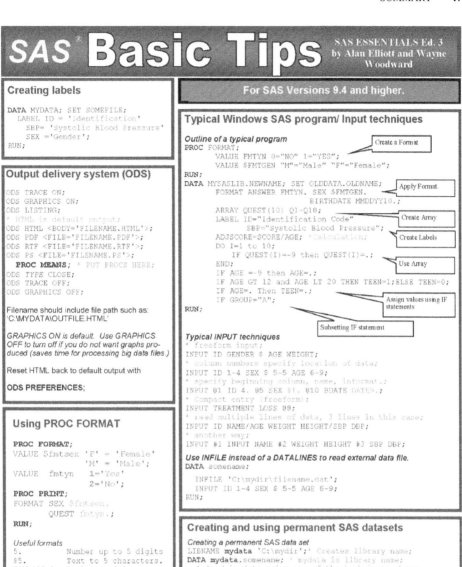

Figure 2.1 SAS Basic Tips, page 1

SAS BASIC TIPS—Page 2

Overview of the DATA STEP

Get data: Create a new SAS data set either by reading a list of data or by using (importing) data that are in another file

Save a dataset: Store data from SAS to a permanent location on your hard drive by naming it with a permanent library name such as MYSASLIB.NEWDATA or
"C:\SASDATA\NEWDATA"

Prepare data for analysis: Data sets must sometimes be manipulated to make the data ready for analysis – including setting missing values, rearranging data, and revising incorrect values.

Rows and Columns within a SAS DATASET

- Each **column** represents a variable and is designated with a variable name (ID, SBP, etc.).
- Each **row**, marked here as OBS1, OBS2, and so on, indicates observations or records.

Common Data Types

- **Numeric Variables** (Default): Designates values that could be used in arithmetic calculations (i.e. no phone number, or Social Security number etc.)
- **Character (Text, String) Variables**: Character variables are used for values that are not used in arithmetic calculations. (i.e. M and F for Gender)
- **Date Variables**: Such as 10/15/09, 01/05/2010, JAN052010, and so on.

Methods of Reading Data into SAS
Freeform Data Entry
```
DATA MYDATA;
INPUT ID $ SBP DBP GENDER $ AGE WT;
DATALINES;
```
Column Input
```
DATA MYDATA;
INPUT ID $ 1-3 SBP 4-6 DBP 7-9 GENDER $ 10
AGE 11-12 WT 13-15;
```
Formatted Input
```
DATA MYDATA
INPUT @1 SBP 3. @4 DBP 3. @7 GENDER $1. @8
WT 3. @12 OWE COMMA9.;
```

Defining the F12 Key

The F12 key command will clear the Log Window and the Results Window and return you to the Editor Window.

1. Press the F9 function key to display the Keys Window. Table 1.2 shows some of the key definitions that appear on your screen.
2. Next to the blank F12 option, enter

```
CLEAR LOG; ODSRESULTS; CLEAR; WPGM;
```

3. Press the Enter key to lock in the new command.

SAS Statements
Each statement begins with an identifying keyword (DATA, PROC, INPUT, DATALINES, RUN, etc.) and ends with a semicolon (;). For example:

DATA TEMP; **SET** MYDATA;
PROC PRINT DATA=TEMP;
RUN;

SAS statements are free-format – they do not have to begin in any particular column.

Tips and Tricks

- **Statements can begin anywhere and end anywhere.** One statement can continue over several lines with a semi-colon at the end of the last line signaling the end of the statement.
- **Several statements may be on the same line**, with each statement terminated with a semicolon.
- **Blanks**, as many as you want but at least one, may be used to separate the components (words) in a SAS program statement.
- **Case (lower- and upper-) doesn't matter in most SAS statements.**
- **Case does matter in data and quoted information.** When you enter character data values (such as M or m for "male"), case will matter when you are sorting or selecting the data.
- **The most common error in SAS programming is a misplaced (or missing) semicolon.** Pay careful attention to the semicolons at the end of the statements; they tell SAS that the statement is completed.
- **A second common error is a missing RUN statement. You must end every program with a RUN statement (in the Windows version of SAS).** If you do not end the program with a RUN statement, your program may not terminate normally, and your last statement may not be completed.
- **A third common error in a SAS program is the presence of unmatched quotation marks.** If you enter a title without matching quotation marks, you will see an error message in your Log file such as "The TITLE statement is ambiguous due to ... unquoted text."
- **A search for errors in a program log should be from the top down.** Fix errors from the top down in your program code, because often the early errors lead to confusion that results in the latter errors.
- **If program errors cause problems that result in SAS's "freezing up" or not completing the steps in your program**, abort SAS's processes by pressing Ctrl-Break and selecting the "Cancel Submitted Statements" option from the dialog box that appears.
- **If you cannot resolve a problem within SAS**, save your files, restart SAS, and reopen the files. Try to find the code that caused the problem, and then re-run your program.

© Alan Elliott and Wayne Woodward, 2022

SAS ESSENTIALS Ed. 3
by Alan Elliott and Wayne Woodward
www.alanellitt.com/sas

Figure 2.2 SAS Basic Tips, page 2

EXERCISES

2.1 **Input data into SAS.**

a. Open the program file EX_2.1.SAS.

```
DATA CHILDREN;
* WT is in column 1-2, HEIGHT is in 4-5 and AGE is in 7-8;
```

```
* Create an INPUT statement that will read in this dataset;
INPUT      ;
DATALINES;
64 57 8
71 59 10
53 49 6
67 62 11
55 51 8
58 50 8
77 55 10
57 48 9
56 42 10
51 42 6
76 61 12
68 57 9
;
Title "Exercise 2.1 - your name";
PROC PRINT DATA=CHILDREN;
RUN;
```

b. Note that the INPUT statement is incomplete. Using one of the input techniques discussed in this chapter, complete the INPUT statement and run the resulting program. Verify that the PROC PRINT represents the data correctly.

2.2 **Complete the INPUT statement using column input.**

The file EX_2.2.SAS contains a partial SAS program that includes data in the following defined columns:

Variable Name	Column	Type
ID	1-5	Text
AGE	6-7	Numeric
GENDER	8	Text (M or F)
MARRIED	9	Text (Y or N)
WEIGHT_IN_POUNDS	10-12	Numeric

a. Complete the SAS code snippet given here using the column input technique.

```
DATA PEOPLE;
INPUT ID $ 1-5 ; * finish INPUT statement;
DATALINES;
0000123MY201
0002143FY154
0004333FN133
0005429MN173
0013249FY114
;
Title "Exercise 2.2 - your name";
PROC PRINT DATA=PEOPLE;
RUN;
```

b. Run the program. Verify that the output lists the data correctly.

CHAPTER 2: GETTING DATA INTO SAS®

2.3 **Complete the INPUT statement using formatted input.**

The SAS program EX_2.3.SAS contains a data file with the following defined columns:

Variable Name	Column	Type
ID	1-5	Text
GENDER	6	Text (M or F)
MARRIED	7	Numeric 0,1
BDATE	8-16	Date

a. Complete the SAS code snippet given below using the formatted input technique. Refer to Table 2.6 to determine which date INFORMAT to use. For the output FORMAT, use the format that outputs dates in the Thursday, November 23, 2023 style.

```
DATA BIRTHDATES;
INPUT ID $ 1-5 ;          * finish INPUT statement;
FORMAT BDATE someformat.; * finish OUTPUT FORMAT statement;
DATALINES;
00001M112JAN1999
00021F003MAR1989
00043F018JUL1991
00054M022DEC1998
00132F110JUL1992
;
Title "Exercise 2.3 - your name";
PROC PRINTDATA=BIRTHDATES;
RUN;
```

2.4 **Read an external data file.**

The file BPDATA.DAT is a raw (ASCII text) data file that contains data with the following defined columns:

Variable Name	Column	Type
ID	1	Text
SBP	2-4	Numeric
DBP	5-7	Numeric
GENDER	8	Text (M or F)
AGE	9-10	Numeric
WEIGHT	1113	Numeric

a. Complete the following SAS code snippet (EX_2.4.SAS) to read in this dataset. The data are in the file named C:\SASDATA\BPDATA.DAT.

```
DATA BPDATA;
INFILE 'inputfilename'; * Finish the INFILE statement;
INPUT ID$ 1 SBP 2-4 ;    * Finish the input statement;
```

```
Title "Exercise 2.4 - your name";
PROC MEANS;
RUN;
```

b. Make sure you include your name in the title statement.

c. Run the program and observe the results.

2.5 **Going deeper: read a large external file.**

In this exercise, you will read data from an external raw data file (TRAUMA.CSV) in comma-separated-values (CSV) format that contains thousands of records. The first few records look like this:

```
SUBJECT,AGE,GENDER,PLACE,STATUS
1868,12.1,Female,Street,Alive
1931,18.7,Female,Street,Alive
1950,16.6,Female,Street,Alive
1960,8.5,Female,Street,Alive
2019,6.7,Male,Unknown,Alive
2044,7.8,Male,Street,Alive
```

Note that the first line (row) in this file contains a header with variable names.

a. Complete the SAS code snippet (EX_2.5.SAS):

```
DATA TRAUMA2;
INFILE 'filename' DLM= FIRSTOBS= ; * Finish the INFILE
statement;
INPUT SUBJECT $ AGE GENDER $ PLACE $ STATUS $;
Title "Exercise 2.5 - your name";
PROC MEANS;
RUN;
```

Finish this incomplete code by doing the following:

I. Enter the appropriate filename in the INFILE statement.

II. Place an appropriate value after the DLM = statement.

III. Place an appropriate value after the FIRSTOBS= statement to begin reading with record 2. Note that in the INPUT statement, the only numeric variable is AGE.

IV. Run the completed program.

You should get the following output (with your name in the title):

Exercise 2.5 – Name
The MEANS Procedure

Analysis Variable: AGE				
N	Mean	Std Dev	Minimum	Maximum
16242	10.6396010	5.8030633	0	21.0000000

V. How many records are there in this dataset?

b. Revise the program slightly: This time includes an OBS= statement so the program will read in only the first 25 records (not including the header line). Change the PROC MEANS to PROC PRINT. Run the revised program. The **Results Viewer** should include a listing of the first 25 records in the dataset.

2.6 **Going Deeper: Reading Gold.**

In this exercise, you will read Yearly Gold Data information from an external raw data file (GOLD.DAT), a column-separated data file. The first few records look like this:

```
Year, Open, High, Low
1980  $559.00    $843.00    $474.00
1981  $592.00    $599.25    $391.75
1982  $399.00    $488.50    $297.00
1983  $452.75    $511.50    $374.75
```

Note that the data values contain dollar signs. One way to read in the data is by using the formatted INPUT statement.

```
INPUT @1 YEAR 4. @5 OPEN DOLLAR9.2 @15 HIGH DOLLAR9.2 @28 LOW
DOLLAR9.2;
```

a. Complete the SAS code snippet (EX_2.6.SAS) using the INPUT statement above, and run the program:

```
DATA GOLD;
  INFILE 'C:\SASDATA\GOLD.DAT' FIRSTOBS=2;
  INPUT @1 YEAR ;
  RUN;
PROC PRINT DATA=GOLD;RUN;
TITLE;FOOTNOTE;
```

b. Note that because the data are complete, and there is a comma between each value, you don't really need the @1, @5, etc., pointers. Eliminate the "@" pointers from the INPUT statement and run the program again. Did you get the same output?

3

READING, WRITING, AND IMPORTING DATA

LEARNING OBJECTIVES

- To be able to work with SAS® libraries and permanent datasets
- To be able to read and write permanent SAS datasets
- To be able to interactively import data from another program
- To be able to define SAS libraries using program code
- To be able to import data using code
- To be able to discover the contents of a SAS dataset
- To be able to understand how the Data Step Reads and Stores Data.

The Era of "Big Data" has arrived. Organizational datasets are growing by leaps and bounds. Institutions, government agencies, and businesses gather information from individual transactions, which can result in hundreds of thousands of data values each day. Whether the dataset you are working with has millions of records or a few hundred isn't the point of this chapter. The point is that once you read in your data, you will want to store it in a way that makes it easy to retrieve for analysis. That means that you may want to store more than raw data values. You'll want your dataset to include variable names, labels for your variables, manipulations you've done to get it ready for analysis, and other information.

SAS® Essentials: Mastering SAS for Data Analytics, Third Edition. Alan C. Elliott and Wayne A. Woodward.
© 2023 John Wiley & Sons, Inc. Published 2023 by John Wiley & Sons, Inc.
Companion website: https://www.alanelliott.com/SASED3/

3.1 WORKING WITH SAS LIBRARIES AND PERMANENT DATASETS

SAS allows you to create permanent datasets that, once created, are easy to access and use. Keep in mind that a SAS dataset is different from a SAS program (code) file or a raw data file. The SAS code files, which we used in Chapter 2, contain a listing of SAS commands and use raw data values (either within the code file or in an external file which is accessed with an INFILE statement). The SAS code files and associated raw data files can be used to create a SAS dataset, but the code file with raw data is not a SAS dataset. Keep the following points in mind. SAS datasets

- are created by a DATA statement (there are other ways to create a SAS dataset that we'll learn later in this chapter)
- are an internal representation of the data created by the DATA statement
- contain more than the data values – they can contain variable names, labels, the results of codings, calculations, and variable formats
- are referred to by a name that indicates whether the dataset is temporary or permanent.

All of the SAS datasets created in Chapter 2 were temporary. (Temporary datasets are also referred to as WORK datasets.) That is, when the SAS program is exited (i.e., you exit the SAS program using **File → Exit**), all temporary datasets are lost. This chapter shows you how to create a SAS dataset that is permanently stored in a folder on your hard disk (or flash drive, network drive, or other storage mediums).

The technique for saving a SAS dataset as a file in a folder location is different from what is used in most Windows applications. Instead of using the **File → Save** paradigm, you store a SAS dataset using code in a SAS program. To illustrate how this works, you need to remember the basic differences between temporary SAS datasets and permanent SAS datasets:

- **A temporary SAS dataset** is named with a single-level name, such as MEASLES or MAR2023. It is created in a DATA statement and is available for analysis as long as you have not exited the SAS environment. (Even if you run other code files and create other SAS datasets, the temporary SAS datasets are available for use.) When you exit the SAS software program, all temporary datasets are erased.
- **A permanent SAS dataset** is a file saved on your hard disk. It can be referred to with its two-part name, such as RESEARCH.DATA1 or MYSASLIB.COVID2022 or with a Windows-type specification such as "C:\SASDATA\DATA1" or "H:\NETWORK\MYFILES\COVIOD2022."
- Figure 3.1 provides a graphical representation of how reading data into SAS works by creating either a temporary (WORK) or a permanent dataset.

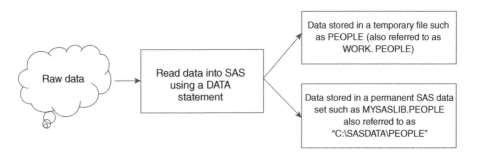

Figure 3.1 Reading and storing data in SAS

The following few sections present information on how to create permanent SAS datasets using two techniques. The first uses Windows file specifications, and the second uses SAS libraries.

3.2 CREATING PERMANENT SAS DATASETS USING THE WINDOWS FILE NAME TECHNIQUE

The Windows file specification technique for creating a permanent SAS dataset is straightforward when using SAS in a Windows operating system. Instead of using a temporary SAS dataset name such as SOMEDATA or MEASLES in the SAS DATA statement, you can use a Windows file name. For example, instead of using

```
DATA PEOPLE;
```

use

```
DATA "C:\SASDATA\PEOPLE";
```

How you implement this in a SAS program is illustrated in Figure 3.1.
Consider the following SAS code (WRITE.SAS):

```
DATA "C:\SASDATA\PEOPLE";
INPUT ID $ 1-3 SBP 4-6 DBP 7-9 GENDER $ 10 AGE 11-12 WT 13-15;
DATALINES;
001120 80M15115
002130 70F25180
003140100M89170
004120 80F30150
005125 80F20110
;
RUN;
PROC MEANS;
RUN;
```

This code creates a dataset called PEOPLE on your hard drive in the C:\SASDATA\ folder, and the Window file name of this permanent file is PEOPLE.SAS7BDAT. To summarize:

- The statement DATA PEOPLE creates an internal SAS dataset that is *temporary*. (It vanishes when you exit the SAS program.)
- The statement DATA "C:\SASDATA\PEOPLE" creates a *permanent* SAS dataset stored on your hard drive.

In fact, you can substitute the Windows file name specification anywhere using the SAS dataset name specification. This is illustrated in several of the upcoming examples.

CITRIX Implementation of SAS: If you are using a server version of SAS, the reference to the file's location on the hard disk will be different from those shown in these examples. For example, when using a Citrix implementation of SAS on Windows, with your data stored in C:\SASDATA, you must use the following code rather than the C:\SASDATA\ folder reference shown in the example:

```
DATA "\\CLIENT\C$\SASDATA\PEOPLE";
```

This code replaces the `DATA` statement in the example where the server address `\\CLIENT\C$` is used instead of `C:`. This is because the server folder access code you use may differ depending on your server implementation of SAS. (The exact folder reference may be different for different server setups. Check your specific server implementation.)

We won't repeat this information for other examples. So, you must take care to replace the C: with the appropriate code if necessary for your server installation of SAS.

HANDS-ON EXAMPLE 3.1

This example illustrates how to write data to a permanent SAS dataset.

1. Open the program file `WRITE.SAS`. Note that the data statement contains a Windows file designation in quotation marks:

```
DATA "C:\SASDATA\PEOPLE";
```

2. Because the dataset name in SAS is `C:\SASDATA\PEOPLE`, it follows that the data are stored in the `C:\SASDATA` folder. To verify this, open your `C:\SASDATA` folder using Windows Explorer. Verify that the `C:\SASDATA` folder contains a file named `PEOPLE.SAS7BDAT`. The .SAS7BDAT file name extension indicates to the Windows operating system the type of file, and in this case, it indicates that this file is a SAS dataset. (To show file name extensions in Windows 8 and 10, you may need to click the "View" tab at the top of the Windows Explorer screen and activate the "File name extensions" box in the Show/hide section.)

3.3   CREATING PERMANENT SAS DATASETS USING A SAS LIBRARY

In the previous section, you created a SAS dataset to a Windows folder (`C:\SASDATA`) by specifying a Windows file name (`C:\SASTATA\PEOPLE`), which indicated the complete file name for the `PEOPLE.SAS7BDAT` file. That technique works only in the Windows environment.

Another technique for reading and writing SAS datasets is to use the SAS Library technique. Before using a SAS Library, you have to "create" one. The following two sections provide two methods for creating a SAS Library.

Before we learn how to set up a SAS Library name, this section shows why it comes in handy. The SAS Library technique involves creating a name that refers to a drive location called (in SAS terminology) a library name. You might think of the library name as a nickname for the library. For example, you could create a library name such as `MYWORK` that refers to a simple location on your computer disk such as

```
C:\LOTSOFFILES\
```

Or, you could create the `MYWORK` library name that refers to a complicated network folder such as

N:\MYNET\ACCOUNTING\STATMENTS\MYFOLDER\RESEARCH\

Once your library name is set up, you can use the short name to refer to your disk location rather than having to use some complex Windows path name every time. With a library name, you could use the file designation:

MYWORK.JAN2022

to refer to a SAS data file named JAN2022.SAS7BDAT rather than the more complicated name:

"N:\MYNET\ACCOUNTING\STATMENTS\MYFOLDER\RESEARCH\JAN2022"

For the examples in this book, we'll create the following SAS Library name and use it for the upcoming exercises:

- The SAS library name: MYSASLIB
- The location of the Windows folder: C:\SASDATA

Therefore, when the library name MYSASLIB is used in a filename, it will refer to a file located in the C:\SASDATA folder. Note that if you've chosen to place the sample files for this book into a different folder than C:\SASDATA, you need to make appropriate adjustments to the library definition in the Hands-On Examples and chapter exercises.

The following two sections describe methods for creating the relationship linking a Windows folder location to a SAS Library name. Two important points to grasp are the following:

- *Every SAS library has a name* such as MYSASLIB, RESEARCH, MYDATA, and so on.
- *Every SAS library name is a name* that points to a specific folder such as C:\SASDATA, N:\NETWORK\DRIVE, I:\MYFLASH\DRIVE, and so on.

Within the SAS program, every SAS dataset has two parts to its name. For example, the SAS dataset referred to as MYSASLIB.MAY2023 consists of the following:

- the *library* named MYSASLIB;
- the *dataset* named MAY2023.

Even temporary files can be referred to with a (temporary) library name, WORK. Therefore, a file named WORK.LOTSADATA consists of:

- the *library* named WORK (refers to the temporary library)
- the *dataset* name is LOTSADATA.

> This SAS naming convention is used for SAS datasets in all types of computers and operating systems – mainframes, Windows, UNIX, and so on. Once you learn how to use SAS in Windows, you can easily transfer that skill to any other computer platform.

To summarize, once you've created a SAS library named MYSASLIB that points to C:\SASDATA, the dataset you refer to as MYSASLIB.MAY2024 is stored on your hard drive using the name C:\SASDATA\MAY2024.SAS7BDAT.

58 CHAPTER 3: READING, WRITING, AND IMPORTING DATA

The following section illustrates how to create a SAS library name using two methods. The first method creates a SAS library name using a Dialog Box, and the second method creates the library name in SAS code.

3.4 CREATING A SAS LIBRARY USING A DIALOG BOX

An easy way to create a SAS library in the Windows version of SAS (with a custom name of your own choosing) is to use the **New Library dialog box**. To display the **New Library dialog box**:

- Click the **Explorer** tab at the bottom of the left window in the main SAS screen to make the SAS **Explorer Window** active
- Select **File → New**
- Click on the **Library** icon
- Click **OK**.

The **New Library** dialog box is shown in Figure 3.2.

- In the **Name** field, enter the name you want to assign to the library. In this book, the shortcut name MYSASLIB is used.
- In the **Path** field, enter the location of the directory. In this book, the files are assumed to be in the C:\SASDATA folder on the hard drive.
- Notice the option **Enable at startup**. If you intend to use this library name repeatedly, you should choose this option by checking the box next to it. Choosing the **Enable at**

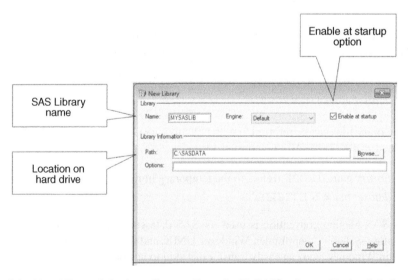

Figure 3.2 New Library dialog box. (Source: Created with SAS® software, Version 9.4. Copyright 2012, SAS Institute Inc., Cary, NC, USA. All Rights Reserved. Reproduced with permission of SAS Institute Inc., Cary, NC)

CREATING A SAS LIBRARY USING A DIALOG BOX

startup option tells SAS to remember the library name and location and to reactivate it each time the SAS program is started.

- You will typically not change anything in the **Engine drop-down** or the **Options** field.

Once you've created the MYSASLIB library, all SAS data files in C:\SASDATA become a part of the library.

HANDS-ON EXAMPLE 3.2

In this example, you will define a SAS library and give it a name.

1. Make the SAS Explorer window active. (Click the **Explorer** tab at the left bottom on the main SAS screen. If it is not there, click on the **View Menu** and open **Explorer**.) Select **File → New**, select the **Library** option, and click **OK**. The dialog box shown in Figure 3.2 appears.
2. Enter MYSASLIB in the **Name** field and C:\SASDATA in the **Path** field, or you can use a path to a network or any other location where your SAS dataset is stored. Check the **Enable at startup** option. Click **OK** to save the information.
3. On the left side of your SAS screen, click the **Explorer** tab. You should see a Libraries icon that resembles a filing cabinet, as shown in Figure 3.3. (If you don't see the Libraries icon, you might need to click **View → Up One Level** to get to this view.) Double-click the **Libraries** icon.

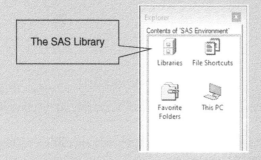

Figure 3.3 The SAS Library icon

4. A new window called **Active Libraries** appears. This window shows which SAS library names have been defined. If you performed the previous steps correctly, you will see a library named **Mysaslib**. Double-click the **Mysaslib** icon to display the contents of the **Mysaslib** window. In that window, you will see a list of the permanent SAS datasets (i.e., files that have a .SAS7BDAT extension) that are in the C:\SASDATA folder. One of the permanent SAS datasets listed here (that you downloaded from the book's website – you did download them didn't you?) is named **Somedata**.
5. Double-click the **Somedata** icon. A window such as the one illustrated in Figure 3.4 is displayed.

(Continued)

(Continued)

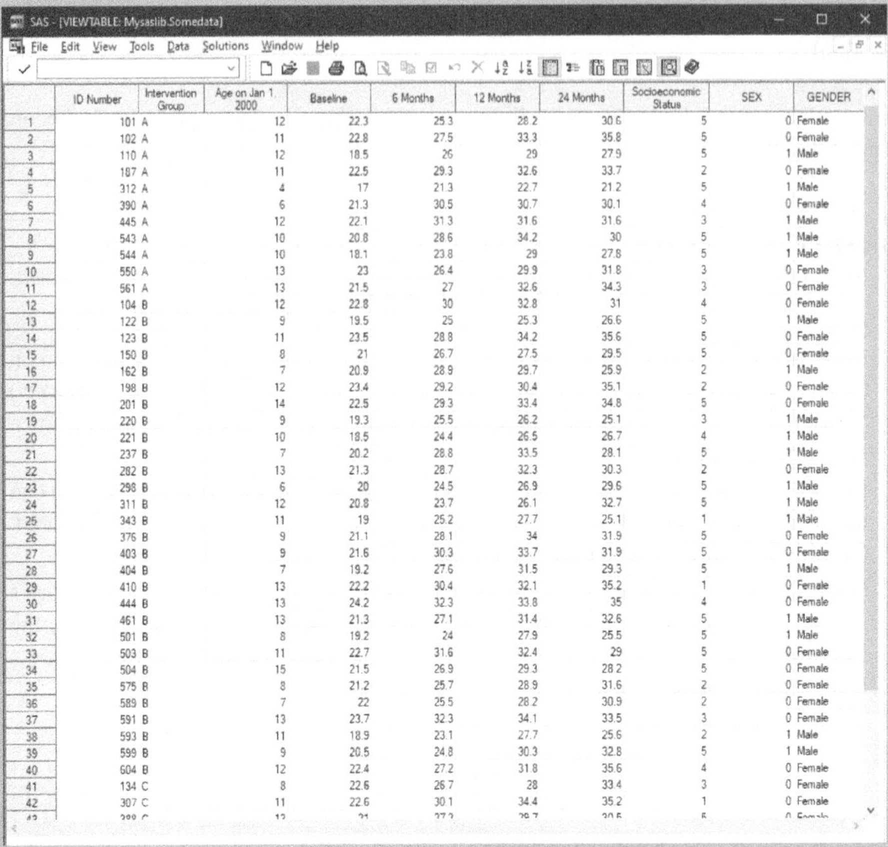

	ID Number	Intervention Group	Age on Jan 1 2000	Baseline	6 Months	12 Months	24 Months	Socioeconomic Status	SEX	GENDER
1	101	A	12	22.3	25.3	28.2	30.6	5	0	Female
2	102	A	11	22.8	27.5	33.3	35.8	5	0	Female
3	110	A	12	18.5	26	29	27.9	5	1	Male
4	187	A	11	22.5	29.3	32.6	33.7	2	0	Female
5	312	A	4	17	21.3	22.7	21.2	5	1	Male
6	390	A	6	21.3	30.5	30.7	30.1	4	0	Female
7	445	A	12	22.1	31.3	31.6	31.6	3	1	Male
8	543	A	10	20.8	28.6	34.2	30	5	1	Male
9	544	A	10	18.1	23.8	29	27.8	5	1	Male
10	550	A	13	23	26.4	29.9	31.8	3	0	Female
11	561	A	13	21.5	27	32.6	34.3	3	0	Female
12	104	B	12	22.8	30	32.8	31	4	0	Female
13	122	B	9	19.5	25	25.3	26.6	5	1	Male
14	123	B	11	23.5	28.8	34.2	35.6	5	0	Female
15	150	B	8	21	26.7	27.5	29.5	5	0	Female
16	162	B	7	20.9	28.9	29.7	25.9	2	1	Male
17	198	B	12	23.4	29.2	30.4	35.1	2	0	Female
18	201	B	14	22.5	29.3	33.4	34.8	5	0	Female
19	220	B	9	19.3	25.5	26.2	25.1	3	1	Male
20	221	B	10	18.5	24.4	26.5	26.7	4	1	Male
21	237	B	7	20.2	28.8	33.5	28.1	5	1	Male
22	282	B	13	21.3	28.7	32.3	30.3	2	0	Female
23	298	B	6	20	24.5	26.9	29.6	5	1	Male
24	311	B	12	20.8	23.7	26.1	32.7	5	1	Male
25	343	B	11	19	25.2	27.7	25.1	1	1	Male
26	376	B	9	21.1	28.1	34	31.9	5	0	Female
27	403	B	9	21.6	30.3	33.7	31.9	5	0	Female
28	404	B	7	19.2	27.6	31.5	29.3	5	1	Male
29	410	B	13	22.2	30.4	32.1	35.2	1	0	Female
30	444	B	13	24.2	32.3	33.8	35	4	0	Female
31	461	B	13	21.3	27.1	31.4	32.6	5	1	Male
32	501	B	8	19.2	24	27.9	25.5	5	1	Male
33	503	B	11	22.7	31.6	32.4	29	5	0	Female
34	504	B	15	21.5	26.9	29.3	28.2	5	0	Female
35	575	B	8	21.2	25.7	28.9	31.6	2	0	Female
36	589	B	7	22	25.5	28.2	30.9	2	0	Female
37	591	B	13	23.7	32.3	34.1	33.5	3	0	Female
38	593	B	11	18.9	23.1	27.7	25.6	2	1	Male
39	599	B	9	20.5	24.8	30.3	32.8	5	1	Male
40	604	B	12	22.4	27.2	31.8	35.6	4	0	Female
41	134	C	8	22.6	26.7	28	33.4	3	0	Female
42	307	C	11	22.6	30.1	34.4	35.2	1	0	Female

Figure 3.4 SAS Viewtable

6. This spreadsheet window is called the SAS Viewtable. It displays the contents of a SAS dataset.
7. Close the `MYSASLIB.SOMEDATA` dataset by selecting **File → Close** or by clicking the **x** at the top right of the window.
8. Return to the SAS Editor window (by clicking on an editor tab at the bottom of your SAS program), select **File → Open Program** . . . and open the program file `LIBNAME1.SAS`, which contains the following short SAS program:

```
PROC MEANS DATA=MYSASLIB.SOMEDATA;
RUN;
```

9. Run this program. In the **Results Viewer**, you'll see that SAS has calculated and reported descriptive statistics for the numeric values in the `MYSASLIB.SOMEDATA` dataset.

Any SAS data file (with a .SAS7BDAT extension) that you copy into the Windows folder linked to a SAS library name automatically becomes a part of that SAS library. Thus, if you acquire a .SAS7BDAT file from some source, all you have to do to make that file available in the SAS MYSASLIB library is to copy that file (using Windows Explorer) to the C:\SASDATA folder.

3.5 CREATING A SAS LIBRARY USING CODE

Another way to create a SAS Library name is by using code. Creating a library within the code makes it active only while the SAS software is running. When you exit the SAS software program (**File → Exit**), the library name is "forgotten" by SAS, although any permanent dataset you've created is still stored in your folder referred to by the library. You may use these permanent SAS datasets again by recreating a library name pointing to the Windows folder where the data file is stored.

The following SAS code creates a SAS library name called MYLIB2 at the location C:\SASDATA:

```
LIBNAME MYLIB2 "C:\SASDATA";
```

When SAS reads this code at the beginning of your SAS program, the library name (MYLIB2) specified by the LIBNAME statement is created and can subsequently be used as a part of a permanent SAS dataset name. For example, suppose that you have a dataset named SURVEY.SAS7BDAT in the folder C:\SASDATA. Once you define the library name MYLIB2 pointing to that folder, you can refer to that dataset using the SAS dataset name MYLIB2.SURVEY.

Note that for illustrative purposes, we are creating the library named MYLIB2 that has a different name than MYSASLIB library previously created, but they are both name s for the C:\SASDATA folder. There is no conflict in creating two SAS library names that point to the same folder on your drive, although in practice, you would not normally have two such library names.

HANDS-ON EXAMPLE 3.3

In this example, you will define a SAS library within code and use the library name to read in a SAS data file.

1. In SAS **Explorer**, display the window named **Contents in the "SAS Environment."** Refer to previous examples if you don't remember how to display this window. Double-click the **Libraries** icon, shown previously in Figure 3.3.
2. A new window named **Active Libraries** appears. Examine the active library names. There should *not* be a library named **Mylib2**.

(Continued)

> (*Continued*)
>
> 3. In the Editor window, open the program file LIBNAME.SAS, which contains the following code:
>
> ```
> LIBNAME MYLIB2 "C:\SASDATA";
> RUN;
> ```
>
> 4. Run this program. The LIBNAME statement creates the SAS library name MYLIB2 but does not create any output. Examine the **Active Libraries** window in the **SAS Explorer Window**. There is now a library icon named **Mylib2**. The folder represented by the filing cabinet icon indicates that SAS created the MYLIB2 library.
> 5. Double-click the **Mylib2** icon and verify that the library includes one or more SAS datasets (including the datasets **Somedata**, **Cars**, and others). Note that this new library contains the same datasets as the MYSASLIB library because they are both defined as name s for "pointing to" C:\SASDATA.

In the previous example, the MYLIB2 library name continues to be active and available for use until you exit the SAS program. However, the library name MYLIB2, created with the dialog box, using the **Enable at startup** option, is remembered by SAS each time you enter the SAS environment.

3.6 USING DATA IN PERMANENT SAS DATASETS

After you have created a SAS library (either a permanent or a temporary one), you can access data in that library within SAS procedures or as input into other DATA steps. In previous examples, we created a permanent library named MYSASLIB and a temporary library named MYLIB2. They both refer to the folder location C:\SASDATA. Suppose that there is a SAS dataset in that folder named SOMEDATA.SAS7BDAT. To access that dataset in SAS in a PROC MEANS statement, you can use either the full Windows file path or either one of the library prefixes. For example, the following are three different statements that all access the same dataset:

```
PROC MEANS DATA='C:\SASDATA\SOMEDATA';RUN;
```

or

```
PROC MEANS DATA=MYSASLIB.SOMEDATA;RUN;
```

or

```
PROC MEANS DATA=MYLIB2.SOMEDATA;RUN;
```

HANDS-ON EXAMPLE 3.4

In this example, you will read data from an existing SAS dataset.

1. Open the program file `READFILE.SAS`. The entire program consists of a single line of code. (This exercise illustrates that when you already have a dataset created (in this case SOMEDATA), you do not need to use a DATA statement in your program.)

   ```
   PROC MEANS DATA='c:\SASDATA\SOMEDATA';RUN;
   ```

   Note that the DATA= statement includes the information

   ```
   DATA='c:\SASDATA\SOMEDATA';
   ```

2. Run this program and observe the output, a table of descriptive statistics created by the `PROC MEANS` statement. The data used in that procedure is from the SAS data file named `SOMEDATA.SAS7BDAT`. (This method of reading SAS datasets works only in the Windows environment.)

3. Change the statement in `PROC MEANS` to `DATA= MYSASLIB.SOMEDATA`. Note that you no longer put the dataset designation name in quotes. Rerun the program to verify that the new dataset designation works and produces the same results as those created in Step 2.

4. Change the file reference to `DATA=MYLIB2.SOMEDATA` and rerun the program. Again you should get the same results (as long as you do not exit SAS and lose the `MYLIB2` library name).

3.7 IMPORTING DATA FROM ANOTHER PROGRAM

In Chapter 2, we showed how to read data from a raw (ASCII) text file using the `INFILE` and `INPUT` statements. In this section, we'll go into more detail about importing data, and we'll do it by describing three methods:

- Importing data using the SAS Import Wizard.
- Importing using `PROC IMPORT`.
- More about importing data using an `INFILE` statement.

3.7.1 Importing Data Using the SAS Import Wizard

The SAS Import Wizard can be used to import data from a number of formats including Microsoft Access, Microsoft Excel, Comma Separated Value (CSV) files, SPSS, STATA, and others. To illustrate the wizard, we'll first look at importing a CSV file into SAS.

These files are characterized by having commas as delimiters. Ideally, this type of file has the following characteristics:

- The first row contains the names of the SAS variables, each separated by a comma, and each adhering to SAS naming conventions. (These were discussed in Chapter 2.)
- Beginning on the second line of the file, each line contains the data values for one subject, where each value is separated by a comma. (Other delimiters such as semicolons are sometimes used.)
- Two commas in a row indicate a missing value. You could also designate numeric data values that are unknown or missing with a dot (.) or a missing value code (such as -99). Missing character values can be represented by a double quote such as "".

If the CSV file does not satisfy these criteria, you may need to edit the file (using a program such as Notepad) to make it conform before proceeding to import. Figure 3.5 shows the first few lines of the CSV file CARSMPG.CSV, which is ready to be imported. Note that line 1 contains the 13 names of the variables separated by commas and each following line also contains 13 values each separated by a comma, consistent with the data type of the variable. (Note also that in this case, it is okay that some values contain blanks, such as the blank between Civic and Hybrid because it is the comma that marks a new value, not a blank.)

To import these data, you can use the SAS **Import Wizard** by selecting **File → Import Data**. The initial screen of the **Import Wizard** is displayed in Figure 3.6.

The **Import Wizard** prompts you to enter the following information:

- **Data source** (type of file to import): In this case, from the list, select the **Comma Separated Values (*.csv)** option.
- **Where the file is located:** In this case, C:\SASDATA\CARSMPG.CSV).
- **The SAS destination**: Specify the SAS library where you want to store the imported data; select the WORK library to import the data into temporary storage or a library name to store the data in a permanent file.

```
BRAND,MODEL,MINIVAN,WAGON,PICKUP,AUTOMATIC,ENGINESIZE,CYLINDERS,CITYMPG,HWYMPG,SUV,AWD,HYBRID
TOYOTA,Prius,0,0,0,1,1.5,4,60,51,0,0,1
HONDA,Civic Hybrid,0,0,0,1,1.3,4,48,47,0,0,1
HONDA,Civic Hybrid,0,0,0,1,1.3,4,47,48,0,0,1
HONDA,Civic Hybrid,0,0,0,0,1.3,4,46,51,0,0,1
HONDA,Civic Hybrid,0,0,0,0,1.3,4,45,51,0,0,1
VOLKSWAGEN,Golf,0,0,0,0,1.9,4,38,46,0,0,0
VOLKSWAGEN,Jetta,0,0,0,0,1.9,4,38,46,0,0,0
VOLKSWAGEN,New Beetle,0,0,0,0,1.9,4,38,46,0,0,0
FORD,Escape HEV 2WD,0,0,0,1,2.3,4,36,31,1,0,1
HONDA,Civic,0,0,0,0,1.7,4,36,44,0,0,0
```

Figure 3.5 CSV data ready for import

IMPORTING DATA FROM ANOTHER PROGRAM   65

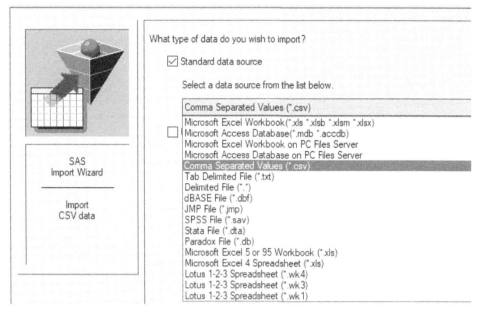

Figure 3.6 Initial screen of the SAS Import Wizard

HANDS-ON EXAMPLE 3.5

In this example, you will use the SAS import wizard to import data from a CSV file.

1. Begin the SAS **Import Wizard** by selecting **File → Import data**.
2. From the **Select a data source** drop-down list, select the type of file you want to import. See Figure 3.6. In this case, select **Comma Separated Values (*.csv)**. After you've selected the type of file to import, click **Next**.
3. When prompted to specify "Where is the file located" either type in or browse to C:\SASDATA\CARSMPG.CSV as shown in Figure 3.7.
4. Click the **Options**... button: the **Options** dialog box allows you to specify whether the names of the variables are on the first line of the CSV file, and on which line the data values begin. For this example, leave the "Get variable names from first row" checked, and first row of dataset at 2. Observe that the delimiter is set as a comma. Also note that the number of rows to guess is set at the SAS default, which is 0. That is, SAS looks at the first 20 values in a column to determine the column's data type (character, numeric, etc.). Note that character data is sometimes referred to as text or string data. Also, numeric data may be referred to as quantitative data. For now, accept all defaults and click **OK**.
5. Click **Next**. The **Library** and **Member** dialog boxes allow you to indicate where the imported data should be stored. As discussed earlier in this chapter, every SAS dataset is stored in a location called a library. You can select either the temporary library

(Continued)

(*Continued*)

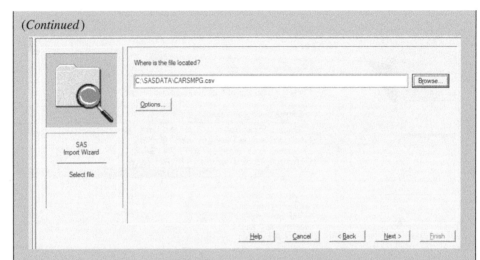

Figure 3.7 Selecting a CSV file name

(WORK) or a permanent library name. In this case, select the WORK library and name the dataset (Member name) MPG_FOR_CARS as shown in Figure 3.8.

6. Click **Next**. SAS asks if you want to create a SAS program file that contains the code that could be used to create this dataset. Enter C:\SASDATA\IMPORTCARS. This has created a SAS program named C:\SASDATA\IMPORTCARS.SAS. (Optional: Check Windows Explorer to verify that this program file exists.) We will use this program in Hands-On Example 3.6. Click **Finish** to complete the Wizard. The data in the CARSMPG.CSV file are imported into SAS and are assigned the SAS dataset name MPG_FOR_CARS in the WORK library. To verify that the SAS dataset has been created:

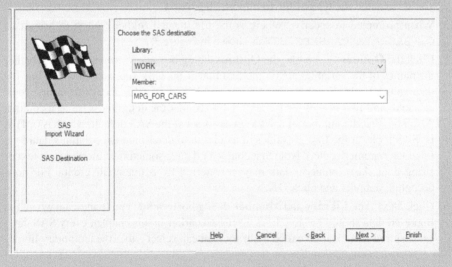

Figure 3.8 Selecting the library and member name

(a) Click the **Explorer** tab at the bottom left of the SAS screen. (If necessary, select **View → Up one level** until you are at the **Contents of SAS Environment Window**.)

(b) Double-click the **Libraries** icon, then the **Work** icon. You should see a dataset table icon named **Mpg_for_cars**.

(c) Double-click the **Mpg_for_cars** table icon, and a table appears that contains the data imported from the CARSMPG.CSV file. This dataset is now ready to use in SAS. Note that because it is in the **Work Library**, it is a temporary dataset.

7. In the Editor window, enter the following code:

   PROC PRINT DATA=MPG_FOR_CARS;RUN;

   Run the program. The following are the first few records of the listing in the **Results Viewer** as shown in Figure 3.9.

8. Re-import this CSV data file by again selecting
   - **File → Import data**
   - Comma separated value (.csv) data source
   - File location C:\SASDATA\CARSMPG.CSV
   - This time select the library named MYSASLIB (instead of WORK)
   - Name the SAS dataset (member) AUTOMPG.
   - Finish

   When you have imported the CSV file into the SAS dataset, its name becomes MYSASLIB.AUTOMPG, and it is stored in a permanent file (on your hard drive as the file named C:\SASDATA\AUTOMPG.SAS7BDAT). (Optional: use Windows Explorer to verify that this file is on your hard drive.)

9. What is the difference between the SAS dataset MPG_FOR_CARS and MYSASLIB.AUTOMPG?

   (MPG_FOR_CAR is a temporary dataset also known as WORK.MPG_FOR_CARS; MYSASLIB.AUTOMPG is a permanent dataset.) If you didn't know the answer, go back and re-read the sections on temporary and permanent SAS datasets.

Obs	BRAND	MODEL	MINIVAN	WAGON	PICKUP	AUTOMATIC	ENGINESIZE	CYLINDERS	CITYMPG	HWYMPG	SUV	AWD	HYBRID
1	TOYOTA	Prius	0	0	0	1	1.5	4	60	51	0	0	1
2	HONDA	Civic Hybrid	0	0	0	1	1.3	4	48	47	0	0	1
3	HONDA	Civic Hybrid	0	0	0	1	1.3	4	47	48	0	0	1
4	HONDA	Civic Hybrid	0	0	0	0	1.3	4	46	51	0	0	1
5	HONDA	Civic Hybrid	0	0	0	0	1.3	4	45	51	0	0	1
6	VOLKSWAGEN	Golf	0	0	0	0	1.9	4	38	46	0	0	0
7	VOLKSWAGEN	Jetta	0	0	0	0	1.9	4	38	46	0	0	0
8	VOLKSWAGEN	New Beetle	0	0	0	0	1.9	4	38	46	0	0	0
9	FORD	Escape HEV 2WD	0	0	0	1	2.3	4	36	31	1	0	1
10	HONDA	Civic	0	0	0	0	1.7	4	36	44	0	0	0
11	VOLKSWAGEN	Jetta Wagon	0	1	0	0	1.9	4	36	43	0	0	0
12	HONDA	Civic	0	0	0	1	1.7	4	35	40	0	0	0
13	TOYOTA	Echo	0	0	0	0	1.5	4	35	42	0	0	0
14	VOLKSWAGEN	Jetta	0	0	0	1	1.9	4	35	42	0	0	0
15	VOLKSWAGEN	New Beetle	0	0	0	1	1.9	4	35	42	0	0	0
16	FORD	Escape HEV 4WD	0	0	0	1	2.3	4	33	29	1	0	1

Figure 3.9 MPG_FOR_CARS dataset n SAS Viewer

3.7.2 Importing Data Using SAS Code

Instead of using the **Import Wizard** to import CSV (or other types of data files), you can import data using SAS code. This technique is handy when you are importing a number of similar files. For example, suppose that your CSV dataset changes over time and you need to re-import it whenever it changes. When you run the SAS program, it reads the new data from the CSV file and creates a SAS dataset that you can use in an analysis.

In Hands-On Example 3.5 involving the **Import Wizard**, we requested that SAS save the code used to import the CARSMPG dataset. The **Import Wizard** created PROC IMPORT code to perform the import procedure, and because we requested it, it "wrote" that code to a file we named IMPORTCARS.SAS. The code created by the **Import Wizard** is as follows:

```
PROC IMPORT OUT= WORK.MPG_FOR_CARS
            DATAFILE= "C:\SASDATA\CARSMPG.CSV"
            DBMS=CSV REPLACE;
            GETNAMES=YES;
            DATAROW=2;
RUN;
```

This code can also be found in program file IMPORTEXAMPLE.SAS

(It may be slightly different depending on your version of SAS.) If you submit this code from the SAS editor, it performs the exact import procedure performed by the **Import Wizard**. A simplified syntax for PROC IMPORT is as follows:

```
PROC IMPORT OUT=dataset
            DATAFILE="filename"
      DBMS=dbmsidentifier <REPLACE>;
            <GETNAMES=YES>; *Default is YES;
            <DATAROW=n>; *Defaultfor n is 2;
            <MIXED=YES>; *Default is NO;
```

PROC IMPORT is a SAS procedure you use to import data from a variety of file formats. There are some advantages to using the PROC IMPORT code rather than using the wizard. Some of those will be discussed in this section. The following is an explanation of the syntax of the PROC IMPORT code used to import CARSMPG:

- The OUT= option defines the name of the SAS dataset where the resulting file will be stored. In this case, because the file is WORK.MPG_FOR_CARS, it is stored in the WORK (temporary) library.
- The DATAFILE= option indicates the location of the original CSV file.
- The DBMS= option indicates the type of data source file to be imported. Table 3.1 lists the types of files SAS can import. In this case, CSV is used because the file to be imported is in CSV format.
- The REPLACE option instructs SAS to replace a previously created dataset that has the same name.
- GETNAMES=YES indicates that the names of the variables appear on the first row of the file. (A NO option would indicate that the names are not on the first row. The GETNAMES option is not available when DBMS=EXCELCS.)

TABLE 3.1 SAS DBMS Data Sources

DBMS Identifier	Data Source	File Extension
XLS	Microsoft Excel 97, 2000, 2002, 2003, or 2007	.XLS
XLSX	Excel 2000 and later	.XLSX
EXCEL	Microsoft Excel 97, 2000, 2002, 2003, or 2007	.xls .xlsb .xlsm .xlsx
EXCELCS	Excel 2007 and higher (if you have the free SAS PC File Server add-on installed.)	.XLSX
DLM	Delimited file (default delimiter is a blank)	.*
CSV	Delimited file (comma-separated values)	.CSV
TAB	Delimited file (tab-delimited values)	.TXT
ACCESS	Microsoft Access database	.MDB
ACCESSCS	Access database (if you have the free SAS PC File Server add-on installed.)	.MDBX
WK1, WK3, or WK4	Lotus 1, 3, or 4 spreadsheet	.WK1 .WK3 .WK4
DBF	dBASE file	.DBF

- `DATAROW=2` indicates that the data values are to be imported starting with the second row.
- `MIXED` = indicates how to "guess" at the appropriate informat for a variable. By default, SAS makes an educated guess based on the first 8 rows of a column. `MIXED=YES` tells the program to look for multiple formats in a column. If found, the column will be read in as a character variable.
- `SCANTEXT = YES` instructs SAS to scan the column for the value with the largest number of characters and use that as the column width.
- `SHEET="Sheet1$"` If you have multiple sheets in an Excel spreadsheet, use this to indicate which sheet to read.
- `USEDATE=YES` instructs SAS to use date formats settings from the Excel file.
- `USEDATE=NO` indicates the date should be read as a character variable.
- `RANGE = "Sheet1.$A1:G33";` tells SAS to import the cells from Sheet1 from the range indicated
- Note: There are other options that are used for different DBMS specifications. Some of these will be discussed in the upcoming examples. Similar DBMS options are available for importing Microsoft Access files. See SAS documentation for more information.

HANDS-ON EXAMPLE 3.6

In this example, you will use SAS code to import data into SAS.

1. Open the program file `IMPORTEXAMPLE.SAS`, which contains the SAS code used to import the `CARSMPG.CSV` data discussed earlier.
2. To illustrate how you can use this import example as a pattern for a different import, change the `OUT= WORK.MPG_FOR_CARS` to `OUT=MYSASLIB.RESULTS` and the `DATAFILE="C:\SASDATA\CARSMPG.CSV"` to `DATAFILE=`

(Continued)

70 CHAPTER 3: READING, WRITING, AND IMPORTING DATA

> (*Continued*)
>
> "C:\SASDATA\SURVEY.CSV". The GETNAMES and DATAROW statements do not need to be changed.
> 3. Add a line at the bottom of the code that reads
>
> PROC PRINT DATA=MYSASLIB.RESULTS;RUN;
>
> 4. Submit the code. It should produce a listing of the Survey data (as MYSASLIB.RESULTS) in the **Results Viewer**.
> 5. As a second example, import an Excel file (.XLS) by opening the SAS program IMPORTEXAMPLE2.SAS
>
> ```
> PROC IMPORT OUT= WORK.FROMXL
>             DATAFILE= "C:\SASDATA\EXAMPLE.XLSX"
>             DBMS=XLSX REPLACE;
>       SHEET="Database";
>       GETNAMES=YES;
> RUN;
> PROC PRINT DATA=FROMXL;RUN;
> ```
>
> 6. Note the option SHEET=. You must know the name of the worksheet in the Excel file, and specify it in this option. Optional: Open the file EXAMPLE.XLSX to verify that the file consists of one sheet, and it is named Database. The GETNAMES=YES indicates that the first row of the Excel file contains the names of the variables.
> 7. Run this code. It should produce a listing of the Example data. (Notice that because this is an XLSX file, we used the XLSX DBMS identifier. If it had been an XLS type file, we would have used the XLS DBMS identifier. An example of how to do this is in the exercises at the end of the chapter.)

3.7.3 Exporting Data Using SAS Code

Exporting data from a SAS data file into another file type (such as a CSV file) is similar to importing. In this case, you use the PROC EXPORT procedure. A simplified syntax for exporting is as follows:

```
PROC EXPORT DATA=dataset
            OUTFILE="fileneme" or OUTTABLE="tablename"
            DBMS=dbmsidentifier
            <LABEL><REPLACE>;
```

The DBMS options are listed in Table 3.1.

> **HANDS-ON EXAMPLE 3.7**
>
> 1. Open the program file EXPORT.SAS. This code exports the SAS dataset named SOMEDATA to a CSV file.

```
PROC EXPORT DATA=MYSASLIB.SOMEDATA
            OUTFILE="C:\SASDATA\EXPORTED.CSV"
            DBMS=CSV
            REPLACE;
RUN;
```

Run this code and observe (in Windows File Explorer) that the file

`C:\SASADATA\EXPORTED.CSV was created.`

2. Open `EXPORTED.CSV` in Notepad or Word and notice that the first line in the file includes the variable names. The remainder of the file includes the data values, separated by commas.
3. Close the `EXPORTED.CSV` file. In your SAS code, add a `LABEL` statement before the `REPLACE` option. Run this revised code. Open `EXPORTED.CSV` in Excel. This time notice that the first row contains the SAS labels for the columns rather than the variable names (Creating and using labels is introduced and discussed in Chapter 4.)
4. Change the `OUTFILE` name to `C:\SASDATA\EXPORTED.XLSX`, change the `DBMS` type to `XLSX`, and run the program. Verify that an Excel file containing the data was created.

3.8 DISCOVERING THE CONTENTS OF A SAS DATASET

Data can come from a number of sources. A popular way to provide data, by e-mail or download, is in a SAS dataset. (These datasets have a .SAS7BDAT file extension.) If you acquire a SAS dataset from another source, you can find out about its contents in two ways:

- Copy the `SAS7BDAT` data file into a folder that is linked to a SAS library (such as `C:\SASDATA`). Use SAS **Viewdata** to view the contents of the dataset.
- Use `PROC DATASETS` to list information about the variables in the dataset.

We've already illustrated how to use **Viewdata** and `PROC PRINT` to see the actual contents of a dataset. The `PROC DATASETS` technique does not show you the actual data. Instead, it provides details about the variables in the dataset, including their names, formats, and labels. For example, the following SAS program (`CONTENTS.SAS`) displays information about the `SOMEDATA` dataset in the `MYSASLIB` library:

```
PROC DATASETS;
CONTENTS DATA= 'C:\SASDATA\SOMEDATA';
RUN;
```

When you run this program, SAS produces a listing of the contents of the dataset, including variable names and other information. It is a handy way to discover what information is in a particular SAS dataset. This is illustrated in Hands-On Example 3.8.

HANDS-ON EXAMPLE 3.8

In this example, you will discover the contents of a SAS data file using PROC DATASETS.

1. Open the program file CONTENTS.SAS. Run the program.
2. Examine the output. The first two sections of the output contain technical information, and the third section (see Table 3.2) contains a list of variables in the dataset. Notice that this SAS dataset description contains:
   - **Variable**: Names are listed in alphabetical order.
   - **Type**: The variable type – Num (numeric), Char (Character/Text), Date, and other SAS data types.
   - **Len**: The number of characters or digits used to store the information.
   - **Label**: A description of the variable (listed only if defined).
3. Change the DATA statement to read

   CONTENTS DATA= MYSASLIB._ALL_;

Run the revised program. The _ALL_ keyword instructs SAS to display information about all SAS datasets in the MYSASLIB library.

TABLE 3.2 Example PROC CONTENTS Output

	Alphabetic List of Variables and Attributes			
#	Variable	Type	Len	Label
3	AGE	Num	8	Age on January 1, 2000
10	GENDER	Char	6	—
2	GP	Char	1	Intervention Group
1	ID	Num	8	ID Number
9	SEX	Num	8	—
8	STATUS	Num	8	Socioeconomic Status
4	TIME1	Num	8	Baseline
5	TIME2	Num	8	6 Months
6	TIME3	Num	8	12 Months
7	TIME4	Num	8	24 Months

3.9 GOING DEEPER: UNDERSTANDING HOW THE DATA STEP READS AND STORES DATA

To understand how SAS works, it can be helpful to "look under the hood" and see what is happening in all those bits and bytes as SAS reads and processes data. Knowing how SAS handles data is a little bit like knowing how the motor in your car works. You can usually drive around okay without knowing anything about pistons, but sometimes, it is good to know what that knocking sound means.

GOING DEEPER: UNDERSTANDING HOW THE DATA STEP READS AND STORES DATA

Having an understanding of the way the DATA step works can help you create programs that avoid problems (programming errors) and take advantage of the internal workings of the program. For example, consider the following simple program.

```
DATA NEW;
INPUT ID $ AGE TEMPC;
TEMPF=TEMPC*(9/5)+32;
DATALINES;
0001 24 37.3
0002 35 38.2
;
RUN;
```

In this program, SAS reads in two lines of data. In addition to reading the data, it also performs a calculation. The TEMPF= statement calculates Fahrenheit temperature from the Celsius temperature represented by the variable TEMPC. To read in these data and do this calculation, SAS goes through three stages represented by the graph in Figure 3.10:

COMPILE SAS reads the syntax of the SAS program to see if there are any errors in the code. If there are no errors found, SAS "compiles" this code – that is, it transforms the SAS code into a code used internally by SAS. (You don't need to know this internal code.)

EXECUTE If the code syntax checks out, SAS begins performing the tasks specified by the code. For example, the first line of code is DATA NEW, so during the execution phase, SAS creates a "blank" dataset (no data in it) named NEW that it will use to put the data into as it is read.

OUTPUT SAS reads in each data line. It interprets this line of data into the values for each variable and stores them into the dataset (NEW in our case) one line at a time until all data have been output into the dataset.

More specifically, once SAS passes the compile phase, it creates an area of memory called the input buffer. This buffer is the location where the data values are first placed before they

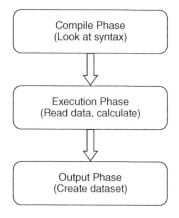

Figure 3.10 The SAS INPUT process

are separated into values associated with a particular variable. In this instance, the first data line contains the following information:

```
0001 24 37.3
```

SAS reads this line of data and places each character into the buffer, as illustrated in the following:

1	2	3	4	5	6	7	8	9	10	11	12
0	0	0	1		2	4		3	7	.	3

SAS also creates another piece of memory called a Program Data Vector (PDV), as shown in the following:

_N_	_ERROR_	ID	AGE	TEMPC	TEMPF
1	0				

The PDV contains the names of the variables specified in the INPUT statement, plus two temporary variables, _N_ and _ERROR_. The variable _N_ is the number of the data line being entered (set to 1 in this case for the first line) and _ERROR_ to indicate if an error in reading data has been encountered. It is initially set at 0 (which means no error detected.)

The INPUT specification tells SAS how to move the characters in the input buffer into the PDV. In this case, the freeform data entry tells SAS to read characters in the input buffer, starting with the far left and going to the right, until it sees a blank. When a blank is encountered, SAS moves the data up to that point into the first variable specified in the PDV. In this case, it moves the "0001" into the memory space in the PDV allocated for ID. When all of the information in the first line has been read, the PDV looks like the following:

_N_	_ERROR_	ID	AGE	TEMPC	TEMPF
1	0	0001	24	37.3	

Notice that the values of each variable read from the data line are now in the PDV. However, the TEMPF variable is set at missing. SAS continues to read the code and sees that the DATA step also includes a calculation

```
TEMPF=TEMPC*(9/5)+32;
```

Using the data already in the PDV for TEMPC (which is 37.3), it uses that information to calculate TEMPF. Once the calculation is performed, that value is also placed in the PDV, and it looks like the following:

_N_	_ERROR_	ID	AGE	TEMPC	TEMPF
1	0	0001	24	37.3	99.14

After all of the data for a line are read in, all of the calculations specified in the DATA step have been performed, and there are no errors detected, SAS outputs the PDV to the dataset (NEW in our example). In this case,

ID	AGE	TEMPC	TEMPF
0001	24	37.3	99.14

Notice at this stage that only the data values appear in the dataset, and the two temporary variables, _N_ and _ERROR_, are not included. Therefore:

- The _N_ and _ERROR_ variables are available from one iteration of the dataset to the next.
- However, they are not saved in the output dataset.
- And they cannot be renamed, kept, or dropped.

The value of understanding how SAS handles data as it is read in from raw data sources is that it provides you with insight in writing code to successfully input the data in the way you want it to be read. The brief discussion here is only an introduction to this process. If you want to delve deeper into the inner workings of the DATA step, refer to SAS documentation.

3.10 SUMMARY

This chapter defined the difference between temporary and permanent SAS datasets and illustrated several methods for importing datasets into SAS using either the SAS Wizard or PROC IMPORT. Finally, the way SAS "thinks" as it is inputting data is explained.

EXERCISES

3.1 **Create a SAS Library Reference.**

You should have already created a SAS library named MYSASLIB in the C:\SASDATA folder (or whatever is appropriate to your SAS setup). If you have not done so, use the dialog box or LIBNAME statement described in this chapter.

a. Create the following SAS libraries using the **New Library** dialog box technique: MYLIB1, MYLIB2, MYLIB3, MYLIB4, MYLIB5. Make all of these library definitions point to the folder C:\SASDATA (or whatever is appropriate to your SAS setup). Verify that the libraries appear in the SAS Explorer.

b. Create the following SAS libraries using the LIBNAME command in code: MYLIBA, MYLIBB, MYLIBC, MYLIBD, MYLIBE (or whatever is appropriate to your SAS setup). Verify that the libraries appear in the SAS Explorer.

3.2 **Create a permanent SAS dataset.**

a. Open the program file EX_3.2.SAS.

```
DATA WEIGHT;
INPUT TREATMENT LOSS @@;
DATALINES;
1 1.0 13.0 1 -1.0 1 1.5 1 0.5 1 3.5
2 4.5 2 6.0 2 3.5 2 7.5 2 7.0 2 6.0 2 5.5
3 1.5 3 -2.5 3 -0.5 3 1.0 3 .5
;
PROC PRINT;
RUN;
```

This program creates a temporary dataset called WEIGHT (or WORK.WEIGHT). Submit this code and use the SAS Explorer (WORK library) to verify that the dataset was created.

b. Change the SAS program so that it creates a permanent SAS dataset named WEIGHT1 in the C:\SASDATA folder *using the Windows folder name in the DATA statement*:

```
DATA "C:_____";
```

(or the appropriate statement for your particular SAS implementation). View the **Log Window** to verify that the permanent dataset was created.

c. Change the SAS program so that it creates a permanent SAS dataset named MYSASLIB.WEIGHT2 (in the previously created SAS library MYSASLIB) using the two-level SAS dataset name:

```
DATA _____._____
```

   I. To verify that the permanent dataset was created, open it from the MYSASLIB library into the **Viewtable**.
   II. Use **File → Print** to print a copy of this (entire) table.

d. Using the following code:

```
PROC DATASETS;
CONTENTS DATA= MYSASLIB.WEIGHT2;
RUN;
```

produce a listing of the contents of the SAS dataset MYSASLIB.WEIGHT2. Change the DATA= statement to

```
CONTENTS DATA=MYSASLIB._ALL_;
```

to display the contents of all datasets in the library.

3.3 **Use SAS code to import XLSX data.**

   a. Write the SAS code to import data from the Excel file FAMILY2.XLSX using these criteria:

      I. Create a SAS dataset named INCOME in the WORK library using the OUT= statement.

II. Use DATAFILE= to specify the location of the original Excel file (XLSX type) on your hard disk.

III. Use DBMS= to specify the data source (see the chart for DBMS options earlier in the chapter for this type of file.)

IV. Use SHEET= to designate "Data" as the name of the Excel table to import.

V. If appropriate, use the GETNAMES= option to indicate that the variable names are on the first row of data.

b. Once the data are imported, view the contents of WORK.INCOME using Viewdata. Close the Viewdata data grid.

c. Calculate descriptive statistics for the data using this code:

```
TITLE "PUT YOUR NAME HERE";
PROC MEANS DATA=INCOME;
RUN;
```

3.4 **More about SAS libraries.**

Suppose that your SAS datasets are stored in the physical directory (folder) named C:\RESEARCH on your hard drive.

a. If you want to create a shortcut library name in SAS named CLINICAL, what LIBNAME statement would specify your data's location? (Fill in the blanks.)

LIBNAME _____ 'C:_____';

b. If you want to use the data in a file named C:\RESEARCH\JAN2009.SAS7BDAT in your SAS program, fill in the blanks in the following DATA = statement:

```
LIBNAME CLINICAL 'C:\RESEARCH';
PROC MEANS DATA = _____._____;
```

c. Use the following information to create a LIBNAME statement:
   Hard drive location is O:\CLINICAL
   SAS library name is YOURNAME

LIBNAME _____ '_____';

d. If you have a SAS library named CLINICAL and you want to create a permanent dataset named MAY2007 that is stored in the C:\RESEARCH folder on your hard drive, what is the name of the file used in SAS code?

_____._____

e. In step d of this exercise, what is the name of the SAS data file on your hard drive?

_____:\RESEARCH\MAY2007._____

3.5 **Going deeper: How SAS reads data.**
In the Compile, Execute, and Output phases in the Data Step, where in the process do these tasks take place?

a. SAS looks at the code to see if there are any errors.
   _____phase
b. SAS reads the data and performs calculations
   _____phase
c. SAS creates the dataset
   _____phase
d. SAS identifies type and length of each variable
   _____phase
e. SAS creates the Input Buffer and puts the contents of data record one in the buffer
   _____phase
f. SAS creates the PDV
   _____phase
g. The initial values in the PDV are

   _N_ = ____
   ERROR = ____

h. After SAS has filled the PDV and there are no more lines of code in the DATA step, it sends the contents of the PDV to the _____ buffer.

3.6 **Export a data set to Excel using PROC EXPORT.**
   a. Use the following data snippet (from EX_3.6.SAS) to export the CARS dataset into an Excel file named CARSDATA.XLSX.
   b. Open the file in Excel to verify that it was correctly created.
   c. What is the sheet name given to the sheet created in Excel?

4

PREPARING DATA FOR ANALYSIS

LEARNING OBJECTIVES

- To be able to label variables with explanatory names
- To be able to create new variables
- To be able to use SAS® IF-THEN-ELSE statements
- To be able to use DROP and KEEP to select variables
- To be able to use the SET statement
- To be able to use PROC SORT
- To be able to append and merge datasets
- To be able to use PROC FORMAT
- Going Deeper: To be able to find first and last values in a group

You've entered your data into SAS, but before you can use them in SAS, you typically need to make corrections, perform calculations, and otherwise prepare your data for analysis. This step is sometimes overlooked in classroom settings where you often use ready-to-analyze textbook data. When you gather your own data, or even when you acquire data from a database, the data may require some changes before you can use them in an analysis.

SAS® Essentials: Mastering SAS for Data Analytics, Third Edition. Alan C. Elliott and Wayne A. Woodward.
© 2023 John Wiley & Sons, Inc. Published 2023 by John Wiley & Sons, Inc.
Companion website: https://www.alanelliott.com/SASED3/

80 CHAPTER 4: PREPARING DATA FOR ANALYSIS

> All of the data manipulation statements discussed in this chapter appear in your program code within the DATA step, except for PROC FORMAT. That is, they appear after the DATA statement and before the first RUN or PROC statement. Thus, all of the results of the data manipulations performed by these statements become part of the active dataset.

4.1 LABELING VARIABLES WITH EXPLANATORY NAMES

SAS labels are used to provide descriptive names for variables. Datasets often contain cryptic variable names such as ID, SBP, WT, and so on. For people familiar with the dataset, these make perfect sense. However, if you want to produce output that is readable by others, creating explanatory labels for variables is good practice. In SAS, this can be accomplished by using the LABEL statement. The LABEL statement uses the format:

```
LABEL      VAR1 = 'Label for VAR1'
           VAR2 = 'Label for VAR2';
```

You can use either single or double quotation marks in the LABEL statement, but you must match the type within each definition statement. When SAS prints out information for VAR1, it also includes the label, making the output more readable. The following program (DLABEL.SAS) illustrates the use of labels.

```
DATA MYDATA;
INFILE 'C:\SASDATA\BPDATA.DAT'; * READ DATA FROM FILE;
INPUT ID $ 1 SBP 2-4 DBP 5-7 GENDER $ 8 AGE 9-10 WT 11-13;
LABEL ID = 'Identification Number'
      SBP= 'Systolic Blood Pressure'
      DBP = 'Diastolic Blood Pressure'
      AGE = 'Age on Jan 1, 2022'
      WT = 'Weight';
RUN;
PROC MEANS;
     VAR SBP DBP AGE WT;
RUN;
```

Notice that the LABEL statement is placed within the DATA step. Also, there is only one semicolon (;) in the LABEL statement, at the very end.

The output listings in Tables 4.1 and 4.2 show how the LABEL statement adds information to the output. The first set of output shows PROC MEANS output *without* the LABEL statement. Notice that all variable names are listed just as they were created from the INPUT statement in the DATA statement.

The next set of output from PROC MEANS is from the same data, and in this case, the LABEL statement, as shown in the aforementioned example code, has been used. As a result, more informative labels are included in the output.

Follow Hands-On Example 4.1 to add label definitions for a dataset within the DATA step.

LABELING VARIABLES WITH EXPLANATORY NAMES

TABLE 4.1 OUTPUT without Labels

		The MEANS Procedure			
Variable	N	Mean	Std Dev	Minimum	Maximum
SBP	5	127.0000000	8.3666003	120.0000000	140.0000000
DBP	5	82.0000000	10.9544512	70.0000000	100.0000000
AGE	5	35.8000000	30.2605354	15.0000000	89.0000000
WT	5	145.0000000	31.6227766	110.0000000	180.0000000

TABLE 4.2 OUTPUT with Labels

		The MEANS Procedure				
Variable	Label	N	Mean	Std Dev	Minimum	Maximum
SBP	Systolic Blood Pressure	5	127.0000000	8.3666003	120.0000000	140.0000000
DBP	Diastolic Blood Pressure	5	82.0000000	10.9544512	70.0000000	100.0000000
AGE	Age on Jan 1, 2022	5	35.8000000	30.2605354	15.0000000	89.0000000
WT	Weight	5	145.0000000	31.6227766	110.0000000	180.0000000

HANDS-ON EXAMPLE 4.1

In this example, you will create labels for variables and learn how to output the labels using PROC PRINT.

1. Open the program file DLABEL2.SAS.

```
DATA MYDATA;
INPUT @1 FNAME $11. @12 LNAME $12. @24 BDATE DATE9.;
FORMAT BDATE WORDDATE12.;
LABEL   FNAME="First Name"; * Complete the labels;
DATALINES;
Bill       Smith        08JAN1952
Jane       Jones        02FEB1953
Clyde      York         23MAR1949
;
PROC PRINT;
RUN;
```

2. Add lines to create labels for LNAME (Last name) and BDATE (Birth date). For example, the LABEL statement could be as follows:

```
LABEL FNAME="First Name"
LNAME="Last Name"
BDATE="Birth Date";
```

Don't forget the semicolon (;) after the last label definition.

(*Continued*)

82 CHAPTER 4: PREPARING DATA FOR ANALYSIS

(*Continued*)

3. Run the program. Notice that the label information is not included in this output.
4. Most SAS PROCs output the label information automatically. However, for PROC PRINT, you must add the option LABEL to the statement. Change the PROC PRINT statement in this program to read:

PROC PRINT LABEL;

5. Run the program and observe how the labels are shown in the output. The output is shown in Table 4.3.

TABLE 4.3   Listing Formatted Dates

Obs	FNAME	LNAME	BDATE
1	Bill	Smith	Jan 8, 1952
2	Jane	Jones	Feb 2, 1953
3	Clyde	York	Mar 23, 1949

4.2   CREATING NEW VARIABLES

It is common to calculate new variables in a dataset in preparation for analysis. For example, you may need to convert a temperature reading from Fahrenheit to Celsius, change the measurement from centimeters to inches, or calculate a score from a series of values. One method for creating new variables in your dataset is to calculate them within the DATA step. This section describes some of the techniques you can use.

You can create new variables within a SAS DATA statement by assigning a value to a new variable name or by calculating a new variable using a mathematical or logical expression. When the SAS program is run, data values for the new variables are assigned to each of the records in the currently active dataset.

> If you have experience using Microsoft Excel, you know that when you change any of the numbers that were used to calculate a value, the calculated value changes automatically. The SAS calculations in the DATA step are **not like** those in Excel. They are one-time calculations that take place when you run the SAS program. If you want to recalculate the expression, you must rerun (resubmit) the SAS program.

4.2.1   Creating Numeric Variables in the DATA Step

Within the SAS DATA step, you can calculate a new variable using current variables in your dataset. These calculations use standard arithmetic operators as defined in Table 4.4.

Hands-On Example 4.2 illustrates how you could calculate a new value for each record in your dataset by using a simple arithmetic expression – in this case, by multiplying two variables (WIDTH and LENGTH) that are in the dataset to create a new variable (AREA).

TABLE 4.4 SAS Arithmetic Operators

Symbol	Meaning	Example
+	Addition	SUM= X+Y
–	Subtraction	DIF=X-Y;
*	Multiplication	TWICE=X*2;
/	Division	HALF=X/2;
**	Exponentiation	CUBIC=X**3;

HANDS-ON EXAMPLE 4.2

In this example, you will create a new variable by calculating from existing information in a SAS dataset.

1. Open the program file DCALC.SAS. This file calculates the square footage (AREA) in various rooms of a house and adds up the total. Notice that the example defines labels for the variables, as discussed and recommended in the previous section.

```
DATA ROOMSIZE;
FORMAT ROOM $10.;
INPUT ROOM $ W L;
* The next line calculates a new variable, AREA;
AREA=L*W;
Label L="Length" W="Width" AREA="Sq. Feet";
DATALINES;
LIVING     14 22
DINING     14 12
BREAKFAST  10 12
KITCHEN    12 16
BEDROOM1   18 12
BEDROOM2   12 14
BEDROOM3   13 16
BATH1       8 12
BATH2       7 10
BATH3       6  8
GARAGE     23 24
;
RUN;
PROC PRINT LABEL; SUM AREA;
RUN;
```

Notice the following statement within the DATA step:

AREA = L * W;

   This statement creates a new variable named AREA that is a result of multiplying length by width.

(Continued)

(Continued)

2. Run the program and observe the output in Table 4.5.

    The calculation AREA = L * W in the DATA step creates a new variable AREA in the dataset named ROOMSIZE.

    - The SUM AREA statement in PROC PRINT causes SAS to report the sum of the AREA values.

    TABLE 4.5 Output Showing Results of Calculation

Obs	ROOM	Width	Length	Sq. Feet
1	LIVING	14	22	308
2	DINING	14	12	168
3	BREAKFAS	10	12	120
4	KITCHEN	12	16	192
5	BEDROOM1	18	12	216
6	BEDROOM2	12	14	168
7	BEDROOM3	13	16	208
8	BATH1	8	12	96
9	BATH2	7	10	70
10	BATH3	6	8	48
11	GARAGE	23	24	552
				2146

3. Note the FORMAT statement in the code. What does it do? If you remove it, what happens to the word BREAKFAST in the output? Why?
4. Suppose the ceilings are 8 feet tall, and you want to calculate the total volume of the home (cubic feet for air conditioning purposes.) Calculate the volume (VOL) using W*L*8.
5. Also add a label to display VOL as Volume and add VOL to the SUM statement. Run the program and observe the results.

SAS allows you to perform more than just simple calculations. You can perform virtually any type of algebraic calculation needed. When performing more extensive calculations, you must pay attention to the order of evaluation used in expressions. This order is similar to that used in common algebraic practice and is also commonly used in other computer programs. The priority of evaluation is as follows:

( )   ** *   /   + -

That is, when SAS evaluates an expression, it first performs calculations in parentheses, then exponentiation, then multiplication, division, addition and then subtraction. For example,

ANSWER = A + B * C;

results in a different answer from

```
ANSWER = (A + B) * C;
```

The first example multiplies B * C, and then adds to the product A. The second adds A+B, then multiplies the resulting sum by C. If you are unsure about how an expression will be evaluated, use parentheses to force the evaluation to be as you intend. For example,

```
NEWVAL = (SCORE1-SCORE2)*INDEX**2;
```

gives the instruction to subtract SCORE2 from SCORE1 first, then multiply the result by $INDEX^2$.

> You may find the following mnemonic (which you may have learned in a previous math class) helpful in remembering the order of operations: "**P**lease **E**xcuse **M**y **D**ear **A**unt **S**ally" (**P**arentheses **E**xponents **M**ultiplication **D**ivision **A**ddition **S**ubtraction).

4.2.2 Creating New Variables as Constant Values

Sometimes, it is handy to define a new variable as a constant value. The reason for creating this type of variable is that you might want to use the value in a series of future calculations. For example, suppose that you are about to calculate new values in your dataset with a formula that uses the value of *pi* (π). A statement such as the following could appear within your DATA step (before you use the variable PI in another calculation):

```
PI = 3.1415927;    *Value to be used later;
```

For example, assuming RADIUS is a variable in your dataset, you could define the following expressions:

```
AREA = PI * RADIUS**2;
CIRCUM=2*PI*RADIUS;
```

to calculate the area and circumference of a circle from the measure of its radius without having to repeatedly enter 3.1415927.

> Variable names on the left side of each equation are created as a *result* of the *statement*. Variables used on the right side of any statement must already be defined in the current dataset. If a variable on the right side of the statement is undefined or contains a missing value, the resulting value for the variable on the left side of the equation (the result) is set to missing.

The SAS SUM Statement is a way to add or accumulate values over the iterations of data records read into the dataset. The format is

```
ACCUM+addval
```

Note that, unlike in the previous examples, there is no equal (=) sign. The `addval` (which can be a constant, variable, or expression) is added to the variable ACCUM for each iteration as the data records are read. The initial value of ACCUM is 0. For each iteration in the DATA step, the value `addval` is added (because of the + sign) to ACCUM as if the statement were ACCUM=ACCUM+`addval`.

4.3 USING IF-THEN-ELSE CONDITIONAL STATEMENT ASSIGNMENTS

Another way to create a new variable in the DATA step is to use the IF-THEN-ELSE conditional statement construct. (This is sometimes also referred to as an "IF statement" or an "IF-THEN statement" because you don't always have to include all three parts in a statement.) The syntax of the full IF-THEN-ELSE statement is as follows:

IF *expression* THEN *statement*; ELSE *statement*;

For example:

IF SBP GE 140 THEN HIGHBP=1; ELSE HIGHBP=0;

This statement tells SAS to do the following: if SBP is greater than or equal to 140 (SBP GE 140), then create a new variable named HIGHBP and set the value of that variable to 1; if SBP is less than 140 (IF SBP LT 140), set the value of the variable HIGHBP to 0. In effect, this creates a grouping variable for the dataset using the variable HIGHBP with values 0 and 1. Because the first reference to HIGHBP is as a number, it is added to the list of variables in the dataset as a numeric variable.

The GE operator used in this example is one of several possible conditional operators that can be used in a comparison statement. A list of common SAS comparison operators is given in Table 4.6. You can use either the symbol or the mnemonic in a SAS expression.

You can also put several conditions together in an IF statement using the AND (mnemonic &), OR (mnemonic |), or NOT (mnemonic ^ or ~) logical operators and create expressions such as

IF AGE GT 19 & GENDER="M" THEN GROUP=1;

or

IF TREATMENT EQ "A" OR GROUP >=1 THEN CATEGORY="GREEN";

Note, as is shown in these examples, that you can also mix mnemonic and symbol operators in the same statement. You can also stack IF-THEN-ELSE statements by using an ELSE IF clause, as in:

IF TRT="A" THEN GROUP=1;
ELSE IF TRT="B" OR TRT="C" THEN GROUP=2;
ELSE GROUP=3;

The preceding lines recode the variable (TRT) into a new variable (GROUP). If TRT is equal to A, then GROUP is set to the value 1. If TRT is B or C, then GROUP is set to the value 2. Otherwise, GROUP is set to the value 3. A summary of the SAS logical operators is shown in Table 4.7.

TABLE 4.6   SAS Conditional Operators

Symbol	Mnemonic	Meaning	Example
=	EQ	Equal to	I=5
^=, or ~=	NE	Not equal to	I NE 5
<	LT	Less than	I<5
<=	LE	Less than or equal to	I<=5
>	GT	Greater than	I GT 5
>=	GE	Greater than or equal to	I >= 5
	IN	Is valued TRUE if a specified number (NUM) is in a list	NUM IN (3,4,5)

USING IF-THEN-ELSE CONDITIONAL STATEMENT ASSIGNMENTS   87

TABLE 4.7 SAS Logical Operators

Symbol	Mnemonic	Meaning	Example
&	AND	Both	I>5 AND J<3
|	OR	Either	I>5 OR J<3
^ or ~	NOT	Negate	NOT (I GT J)

HANDS-ON EXAMPLE 4.3

In this example, you will use an IF-THEN-ELSE conditional statement to create a new variable.

1. Open the program file DCONDITION.SAS.

   ```
   DATA MYDATA;
   INPUT ID $ SBP DBP GENDER $ AGE WT;
   IF SBP GE 140 THEN STATUS="HIGH"; ELSE STATUS="OK";
   DATALINES;
   001 120 80 M 15 115
   002 130 70 F 25 180
   003 140 100 M 89 170
   004 180 80 F 30 150
   005 125 80 F 20 110
   ;
   PROC PRINT;
   RUN;
   ```

2. Notice the IF-THEN-ELSE statement. If systolic blood pressure (SBP) is greater than or equal to 140, the program assigns the value HIGH to the variable STATUS. Otherwise, it assigns the value OK to that variable.
3. Run the program and observe the results in Table 4.8.

   TABLE 4.8 Results of IF-THEN-ELSE Statement

Obs	ID	SBP	DBP	GENDER	AGE	WT	STATUS
1	001	120	80	M	15	115	OK
2	002	130	70	F	25	180	OK
3	003	140	100	M	89	170	HIGH
4	004	180	80	F	30	150	HIGH
5	005	125	80	F	20	110	OK

   Notice that the new variable STATUS is assigned either the value OK or the value HIGH depending on the value of SBP. Also notice that because the first use of the variable STATUS was as a character (text) value, it is set to a SAS character type variable.

4.3.1 Using IF-THEN to Assign Missing Values

An `IF-THEN` statement is a common way to specify missing values in a SAS `DATA` step. For example,

```
IF AGE EQ -9 THEN AGE = .;
```

indicates that if the value of `AGE` is equal to −9, the value of `AGE` is set to a missing value . (dot). Because there is no `ELSE` clause, the value of `AGE` is not changed when `AGE` is not equal to −9. To specify a missing character value, use two quotation marks. For example,

```
IF GENDER NE "M" AND GENDER NE "F" THEN GENDER = "";
```

specifies a missing value for `GENDER` (two quotation marks in a row, "") when it is neither `M` nor `F`. A blank (two quotation marks in a row) is a common missing value for character variables.

When recoding missing values using an `IF-THEN` statement, you need to take care that all possibilities are accounted for. For example, suppose that you are recoding AGE into a new variable called `TEEN`. You could use the following code:

```
IF AGE GT 12 AND AGE LT 20 THEN TEEN=1;ELSE TEEN = 0;
IF AGE = . THEN TEEN = .;
```

The use of `IF AGE = . THEN TEEN = .` guarantees that the value of `TEEN` is set to missing for any observation for which AGE is missing. Otherwise, any observation where AGE is less than or equal to `12` and greater than or equal to `20` is coded as `TEEN = 0;`.

HANDS-ON EXAMPLE 4.4

In this example, you will use `IF-THEN` and `IF-THEN-ELSE` statements in the `DATA` step to create a new variable and define a missing value.

1. Open the program file `DMISSING.SAS`.

```
DATA MYDATA;
INPUT ID $ SBP DBP GENDER $ AGE;
IF AGE GT 12 AND AGE LT 20 THEN TEEN=1;ELSE TEEN=0;
IF AGE=. THEN TEEN=.;
DATALINES;
001 120 80 M 15
002 130 70 F .
003 140 100 M 12
004 180 80 F 17
005 144 80 F 23
```

```
006 165 80 M 18
007 121 80 F 19
008 195 80 M 11
009 162 80 M 13
010 112 80 F 17
;
PROC PRINT;
RUN;
```

2. Notice the IF statements that define the new variable TEEN. They also set TEEN to missing (.) if AGE is missing. Run this program to see that the value of TEEN in the second observation is set to missing.
3. Comment out the second IF AGE statement by placing an asterisk (*) before the IF. (The line should be displayed in green signifying that it is now a comment.) Submit the changed code and observe that the value of TEEN for ID 002 was incorrectly given the value 0.
4. Use the second method described previously to do the same task:

```
IF AGE=. THEN TEEN=.;
ELSE IF AGE GT 12 AND AGE LT 20 THEN TEEN=1;
ELSE TEEN=0;
```

Submit the program. Did this version work correctly?

4.3.2 Using IF and IF-THEN To Subset Datasets

Datasets can be quite large. You may have a dataset that contains some group of subjects (records) that you want to eliminate from your analysis. In that case, you can subset the data so it will contain only those records you need. For example, if your analysis concerns pregnancies, you may want to limit your dataset to females.

One method of eliminating certain records from a dataset is to use a subsetting IF statement in the DATA step. The syntax for this statement is as follows:

```
IF expression;
```

where the expression is some conditional statement in terms of the variables in the dataset. For example, to select records containing the value F (only females) from a dataset, you could use this statement within a DATA step:

```
IF GENDER EQ 'F';
```

Note that you can use single or double quotation marks ("F" or 'F' but not "F' or 'F") in this statement.

This subsetting IF statement is similar to a gate. It allows only records that meet a certain criterion to enter into the dataset. In this case, only those records whose GENDER value is recorded as F are retained in the dataset.

The opposite effect can be created by including the statement THEN DELETE at the end of the statement:

IF expression THEN DELETE;

For example, to get rid of certain records (all males) in a dataset, you could use the code

IF GENDER EQ 'M' THEN DELETE;

When data are read into a SAS dataset using this procedure and a record contains M for GENDER, then that record is *not* retained in the SAS dataset. If the only values for GENDER in the dataset are F and M, then these two subsetting strategies will yield the same results.

You are not limited to using the IF statement on character variables. An example that uses the IF statement on a numeric variable is as follows:

IF AGE GT 19;

You can also use combinations of conditions in any of these IF statements, such as

IF AGE GT 19 AND GENDER ="M";

> The statement DELETE *does not* indicate that these records are erased from the raw data file on disk. The records are only eliminated from the SAS dataset created in the DATA step.

HANDS-ON EXAMPLE 4.5

In this example, you use a subsetting IF to select only certain records for inclusion in the MYDATA SAS dataset created in the DATA step.

1. Open the program file DSUBSET1.SAS. In this example, subjects less than or equal to 10 years of age are included in the dataset named MYDATA.

   ```
   DATA MYDATA;
   INFILE 'C:\SASDATA\EXAMPLE.DAT';
   INPUT  ID $ 1-3 GP $ 5 AGE 6-9 TIME1 10-14 TIME2 15-19;
   IF AGE LE 10;
   PROC PRINT;
   RUN;
   ```

   Notice that the subsetting IF statement selects only records for which AGE is less than or equal to 10. (This does not change the contents of the file C:\SASDATA\EXAMPLE.DAT; it affects only the contents of the new SAS dataset MYDATA.)

2. Run this program and observe the results. The data listing includes only the 22 subjects whose age is less than or equal to 10.

4.3.3 Using IF-THEN and DO for Program Control

Another use of the `IF` statement is to control the flow of your SAS program in conjunction with a `DO` statement. In this case, you can cause a group of SAS commands to be conditionally executed by using the following type of code:

```
IF expression THEN DO;
Code to execute;
END;
```

For example, suppose that you want to calculate `BMI` (Body Mass Index) for subjects in a dataset, but the formula is only relevant for subjects older than 19 years of age. Plus, at the same time, you want to assign other values for this same group of subjects. You could use the following code:

```
IF AGE GT 19 THEN DO;
   BMI=(WTLBS/HTINCH**2)*703;
   ISADULT=1;
   INCLUDEINSTUDY="Yes";
END;
```

In this case, when AGE for a subject is older than 19, a series of three lines of code is processed. For other subjects in the dataset where AGE is less than or equal to 19, `BMI`, `ISADULT`, and `INCLUDEINSTUDY` are not assigned any values, and thus they have a missing value. (You might have a second `IF-THEN-DO` statement to assign other values for this case.)

4.3.4 Using @ and IF to Conditionally Read Lines in a Dataset

For big datasets, it is often the case that you don't want to read in all of the data. One method you could use to conditionally read in certain records is to set up a test condition and read in the record only if it meets that condition. To do this, you can use the @ (at) sign in your input statement to read in a portion of a record and put the input on hold while you decide whether or not to read in the complete record. This is referred to as a "trailing at." This is illustrated in Hands-On Example 4.6.

HANDS-ON EXAMPLE 4.6

This example illustrates the use of a trailing at to conditionally read in data.

1. Open the program file `DSUBSET2.SAS`. Note that the first `INPUT` statement ends in a trailing at. The `IF` statement tests for records where the group is A and AGE is greater than or equal to 10 and reads only the records that match the condition. Run the program and observe the results.

   ```
   DATA MYDATA;
   INFILE 'C:\SASDATA\EXAMPLE.DAT';
   INPUT GP $ 5 AGE 6-9 @;
   ```

   (Continued)

(Continued)

```
    IF GP EQ "A" AND AGE GE 10 THEN
    INPUT ID   $ 1-3 GP $ 5 AGE 6-9 TIME1 10-14 TIME2 15-19;
    PROC PRINT DATA=MYDATA;
    RUN;
```

2. Change the IF statement to look for groups A or B using

   ```
   IF (GP EQ "A" OR GP="B") AND AGE GE 10;
   ```

   Resubmit and observe the difference.
3. Change the program to only read records where ID is less than or equal to 100. Resubmit to verify that your changes work.

4.4 USING DROP AND KEEP TO SELECT VARIABLES

Sometimes, you may have a dataset containing *variables* you do not want to keep. The DROP and KEEP statements in the DATA step allow you to specify which variables to retain in a dataset. The general form for these statements is as follows:

```
DROP variables;
KEEP variables;
```

For example,

```
DATA MYDATA;
INPUT A B C D E F G;
DROP E F;
DATALINES;
... etc ...
```

reads in all of the data but then drops the variables E and F so that these variables are no longer in the temporary SAS dataset named MYDATA.

You can also use DROP and KEEP as you are reading in the data. In this case, the statement is included in the DATA statement as an option specified in parentheses. The format for this use is (DROP= list of variables) or (KEEP= list of variables). For example,

```
DATA MYDATA (DROP=E F);
```

Hands-On Example 4.7 illustrates this technique.

HANDS-ON EXAMPLE 4.7

In this example, you will read in a dataset but keep only a selection of the variables.

1. Open the program file DKEEP.SAS.

   ```
   DATA MYDATA;
   INFILE 'C:\SASDATA\EXAMPLE.CSV' DLM=',' FIRSTOBS=2 OBS=26;
   INPUT   GROUP $ AGE TIME1 TIME2 TIME3 TIME4 SOCIO;
   KEEP AGE TIME1 SOCIO;
   PROC PRINT DATA=MYDATA;
   RUN;
   ```

   This SAS program reads the first 25 records from a file with comma-separated values. Using this type of data entry, you must read in all variables. To eliminate variables you do not need, use the KEEP statement (or DROP statement).

2. Run the program. Although the program reads in seven variables, the listing from the PROC PRINT statement includes only those variables you have specified in the KEEP statement.

3. Replace the KEEP statement with the statement

   ```
   DROP GROUP AGE SOCIO;
   ```

   Run the program and observe the results.

4. Change the code so the DROP option appears in the DATA step instead of being its own statement. Comment out the current KEEP statement and use this option instead:

   ```
   DATA MYDATA (DROP=list variables to drop);
   ```

4.5 USING THE SET STATEMENT TO READ AN EXISTING DATASET

Often, you have a "main" dataset that is from some corporate, government, or organization source. If you don't want to change the original dataset, you should work with a copy of it. The SET statement is what you need. The SET statement in a DATA step is used to transfer data from an existing SAS dataset to a new one. When you use a SET statement, there is no need for an INPUT, INFILE, or DATALINES statement. The SET statement is often used when you want to create one or more new SAS datasets by subsetting a larger dataset but at the same time preserve all data in the original dataset.

The general form of the SET statement is as follows:

```
SET SASdataset;
```

94 CHAPTER 4: PREPARING DATA FOR ANALYSIS

For example:

```
DATA NEW; SET OLD;
```

or

```
DATA NEW; SET "C:\SASDATA\OLD";
```

In this example, the SAS dataset named OLD already exists. All variables and observations in the existing SAS dataset (OLD) are automatically passed through the input buffer to the new SAS dataset (NEW) (unless otherwise directed with programming statements). Just as when any new dataset is created, additional variables can be created with assignment statements within the SAS step.

For example, suppose that you already have a dataset named ALL. It contains several variables including the variable GENDER that is coded "M" or "F."

Suppose you want to create a subset of ALL that contains only males. The SAS code used to do this is as follows:

```
DATA MALES;
SET ALL;
IF GENDER ='M';
RUN;
```

In this code, the DATA step creates a new dataset named MALES. The SET statement passes all data from the existing dataset ALL into the new dataset MALES. The IF statement causes the buffer to only keep the records if GENDER ="M". The result is a dataset named MALES that contains only the selected records. Figure 4.1 illustrates this process.

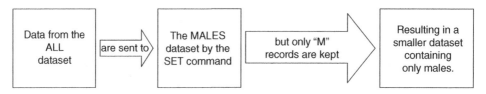

Figure 4.1 Creating a new dataset using the SET statement

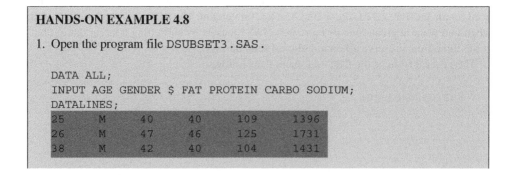

HANDS-ON EXAMPLE 4.8

1. Open the program file DSUBSET3.SAS.

```
DATA ALL;
INPUT AGE GENDER $ FAT PROTEIN CARBO SODIUM;
DATALINES;
25      M       40      40      109     1396
26      M       47      46      125     1731
38      M       42      40      104     1431
```

```
42        M        48        46        123        1711
65        M        41        41        112        1630
68        M        34        33        96         1192
20        F        39        29        118        1454
30        F        40        40        115        1532
60        F        39        40        123        1585
;
DATA MALES;
SET ALL;
IF GENDER ='M';
RUN;
```

Notice that this code already includes one subsetting IF statement in the DATA step to create a new dataset named MALES.

2. At the end of the code, add these program lines to create a FEMALES dataset using the appropriate SET and IF statements.

```
DATA FEMALES;
SET ALL;
IF GENDER ='F';
RUN;
```

3. Add lines that perform a PROC PRINT for MALES and a separate PROC PRINT for FEMALES. Remember to include a RUN statement at the end of the program. The following is the statement for MALES:

```
PROC PRINT DATA=MALES;RUN;
```

Similarly, add this line to the program to print the contents for the FEMALE dataset:

```
PROC PRINT DATA=FEMALES;RUN;
```

4. Run the completed program. How many datasets have been created using this program? What are their names? How are they related?

4.6 USING PROC SORT

The SORT procedure can be used in the DATA step to rearrange the observations in a SAS dataset or create a new SAS dataset containing the rearranged observations. Using PROC SORT you can sort on multiple sort fields and sort in ascending or descending order. The sorting sequence information for SAS datasets is shown in Table 4.9. The data are listed from the smallest to the largest in terms of each character's computer index designation (ASCII code). That is, in terms of sequence, a blank is the smallest character and ~ is the largest.

TABLE 4.9 SAS Sorting Sequence

Sorting sequence for **character variables**	blank!"#$%&©()*+,−./0123456789:;<=>?@ ABCDEFGHIJKLMNOPQRSTUVWXYZ [\]^_abcdefghijklmnopqrstuvwxyz(|)~
Sorting sequence for **numeric variables**	Missing values first, then numeric values
Default sorting sequence	Ascending (or indicate BY DESCENDING)

The syntax for PROC SORT is as follows:

PROC SORT 'options'; BY variable(s);

Options for PROC SORT include

DATA=datasetname;
OUT= outputdatasetname;

For example,

PROC SORT DATA=MYDATA OUT=MYSORT; BY RECTIME;

creates a new sorted SAS dataset named MYSORT (and the records in the original SAS dataset MYDATA remain in their original order). If you *do not include* an OUT= statement, SAS sorts the current SAS dataset. In the aforementioned example, if the OUT= statement is not included, the MYDATA dataset is sorted.

The BY statement in the PROC SORT paragraph must contain one or more variables and optionally the DESCENDING keyword. The BY DESCENDING statement tells SORT to arrange the values from highest to lowest instead of the default lowest to the highest order. For example:

BY AGE;   * Sorts in ascending/alphabetical order;
BY DESCENDING AGE; * Sorts in descending order;

You can also sort several variables at a time. For example,

PROC SORT; BY GROUP SOCIO;

will cause the dataset to be sorted on GROUP within SOCIO.

> Uppercase characters (e.g., "A") come before lowercase characters (e.g., "a"). Thus, when sorting, Z comes before a. If variables such as GENDER are entered without consistency for uppercase and lowercase (e.g., M, F, m, and f), then your sort will not properly place the records into a Male and Female grouping.

> If you *do not include* an OUT= statement, SAS sorts the current SAS dataset. In the aforementioned example, if the OUT= statement is not included, the MYDATA dataset is sorted.

HANDS-ON EXAMPLE 4.9

In this example, you will sort data in ascending and descending order.

1. Open the program file `DSORT1.SAS`.

   ```
   DATA MYDATA;
   INPUT GROUP RECTIME;
   DATALINES;
   1    4.2
   2    3.6
   2    3.1
   1    2.1
   1    2.8
   2    1.5
   1    1.8
   ;
   PROC SORT DATA=MYDATA OUT=S1; BY RECTIME;
   Title 'Sorting Example - Ascending';
   PROC PRINT DATA=S1;
   RUN;
   ```

   Note that this program sorts the data and stores the results in a SAS data file named `S1`. Submit the code and observe the sorted output.

2. Write second `PROC SORT` and `PROC PRINT` statements to sort `MYDATA` descending and save the results in SAS dataset `S2` by using the code

   ```
   PROC SORT DATA=MYDATA OUT=S2; BY DESCENDING RECTIME;
   ```

   Submit the revised program.

   Notice that the two `PROC SORT` routines produce two different results, the dataset `S1` sorted ascending and dataset `S2` sorted descending as shown in Table 4.10. Note also that in order for the "descending table to have a correct title, you will need to add an appropriate `TITLE` statement between `PROC SORT` and `PROC PRINT`.

TABLE 4.10 SAS Output Showing Sort Ascending and Descending

Sorting Example – Ascending			Sorting Example – Descending		
Obs	GROUP	RECTIME	Obs	GROUP	RECTIME
1	2	1.5	1	1	4.2
2	1	1.8	2	2	3.6
3	1	2.1	3	2	3.1
4	1	2.8	4	1	2.8
5	2	3.1	5	1	2.1
6	2	3.6	6	1	1.8
7	1	4.2	7	2	1.5

Other handy statements in PROC SORT are DROP, KEEP, and RENAME. These are illustrated in Hands-On Example 4.10.

HANDS-ON EXAMPLE 4.10

Suppose that you are sorting a dataset for the purpose of merging it (which is discussed in the following section) and you need to change the variable named ID to CASE.

1. Open the program file DSORT2.SAS.

```
PROC SORT DATA="C:\SASDATA\SOMEDATA"
       OUT=ANALYSIS
          (KEEP=ID GP AGE GENDER RENAME=(ID=SUBJECT AGE=DXAGE));
BY ID;
RUN;
PROC PRINT DATA=ANALYSIS;RUN;
```

Notice that there is a set of parentheses following the OUT= statement. Within those parentheses is a KEEP= statement that specifies which variables to keep in the output file ANALYSIS. Inside these parentheses is another statement, RENAME with its own parentheses. Inside the RENAME= parentheses are two statements that rename an old variable with a new name. The syntax of this RENAME statement is

```
RENAME =(oldname1=newname1 oldname2=newname2, etc...)
```

Therefore, the RENAME statement in DSORT2.SAS renames two variables and saves four variables in the output. Submit this code and observe the (partial) results shown in Table 4.11.

However, note that the names to KEEP must be the original variable names, not the new names. Notice that ID and AGE are kept, but in the new dataset, they are named SUBJECT and DXAGE. Run the program and observe the output.

2. Take the variable name GENDER out of the KEEP statement. Rerun the program and see that the results no longer include the GENDER column.
3. Rename the GP variable GROUP. Rerun the code and observe the changed results.

TABLE 4.11 SAS Output Showing Result of KEEP and RENAME in SORT Statement

Obs	SUBJECT	GP	DXAGE
1	101	A	12
2	102	A	11
3	104	B	12
4	110	A	12
5	122	B	9

4.7 APPENDING AND MERGING DATASETS

Other useful SAS data manipulation capabilities are appending and merging datasets.

- *Appending* adds new *records* to an existing dataset.
- *Merging* adds *variables* to a dataset through the use of a key identifier that is present in both datasets (usually an identification code).

4.7.1 Appending Two Datasets

Appending datasets combines records from two or more datasets. (This is sometimes called a vertical merge.) For example, suppose that data are collected at two locations and you want to combine the datasets into one SAS dataset for analysis. The datasets must have at least some of the same variables in common. Appending is accomplished by including multiple dataset names in the SET statement. For example,

```
DATA NEW; SET OLD1 OLD2;
```

creates the dataset NEW, which consists of the records in the OLD1 dataset, as well as the records from the OLD2 dataset.

HANDS-ON EXAMPLE 4.11

In this example, you will append one SAS dataset to another, creating a larger dataset.

1. Open the program file DAPPEND1.SAS.

```
DATA OLD1;
INPUT SUBJ $ AGE YRS_SMOKE;
DATALINES;
001 34 12
003 44 14
004 55 35
006 21 3
;
DATA OLD2;
INPUT SUBJ $ AGE YRS_SMOKE;
DATALINES;
011 33 11
012 25 19
023 65 45
032 71 55
;
RUN;
```

This program creates two datasets, one named OLD1 and the other named OLD2. It does not create any output. Run the program and verify by looking at the contents

(Continued)

(Continued)
of the **Log Window** that the datasets have been created. Observe the following statements in the **Log Window**:

```
NOTE: The data set WORK.OLD1 has 4 observations and 3 variables.
NOTE: The data set WORK.OLD2 has 4 observations and 3 variables.
```

2. Return to the Editor window and add the following statements to the end of this SAS program:

```
DATA NEW;SET OLD1 OLD2; RUN;
PROC PRINT DATA=NEW;RUN;
```

3. Run the SAS program and observe the results. The data from the `OLD2` dataset is appended to the `OLD1` dataset, and the combined datasets are stored in the `NEW` dataset. The results (contents of the `NEW` dataset) are listed using `PROC PRINT`.

4. Open the program file `DAPPEND2.SAS`. Notice that in this program, the `OLD2` dataset contains an additional variable, `MARRIED`.

```
DATA OLD1;
... etc ...
DATA OLD2;
INPUT SUBJ $ AGE YRS_SMOKE MARRIED;
DATALINES;
011 33 11 1
012 25 19 0
023 65 45 1
032 71 55 1
;
RUN;
DATA NEW;SET OLD1 OLD2; RUN;
PROC PRINT DATA=NEW;RUN;
```

5. Run this SAS program and observe the results. Where there is no matching variable in an appended dataset, the data are set to missing in the merged dataset.

4.7.2 Merging Datasets by a Key Identifier

Another way to combine datasets is to merge them by adding *new variables* from one dataset to an existing dataset. (This is sometimes called a horizontal merge or a many-to-many merge.) In order to accomplish this, you must have a matching identifier in each dataset that can be used by SAS to match the records during the merge. For example, suppose that you have data collected on patients taken on two separate occasions, and the data for each collection time are in separate datasets. If you want to combine the data so that you can compare pre- and post-treatment values, the technique for merging the datasets using some key identifier (such as patient ID) is as follows:

1. Sort each dataset by the key identifier.
2. Within a DATA step, use the MERGE statement along with a BY statement to merge the data by the key identifier.

As with the SORT statement, you can RENAME, DROP, and KEEP variables during the MERGE. You can also merge many files at a time. The following shows the syntax for merging four datasets and performing a RENAME, DROP, and KEEP on the third dataset.

```
DATA newdataset;
MERGE data1
      data2
      data3 (RENAME=(oldname=newname)
            DROP=variables or KEEP=variables))
      data4;
BY keyvar
RUN;
```

HANDS-ON EXAMPLE 4.12

In this example, you will merge datasets using a key identifier to match records. In this case, the new dataset will contain more variables than either of the two original datasets.

1. Open the program file DMERGE1.SAS. The first dataset created in this example is named PRE.

   ```
   DATA PRE;
   INPUT CASE $ PRETREAT;
   DATALINES;
   001 1.02
   002 2.10
   etc ...
   *A second dataset in the program is named POST:
   DATA POST;
   INPUT CASE $ POSTREAT;
   DATALINES;
   001 1.94
   002 1.63
   etc ...
   ```

2. Note that each dataset contains a variable named CASE. This is the key identifier that tells SAS which records to match when performing the merge. Before merging the datasets, you must sort them on the same identifier:

(Continued)

(*Continued*)

```
PROC SORT DATA=PRE; BY CASE;
PROC SORT DATA=POST; BY CASE;
```

The code you opened in SAS is not complete. You must complete the code yourself to make it run properly. For example, add the BY CASE; statement to the second PROC SORT.

Also, complete the DATA statement in the program to perform the desired merge:

```
DATA PREPOST;                * Create new dataset;
  MERGE PRE POST; BY CASE;
  DIFF=POSTREAT - PRETREAT;  * Calculate DIFF in new dataset;
RUN;
```

The MERGE statement merges the datasets. Because you are interested in the difference between the PRE and POST values, the DATA step includes the DIFF= calculation. (Notice that the DIFF calculation is appropriate in this case because the code is within a DATA step.)

3. Run the completed program and observe the results shown in Table 4.12.
4. Make a change to the MERGE statement to rename the variable PRETREAT to BASE-LINE.

```
MERGE PRE (RENAME=(PRETREAT = BASELINE))POST;
```

You will also need to change the DIFF= calculation to

```
DIFF= POSTREAT-BASELINE;
```

Resubmit the code and verify that you obtain the expected results.

5. In your SAS code, delete the data line for record 2 in the PRE dataset. Resubmit the program and observe the results. When there is no matching key value, a missing value is placed in the cell, and the calculation based on that value is a missing value.
6. Remove the BY CASE statement after the MERGE statement. Resubmit the program and observe the results. Compare the output to the original output and see that the merge no longer produces the expected results. Forgetting to use the BY statement in a MERGE is a common programming mistake that leads to incorrect results.

TABLE 4.12 Example Output from Merged Datasets

Obs	CASE	BASELINE	POSTREAT	DIFF
1	001	1.02	1.94	0.92
2	002	2.10	1.63	−0.47
3	003	1.88	2.73	0.85
4	004	2.20	2.18	−0.02
5	005	1.44	1.82	0.38
6	011	1.55	1.94	0.39
7	013	1.61	2.25	0.64
8	014	2.61	1.70	−0.91
9	015	1.56	1.78	0.22
10	016	0.99	1.52	0.53
11	022	1.53	1.97	0.44

4.7.3 Few-to-Many Merge

A Few-To-Many merge is used when you have records in one dataset that you want to merge into some table that contains (typically) a smaller number of categories. For example, suppose that you own an auto parts store where you sell a lot of (maybe a hundred thousand) different kinds of products. You sell these products to certain types of buyers (only a few types). Each type of buyer gets a particular discount. Counter customers (the public) get no discount. Other parts stores get a discount of 40%, and auto repair shops get a discount of 33%. This discount schedule may change at any time, so you want to create a program where you can quickly merge the thousands of items to the various buyer categories – thus a Few-to-Many (or some would call it Many-to-Few) match. As another example, a credit card company might be dealing with millions of transactions each month. At the end of the month, you want to match each credit card sale (the many) to the card owner (the fewer) in order to combine those purchases by the card owner to produce a monthly statement.

To continue with the auto parts example, suppose that you want to produce a report of seven transactions (to keep it simple) defined as shown in Table 4.13. The table shows the part, the type of customer, and the retail price. This is information in your "many" dataset. A second table (also shown in Table 4.13) contains the current discounts offered. In this dataset, there are only three records. This is your "few" dataset.

The goal is to merge the current discounts with the parts sold, allowing you to calculate the actual price you charged for the part. To perform the merge that you desire, first sort both files by the key identifier. Notice that BUYERTYPE is in both datasets. This is the key identifier by which you want to sort in preparation for the merge. The following example illustrates this merge.

TABLE 4.13 Data for a Few-to-Many Merge

Transactions (Dataset name SALES)

ITEM	BUYERTYPE	PRICE
CARBCLEANER	REPAIR	2.30
BELT	CONSUMER	6.99
MOTOROIL	CONSUMER	14.34
CHAIN	STORE	18.99
SPARKPLUGS	REPAIR	28.99
CLEANER	CONSUMER	1.99
WRENCH	STORE	18.88

Discount Types (Dataset name DISCOUNT)

BUYERTYPE	DISCOUNT
CONSUMER	0.00
REPAIR	0.33
STORE	0.40

Note that this merge appears to be very similar to merging any two files by a key identifier, and it is. The difference is that in the first dataset, there can be many of the same values for the key identifier (BUYERTYPE) and a list of similarly named unique categories in the second data file.

HANDS-ON EXAMPLE 4.13

1. Open the program file DMERGE2.SAS. The part of the code that performs the Few-to-Many merge is as follows:

```
PROC SORT DATA=SALES; BY BUYERTYPE;RUN;
PROC SORT DATA=DISCOUNT; BY BUYERTYPE; RUN; *COMPLETE THIS;
DATA REPORT;
   MERGE SALES DISCOUNT;BY BUYERTYPE;*COMPLETE THIS;
   FINAL = ROUND(PRICE*(1-DISCOUNT),.01);
RUN;
PROC PRINT DATA=REPORT;
SUM FINAL;
RUN;
```

Note that the code first sorts the data. In the DATA statement, the merge is performed, and at the same time, the final price is calculated. (The ROUND function used in the example limits the number for FINAL to two digits. See Appendix B for more explanation of how to use the ROUND function.)

2. The code in the file has some pieces missing. Complete the code as shown previously.
3. Submit the code and observe the results. The results are shown in Table 4.14. Observe the results. Notice how each ITEM is paired with a DISCOUNT according to the matching BUYERTYPE. The last column (FINAL) contains the calculated actual

TABLE 4.14 Results for the Few-to-Many Merge

Obs	ITEM	BUYERTYPE	PRICE	DISCOUNT	FINAL
1	BELT	CONSUMER	6.99	0.00	6.99
2	MOTOROIL	CONSUMER	14.34	0.00	14.34
3	CLEANER	CONSUMER	1.99	0.00	1.99
4	CARBCLEANER	REPAIR	2.30	0.33	1.54
5	SPARKPLUGS	REPAIR	28.99	0.33	19.42
6	CHAIN	STORE	18.99	0.40	11.39
7	WRENCH	STORE	18.88	0.40	11.33
					67.00

price charged for each item. The `SUM` in `PROC PRINT` produces a final total for all sales. For this report, the store sold $67.00 worth of merchandise.

4. The store is running a sale and is offering consumers a 15% discount. Change the consumer discount from 0 to 0.15 and rerun the program. How much less did the store make?

5. In the `SALES` dataset, add a line in the `DATA` step (before the `DATALINES` statement) that raises all prices by 15%. For example:

```
PRICE=PRICE*1.15;
```

Resubmit the program. How much more money would this scenario bring to the store?

There are other ways to perform appends and merges in SAS. Two other alternatives (not discussed here) are `PROC APPEND` and `PROC SQL`. Also, a version of appending called interleaving is not discussed here because it is seldomly used. The SAS merge process can be tricky, and you must be careful not to violate certain rules. The following list will help you avoid common merging mistakes:

- Be careful if there are duplicate values of your `KEY` variable(s) in datasets to be merged. These types of merges are sometimes referred to as few-to-many, many-to-few, or many-to-many and should be avoided.
- Do not leave off the `BY` statement when merging datasets.
- Make sure your `KEY` variables in datasets to be merged are of the same data type, length, and format.
- Make sure your `KEY` variables uniquely identify or label the records within the dataset. Beware of ambiguous `KEY` variables that might have been miscoded.
- Watch out for variables in datasets that have the same name but mean something different. For example, suppose that you have a variable name `EX` in both datasets you intend to merge. In one dataset it means `EXPENSE` and in another `EXTRA` – they are two different variables with two different meanings. These are sometimes referred to

as overlapping variables. When the datasets are merged, the resulting value of an overlapping variable may not be what you expect.

4.8 USING PROC FORMAT

In Chapter 3, we saw how you can use predefined SAS date formats to output dates in selected formats. For example, the format Worddate12. outputs dates in a Jan 10, 2016 format. The PROC FORMAT procedure allows you to create your own custom formats. These custom formats allow you to specify the information that will be displayed for selected values of a variable. Note that PROC FORMAT (which creates the format definition) is *not* a part of the DATA step. However, it is related to how data are displayed in the output and is therefore related to the topics of this chapter.

Once a FORMAT has been created, it must be applied within a SAS procedure in order to take effect. Thus, the steps for using formatted values are as follows:

1. Create a FORMAT definition using PROC FORMAT.
2. Apply the FORMAT to one or more variables. You can apply a format (once it is defined in PROC FORMAT) in a DATA step or in a data analysis PROC statement.

For example, suppose that your SAS dataset contains a numeric variable named MARRIED coded as 0 and 1. To cause SAS to output the words Yes and No rather than 0 and 1, construct a format definition for MARRIED using the PROC FORMAT statement, as illustrated in the following:

```
PROC FORMAT;
     VALUE FMTMARRIED 0="No"
                      1="Yes";
RUN;
```

Hands-On Example 4.14 illustrates how you would apply the FMTMARRIED format to a variable within a SAS procedure. You can choose any legitimate name for your format and define as many format names as are required for your dataset. Custom formats must be defined in your code before they are used; therefore, you will usually place the PROC FORMAT early in your program listing.

Formats are not limited to numeric variables. For example, if a text variable was coded Y and N for Yes and No to survey answers, you could create the following text format (for the text variable ANSWER). Notice that formats for text variables begin with a dollar sign ($):

```
PROC FORMAT;
     VALUE $FMTANSWER 'Y' = 'Yes'
              'N' = 'No';
   PROC PRINT;
   FORMAT ANSWER $FMTANSWER.;
   RUN;
```

This code causes PROC PRINT to use the formatted values Yes and No instead of Y and N in the output. Note that the format name when specified in a FORMAT statement ends with a dot (.).

HANDS-ON EXAMPLE 4.14

In this example, you will create a format definition (FMTMARRIED) for a numeric variable and apply that definition to the variable MARRIED in PROC PRINT.

1. Open the program file DFORMAT1.SAS.

   ```
   PROC FORMAT;
        VALUE FMTMARRIED 0="No"
                         1="Yes";
   RUN;
   PROC PRINT DATA="C:\SASDATA\SURVEY";
        VAR SUBJECT MARRIED;
        FORMAT MARRIED FMTMARRIED.;
   RUN;
   ```

   Make note of the following:
   - The PROC FORMAT statement defines a format named FMTMARRIED.
   - The FORMAT statement in PROC PRINT tells SAS to apply the FMTMARRIED format definition to the variable MARRIED.
   - Notice that in a FORMAT statement, a SAS format specification ends with a dot (.).
   - When SAS lists the data, as in a PROC PRINT, instead of listing data from the MARRIED variable as 0 or 1, it outputs the data as Yes or No.

2. Run the program and observe the output. The MARRIED column should contain the words Yes and No rather than 1 and 0.

3. Another variable in this dataset is GENDER. This text variable includes the codes M for MALE and F for FEMALE. Change the PROC FORMAT step to also include a definition for a format named $FMTGENDER by adding the VALUE statement.

   ```
   VALUE $FMTGENDER "F" = "Female"
                    "M" = "Male";
   ```

   Change the VAR statement to

   ```
   VAR SUBJECT MARRIED GENDER;
   ```

   and the FORMAT statement to

   ```
   FORMAT MARRIED FMTMARRIED. GENDER $FMTGENDER.;
   ```

4. Submit the new program and observe the output. Does the MARRIED column include Male and Female instead of M and F? See the results in Table 4.15.

(Continued)

> (*Continued*)
>
> **TABLE 4.15 Application of a FORMAT**
>
> | Obs | SUBJECT | MARRIED | GENDER |
> |---|---|---|---|
> | 1 | 1 | Yes | Male |
> | 2 | 2 | Yes | Male |
> | 3 | 3 | No | Female |
> | 4 | 4 | No | Female |
> | 5 | 5 | Yes | Female |
> | 6 | 6 | Yes | Female |
>
> Notice that in the MARRIED column, the data are output as Yes and No rather than as 0 and 1 and GENDER is displayed as Male and Female.

The format name FMTGENDER is arbitrary. However, it is often a good idea to use the prefix FMT in any user-created format name to help you avoid getting format names mixed up with variable names. Format names must follow normal SAS naming conventions, but they may not end in a number.

Formats may also use ranges. For example, suppose that you want to classify your AGE data using the designations Child, Teen, Adult, and Senior. You could do this with the following format:

```
PROC FORMAT;
Value FMTAGE    LOW-12 = 'Child'
                13,14,15,16,17,18,19 = 'Teen'
                20-59 = 'Adult'
                60-HIGH = 'Senior';
```

LOW means the lowest numeric value for this variable in the dataset; HIGH means the highest numeric value. You can also use OTHER= to indicate any other data not specified by the other assignments. Also, you can use lists, as in the Teen specification. (Yes, it would have been simpler to use 13–19, but we did it this way for illustrative purposes.)

You can also assign the same format to several variables. If you have questionnaire data with variable names Q1, Q5, Q7 where each question is coded as 0 and 1 for answers Yes and No, respectively, and you have a format called FMTYN, you could use that FORMAT in a procedure as in the following snippet:

```
    PROC PRINT;
    FORMAT Q1 Q5 Q7 FMTYN.;
RUN;
```

USING PROC FORMAT

HANDS-ON EXAMPLE 4.15

In this example, you will create formats using ranges.

1. Open the program file DFORMAT2.SAS.

   ```
   PROC FORMAT;
       VALUE FMTAGE    LOW-12 = 'Child'
                       13,14,15,16,17,18,19 = 'Teen'
                       20-59 = 'Adult'
                       60-HIGH = 'Senior';
   VALUE FMTSTAT       1='Lower Class'
                       2='Lower-Middle'
                       3='Middle Class'
                       4='Upper-Middle'
                       5='Upper';
   RUN;
   PROC PRINT DATA="C:\SASDATA\SOMEDATA";
   VAR AGE STATUS GP;
   FORMAT AGE FMTAGE. STATUS FMTSTAT.;
   TITLE 'PROC FORMAT Example';
   RUN;
   ```

2. Run the program and observe the output. Notice that the values listed for AGE and STATUS are not the original values. Instead, the formatted values are listed.
3. Add the following format to the program (including changing the FORMAT statement in the PROC PRINT procedure.)

   ```
   VALUE $FMTGP 'A'='Southern Suburbs'
                'B'='Urban'
                'C'='Northern Suburbs';
   ```

4. Run the revised program and observe the output.
   Occasionally, there is some confusion over the difference between labels and formats. Labels provide more explanatory names to *variables* and relate to the names of *variables* only (and not to *data values*). Formats affect how actual *data values* are listed in SAS output tables.

4.8.1 Creating Permanent Formats

In all the previous examples, formats were applied in a PROC step and are considered temporary formats. You may also create permanent formats in a manner similar to that used when creating permanent datasets.

When permanent SAS formats are created, they are stored in a format catalog, either in the default SAS format catalog or in your own specified format catalog. For example, to store a SAS format in a specified permanent library location, you could use code such as

110 CHAPTER 4: PREPARING DATA FOR ANALYSIS

```
PROC FORMAT LIBRARY = MYSASLIB<.name>;
```

In this case, the MYSASLIB refers to a SAS library location you have previously created. (Refer to Chapter 3 for creating a SAS Library if you have not already created the MYSASLIB library.)

If you do not include a name for the format folder, the default format library name FORMATS is used. (In this case, it would be the MYSASLIB.FORMATS library.) That is, when you create a SAS format catalog, a folder icon appears in the designated SAS Library. In this case, it is named FORMATS and appears in the MYSASLIB library. You can verify its existence by examining the MYSASLIB library using SAS Explorer. If you double-click on the FORMATS folder, you will see subfolders named with the names of the formats you have created. For example, the code

```
PROC FORMAT LIBRARY = MYSASLIB;
    VALUE FMTMARRIED 0="No"
                     1="Yes";
    VALUE $FMTGENDER "F" = "Female"
                     "M" = "Male";
RUN;
```

creates two subfolders in the MYSASLIB.FORMATS folder named FMTMARRIED and $FMTGENDER. These folders are illustrated in Figure 4.2.

To examine the contents of a particular format, double-click on the format subfolder. For the FMTMARRIED folder, the resulting information shows the details of that format when it was created, and what version of SAS was used.

Once you have created permanent formats, you can use them in both PROC and DATA step statements. To tell SAS the location of a particular format, use the statement.

```
OPTIONS FMTSEARCH=(proclib);
```

where PROCLIB is the name of the SAS Library where your formats folder is located. For example, if you have previously created and stored the FMTMARRIED and $FMTGENDER formats in your MYSASLIB.FORMATS folder, you could use the following code to access those formats with PROC PRINT (or any PROC.)

```
OPTIONS FMTSEARCH=(MYSASLIB.FORMATS);
PROC PRINT DATA="C:\SASDATA\SURVEY";
    VAR SUBJECT MARRIED GENDER;
    FORMAT MARRIED FMTMARRIED. GENDER $FMTGENDER.;
RUN;
```

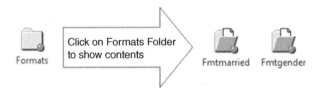

Figure 4.2 FORMAT folders

To discover what formats are in a particular format library, you can use the PROC CATALOG procedure as shown here. This code displays all of the formats stored in the MYSASLIB.FORMATS library.

```
PROC CATALOG CATALOG = MYSASLIB.FORMATS;
CONTENTS;
RUN;
QUIT;
```

HANDS-ON EXAMPLE 4.16

In this example, you will explore creating and using permanent SAS formats.

1. Open the program file DFORMAT3.SAS.

   ```
   PROC FORMAT LIBRARY = MYSASLIB.CAT1; *CREATE A FORMAT CATALOG;
       VALUE FMTMARRIED 0="No"
                       1="Yes";
       VALUE $FMTGENDER "F" = "Female"
                       "M" = "Male";
   RUN;
   ```

   Run this program. Examine the SAS Library named MYSASLIB and note that it contains a folder named CAT1. This folder contains the two formats you just created.

2. Open the program file DFORMAT3A.SAS.

   ```
   OPTIONS FMTSEARCH=(_____);
   PROC PRINT DATA="C:\SASDATA\SURVEY";
       VAR SUBJECT MARRIED GENDER;
       FORMAT MARRIED FMTMARRIED. GENDER $FMTGENDER.;
   RUN;
   ```

   Fill in the blank in the OPTIONS statement with MYSASLIB.CAT1 (the name of the format catalog you just created). Submit this code and verify that the output shows that the formats have been applied in the MARRIED and GENDER columns.

Suppose you have created formats in the past and somehow lost them, or someone sends you a .SASB7DAT file that uses created formats, but they did not send you that library. If you attempt to use that dataset, you will get the following error message in the **Log Window**.

```
ERROR: The format FMTMARRIED was not found or could not be loaded.
```

If this occurs, you must use the following OPTIONS statement (above the code where you refer to the dataset) to tell SAS to access the dataset, or run the procedure without using the defined formats:

```
OPTIONS NOFMTERR;
```

In this case, the output displays the raw values of the variables instead of the assigned format labels.

4.9 GOING DEEPER: FINDING FIRST AND LAST VALUES

A technique that can come in handy to understand the contents of a dataset is to be able to identify the first and last values in the dataset by some grouping variable. For example, in the SOMEDATA dataset, the data are recorded by four Intervention Groups (GP) labeled A, B, C, and D. Suppose that you want to identify the first and last person (ID) in each of those groups. In a SAS DATA step, you identify the first and last values by FIRST.GP and LAST.GP, where GP is the name of the sorted grouping (or key) variable. Hands-On Example 4.17 illustrates how that could be done.

HANDS-ON EXAMPLE 4.17

This example illustrates how to locate the first and last values in a dataset that is organized by some grouping variable.

1. Open the program file DFINDFIRST.SAS. Notice the names FIRST.GP and LAST.GP used in the IF statements. Run this code and observe the results. The first few records are shown in Table 4.16.

```
DATA FIRST;
SET MYSASLIB.SOMEDATA;
BY GP;
```

TABLE 4.16 Finding First and Last in a Group

Obs	ID	GP	WHICHID
1	101	A	ISFIRST
2	102	A	
3	110	A	
4	187	A	
5	312	A	
6	390	A	
7	445	A	
8	543	A	
9	544	A	
10	550	A	
11	561	A	ISLAST
12	104	B	ISFIRST
13	122	B	
14	123	B	

```
IF FIRST.GP THEN WHICHID="ISFIRST";
IF LAST.GP THEN WHICHID="ISLAST";
RUN;
PROC PRINT DATA=FIRST;
VAR ID GP WHICHID;
RUN;
```

   Note that ID 101 is the first subject in group A and is now labeled ISFIRST, and ID 561 is the last subject in group A and is labeled ISLAST.

2. The IF statement created values for WHICHID, but the rest are blank. Correct this by putting the code
     WHICHID="NEITHER"; after the by GP statement. Resubmit and observe how this changes the output.

4.10 SUMMARY

This chapter discussed several techniques for preparing your data for analysis. In the next chapter, we begin the discussion of SAS procedures that perform analyses of the data.

EXERCISES

4.1 **Sort data.**

   Modify the program (EX_4.1.SAS) to sort (alphabetically) first name within last (**last first**) by adding FIRST to the BY variables in the PROC SORT statement.

```
DATA MYDATA;
INPUT @1 LAST $20. @21 FIRST $20. @45 PHONE $12.;
Label LAST = 'Last Name'
      FIRST = 'First Name'
      PHONE = 'Phone Number';
DATALINES;
Reingold            Lucius              201-876-0987
Jones               Pam                 987-998-2948
Abby                Adams               214-876-0987
Smith               Bev                 213-765-0987
Zoll                Tim Bob             303-987-2309
Baker               Crusty              222-324-3212
Smith               John                234-943-0987
Smith               Arnold              234-321-2345
Jones               Jackie              456-987-8077
;
*-------- Modify to sort by first name within last (by last first);
PROC SORT; BY LAST;
PROC PRINT LABEL;
TITLE 'ABC Company';
TITLE2 'Telephone Directory';
RUN;
```

114 CHAPTER 4: PREPARING DATA FOR ANALYSIS

4.2 **Create subset datasets.**

Using the CARS permanent SAS dataset, write SAS code to do the following:

a. Create a subset (into a SAS dataset named SMALL) consisting of all vehicles whose engine size is less than 2.0L. On the basis of this dataset, find the average city and highway miles per gallon for these vehicles.

```
DATA SMALL;SET "C:\SASDATA\CARS";
IF ENGINESIZE LT 2;
```

b. Create a subset (into a SAS dataset named HYBRID) of all hybrid vehicles in the dataset. For these vehicles:

  I. List the BRAND and MODEL names.

  II. Find the average city and highway miles per gallon.

c. Create a subset (into a SAS dataset named AWDSUV) consisting of all vehicles that are both SUVs and have all-wheel drive. For these vehicles:

  I. List the BRAND and MODEL names.

  II. Use PROC MEANS to find the average city and highway miles per gallon.

d. Sort the data in AWDSUV by highway miles per gallon (smallest to largest). List the BRAND, MODEL, and highway miles per gallon for this sorted dataset.

4.3 **Use DROP or KEEP.**

Using the SOMEDATA permanent SAS dataset, write SAS code to do the following:

a. Calculate the changes from baseline (i.e., 6-month reading minus baseline, etc.) for 6, 12, and 24 months. Times are given as:

```
TIME1 is BASELINE
TIME2 is six months
TIME3 is twelve months
TIME4 is twenty-four months
```

To calculate the changes, use the following:

```
CHANGE6=TIME2-TIME1;
CHANGE12=TIME3-TIME1;
etc.
```

Use PROC MEANS to find the average and standard deviations for these new "change variables."

b. Create a SAS dataset that contains only the variables ID number, intervention group, age, and gender and also includes only those subjects in intervention group A who were females. List the ID numbers for these subjects.

4.4 **Use** PROC FORMAT.

a. In the Editor window, open the program file EX_4.4.SAS.

```
PROC FORMAT;
VALUE fmtYN    0 = 'No'
               1 = 'Yes';
```

```
DATA QUESTIONS;
INPUT Q1 Q2 Q3 Q4 Q5;
DATALINES;
1 0 1 1 0
0 1 1 1 0
0 0 0 1 1
1 1 1 1 1
1 1 1 0 1
;
PROC PRINT;
* PUT FORMAT STATEMENT HERE;
RUN;
```

b. Run the program and observe the output. The Q1–Q5 variables are reported as 0 and 1 values.

c. In the PROC PRINT paragraph, include the statement

```
FORMAT Q1-Q5 fmtYN.;
```

d. Run the program and observe the output. How does the output differ from the version without the FORMAT statement?

4.5 **Creating and using Permanent SAS Formats.**

a. Open the program file EX_4.5.SAS. Change the existing FORMAT statement to create a permanent catalog named CAT2 using the statement

```
PROC FORMAT LIBRARY = MYSASLIB.CAT2;
```

b. Submit the program and verify that a catalog named CAT2 has been created in your MYSASLIB library.

c. Open the program file EX_4.5A.SAS.

```
PROC PRINT DATA="C:\SASDATA\SOMEDATA";
VAR AGE STATUS GP;
FORMAT AGE FMTAGE. STATUS FMTSTAT.;
TITLE 'PROC FORMAT Example';
RUN;
```

This program displays a listing of some of the variables in the SOMEDATA dataset. Submit the code above and observe that no output is created. Verify in the **Log Window** that an error occurred indicating that SAS could not find the appropriate formats.

d. Add an OPTIONS statement before the PROC PRINT to indicate the location of the CAT2 format catalog.

```
OPTIONS FMTSEARCH=(MYSASLIB.CAT2);
```

e. Submit the revised code and verify that the program now runs and uses the defined formats to display the AGE and STATUS columns.

116 CHAPTER 4: PREPARING DATA FOR ANALYSIS

4.6 Using the CARS dataset, you want to find which cars by brand have the lowest and highest city miles per gallon.

   a. First, sort the CARS dataset by BRAND and CITYMPG.

   ```
   PROC SORT DATA=MYSASLIB.CARS OUT=MPG;
   BY BRAND CITYMPG;
   RUN;
   ```

   Observe the output using a PROC PRINT statement.

   b. Using FIRST and LAST data step variables within the sorted data, determine which car has the lowest and highest city miles per gallon, and display the results.

   ```
   DATA MPGRANKED; SET MPG; BY BRAND;
   IF FIRST.BRAND THEN MPGRANK="LOWEST";
   IF _____ THEN MPGRANK="HIGHEST";
   RUN;
   PROC PRINT DATA=_____;
   VAR BRAND MODEL CITYMPG MPGRANK;
   RUN;
   ```

   c. Note a problem with MPGRANK. The information HIGHEST is truncated to six characters because the first instance (LOWEST) is six characters long. Correct this problem by using the FORMAT statement

   ```
   FORMAT MPGRANK $7.;
   ```

   in the DATA step (after BY BRAND). Observe how this fixes the problem.

   d. To make your results more compact, use a subsetting IF statement to display only the records that are either LOWEST or HIGHEST using

   ```
   IF MPGRANK NE "";
   ```

   as the last entry in the DATA statement. Observe the results.

5

PREPARING TO USE SAS® PROCEDURES

LEARNING OBJECTIVES

- To be able to use SAS Support Statements
- To be able to use TITLE and FOOTNOTE
- To be able to include comments in your code
- To be able to use RUN and QUIT correctly
- To be able to understand SAS PROC statement syntax
- To be able to use VAR statements
- To be able to use BY statements
- To be able to use ID statements
- To be able to use LABEL statements in a SAS procedure
- To be able to use WHERE statements
- To be able to use PROC PRINT

Plus:

- Going Deeper: To be able to use common System Options
- Going Deeper: To be able to split column titles

SAS® Essentials: Mastering SAS for Data Analytics, Third Edition. Alan C. Elliott and Wayne A. Woodward.
© 2023 John Wiley & Sons, Inc. Published 2023 by John Wiley & Sons, Inc.
Companion website: https://www.alanelliott.com/SASED3/

118   CHAPTER 5: PREPARING TO USE SAS® PROCEDURES

This chapter illustrates several SAS features that allow you to make your output more readable and more organized. It also introduces you to details about SAS procedures (PROCS) that are the core of SAS analysis.

5.1   UNDERSTANDING SAS SUPPORT STATEMENTS

Before using specific SAS procedures for data analysis, you should understand several basic supporting SAS options and statements that are often used in conjunction with SAS data analysis procedures. This chapter introduces you to these options and statements.

5.1.1   Using TITLE and FOOTNOTE Statements

The TITLE statement instructs SAS to place a line of text at the top of each output page or at the beginning of a procedure's output (depending on the output destination). Similarly, a FOOTNOTE statement places text lines at the bottom of each output page. Up to nine title or footnote lines can be specified. When outputting to an HTML destination, there are no pages. So, the title appears before procedure output. When outputting in PDF, for example, the title appears at the top of the page. (Similarly for footnotes.) Output destinations are discussed in Chapter 8. For example:

```
TITLE 'title text';
FOOTNOTE 'footnote text';
```

or

```
TITLEn 'title text';
FOOTNOTEn 'footnote text';
```

where n is a number between 2 and 9. A TITLE or FOOTNOTE statement without a number specifies an initial TITLE or FOOTNOTE line. Subsequent TITLE or FOOTNOTE lines can be defined using TITLE2, TITLE3, and so on up to TITLE9; and FOOTNOTE2, FOOTNOTE3, and so on up to FOOTNOTE9. (There are no blanks between TITLE or FOOTNOTE and the number.) For example:

```
TITLE 'The first line of the title';
TITLE2 'The second line of the title';
TITLE5 'Several lines skipped, then this title on the fifth line';
FOOTNOTE 'This is a footnote';
FOOTNOTE3 'This is a footnote, line 3';
```

You can use either single or double quotation marks when defining a title or footnote, but you must be consistent for each definition. Once you have specified a TITLE or FOOTNOTE, it is used in all subsequent SAS output (even when submitting a new SAS program as long as you do not exit the current SAS software program) until you redefine the TITLE or FOOTNOTE lines. For example, if you include a TITLE2 statement in your code, it cancels any prior TITLE2 and any TITLEn statements where n is greater than 2. Footnotes work the same way.

You can cancel all TITLE and FOOTNOTE lines with the statement

```
TITLE; FOOTNOTE;
```

It is a good programming practice to include this line at the bottom of your SAS code file to prevent stray titles or footnotes from appearing in subsequent SAS output. We've put this code at the end of most of the SAS program example files used in this book.

> **HANDS-ON EXAMPLE 5.1**
>
> In this example, the output is created using two SAS PROCs. A TITLE is defined for the first PROC and changed for the second.
>
> 1. Open the program file DTITLE1.SAS.
>
>    ```
>    DATA MYDATA;
>    INFILE 'C:\SASDATA\EXAMPLE.DAT';
>    INPUT ID 1-3 GP $ 5 AGE 6-9 TIME1 10-14 TIME2 15-19
>        TIME3 20-24;
>    RUN;
>    PROC PRINT;
>    TITLE 'Example SAS programs';
>    TITLE2 'These are data from the example file.';
>    TITLE4 'Using the EXAMPLE dataset.';
>    FOOTNOTE 'This is a footnote';
>    PROC MEANS;
>    TITLE2 'This is output from PROC MEANS.';
>    RUN;
>    TITLE; FOOTNOTE; *Turns off TITLE and FOOTNOTE;
>    ```
>
> 2. Run the program and observe the output. Note that the program has a PROC PRINT and a PROC MEANS statement. A partial sample of the output from PROC PRINT is given in Table 5.1. Note that the title lines 1, 2, and 4 at the top match the titles defined in the code. The footnote appears at the bottom of the procedure.
>
>   **TABLE 5.1 Output Showing a TITLE and FOOTNOTE for PROC PRINT**
>
> | | | | | | | |
> |---|---|---|---|---|---|---|
> | | | | Example SAS programs
These are data from the example file.

Using the EXAMPLE data set. | | | |
> | Obs | ID | GP | AGE | TIME1 | TIME2 | TIME3 |
> | 1 | 101 | A | 12 | 22.3 | 25.3 | 28.2 |
> | 2 | 102 | A | 11 | 22.8 | 27.5 | 33.3 |
> | 3 | 104 | B | 12 | 22.8 | 30.0 | 32.8 |
> | 4 | 110 | A | 12 | 18.5 | 26.0 | 29.0 |
> | 5 | 122 | B | 9 | 19.5 | 25.0 | 25.3 |
> | 6 | 123 | B | 11 | 23.5 | 28.8 | 34.2 |
> | 7 | 134 | C | 8 | 22.6 | 26.7 | 28.0 |
> | 8 | 150 | B | 8 | 21.0 | 26.7 | 27.5 |
> | 9 | 162 | B | 7 | 20.9 | 28.9 | 29. |
> | | | | This is a footnote | | | |
>
> *(Continued)*

(*Continued*)
3. Observe the second output table, the one produced by PROC MEANS (see Table 5.2). Note that for this output, the title in line 2 has changed according to the TITLE2 statement before the PROC MEANS in the code, and all higher titles (TITLE4) are canceled.

TABLE 5.2 Output Showing a TITLE and FOOTNOTE for PROC MEANS

Example SAS programs					
:--:					
This is the output from PROC Means.					
The MEANS Procedure					
Variable	N	Mean	Std. Dev.	Minimum	Maximum
ID	50	374.2200000	167.4983143	101.0000000	604.0000000
AGE	50	10.4600000	2.4261332	4.0000000	15.0000000
TIME1	50	21.2680000	1.7169551	17.0000000	24.2000000
TIME2	50	27.4400000	2.6590623	21.3000000	32.3000000
TIME3	50	30.4920000	3.0255942	22.7000000	35.9000000
This is a footnote					

4. Put a TITLE6 line in the code that reads 'This is Title 6' after the TITLE2 and before the RUN statement for PROC MEANS. Submit the changed code and note where this title appears. Is the previous title at line 4 present this time? Why?

It is important to realize that TITLEs and FOOTNOTEs are remembered as long as the SAS software program is active (i.e., you have not exited the SAS software program). Therefore, if you create a title using one program, be sure to change or erase the TITLE when beginning a different analysis, such as starting a new procedure or performing an analysis on a different set of data. Otherwise, you'll end up having an incorrect TITLE and/or FOOTNOTE on your output.

Therefore, either erase titles and footnotes at the end of the program or at the beginning of a program if there may be previous titles created in an earlier program.

Customizing TITLEs and FOOTNOTEs: There are a number of options that can be used with the TITLE or FOOTNOTE statements to customize the look of your title. Some of these options include specifying color with a C= or COLOR= option. For example:

TITLE C=BLUE "*This is a title*";

causes the title to be displayed in blue. A summary of some common options is given in Table 5.3.

HANDS-ON EXAMPLE 5.2

This example illustrates some options for the `TITLE` statement.

1. Open the program file `DTITLE2.SAS`.

   ```
   TITLE J=RIGHT
         FONT= 'SWISS' COLOR=RED BOLD BCOLOR=YELLOW "EXAMPLE "
         C=BLACK BOLD ITALIC "TODAY IS &amp;SYSDATE9";
   TITLE2 J=CENTER C="CX3230B2" "THIS IS BRILLIANT BLUE.";
   FOOTNOTE J=CENTER COLOR=PURPLE UNDERLINE=1
            HEIGHT=8PT "8 POINTS"
            HEIGHT=16PT "16 POINTS"
            HEIGHT=24PT "24 POINTS";
   ;
   PROC MEANS DATA=MYSASLIB.PEOPLE;
   RUN;
   ```

2. Note how `TITLE` and `FOOTNOTE` options are used in this program. Run the program and observe the results as shown in Figure 5.1.

Figure 5.1 Illustrating options for titles and footnotes

3. Change the font specification to Times New Roman, add an underline to `TITLE2`, and change the `FOOTNOTE` to `TITLE3`. Submit the code and note the changes.

5.1.2 Including Comments in Your SAS Code

It is a good programming practice to include explanatory comments in your code. This allows you to understand what you have done when you go back into your code the next day (or the next year). Moreover, if your code is used by other SAS programmers, this will help them understand what your code does.

Comment statements can be used almost anywhere in SAS code to document the job, and you can use any number of comment statements in your code. There are two ways to include comments in your code. In this first example, a comment begins with an asterisk (*) and ends with a semicolon.

TABLE 5.3 Common Options for TITLEs and FOOTNOTEs

Title Option	Meaning
C=*color* or COLOR= *color*	Specifies color. See Appendix A for a summary of colors available
C="*RGBCode*"	An RGB code allows you to specify over 16 million different colors using an RGB coding system. See Appendix A for more information. For example, C="CX803009" specifies vivid reddish orange. Colors may also be defined using a Hex code. Companies often have defined colors to be used in official documents, and using the corresponding RBG or Hex code is a way to match those colors
BCOLOR= *color*	Specifies background color (See Appendix A for color options.)
J=*type*	Justification. Options are left, right, or center
FONT=*fontname*	Specifies the font for text. See Appendix A for more information. Example FONT= 'SWISS'
H=*n* or HEIGHT=*n*	Specifies the height of the text. See Appendix A for more information. The default height is 1. H=16 pt means 16 points
BOLD	Specifies bold
ITALIC	Specifies italic
UNDERLIN=*n*	0 means none, 1, 2, or 3 means underline and may specify thickness in some instances (Yes, this is the spelling used by SAS.)

```
*This is a message
It can be several lines long
But it always ends with an;
**************************************************************
   * Boxed messages stand out more, still end in a semicolon *
**************************************************************;
DATA MYDATA; * You can put a comment on a line of code;
```

Another useful technique is to begin a comment with /* and end with */. This technique is useful when you want to include statements that contain a semicolon. For example, you can comment out several lines of code that you do not want to currently use, but that you want to retain for possible future use or reference. Here are two examples of this comment technique:

```
/*This is a SAS comment*/
/* Use this comment technique to comment out lines of code
PROC PRINT;
PROC MEANS;
The next line ends the comment - the PROCS were ignored
*/
```

In your own programming, it is a good idea to begin every program with a note about what it does, when it was written, and who wrote it. It may not seem important at the moment, but 6 months or a year later, the note can save your valuable time.

Note that you can put a comment using either method on a line of code, but you cannot put a comment on a line of data values that follow the DATALINES statement.

- Comments are a way to cause SAS to ignore some part of your program when it is submitted. This allows you to run part of your code without having to erase code that might come in handy again, but you don't want to run that code every time you run your program. Examples include a format definition or the creation of a dataset.
- Another way to run only certain lines of code in your program is to highlight the code you want and click the running man icon (or choose Run from the menu.). It's that simple. SAS will only act on the highlighted code.

5.1.3 Using RUN and QUIT Statements

The RUN statement causes previously entered SAS statements to be executed. It is called a boundary statement. For example:

```
PROC PRINT;
PROC MEANS;
RUN;
```

Include a RUN statement after major sections in your SAS code. A RUN statement *must* follow the final procedure section (e.g., PROC MEANS) in your SAS code. It is often the last statement in a Windows SAS program.

Another boundary statement is the QUIT statement. It is sometimes used in conjunction with a RUN statement to cease an active procedure. For example, you might see the following code:

```
PROC REG;
RUN;
QUIT;
```

The REG (regression) procedure, and others, need a QUIT to instruct them to cease operation. As procedures are described in the text, those that require a QUIT statement will be noted. If you are unsure, it never hurts to include both a RUN and a QUIT at the end of any SAS job.

If you fail to include a RUN statement at the end of your SAS job (or if SAS runs into an error and never sees the RUN statement), the SAS processor may continue to run in the background. This can cause unpredictable problems. If this occurs, press Ctrl-Break. An option will appear allowing you to "Cancel Submitted Statements."

5.2 UNDERSTANDING PROC STATEMENT SYNTAX

Although there are scores of SAS PROCs (procedures), the syntax is consistent across all of them. Once you learn how SAS PROCs work in general, you should be able to use PROCs that are not covered in this book. This section discusses common syntax used in most analytical PROCs. *Options* and *Statements* that are specific to a PROC are discussed in the context of that PROC. Table 5.4 lists often-used statements and options common to most SAS PROCs.

TABLE 5.4 Common Options and Statements for SAS PROCS

Common Options for SAS PROCS

PROC Option	Meaning
DATA=*datasetname*	Specifies which dataset to use in the analysis
OUT=*datasetname*	Creates an output dataset containing results from the procedure

Common statements for SAS PROCS

PROC statement	Meaning
VAR *variable(s)*;	Instructs SAS to use only the variables in the list for the analysis
BY *variable(s)*;	Causes SAS to repeat the procedure for each different value of the named variable(s). (The dataset must first be sorted by the variables listed in the BY statement.)
ID *variable(s)*;	Instructs SAS to use the specified variable as an observation identifier in a listing of the data
LABEL var='*label*';	Assigns a descriptive label to a variable. (This statement can also appear in the DATA step.)
WHERE *condition*;	Specifies a condition for which only those records meeting the condition are used in the procedure

A simplified syntax for the SAS PROC statement is as follows:

```
PROC name <options>;
   <Statements/statementoptions;>
RUN;
```

In a SAS PROC statement:

- Options appear before the first semicolon. Typical options include DATA=, NOPRINT, and others (these are explained in the following sections).
- Statements (procedure statements) appear after the first semicolon. When present, these statements usually specify choices within the procedure (i.e., a VAR statement indicates which variables to use). You may sometimes use multiple statements in a PROC procedure.
- *Statementoptions* appear after a slash (/) and specify options for the choices in the Statement. For example, the *statmentoption*/CHISQ may instruct SAS to output a Chis-quare type analysis).

The most commonly used *option* within the PROC statement is the DATA= option. For example:

```
PROC PRINT DATA=MYDATA;
RUN;
```

The DATA= *option* tells SAS which dataset to use in the analysis. If there has been only one dataset previously created (within a session), there is no need for this statement.

However, if you've created more than one dataset during a particular SAS session, it's a good idea to include a DATA= *option* to specify which dataset SAS is to use. If you do not specify a DATA= *option*, SAS will use the most recently defined dataset for the specified procedure.

Procedure *Statements* are often required to specify information about how an analysis is to be performed. For example, in the code

```
PROC PRINT;
VAR ID GROUP TIME1 TIME2;
RUN;
```

the VAR statement specifies that only the listed variables are to be used in the procedure. Statements can themselves have options. For example:

```
PROC FREQ;
   TABLES GROUP*SOCIO/CHISQ;
RUN;
```

In this example, the TABLES statement of the PROC FREQ (Frequencies) procedure (which is covered in detail in Chapter 11) instructs SAS to create a table based on the data from the variables GROUP and SOCIO. The statement option /CHISQ instructs SAS to calculate and print chi-square and other statistics for this table.

Using common SAS syntax, you can build statements to perform many types of analyses. Because each PROC is described in future chapters, additional options and statements will be described that are unique to that procedure. However, some statements can be used with (almost) every SAS procedure.

5.2.1 Using the VAR Statement in a SAS Procedure

The VAR statement specifies which variables in the selected SAS dataset are to be used by the procedure. The syntax of the statement is as follows:

```
VAR variable list;
```

An example is as follows:

```
PROC MEANS;
   VAR HEIGHT WEIGHT AGE;
RUN;
```

Note that variables listed in the VAR statement are separated by spaces, and not by commas. This statement tells SAS to perform PROC MEANS only on the three listed variables. If you have several variables in your dataset, this is a way to limit your analysis only to those of interest.

> To save time and typing, you can use a list such as Q1-Q50 (using a single dash) to indicate 50 different sequential variable names with identical prefixes and a number suffix (Q1, Q2, Q3, etc.) in a VAR statement where the variable names are from a

consecutive numerical list. SAS also understands the variable list ID -- SBP (two dashes between variable names) to indicate all variables between ID and SBP (inclusive) in the order defined in the dataset.

Although not discussed in detail here, you can also control elements of the variable labels including color, justification, etc. using /STYLE *statementoption*. This is illustrated in Section 5.7.

5.2.2 Using the BY Statement in a SAS Procedure

The BY statement is a powerful and handy method that allows you to quickly analyze subsets of your data. To use the BY statement, you must have a grouping variable such as GENDER, GROUP, RACE, and so on. Then you must sort your data (using PROC SORT) by the variable on which you wish to subset (sorting has been discussed in Chapter 4). Finally, when you add the BY statement, SAS performs the analysis separately for the groups specified in the BY statement variable. The syntax for the BY statement is as follows:

```
BY variable list;
```

Please Note: You should sort a SAS dataset using PROC SORT on the grouping variable before you perform an analysis using the BY statement.

HANDS-ON EXAMPLE 5.3

In this example, you will calculate means BY group.

1. Open the program file DSORTMEANS.SAS.

```
DATA MYDATA;
INFILE 'C:\SASDATA\EXAMPLE.DAT';
INPUT ID 1-3 GP $ 5 AGE 6-9 TIME1 10-14 TIME2 15-19
   TIME3 20-24 STATUS 31;
RUN;
* FIRST TIME ALL DATA AT ONCE;
PROC MEANS;
   VAR TIME1 TIME2;
RUN;*NOTE USES MYDATA BY DEFAULT;
* SECOND TIME MEANS BY GROUP;
PROC SORT DATA=MYDATA OUT=SORTED;BY GP; RUN;
PROC MEANS DATA=SORTED;
   VAR TIME1 TIME2;
   BY GP;
RUN;
```

2. Run the program. Observe that the first PROC MEANS statement produces the output as given in Table 5.5.

TABLE 5.5 Output from PROC MEANS

Output with no sorting The MEANS Procedure					
Variable	N	Mean	Std. Dev.	Minimum	Maximum
TIME1	50	21.2680000	1.7169551	17.0000000	24.2000000
TIME2	50	27.4400000	2.6590623	21.3000000	32.3000000
Output following SORT BY GP The MEANS Procedure GP=A					
Variable	N	Mean	Std. Dev.	Minimum	Maximum
TIME1	11	20.9000000	2.0813457	17.0000000	23.0000000
TIME2	11	27.0000000	2.9295051	21.3000000	12.3000000
GP=B					
Variable	N	Mean	Std. Dev.	Minimum	Maximum
TIME1	29	21.1862069	1.5853625	18.5000000	24.2000000
TIME2	29	27.4344828	2.6357807	23.1000000	32.3000000
GP=C					
Variable	N	Mean	Std. Dev.	Minimum	Maximum
TIME1	10	21.9100000	1.6649658	19.0000000	23.9000000
TIME1	10	21.9100000	1.6649658	19.0000000	23.9000000
TIME2	10	27.9400000	2.6137245	22.5000000	31.1000000

This output reports means for the entire dataset (50 records).

3. Observe that when the BY statement is used (after the data are sorted by GP), the means are calculated separately for each value of the grouping variable GP, which has three levels (A, B, and C) as given in Table 5.5.
4. Change the grouping variable used in the SORT and PROC statements to STATUS and rerun the program. Note that the program creates five tables, one for each value of the variable STATUS.

5.3 USING THE ID STATEMENT IN A SAS PROCEDURE

The ID statement provides you with a way to increase the readability of your output. It allows you to specify a variable to be used as an observation (record, subject) identifier. For example, in a PROC PRINT procedure, you can use an ID statement that specifies a variable to be displayed at the far left of your listing.

HANDS-ON EXAMPLE 5.4

In this example, you will learn how to use the `ID` statement.

1. Open the program file `D_ID.SAS`.

```
DATA WEIGHT;
INFORMAT MDATE MMDDYY10.;
FORMAT MDATE DATE9.; *OUTPUT FORMAT;
INPUT MDATE RAT_ID $ WT_GRAMS TRT $ PINKEYE $;
DATALINES;
02/03/2009 001 093 A Y
02/04/2009 002 087 B N
02/04/2009 003 103 A Y
02/07/2009 005 099 A Y
02/08/2009 006 096 B N
02/11/2009 008 091 B Y
;
RUN;
PROC PRINT DATA=WEIGHT;
RUN;
```

This program lists information about results of an experiment involving six rats.

2. Run the program and observe the (partial) results given in Table 5.6 which shows the first six lines of the output. Note that the `RAT_ID` is the second column of the data, and the data in the `Obs` column are the sequential record numbers from the dataset.

TABLE 5.6 Output before Defining ID Column

Obs	MDATE	RAT_ID	WT_GRAMS	TRT	PINKEYE
1	03FEB2009	001	93	A	Y
2	04FEB2009	002	87	B	N
3	04FEB2009	003	103	A	Y
4	07FEB2009	005	99	A	Y
5	08FEB2009	006	96	B	N
6	11FEB2009	008	91	B	Y

3. Revise the `PROC PRINT` line to include the following statement and then rerun the program. Observe the output in Table 5.7.

```
PROC PRINT DATA=WEIGHT;ID RAT_ID;
```

The output now places the `RAT_ID` variable in the leftmost column, replacing the generic `Obs` column. The `ID` statement places a specified variable in the first column, usually to make it easier to identify your data records.

TABLE 5.7 Output after Defining ID Column

RAT_ID	MDATE	WT_GRAMS	TRT	PINKEYE
001	03FEB2009	93	A	Y
002	04FEB2009	87	B	N
003	04FEB2009	103	A	Y
005	07FEB2009	99	A	Y
006	08FEB2009	96	B	N
008	11FEB2009	91	B	Y

5.4 USING THE LABEL STATEMENT IN A SAS PROCEDURE

In Chapter 4, we learned how to use the LABEL statement within the DATA step to create labels for several variables. Another version of the LABEL statement allows you to create labels for variable names within a procedure. For example, suppose in Hands-On Example 5.4 you want the column for the variable TRT to read "Treatment." For this, include the statement

```
LABEL TRT='Treatment';
```

after the PROC PRINT statement and before the RUN statement. Moreover, you can include a LABEL option in the PROC PRINT statement to indicate that you want the printout to use defined labels.

```
PROC PRINT LABEL DATA=WEIGHT;
   ID RAT_ID;
   LABEL TRT='Treatment';
RUN;
```

> Please Note: In most other procedures, you can define labels in the DATA step or in the PROC and they will appear appropriately in the output. However, in the PROC PRINT procedure, you must include the LABEL option (i.e., PROC PRINT LABEL;) in order for labels to appear in the output.

HANDS-ON EXAMPLE 5.5

In this example, you will learn how to use the LABEL statement within a procedure.

1. Open the program file D_ID.SAS. Assuming that you still have the ID statement in place from Hands-On Example 5.4, add a LABEL option to the PROC PRINT paragraph and a label statement specifying a label for the variable TRT.

   ```
   PROC PRINT LABEL DATA=WEIGHT;
      ID RAT_ID;
      LABEL TRT='Treatment';
   RUN;
   ```

   (Continued)

(*Continued*)

2. Run the revised program. Observe the new column label for `TRT` in Table 5.8.
3. Remove the `LABEL` option in the `PROC PRINT` statement and rerun the program. Observe the difference in the output (the `TRT` column).
4. After the last `RUN` statement, add the following `PROC MEANS` with `LABEL` statement for `WT_GRAMS` and `MDATE`:

```
PROC MEANS DATA=WEIGHT;
   LABEL WT_GRAMS="Weight in Grams"
   MDATE="MEDOBS Date";
RUN;
```

Run the revised program and observe the output given in Table 5.8. Note that labels are included for the `WT_GRAMS` and `MDATE` variables, and that you were not required to use the `LABEL` option when using `PROC MEANS`.

TABLE 5.8 Using Revised Labels in PROC PRINT

The MEANS Procedure						
Variable	Label	N	Mean	Std. Dev.	Minimum	Maximum
MDATE	MEDOBS Date	6	17934.17	3.0605010	17931.00	17939.00
WT_GRAMS	Weight in Grams	6	94.8333333	5.7416606	87.0000000	103.0000000

5.5 USING THE WHERE STATEMENT IN A SAS PROCEDURE

The `WHERE` statement allows you to specify a conditional criterion for which output will be included in an analysis. This is different from the subsetting `IF` statements described in Chapter 4. In that chapter, the subsetting `IF` statement in the `DATA` step dictates which records are kept in a dataset for subsequent analyses on that particular dataset. The `WHERE` statement affects only the data used for a particular procedure. For example, in the `ID` example, suppose you want to output data only for treatment A in the `PROC PRINT` statement. Use the statement

```
WHERE TRT="A";
```

in the `PROC PRINT` (or any) procedure to limit the analysis to only those records that meet the criteria in the `WHERE` statement. Subsequent procedures within the same SAS program will *not* be subject to the `WHERE` restriction unless it is invoked again for that procedure.

> **HANDS-ON EXAMPLE 5.6**
>
> In this example, you will learn how to use the WHERE statement.
>
> 1. In the SAS file D_ID.SAS used in Hands-On Examples 5.4 and 5.5, modify the PROC PRINT statement to read:
>
> ```
> PROC PRINT LABEL DATA=WEIGHT;
>     ID RAT_ID;
>     LABEL TRT='Treatment';
>     WHERE TRT="A";
> RUN;
> ```
>
> 2. Run the program and observe the output given in Table 5.9. Note that the WHERE statement only changed the output for the PROC PRINT (where it was used) and not the PROC MEANS procedure.
>
> Only the records that match the criterion specified by the WHERE statement are included in the output.
>
> **TABLE 5.9 Selection Using the Where Statement**
>
> | RAT_ID | MDATE | WT_GRAMS | Treatment | PINKEYE |
> |---|---|---|---|---|
> | 001 | 03FEB2009 | 93 | A | Y |
> | 003 | 04FEB2009 | 103 | A | Y |
> | 005 | 07FEB2009 | 99 | A | Y |

5.6 USING PROC PRINT

Although several previous examples have used a simple version of the PROC PRINT procedure, a number of options for this procedure have not been discussed. As demonstrated in previous examples, PROC PRINT outputs a listing of the data values in a SAS dataset. Using the options described in Table 5.10, you can enhance that listing to create a more useful report. Some of the options listed below duplicate those already discussed but are included for the sake of completeness.

Following are several examples illustrating the use of these options. For example, the N option instructs SAS to print the number of observations and allows you to specify a label to be printed such as:

```
N='Observations ='
```

The OBS option allows you to specify a label at the top of the Obs column in the data listing. For example:

```
OBS='Subjects'
```

TABLE 5.10 Common Options and Statements for PROC PRINT

Common Options for PROC PRINT

PROC PRINT Option	Meaning
`DATA=`*datasetname*	Specifies which dataset to use
`DOUBLE`	Double-spaces the output for ODS output formats such as RTF and PDF (This option has no effect in default SAS output, which is HTML.)
`LABEL`	Instructs SAS to use variable labels as column headings
`N <= '`*label*`'>`	Includes the number of observations in the listing and optionally specifies a label
`OBS='label'`	Specifies a label for the `Obs` column
`NOBS`	Suppresses the output column that identifies observations by a number
`ROUND`	Rounds numbers to two decimal places unless they are otherwise formatted
`SPLIT='char'`	Defines a split character that is used to split label names for column headings. See the section "Splitting Column Titles" in this chapter
`(FIRSTOBS=`*n1* `OBS=`*n2*`)`	Causes `PROC PRINT` to only use selected records in a dataset from *n1* as specified by `FIRSTOBS=` to the *n2* as specified by `OBS=`)
`STYLE(item)=[options]`	There are many ways to customize your PROC PRINT table. Some of these are illustrated in Hands-On Example 5.7. More information about customizing (ODS) output is given in Chapter 8.

Common statements for PROC PRINT

PROC PRINT statement	Meaning
`VAR` *variable list*`;`	Specifies which variables to include in the listing and in which order
`SUM` *vars*`;`	Specifies variables to report sum
`BY, ID, WHERE`	See Table 5.4

The `SUM` statement specifies that a sum of the values for the variables listed is to be reported. For example:

```
SUM COST;
```

The following SAS program (`APRINT1.SAS`) utilizes several of the `PROC PRINT` options and the `SUM` statement:

```
PROC PRINT DATA=MYSASLIB.SOMEDATA
    N= 'Number of Subjects is: '
    OBS='Subjects';
SUM TIME1 TIME2 TIME3 TIME4;
```

TABLE 5.11 PROC PRINT Using SUM, OBS, and N

Subjects	ID	GP	AGE	TIME1	TIME2	TIME3	TIME4	STATUS	SEX	GENDER
1	101	A	12	22.3	25.3	28.2	30.6	5	0	Female
2	102	A	11	22.8	27.5	33.3	35.8	5	0	Female
3	110	A	12	18.5	26.0	29.0	27.9	5	1	Male
4	187	A	11	22.5	29.3	32.6	33.7	2	0	Female
49	566	C	10	23.9	30.7	31.8	34.4	5	0	Female
50	578	C	12	21.6	28.1	30.9	34.1	5	0	Female
				1063.4	**1372.0**	**1524.6**	**1541.9**			

Number of Subjects is: 50

A partial listing of this output is displayed in Table 5.11 showing how the listing has changed because of the options included in the `PROC PRINT` statement

```
N = 'Number of Subjects is:'
```

produces the statement at the bottom of the printout reporting the number of subjects. Moreover, the option

```
OBS='Subjects'
```

causes the label above the record numbers to be `Subjects` rather than `Obs`. Finally, the statement

```
SUM TIME1 TIME2 TIME3 TIME4;
```

causes the totals for the four variables to be reported at the bottom of the report.

HANDS-ON EXAMPLE 5.7

1. Open the program file `APRINT1.SAS`.

   ```
   PROC PRINT DATA=MYSASLIB.SOMEDATA
     N ='Number of Subjects is:'
     OBS='Subjects';
   SUM TIME1 TIME2 TIME3 TIME4;
   TITLE 'PROC PRINT Example';
   RUN;
   ```

   Note the `N=` and `OBS=` options and the `SUM` statement. Run the program and observe their effect on the output.

2. Modify the program by adding

   ```
   PROC SORT DATA=MYSASLIB.SOMEDATA;BY GP;
   ```

   before the `PROC PRINT`.

   (*Continued*)

> (*Continued*)
>
> 3. Modify the SUM line as
>
>   SUM TIME1 TIME2 TIME3 TIME4;BY GP;
>
> 4. Run the program. How does this change the output? In particular, note the report for the number of subjects at the end of each Intervention Group tables.
> 5. To illustrate how options can be used to customize the look of your table, add the following statements before the N= option. Note the style option is a way to specify certain output colors and formats. Other options control the color and font attributes of the header, the OBS output, and the grand total. So, the code:
>
>   STYLE(COL)=[BACKGROUNDCOLOR=LIGHTYELLOW]
>
>   instructs SAS to output the columns (col) with a background color light yellow. Styles are covered in more detail in Chapter 8. For this example, add these statements (before the N=*option*.)
>
>    ```
>    STYLE(COL)=[BACKGROUNDCOLOR=LIGHTYELLOW]
>    STYLE(HEADER)= [BACKGROUND=LIGHTGREEN FONT_WEIGHT=BOLD]
>    STYLE(OBS)= [BACKGROUND=LIGHTGRAY]
>    STYLE(GRANDTOTAL)=[FOREGROUND=DARKBLUE]
>    GRANDTOTAL_LABEL='Grand Total'
>    ```
>
>   Run the program and observe the results. Refer to Appendix A for other color combinations. Also, there are other options for styles not covered here. Refer to the SAS documentation for additional options.

5.7 GOING DEEPER: SPLITTING COLUMN TITLES IN PROC PRINT

The statements illustrated for PROC PRINT are the most commonly used, but there are other statements available that can enhance your listing even more. See the SAS documentation.

One handy statement is SPLIT= that allows you to specify how labels are split into column titles. Normally, SAS splits titles at blanks when needed to conserve space in a report. For example, the following SAS program (APRINT2.SAS) defines labels for a number of variables. (It also uses the PROC PRINT options FIRSTOBS=11 and OBS=20 to limit the number of records listed.)

```
PROC PRINT LABEL DATA="C:\SASDATA\TRAUMA"
     (FIRSTOBS=11 OBS= 20);
VAR INC_KEY AGE GENDER ISS INJTYPE DISSTATUS;
LABEL INC_KEY='Subject ID '
         AGE='Age in 2014 '
         GENDER='Gender '
     ISS='Injury Severity Score '
     INJTYPE='Injury Type '
     DISSTATUS='Discharge Status';
RUN;
```

When this code is run, it produces the output given in Table 5.12. Observe how certain labels are split across two lines. SAS does this, so columns are not too wide. This is a good default feature. However, SAS allows you to be more specific. You can tell SAS where you want the labels to be split using the SPLIT= option.

TABLE 5.12 Example of Column Titles in PROC PRINT

		Trauma Data Report				
Obs	Subject ID	Age in 2014	Gender	Injury Severity Score	Injury Type	Discharge Status
11	603108	15.6	Female	6	Blunt	Alive
12	603110	14.3	Female	17	Blunt	Alive
13	603112	12.0	Female	5	Blunt	Alive
14	603121	13.6	Male	21	Blunt	Alive
15	603122	14.1	Male	1	Blunt	Dead
16	603129	12.7	Male	24	Blunt	Alive
17	603131	6.9	Female	1	Penetrating	Alive
18	603133	16.1	Male	5	Blunt	Alive
19	603138	14.4	Female	5	Blunt	Alive
20	603142	10.9	Male	9	Penetrating	Alive

HANDS-ON EXAMPLE 5.8

1. Open the program file APRINT3.SAS. Run this program and observe the results in Table 5.13.

TABLE 5.13 Example Revised of Column Titles in PROC PRINT

			Trauma Data Report			
Observation Number	Subject ID	Age in 2014	Gender	Injury Severity Score	Injury Type	Discharge Status
===========	===========	===========	===========	===========	==========	=============
11	603108	15.6	Female	6	Blunt	Alive
12	603110	14.3	Female	17	Blunt	Alive
13	603112	12.0	Female	5	Blunt	Alive
14	603121	13.6	Male	21	Blunt	Alive
15	603122	14.1	Male	1	Blunt	Dead
16	603129	12.7	Male	24	Blunt	Alive
17	603131	6.9	Female	1	Penetrating	Alive
18	603133	16.1	Male	5	Blunt	Alive
19	603138	14.4	Female	5	Blunt	Alive
20	603142	10.9	Male	9	Penetrating	Alive
			N = 10			

```
TITLE "Trauma Data Report";
PROCPRINT DATA="C:\SASDATA\TRAUMA"
   (FIRSTOBS=11 OBS=20)
   SPLIT='*'
   N OBS='Observation*Number*===========';
```

(Continued)

(*Continued*)

```
VAR INC_KEY AGE GENDER ISS INJTYPE DISSTATUS;
LABEL INC_KEY='Subject*ID*============'
      AGE='Age in*2014*============'
      GENDER='Gender* *============'
      ISS='Injury Severity*Score*============'
      INJTYPE='Injury*Type*==========='
      DISSTATUS='Discharge*Status*==============';
RUN;
```

Note the changes from the output previously shown in Table 5.12. The `SPLIT="*"` statement tells SAS that the asterisk (*) is the delimiter that will be used to split labels into two or more lines. The double underline (created by repeated equal signs (=)) provides a break between the label and the column contents. The statement

```
N OBS = 'Observation*Number*===========';
```

tells SAS to report the total number of observations at the bottom of the listing (N) and to use the specified label for the OBS column. Note that the label contains three sections, each section splits by the delimiter*. Thus, the label displayed will be three lines long:

```
Observation
    Number
===========
```

Labels for the variables are similarly defined where the * tells SAS where to split the labels as they are used in the columns for the listing. The label GENDER has a * * (note the blank between the *) indicating a skip to make the underline for Gender line up with the other labels. Compare this listing with the previous listing in Table 5.12.

Another feature to note is that labels (like the data values) are either right or left justified depending on whether the column represents numeric or text data.

2. Revise the code by changing the split character to a tilde (~) and change the underlines to single dashes. Rerun the code to make sure it works as expected.
3. GOING DEEPER: You can also control the justification and colors of the column labels with a STYLE option in the VAR statement. Change the VAR statement to the following by adding a STYLE statement option:

```
VAR INC_KEY AGE GENDER ISS INJTYPE DISSTATUS
/ STYLE(HEADER)={COLOR=RED background=YELLOW JUST=C};
```

Resubmit this code and observe how the STYLE(HEADER) option allows you to specify the characteristics of the label text. In this example, it sets color, background, and justification. Change the location in STYLE() from HEADER to DATA and resubmit again to see how to apply these options to the data portion of the output.

5.8 GOING DEEPER: COMMON SYSTEM OPTIONS

Although not a part of a `PROC` statement, System Options can be used to customize the way output is displayed or how data in a dataset are used. System Options must be placed in the code in such a way that they are operated on before they are needed by any following code. Usually, these options are placed at the top of the code file. We'll use System Options in some examples throughout the text as needed, and they will be described when used. This section introduces some commonly used options. System Options are specified using the `OPTIONS` statement. The syntax for the `OPTIONS` statement is

```
OPTIONS option1 option2 . . .;
```

Any number of options may be included in a single statement. The System Options listed in Table 5.14 are ones that are commonly used. Some of those options will be used in upcoming examples.

TABLE 5.14 Common System Options

Common System Options	Meaning
`FIRSTOBS=n` and `OBS=n;`	Specifies the first observation to be used in a dataset (`FIRSTOBS=`) and the last observation to be used (`OBS=`).
	For example
	`OPTIONS FIRSTOBS=2 OBS=21;`
	causes SAS to use data records 2 through 21 in any subsequent analysis. When this option is set, it is usually a good idea to reset the values to
	`OPTIONS FIRSTOBS=1; OBS=MAX;`
	at the end of the program so subsequent analyses are not limited by the same options.
`YEARCUTOFF=year`	Specifies the cut-off year for two-digit dates in a 100-year span starting with the specified date. For example, if `YEARCUTOFF=1920`, then the date 01/15/19 would be considered 2019 while 01/15/21 would be seen as 1921 (Warning: SAS changes the default `YEARCUTOFF` occasionally. At the time of this writing, for SAS 9.4, the cut-off year was 1940.) Always designing your dataset with 4-digit years will avoid this problem.
`PROBSIG=n`	Specifies the number of decimals used when reporting p-values. For example, PROBSIG=3 would cause p-values to be reported to three decimal places.
`LINESIZE= n` and `PAGESIZE= n`	Controls the number of characters in an output line (LINESIZE) or the number of lines on a page (PAGESIZE) for RTF and PDF output.
`NONUMBER`	Specifies that no page numbers will be included in RTF or PDF output.
`NODATE`	Specifies that no date will be included in RTF or PDF output.
`ORIENTATION=option`	Specifies paper orientation. Options are `PORTRAIT` or `LANDSCAPE` for RTF or PDF output.
`NOCENTER`	Left justifies output (default is centered)

> **HANDS-ON EXAMPLE 5.9**
>
> 1. Open the program file `SYSOBS.SAS`.
>
> ```
> OPTIONS FIRSTOBS=11 OBS=20;
> PROC PRINT LABEL DATA="C:\SASDATA\SOMEDATA";
> RUN;
> OPTIONS FIRSTOBS=1 OBS=MAX;
> ```
>
> Run this program and observe that only the records 11 through 20 are displayed in the `PROC PRINT` listing.
>
> 2. Change `FIRSTOBS` to 5 and `OBS` to 25. Rerun and observe which records are displayed.
>
> Note that the last line in the code resets `FIRSTOBS` and `OBS` to 1 and `MAX`. It is important to reset these values; otherwise, any following SAS program may not use all of the records in a specified dataset.

System Options should not be confused with similar-looking options that are available as a part of a procedure. System Options affect *all* the following programs you may run during the current SAS session unless you reset the option. Procedure options only affect that procedure. For example, in `PROC PRINT`, you can use

```
PROC PRINT DATA=MYDATA (FIRSTOBS=11 and OBS=20);
```

to limit the number of records processed in that single procedure, but this does not affect any following procedures. On the other hand, if you use

```
OPTIONS FIRSTOBS=11 and OBS=20;
PROC PRINT DATA=MYDATA;
```

these statements limit the use of *all* the following datasets until you reset the settings with

```
OPTIONS FIRSTOBS=1 OBS=MAX;
```

or until you exit the SAS Software program.

5.9 SUMMARY

This chapter introduced you to the syntax of SAS procedures in preparation for using specific PROCs discussed in the remainder of the book. It also introduced `PROC PRINT` and illustrated some of the common options used for this procedure.

EXERCISES

5.1 **Add titles and footnotes to output.**

Open the program file `FIRST.SAS`:

a. Using the `TITLE` and `FOOTNOTE` statements, place five lines of titles at the beginning and two footnotes (i.e., footnote lines) at the end of the output.

b. Run the program and observe the output.

5.2 **Use the `ID` statement.**

Working with the `C:\SASDATA\CARS` dataset, use the `ID` statement in `PROC PRINT` to create a listing in which the `MODEL` variable is listed in the first column of the output.

5.3 **Use the `LABEL` statement.**

a. Open the program file `DLABEL.SAS`. Add a blank line between the `PROC MEANS` statement and the `RUN` statement. Cut and paste the `LABEL` statement (from the word `LABEL` to the semicolon) from its current location to just before the final `RUN` statement (this places it in the `PROC MEANS` paragraph). Run the program.

b. Do the labels appear in the output?

c. After the final `RUN` statement in the program, add a new line:

```
PROC PRINT;RUN;
```

Rerun the program. Do the labels appear in the output? Why or why not?

d. Change the `PROC PRINT` statement to

```
PROC PRINT LABEL;RUN;
```

Rerun the program. Do the labels appear? Why or why not? If you don't remember, go back to Section 5.4 to find out.

5.4 **Split Labels**

Open the program file `APRINT1.SAS`.

a. Add a `LABEL` statement to define labels for the `TIME` variables as

```
TIME1 = "Baseline Value"
TIME2 = "Observation at 60 Seconds"
TIME3 = "Observation at 120 Seconds"
TIME4 = "Observation at 5 Minutes"
```

b. Select a delimiter to be used to split the labels into several lines, and alter the `LABEL` statements to include the delimiter plus an underline. For example:

```
LABEL TIME1="Baseline*Value*--------"
```

c. Using the `SPLIT=` statement, complete the changes in the code to cause the labels to be split when the listing is printed.

d. Submit the code and verify that the listing includes labels split to your specification.

5.5 Create a report using the techniques you've learned so far.
   a. Using PROC IMPORT, import the file SURVEY.CSV naming it MYSASLIB.CLINIC (a permanent filename)
   b. Create a new data set named WORK.ANSWERS from the imported data set.
   c. Create a Yes/No FORMAT and apply it to MARRIED (1 means yes, 0 means no)
   d. Create Labels for the variables EDU (Education) and TEMP (Temperature)
   e. Plus – include a two-line Title
      - Line 1: Example of a Simple SAS Report
      - Line 2: Programmed by Your Name
   f. Use a PROC PRINT statement displaying labels and formats in the output.

6

PREPARING DATA FOR ANALYSIS

LEARNING OBJECTIVES

- To be able to use SAS Functions
- To be able to TRANSPOSE datasets
- To be able to perform data recoding using SELECT
- To be able to use SAS programming techniques to clean up a messy dataset

The SAS programming language is extensive and flexible, and there are many features of the language that were not discussed in the previous chapters of this book. The next two chapters are included for those who want to go deeper into SAS programming. This chapter introduces several topics that are commonly used in preparing data for analysis not covered in previous chapters. Additional programming topics including arrays and macros are covered in Chapter 7.

6.1 USING SAS FUNCTIONS

Sophisticated calculations can be created using SAS functions. These functions include methods for manipulating arithmetic, character, and date data. The format for the use of SAS functions is

```
RESULT = function(argument1,argument2, etc.);
```

SAS® Essentials: Mastering SAS for Data Analytics, Third Edition. Alan C. Elliott and Wayne A. Woodward.
© 2023 John Wiley & Sons, Inc. Published 2023 by John Wiley & Sons, Inc.
Companion website: https://www.alanelliott.com/SASED3/

Where `argument1` etc. is information (numeric or string) you "send" to the function for it to act upon. Functions can require one or more arguments. Some require no arguments.

For example, a few mathematical functions already built in SAS are as follows:

```
S = ABS(X);          Absolute value
S = FACT(X);         Factorial
S = INT(X);          Integer portion of a value
S = LOG(X);          Natural log
S = SQRT(X);         Square root
```

Some of the other functions available in SAS are `SUBSTR`, `LAG`, `COS`, `SIN`, `ARCS`, `LOG10`, `UNIFORM`, `NORMAL`, `SUM`, and `MEAN`. A list of commonly used SAS functions and additional examples can be found in Appendix B. These include functions of the following types:

- Arithmetic/Mathematical
- Trigonometric
- Date and Time
- Character
- Truncation
- Special Use and Miscellaneous
- Financial
- Access Previous Observations (Lags)

Functions can also be used as a part of a more extensive calculation. For example,

```
C = MEASURE + SQRT(A**2 + B**2);
```

would calculate the square of A and then the square of B, then add those two numbers, take the square root of that value, add that number to the value of `MEASURE`, and assign the result to the variable named `C`.

As mentioned previously, functions may require no argument, one argument, or multiple arguments. A few multi-argument mathematical functions in SAS are the following:

```
S = MAX(x1,x2,x3,...);      Maximum value in the list of arguments
S = MIN(x1,x2,x3,...);      Minimum value in the list of arguments
S = SUM(x1,x2,x3,...);      Sum of non-missing values in a list of arguments
S = MEDIAN(x1,x2,x3,...);   Median of a list of non-missing values
S = ROUND(value, round);    Rounds value to nearest round-off unit
```

These functions allow missing values in the list. Thus, the `SUM` function

```
TOTAL=SUM(TIME1,TIME2,TIME3,TIME4);
```

can create a different answer than

```
TOTAL= TIME1+TIME2+TIME3+TIME4;
```

It is important to understand that in the second instance, if the value for TIME2 were missing for a subject, the value of TOTAL would be set at missing (.). However, in the assignment using the SUM function, the value of TOTAL would contain the sum of the other time values even if TIME2 were missing. You must ultimately decide which way you want to treat missing values.

The round-off unit in the ROUND function determines how the rounding will be performed. The default is 1. A round-off value of 0.01 means to round it off to the nearest 100th, and a round-off value of 5 means to round it off to the nearest 5. Here are a few examples:

ROUND(3.1415,.01)	Returns the value 3.14
ROUND(107,5)	Returns the value 105
ROUND(3.6234)	Returns the value 4 (an integer)

A function that takes no argument is TODAY(). For example, the code

```
NOW=TODAY();
```

puts the current date value from the computer's clock into the variable named NOW.

When arguments are a list of values, such as in MAX or MIN, you can specify the list as variables separated by commas, or as a range preceded by the word OF. For example, if these variables have the following values:

```
X1 = 1;   X2 = 2;   X3 = 13;   X4=10;
```

Then

MAX(1,2,3,4,5)	Returns the value 5
MAX(X1,X2,X3,X4)	Returns the value 13
MAX(OF X1-X4)	Returns the value 13

Note that the designation OF X1-X4 is interpreted by SAS as all same-named consecutively numbered variables from X1 to X4 in the MAX() function example earlier. If there are missing values in the list, they are ignored. Other similar functions are illustrated here:

MIN(OF X1-X4)	Returns the value 1
SUM(OF X1-X4)	Returns the value 26
MEDIAN(OF X1-X4)	Returns the value 6 (the average of 2 and 10)
NMISS(OF X1-X4)	Returns 0 (number of missing values)
N(OF X1-X4)	Returns 4 (number of non-missing values)

There is also an extensive collection of functions available to allow you to manipulate character and date values. As an example of a date function, the INTCK function counts the number of intervals (INTerval Count) between two dates or times. The result is an integer value. The format for this function is

```
INTCK('interval',from,to)
```

144 CHAPTER 6: PREPARING DATA FOR ANALYSIS

The `from` and `to` variables are SAS date values. The `'interval'` argument (the quotes are required) can be one of the following lengths of time:

```
DAY
WEEKDAY
WEEK
TENDAY
SEMIMONTH
MONTH
QTR
SEMIYEAR
YEAR
```

Thus, the following function counts the number of years between the date variables `begdate` and `enddate` because the date interval indicated is `'YEAR'`.

```
INTCK('YEAR',begdate,enddate)
```

The following code counts the number of months between these two dates because the date interval indicated is `'MONTH'`.

```
INTCK('MONTH',begdate,enddate)
```

Another handy date function is the `MDY` function. This allows you to convert numbers into a SAS date value. For example,

```
BDATE = MDY(10,12,1989);
```

converts the date October 12, 1989 into a SAS date value. Dates must be in SAS date format in order for you to use other SAS date functions or to output dates using SAS format specifications. Or if in a dataset you have these values

```
    MTH=5  DAY=9  YR=2016
```

then the `MDY` function

```
    VISITDATE = MDY(MTH,DAY,YR);
```

creates a SAS date variable named `VISITDATE` with the value May 9, 2016. For a more comprehensive list of functions, refer to Appendix B in this book.

HANDS-ON EXAMPLE 6.1

This example illustrates a method for calculating the difference between two dates (AGE) using the `INTCK` function. Open the program file `DDATES.SAS`.

```
DATA DATES;
INPUT @1 BDATE MMDDYY8.;
```

```
TARGET=MDY(08,25,2023);         * Uses MDY() function;
AGE=INTCK('YEAR',BDATE,TARGET);  * INTCK function;
DATALINES;
07101952
07041776
01011900
;
PROC PRINT DATA=DATES;
FORMAT BDATE WEEKDATE. TARGET MMDDYY8.;
RUN;
```

The SAS code involves the following steps:

1. Read in BDATE as a SAS date value using the INFORMAT MMYYDD8. specification in the INPUT statement.
2. Convert the target date 08/25/2023 into a variable named TARGET using the MDY function.
3. Use the INTCK function to count the number of years between each date (BDATE) in the dataset and the TARGET date and assign this value to the variable AGE.
4. Assign the count of years to the variable named AGE.

Here are the steps:

1. Note the statement.

```
INPUT @1 BDATE MMDDYY8.;
```

This statement reads in a variable named BDATE as a SAS date value starting in column 1 (specified by the pointer @1) using the format MMDDYY8. The SAS date variable TARGET is created with the MDY function, using constant values for month, day, and year. To display the dates in a readable manner, a FORMAT statement is included in the PROC PRINT paragraph.

2. Run the program and observe the output in Table 6.1.

Note that the program calculates AGE as the difference in years (the count of the "year" intervals) between the TARGET date and the date value for each record in the SAS dataset.

TABLE 6.1 Results of the INTCK Function Calculation

Obs	BDATE	TARGET	AGE
1	Thursday, July 10, 1952	08/25/23	71
2	Thursday, July 4, 1776	08/25/23	247
3	Monday, January 1, 1900	08/25/23	123

(*Continued*)

> (*Continued*)
>
> 3. To illustrate the TODAY() function, change the TARGET date to today's date using the code
>
>    ```
>    TARGET=TODAY();
>    ```
>
>   Rerun the program and observe the results.
>
> 4. Another function you can use to find the difference between two dates is YRDIF. Change the AGE=INTCK statement to the following:
>
>    ```
>    AGE=YRDIF(BDATE,TARGET,'ACT/365');
>    ```
>
>   This function finds the difference in two dates using the method of dividing the number of days by 365. Run the program. See Appendix B for more information about this and other functions.
>
> 5. For you history buffs: If you include the date
>
>    ```
>    09041752
>    ```
>
>   (September 4, 1752) into the list of data values and run the program, it indicates that this date fell on a Monday (in the current Gregorian calendar). However, this date did not exist. When Britain adopted the Gregorian calendar, the day after September 2, 1752 was September 14, 1752. Thus, if SAS is used to calculate dates before September 2, 1752, care must be taken because SAS does not convert dates to the older Julian calendar designation.

6.2 USING PROC TRANSPOSE

PROC TRANSPOSE allows you to restructure the values in your dataset by transposing (or reorienting) the data. This is typically performed when your data are not in the structure required for an analysis. For example, if your data are listed with subjects as columns, you can transpose the data so the subjects represent rows, which is the correct structure for SAS data. The simplified syntax for PROC TRANSPOSE is:

```
PROC TRANSPOSE DATA=input-data OUT=output-data;
<PREFIX=prefix>;
<BY <variables>
VAR variables;
```

- The *input* dataset is the dataset to be transposed.
- The *output* dataset is the resulting dataset.
- PREFIX specifies a prefix to the names of variables created in the transposition. The default names are COL1, COL2, and so on.
- The BY variable, when specified, causes the transpose to be performed once per category specified by the BY group's categorical variable.
- The VAR statement specifies which variables are to be transposed.

HANDS-ON EXAMPLE 6.2

Suppose you have been given a dataset in which each line represents a variable and each subject's data are in a single column. This example illustrates how to use PROC TRANSPOSE to restructure this dataset. Note that because some of the data are of character type, you must indicate that with a $ after each variable name in the INPUT statement.

1. Open the program file DTRANSPOSE1.SAS.

```
DATA SUBJECTS;
INPUT SUB1 $ SUB2 $ SUB3 $ SUB4 $;
DATALINES;
12 21 13 14
13 21 12 14
15 31 23 23
15 33 21 32
M  F  F  M
;
PROC TRANSPOSE DATA=SUBJECTS OUT=TRANSPOSED;
   VAR SUB1 SUB2 SUB3 SUB4;
RUN;
PROC PRINT DATA=TRANSPOSED;
RUN;
```

2. Run the program. The OUT=TRANSPOSED statement creates a SAS dataset named OBS_TRANSPOSED produced by PROC TRANSPOSE. The resulting dataset is given in Table 6.2.

TABLE 6.2 Example of Transposed Data

Obs	_NAME_	COL1	COL2	COL3	COL4	COL5
1	SUB1	12	13	15	15	M
2	SUB2	21	21	31	33	F
3	SUB3	13	12	23	21	F
4	SUB4	14	14	23	32	M

3. You can control the names of the columns with a PREFIX statement. Change the PROC TRANSPOSE statement to

```
PROC TRANSPOSE DATA=SUBJECTS OUT=TRANSPOSED PREFIX=INFO;
```

Rerun the program and observe the names of the columns.

4. The solution isn't exactly what you want. One way you could fix that is to use the RENAME statement to give each column a new name. Enter the following code. Note how each variable is renamed to a better name, particularly GENDER:

(*Continued*)

(*Continued*)

```
DATA NEW;SET TRANSPOSED;
RENAME INFO1=T1 INFO2=T2 INFO3=T3 INFO4=T4 INFO5=GENDER;
RUN;
PROC PRINT DATA=NEW;
RUN;
```

Submit this code and verify that it provides better names for the dataset.

Another possibility when transposing data is to use an available label to name new variables. Consider the following code (DTRANSPOSE1A.SAS):

```
DATA SUBJECTS;
INPUT LAB $ SUB1 $ SUB2 $ SUB3 $ SUB4 $;
DATALINES;
BASELINE 12 21 13 14
TIME1 13 21 12 14
TIME2 15 31 23 23
TIME3 15 33 21 32
GENDER M F F M
;
PROC TRANSPOSE DATA=SUBJECTS OUT=TRANSPOSED;
  ID LAB;
  VAR SUB1 SUB2 SUB3 SUB4;
RUN;
PROC PRINT DATA=TRANSPOSED;
RUN;
```

In this version of the program, there is a variable called LAB that provides a label for columns. The ID statement in PROC TRANSPOSE allows you to specify LAB for the new variable names.

HANDS-ON EXAMPLE 6.3

1. Open the file DTRANSPOSE1A.SAS. Submit the code and observe the output given in Table 6.3. The following is a snippet of the code.

```
PROC TRANSPOSE DATA=SUBJECTS OUT=TRANSPOSED;
  ID LAB;
  VAR SUB1 SUB2 SUB3 SUB4;
RUN;
```

Because of the ID statement in PROC TRANSPOSE, the values for ID become the column names.

2. Change the values in the dataset from BASELINE, TIME1, and so on to READ1 through READ4. Leave GENDER as is. Resubmit the code and note the changed variable names.

TABLE 6.3 Transposed Data with ID

Obs	_NAME_	BASELINE	TIME1	TIME2	TIME3	GENDER
1	SUB1	12	13	15	15	M
2	SUB2	21	21	31	33	F
3	SUB3	13	12	23	21	F
4	SUB4	14	14	23	32	M

6.2.1 Using TRANSPOSE to Deal with Multiple Records per Subject

As a second example illustrating the use of PROC TRANSPOSE, suppose you have data that have several (and possibly differing numbers of) observations per subject, but you want to analyze the data by observation (a set of observations per row). You can use PROC TRANSPOSE to transpose the data by a key variable. For example, Table 6.4 is a partial list

TABLE 6.4 Data from the COMPLICATIONS Dataset

Obs	SUBJECT	COMPLICATION
1	1001	Heart Attack
2	1033	Heart Attack
3	1048	Heart Attack
4	1054	Heart Attack
5	1073	Heart Attack
6	1135	Wound Infection
7	1149	Compartment Syndrome
8	1368	Pneumonia
9	1379	Heart Attack
10	1536	Heart Attack
11	1545	Heart Attack
12	1615	Heart Attack
13	1643	Pancreatitis
14	1645	Pneumonia
15	1796	Coagulopathy
16	1859	Pneumonia
17	1921	Pneumonia
18	1921	Renal Failure
19	2023	Pneumonia
20	2039	Coagulopathy

of records in a simulated SAS dataset named COMPLICATIONS. These are observed complications for accident victims. Note that some subjects have more than one complication (subject 1921, for example).

> **HANDS-ON EXAMPLE 6.4**
>
> This example uses PROC TRANSPOSE to rearrange the COMPLICATIONS dataset so that there is only one subject per row and to output a dataset for subjects with three or more complications. Notice that data are already sorted by SUBJECT, but if this was not the case, you would need to do a PROC SORT by SUBJECT before doing PROC TRANSPOSE.
>
> 1. Open the program file DTRANSPOSE2.SAS.
>
> ```
> PROC TRANSPOSE DATA=MYSASLIB.COMPLICATIONS
>     OUT=COMP_OUT PREFIX=COMP;
> (Continued)
>     BY SUBJECT;
>     VAR COMPLICATION;
> RUN;
> DATA MULTIPLE;SET COMP_OUT;
> DROP _NAME_ ;
> IF COMP3 NE '' ;
> RUN;
> PROC PRINT DATA=MULTIPLE; VAR SUBJECT COMP1-COMP4;
> FORMAT SUBJECT 10. COMP1-COMP4 $10.;
> RUN;
> ```
>
> The PROC TRANSPOSE step outputs a dataset named COMP_OUT. The BY statement indicates that the transpose is by SUBJECT, so there will be one subject per row. The PREFIX= option specifies the prefix for the new variables created for complications. That is, the first complication for a subject is named COMP1, the second COMP2, and so on. The VAR statement indicates which variable (COMPLICATION) will be used as the output variable (named with the COMP prefix).
>
> A second dataset is created (MULTIPLE) that contains only 38 records where the value for COMP3 is not blank. Thus, only subjects that have three or more complications are retained in the MULTIPLE dataset. For example, subject 1921 has only two complications, so this subject is not included in the MULTIPLE dataset, but subject 2076 does appear. The results are reported using a PROC PRINT statement.
>
> 2. Run this program. The abbreviated output given in Table 6.5 presents the first few records. Note that each of the subjects listed has at least three complications.
> 3. Change the code to select only subjects with four or more complications, that is, change the IF statement and change COMP4 to COMP5 in two places. Resubmit and confirm that the change worked.

TABLE 6.5 Output from PROC TRANSPOSE

Illustrates PROC TRANSPOSE, compressing multiple subject records to a single record. Reports subjects having 3 or more complications.

Obs	SUBJECT	COMP1	COMP2	COMP3	COMP4
1	2076	Pneumonia	Heart Atta	Renal Fail	
2	3585	DVT (Lower	Pneumonia	Renal Fail	
3	3630	DVT (Lower	Heart Atta	Pancreatit	Pneumonia
4	4585	Compartmen	Pneumonia	Skin Break	
5	4599	Aspiration	Pneumonia	Renal Fail	
6	4760	Acute Resp	Pneumonia	Renal Fail	

6.3 THE SELECT STATEMENT

The IF statement allows you to conditionally perform certain SAS calculations and assignments and is handy for recoding a small number of values. For more complex recoding, or recoding with more categories, the SELECT statement is often easier and more efficient to use. The SELECT statement evaluates the value of a variable and creates new assignments based on those values. Syntax (simplified) is as follows:

```
SELECT <(select-expression)> ;
     WHEN-1  statement;
     ...    WHEN-n statement;
     < OTHERWISE statement;>
```

For example, suppose you want to calculate NEWVAL according to some specific values of the variable OBSERVED. That is, if OBSERVED=1, you want to set NEWVAL=AGE+2. If OBSERVED=2 or 3, you want to set NEWVAL=AGE+10, and so on. A SELECT statement to perform this recoding would be as follows:

```
SELECT (OBSERVED);
   WHEN (1) NEWVAL=AGE+2;
   WHEN (2, 3) NEWVAL=AGE+10;
   WHEN (4,5,6) NEWVAL=AGE+20;
OTHERWISE NEWVAL=0;
END;
```

Another way to use SELECT is without using the specific SELECT(*expression*) style as in the aforementioned. In this case, the WHEN statements include conditional expressions that should be in parentheses. For example:

```
SELECT;
   WHEN (GP='A') STATUS2=1;
```

152 CHAPTER 6: PREPARING DATA FOR ANALYSIS

```
WHEN (GP='B' AND SEX=1) STATUS2=2;
WHEN (GP='C' AND SEX=0) STATUS2=3;
OTHERWISE STATUS2=0;
END;
```

HANDS-ON EXAMPLE 6.5

As an example using the SELECT statement, suppose you want to recode the numeric variable STATUS into a character variable called ECONOMIC, and combine some of the groups.

1. Open the program file DSELECT.SAS. Note the SELECT statement. Submit the code and verify that it creates a new variable named ECONOMIC. Partial output is given in Table 6.6.

```
DATA MYDATA;SET "C:\SASDATA\SOMEDATA";
FORMAT ECONOMIC $7.;
SELECT(STATUS);
   WHEN (1,2) ECONOMIC="LOW";
   WHEN (3) ECONOMIC="MIDDLE";
   WHEN (4,5) ECONOMIC="HIGH";
   OTHERWISE ECONOMIC="MISSING";
END;
PROC PRINT DATA=MYDATA;
RUN;
```

TABLE 6.6 Example of SELECT Statement

This example illustrates the Select Statement										
Obs	ID	GP	AGE	TIME1	TIME2	TIME3	TIME4	STATUS	SEX	GENDER	ECONOMIC
1	101	A	12	22.3	25.3	28.2	30.6	5	0	Female	HIGH
2	102	A	11	22.8	27.5	33.3	35.8	5	0	Female	HIGH
3	110	A	12	18.5	26.0	29.0	27.9	5	1	Male	HIGH
4	187	A	11	22.5	29.3	32.6	33.7	2	0	Female	LOW
5	312	A	4	17.0	21.3	22.7	21.2	5	1	Male	HIGH
6	390	A	6	21.3	30.5	30.7	30.1	4	0	Female	HIGH

2. Using another SELECT statement without the SELECT(*expression*) style, add new code to the DATA step that does the following recode:
   - For ages 12 and under set AGEGROUP="CHILD".
   - For ages 13 to 19 set AGEGROUP="TEEN".
   - For ages 20 and above set AGEGROUP="ADULT".
   - OTHERWISE set AGEGROUP="NA".

   Here is the partial code to be placed after the END statement:

```
SELECT;
WHEN (AGE LE 12) AGEGROUP="CHILD";
WHEN (AGE GT 12 AND AGE LE 19) AGEGROUP="TEEN";
```

You finish the code.

> 3. Submit the code and verify that your new SELECT statement works. Note that although both STATUS and AGE are numeric, STATUS is categorical and lends itself to using SELECT with the SELECT(*expression*) style, whereas AGE is a continuous numeric variable that lends itself to using a SELECT statement without the SELECT(*expression*) style.

6.4 GOING DEEPER: CLEANING A MESSY DATASET

Many of the features in the SAS language are helpful in cleaning up messy data. By messy we mean datasets that are not yet ready for analysis. Most data analysts experience problems dealing with files that contain data that have coding problems and must be fixed before a proper analysis is possible. This section walks you through a case study using a dataset with problems and illustrates how they might be corrected.

The dataset in question is named MESSYDATA and represents an actual dataset (with some information changed to protect the innocent) that was created from a survey taken at a clinic. Originally entered into an Excel spreadsheet, it was imported into SAS using PROC IMPORT. Information about this process has been discussed in Chapter 3. In this case study, we start with the MESSYDATA.SA7BDAT file, which contains a number of problems that should be fixed before performing any type of analysis.

We use this case study to provide a series of actions you can take to correct errors in the data. Of course, if you do not first clean up your dataset, any analysis may be wrong, may reach incorrect conclusions, and may cost you time and money if you have to make the corrections later.

A portion of the original MESSYDATA dataset is displayed in Table 6.7.

A few of the problems you might quickly note include the following:

- Line 17 is blank.
- There are non-date values in the DateLeft column.
- There is a non-number in the Age column (>29).
- Values of GENDER are mixed upper and lower cases.
- There are multiple answers in columns that should have only one answer.

It is easy to see that some of these errors should have been dealt with in the design of the survey or in the data entry process. Nevertheless, this dataset represents an accurate view of typical problems an analyst will see.

You can correct these issues using SAS code. One reason this is a good idea is that it leaves an audit trail of changes. You should always make changes in a *copy* of the original dataset so that the original data will not be changed in any way. Again, this is a good practice. If you alter the original data file, and accidentally change something that should not have been changed, it might be difficult to get the original dataset back.

Additionally, some projects (such as those funded by grants) may make it a requirement to keep an accurate record of all changes in a data set. Making changes in SAS code may provide an adequate record of changes.

A number of functions are utilized in the data clean-up illustrated in the following sections. Please refer to Appendix B for a specific description of the functions.

6.4.1 Fix Labels, Rename Variables

The first step in creating a clean dataset is to attach labels to the variables. This helps create a more understandable output. Of course, you need to know what each variable means in

TABLE 6.7 A Messy Dataset

Subject	DateArrived	TimeArrive	DateLeft	TimeLeft	Married	Single	Age	Gender	Education	Race	How_Arrived	Top_Reason	Arrival	Satisfaction
1	2/7/2005	11:18:00 A	2/7/2005	2:15:00 PM		1	0.29	M	12	H	Bus	1	101.5	84.7
2	3/10/2005	1:23:00 PM	3/11/2005	12:40:00 AM	1		0.21	M	12	AA	Car	1	98.6	82.8
3	3/2/2005	1:08:00 PM	3/2/2005	5:07:00 PM	0		1.21	F	12	AA	Car	2	98.2	68.2
4	5/2/2005	11:46:00 A	5/2/2005	9:45:00 PM		1	1	F	11	AA	Car	1	100	70
5	1/31/2005	12:09:00 P	2/1/2005	1:30:00 AM		1	0.29	F	12	NA	Car	1	101	81.4
6	2/1/2005	2:27:00 PM	2/1/2005	2:40:00 PM	0		0.47	F	12	H	Car	3	37.2	87.7
7	2/3/2005	11:50:00 A	2/3/2005	9:08:00 PM		1	0.45	F	10	AA	Bus	1,3	99.8	68.7
8	2/4/2005	10:47:00 A	#VALUE	11:50:00 PM	0		0.21	M	12	H	Car	1	99.5	89.2
9	2/5/2005	11:42:00 A	2/5/2005	12:51:00 PM		1	0.46	F	14	MEX	Car	2	97.1	98.7
10	1/30/2005	12:53:00 P	1/30/2005	11:40:00 PM	1		1.28	M	12	C	Car	3	36.7	85.6
11	2/6/2005	9:22:00 AM	2/6/2005	2:40:00 PM	0	0	0.34	F	9	H	Car	1	101	73.6
12	1/30/2005	2:30:00 AM	Missing	8:00:00 PM		1	1.20	F	11	H	Bus	3	37.2	79.4
13	2/6/2005	7:10:00 AM	2/7/2006	12:20:00 AM	0	1	0.35	M	16	AA	Car	1	97.8	86.9
14	2/6/2005	9:35:00 AM	2/6/2005	7:25:00 PM		1	0.33	F	12	H	Walk	1	37.5	87.6
15	2/7/2005	9:50:00 AM	2/7/2005	8:30:00 AM		1	0.27	F	12	AA	Car	3	101.8	79.1
16	2/7/2005	10:55:00 A		11:20:00 AM									100.2	81
17														
18	2/7/2005	12:40:00 P	2/7/2005	3:15:00 PM		1	0.36	F	12	H	Bus, Walk	1	97.1	.99
19	5/9/2005	1:20:00 PM	5/9/2005	7:20:00 PM	0		1.34	F	12	AA	Car	1	38	81
20	2/7/2005	2:40:00 PM	2/8/2005	2:00:00 AM		1	0.26	M	12	X	Car	1.5	97.8	86.3
21	2/9/2005	12:00:00 A	2/9/2005	7:45:00 PM		1	0.44	M	9	H	Car	2	99.2	87
22	1/21/2005	4:35:00 PM	3/21/2005	1:36:00 PM	0	0	1.29	F	16		Car	1	36.8	86.8
23	1/31/2005	8:20:00 AM	2/1/2005	1:50:00 AM	1		0.53	M	7	C	Car	3	37.6	84.8
24	1/25/2005	10:25:00 A	10/28/2001	12:00:00 AM	1		0.45	F	12	A	Car	3	99.8	85.4
25	1/25/2007	12:10:00 P	1/25/2005	7:45:00 PM		1	0.22	f	10	M	Car	2	97.4	85.4
26	1/25/2005	2:44:00 PM	1/25/2005	3:25:00 PM	11		0.22	F	9	C	Car	3	98.6	84.6
27	1/25/2005	10:29:00 A	1/25/2005	11:30:00 PM	1		0.42	M	10	H	Bus	1	99.2	86.1
28	1/25/2005	2:52:00 PM	1/26/2005	12:30:00 AM	1		0.39	M	8	H	UKN	3	98.8	86.1
29	1/25/2005	12:30:00 P	1/26/2005	3:20:00 AM	0		1.48	M	9	AA	Bus	2	99.7	72
30	1/25/2005	2:25:00 PM	1/26/2005	2:25:00 AM	1		0.39	M	13	AA	Car	1	98.3	83.5
31	1/25/2005	11:00:00 A	1/26/2005	12:45:00 AM		1	0.28	f	12	H	Car	2	37.8	86
32	1/25/2005	2:20:00 PM	1/26/2005	2:45:00 AM	0		1.21	F	10	AA	Car	1	101.9	74
33	1/30/2005	1:38:00 PM	1/31/2005	3:10:00 AM	0	0	1.22	F	12	C	Car	3&1	98.5	86.5
34	1/26/2005	7:55:00 PM	1/26/2005	10:40:00 PM		1	1.33	F	12	AA	NA	1,2,3	97	74.5
35	2/13/2005	7:15:00 PM	2/14/2005	6:30:00 PM		1	0.45	F	9	H	Walk, Car	1	97.7	86.7
36	1/27/2005	12:45:00 P	1/27/2005	2:40:00 PM	0		0.31	F	12	W	Car	3	36.6	84.5
37	1/27/2005	9:15:00 AM	1/27/2005	5:10:00 PM	0	0	1.65	M	11	AA	Car	3	1018	84.1
38	1/27/2005	9:30:00 AM	1/27/2005	5:15:00 PM	0	1	1.28	F	7	H	Car	2	101.3	67.8
39	1/27/2005	9:52:00 AM	1/27/2005	6:25:00 PM	1		0.49	M	12	H	Car	1	98.6	88.7
40	1/28/2005	Noon	1/28/2005	6:15:00 PM	1	0	1.29	f	12	AA	Car	2	97.4	58.6

order to assign the variable labels, and this should be provided in the form of some type of data dictionary, as illustrated in Table 6.8.

TABLE 6.8 Data Dictionary for Messy Dataset

Variable	Variable Name	Type	Label
1	Subject	Char	Subject ID
2	DateArrived	Date	Date Arrived
3	TimeArrived	Time	Time Arrived
4	DateLeft	Date	Date Left
5	TimeLeft	Time	Time Left
6	Married	Num	Married?
7	Single	Num	Single?
8	Age	Num	Age Jan 1, 2014
9	Gender	Char	Gender
10	Education	Num	Years of Schooling
11	Race	Char	Race
12	How_Arrived	Char	How Arrived at Clinic
13	Top_Reason	Num	Top Reason for Coming
14	Arrival	Num	Temperature
15	Satisfaction	Num	Satisfaction Score

You may note in Table 6.7 that there are some variables whose values are suspect, and some that are not of the right data type. That will be fixed later. In this example, the program assumes that you have a SAS library named `MYSASLIB`. (A `LIBNAME` statement is included in the code to help you set up this library if needed.) The `PROC PRINT` statement lists the first 10 observations to make it easy to verify that the changes made took place.

HANDS-ON EXAMPLE 6.6

1. Open the program file `MESSY1.SAS` (Only a portion of the code for this exercise is shown here.). In this program, only some of the labels have been assigned. The variable `ARRIVAL` is replaced with `TEMP` to avoid confusion with other variables. Submit this code and observe the results given in Table 6.9.

```
DATA MYSASLIB.CLEANED;SET MYSASLIB.MESSYDATA;
LABEL
     EDUCATION='Years of Schooling'
     HOW_ARRIVED='How Arrived at Clinic'
     TOP_REASON='Top Reason for Coming'
     SATISFACTION='Satisfaction Score';
TEMP=ARRIVAL;
DROP ARRIVAL;
LABEL TEMP='Arrival Temperature';
RUN;
PROC PRINT LABEL
     DATA=MYSASLIB.CLEANED
```

(*Continued*)

(Continued)

```
        (FIRSTOBS=1 OBS=10);
    VAR SUBJECT EDUCATION
        TOP_REASON SATISFACTION;
RUN;
```

TABLE 6.9 First Pass to Fix Messy Data

Obs	SUBJECT	Years of Schooling	Arrival Temperature	Top Reason for Coming	Satisfaction Score
1	00001	12	101.5	1	84.7
2	00002	12	98.6	1	82.8
3	00003	12	98.2	1	68.2
4	00004	11	100.0	2	70
5	00005	12	101.0	1	81.4
6	00006	12	37.2	3	87.7
7	00007	10	99.8	1,3	68.7
8	00008	12	99.5	1	89.2
9	00009	14	97.1	2	88.7
10	00010	14	36.7	1	85.6

2. Enter the remainder of the Labels (see Table 6.8) and submit the program. Note that the resulting program is stored in the file `MYSASLIB.CLEANED`.

   Note: It is possible to rename the `ARRIVAL` variable using the code.

```
RENAME ARRIVAL=TEMP;
```

However, because we will use `TEMP` in some later calculations within the same dataset, it is better in this case to create a new variable. The reason for this is that if `RENAME` is used, the variable `TEMP` can't be used in latter calculations within the same DATA step.

6.4.2 Fix Case Problems, Allowed Categories, and Delete Unneeded Lines

A second common fix to apply to data values is to make sure that case is consistent within a variable. For example, if `GENDER` is recorded as `F` and `f`, the computer sees that as two separate categories and any summary by gender will not be correct. To correct case problems, you can use the `UPCASE()` or `LOWCASE()` function to convert data values to all upper or all lower case.

A second common fix is to verify that all items in a categorical variable are allowable. For example, in the `HOW_ARRIVED` variable, only `CAR`, `BUS`, or `WALK` is acceptable. If data values do not equal any of the allowed values, we'll set the value to missing. In the case of `HOW_ARRIVED` missing is a blank value, represented by `" "`. The `IN` function can check to see if a variable contains one of the values in a set of specified values. For example:

```
IF HOW_ARRIVED NOT IN ('CAR','BUS','WALK') THEN HOW_ARRIVED="";
```

GOING DEEPER: CLEANING A MESSY DATASET   157

A third easy-to-perform check is to delete irrelevant records. In this dataset, if a line does not contain a Subject ID, we want to eliminate that record. This is done with an IF statement.

```
IF SUBJECT ="" THEN DELETE;
```

These fixes are illustrated in the next example.

HANDS-ON EXAMPLE 6.7

1. Open the program file `MESSY2.SAS`. Note the code below that is added to the end of the `LABEL` statements you added to `MESSY1.SAS` while working Hands-On Example 6.6. Submit the code and observe the results. A portion of the resulting dataset (with selected variables) is shown in Table 6.10. You should note that all character variables are in upper case, and some of the `HOW_ARRIVED` values (SUBJECTs 17 and 24) have been set to blank.

```
* Fix 1 - Convert character values to all uppercase;
GENDER=UPCASE(GENDER);
RACE=UPCASE(RACE);
HOW_ARRIVED=UPCASE(HOW_ARRIVED);
* Fix 2 Set bad unknown values to Missing;
IF HOW_ARRIVED NOT IN ('CAR','BUS','WALK') THEN HOW_ARRIVED="";
* FIX 2 - Get rid of empty rows;
IF SUBJECT="" THEN DELETE;
```

TABLE 6.10 Some Corrected Variables in the Messy Dataset

Obs	Subject ID	Gender	Race	How Arrived at Clinic
1	00001	M	H	BUS
2	00002	M	AA	CAR
3	00003	F	AA	CAR
4	00004	F	AA	CAR
5	00005	F	NA	CAR
6	00006	F	H	CAR
7	00007	F	AA	CAR
8	00008	M	H	BUS

2. Add a new line using the `IN()` function to check for correct values in GENDER (M or F) and in RACE (H, AA, and C). Specifically, add the statement

```
IF GENDER NOT IN('M','F') THEN GENDER="";
```

immediately following the last `IF` statement in the new code above. Submit the new code and observe the results.

(Continued)

158 CHAPTER 6: PREPARING DATA FOR ANALYSIS

(*Continued*)

3. Upon further examination, suppose you verify with the survey coordinator that an answer of "MEX" or "M" for a race should be corrected to "H", "W" should be corrected to "C", and "A" should be corrected to "AA". If RACE is "X", "NA" or ".", then set it to blank (missing). The code to correct two of these items (which must appear after the UPCASE fixes and before the IN() function fixes) is

```
IF RACE="MEX" OR RACE="M" THEN RACE="H";
IF RACE="X" OR RACE="NA" THEN RACE="";
```

Enter these fixes. Submit the code and verify that the fixes work (and as always, we recommend checking the **Log Window** for error messages).

6.4.3 Check and Fix Incorrect Categories, Fix Duplicated Variables

Two troubling variables are MARRIED and SINGLE. The survey asked respondents their marital status, and the information was recorded in the dataset where 1 means yes and 0 means no. Technically, these two variables should be the opposite of each other, and you should only require one of them in the dataset. However, if you look at the frequencies of each using PROC FREQ, you will discover that the two variables are not providing consistent information. To check these and other categorical variables, use the PROC FREQ procedure. This is illustrated in Hands-On Example 6.8.

HANDS-ON EXAMPLE 6.8

This example checks for incorrect entries using PROC FREQ and fixes the problems.

1. Open the program file DISCOVER1.SAS.

```
PROC FREQ DATA=MYSASLIB.CLEANED;
TABLES MARRIED SINGLE TOP_REASON RACE GENDER HOW_ARRIVED;
run;
```

Submit the code and observe the results in Table 6.11. (Only some of the output is shown here.) Note the following:
- There are codes for MARRIED that are incorrect.
- The two tables MARRIED and SINGLE are not the opposite of one another as you would expect. Typically, you'd only need one table or the other. In this case, suppose you choose to use the SINGLE table, so the MARRIED variable can be deleted.

- Several values for TOP_REASON are incorrect.
- RACE still contains some incorrect values. (The previous fix didn't correct all the problems.)

TABLE 6.11 Frequencies for Several Categorical Variables

Married?				
Married	Frequency	Percent	Cumulative Frequency	Cumulative Percent
0	38	48.10	38	48.10
1	39	49.37	77	97.47
9	1	1.27	78	98.73
11	1	1.27	79	100.00

Single?				
Single	Frequency	Percent	Cumulative Frequency	Cumulative Percent
0	43	54.43	43	54.43
1	36	45.57	79	100.00

Top Reason for Coming				
Top_Reason	Frequency	Percent	Cumulative Frequency	Cumulative Percent
1	44	55.70	44	55.70
1,2,3	1	1.27	45	56.96
1,3	1	1.27	46	58.23
1.5	1	1.27	47	59.49
2	19	24.05	66	83.54
3	12	15.19	78	98.73
3&1	1	1.27	79	100.00

Race				
Race	Frequency	Percent	Cumulative Frequency	Cumulative Percent
A	1	1.32	1	1.32
AA	37	48.68	38	50.00
C	5	6.58	43	56.58
H	32	42.11	75	98.68
W	1	1.32	76	100.00
Frequency Missing = 3				

- `GENDER` appears to be okay now. Note that it contains two missing values as a result of the fixes in the previous example. (The associated `PROC FREQ` output is not shown here.)
- All categories in `HOW_ARRIVED` are okay. (Not shown here.)

With this information, you can continue cleaning the dataset.

(*Continued*)

160 CHAPTER 6: PREPARING DATA FOR ANALYSIS

(*Continued*)

2. Open the program file MESSY3.SAS. Observe the code listed under the Step 3 comment lines. The code already includes the following fixes:
   - The MARRIED variables are eliminated.
   - Additional RACE fixes are included.

   ```
   DROP MARRIED;
   IF RACE NOT IN ('AA','H','C') THEN RACE="";
   ```

   To fix the TOP_REASON problems, use an IF statement shown below in Step 3 of the code. Note that TOP_REASON is a character variable.

   ```
   IF TOP_REASON NE "1" And
      TOP_REASON NE "2" AND
      TOP_REASON NE "3" THEN TOP_REASON="";
   ```

   Submit the updated code and observe the results.

3. Go back to DISCOVER1.SAS, take out the MARRIED variable from the TABLE statement and resubmit it to verify that all changes have been made. Note that for character variables, the missing values are noted at the bottom of the table, and for the numeric variable (TOP_REASON) the missing values category (.) is listed in the table.

6.4.4 Check and Fix Out-of-Range Numeric Variables

Now that categorical variables have been cleaned, it is time to turn attention to the numeric measurement variables: AGE, ARRIVAL (Temperature), EDUCATION, and SATISFACTION. One way to take an initial look at numeric variables is with PROC MEANS which is discussed in Section 9.1. Specifically, using PROC MEANS you are looking for unusual minimum and maximum values. The following example illustrates how to find problem values using PROC MEANS and how to correct them.

HANDS-ON EXAMPLE 6.9

This example corrects problems associated with numeric variables.

1. Open the program file DISCOVER2.SAS and submit it to create the table of means shown in Table 6.12. (The output depends on you doing the previous exercises correctly, so if you have not done them your results may differ.)

   ```
   PROC MEANS MAXDEC=2 DATA=CLEAN.CLEANED;RUN;
   ```

   When you run PROC MEANS, it displays the means of all numeric variables. Note in this case that AGE is not included in the output. On further inspection, you can see that AGE is a character-type variable in the dataset. (A common problem when importing data from Excel or CSV files.) One way to convert this variable to numeric is with the INPUT() function (see Appendix B). Also note problems with some minimum and maximum values. On further inspection, you decide these are the items that need to be fixed:

- Set the value 99 for EDUCATION as a missing value.
- Some temperatures were taken as Celsius and some as Fahrenheit. They should all be Fahrenheit. In consultation with the physician in the clinic, you decide that any temperature under 45 should be considered Celsius and should be converted to Fahrenheit.
- By looking at records, you find that the temperature coded as 1018 should have been 101.8.
- Set the value -99 for SATISFACTION as a missing value.

TABLE 6.12 Checking Means for Several Variables

Variable	Label	N	Mean	Std Dev	Minimum	Maximum
Single	Single?	79	0.46	0.50	0.00	1.00
Education	Years of Schooling	78	14.64	17.16	6.00	99.00
Satisfaction	Satisfaction Score	79	77.92	21.70	-99.00	94.10
TEMP	Arrival Temperature	79	96.04	108.39	36.30	1018.00

Convert AGE to numeric. Give it the new name AGEN and assign a label. Note that changes need be made to SINGLE.

2. To fix the problems mentioned above, open the program file MESSY4.SAS. Observe the code listed under Part 4 in the code, where each of the identified fixes is made. Submit the code and observe the output.

```
IF EDUCATION=99 THEN EDUCATION=.;
IF TEMP LT 45 THEN TEMP=(9/5)*TEMP+32;
IF TEMP=1018 THEN TEMP=101.8;
IF SATISFACTION=-99 THEN SATISFACTION=.;
* Convert AGE from character to Numeric;
AGEN=INPUT(AGE,5.);
DROP AGE;
LABEL AGEN="Age Jan 1, 2014";
```

3. To verify that the changes have been made, rerun the code in DISCOVER2.SAS. Now that AGEN is a numeric variable, note that the minimum and maximum are 0 and 220. In checking with the researcher, you conclude that ages should be between 10 and 99.

4. Go back to MESSY4.SAS and make this additional fix. Note that you are now using the numeric variable AGEN rather than the previous character variable AGE.

```
IF AGEN LT 10 or AGEN GT 99 THEN AGEN=.;
```

Run the MESSY4.SAS code, then run the DISCOVER2.SAS code to make sure the correction took place.

When you convert character data to numeric, as in the case of AGE to AGEN, any values that cannot be converted to numeric are made missing (.). Therefore, the entries for AGE such as >29 are recoded as missing values as a result of this conversion. Note in the **Log Window** that SAS indicates an "invalid argument" warning because of the non-numeric AGE entries for subjects 21 and 67 during the conversion to the variable AGEN.

6.4.5 Correct Date and Time Values

The date and time the subject arrived and left the clinic are needed in order to calculate how long it took to serve each patient. However, these values are currently of character type. The following example illustrates how to convert the character variables to SAS date and time values, and how to combine them into a single DATETIME value.

This conversion uses the INPUT() function to convert from character to date and time format. For example, to convert the DATEARRIVED variable:

```
DATEARRIVED2=INPUT(TRIM(DATEARRIVED),MMDDYY10.);
```

The new (converted) variable must have a new name. In this case, DATEARRIVED2 as you cannot convert a variable to a different type and keep the same name. In this conversion, the DATEARRIVED character value is first trimmed using the TRIM() function. The resulting trimmed character value is then converted using the INPUT)() function with the INFORMAT MMDDYY10. and stored in the new variable named DATEARRIVED2.

The TIME value is a little trickier. Because the character value for TIME includes AM or PM, this information must be stripped off before it can be converted. To do this, the blank before the AM or PM is located using the FIND() function:

```
I= FIND(TIMEARRIVE," ");
```

However, you want to know if TIME was PM or AM because that information is needed to convert TIME to 24-h style (military-type) time. Therefore, perform a second FIND and look for P as that will uniquely indicate PM. Use this code:

```
P= FIND(TIMEARRIVE,"P");
```

If P is greater than 0, then a P is present in the statement, and you know that the time represents a time value after noon. Otherwise, the time represents a time before noon (AM).

The result of these two FIND() functions is a value for I representing the location of the blank before AM or PM and P representing whether the time is before or after noon.

The I value is used with a SUBSTR() function to extract a sub-string that contains only the time information:

```
TIMEARRIVE=SUBSTR(TIMEARRIVE,1,I-1);
```

Use an INPUT() function converting the resulting character value for TIMEARRIVE to a new DATE-type variable called TIMEARRIVET by applying the time format TIME8.

```
TIMEARRIVET=INPUT(TRIM(TIMEARRIVE),TIME8.);
```

You're not finished yet! Because time is recorded as AM or PM, you may need to adjust the time. If the time is in the afternoon (if P=1), you want to add 12 hours to the recorded time value. Because the time value is in seconds, you should calculate that the number of seconds in 12 hours is $12 \times 60 \times 60 = 43{,}200$, so the conversion statement is

```
IF P>0 THEN TIMELEFTT=TIMELEFTT+43200;
```

Finally, the DHMS() function is used to combine the date and time values using the code:

```
ARRIVEDT=DHMS(DATEARRIVED2,0,0,TIMEARRIVET)
```

The arguments for the DHMS() function are DATE, HOUR, MINUTE, and SECONDS. In this case, we use

- DATEARRIVED2 as the DATE value
- TIMEARRIVET as the SECONDS value
- Because the seconds incorporate hours and minutes, we enter 0 for the HOUR and 0 for the MINUTE arguments.

Appropriate formats are also used so that date and time values are displayed appropriately. Unneeded variables are dropped. The following example illustrates how this is done.

HANDS-ON EXAMPLE 6.10

This example corrects problems associated with numeric variables.

1. Open the program file MESSY5.SAS and observe the code after the Part 5 comment.

   ```
   FORMAT ARRIVEDT DATETIME18.;
   DATEARRIVED2=INPUT(TRIM(DATEARRIVED),MMDDYY10.);
   I= FIND(TIMEARRIVED," ");
   P=FIND(TIMEARRIVED,"P");
   TIMEARRIVED=SUBSTR(TIMEARRIVED,1,I-1);
   TIMEARRIVEDT=INPUT(TRIM(TIMEARRIVED),TIME8.);
   IF P>0 AND TIMEARRIVEDT LT 43200 THEN
   TIMEARRIVEDT=TIMEARRIVEDT
   +43200;
   ARRIVEDT=DHMS(DATEARRIVED2,0,0,TIMEARRIVEDT) ;
   Label ARRIVEDT="Date & Time Arrived";
   ```

   Run this program and observe the output in Table 6.13.
   Table 6.13 shows relevant variables for the first few converted records. Note that the original values are left justified in the cell, indicating character values and the converted values are right justified indicating numbers/dates. The ARRIVEDT (Date & Time Arrived) value includes the combined date and time values. Also note that the time arrived for the second observation, which was 1:23 PM was converted to the 24-h time 13:23.

   (Continued)

(Continued)

TABLE 6.13 Converting Date and Time Values

Obs	Subject ID	Date Arrived	Time Arrived	DATEARRIVED2	TIMEARRIVET	Date & Time Arrived
1	00001	2/7/2005	11:18:00 A	07FEB2005	11:18:00	07FEB05:11:18:00
2	00002	3/10/2005	1:23:00 PM	10MAR2005	13:23:00	10MAR05:13:23:00
3	00003	3/2/2005	1:08:00 PM	02MAR2005	13:08:00	02MAR05:13:08:00
4	00004	5/2/2005	11:46:00 A	02MAY2005	11:46:00	02MAY05:11:46:00
5	00005	1/31/2005	12:09:00 P	31JAN2005	24:09:00	01FEB05:00:09:00
6	00006	2/1/2005	2:27:00 PM	01FEB2005	14:27:00	01FEB05:14:27:00
7	00007	2/3/2005	11:50:00 A	03FEB2005	11:50:00	03FEB05:11:50:00
8	00008	2/4/2005	10:47:00 A	04FEB2005	10:47:00	04FEB05:10:47:00
9	00009	2/5/2005	11:42:00 A	05FEB2005	11:42:00	05FEB05:11:42

2. Using the conversion code used to convert DATEARRIVED and TIMEARRIVE, perform a similar conversion for DATELEFT and TIMELEFT.

    ```
    DATELEFT2=INPUT(TRIM(_____),MMDDYY10.);
    ```

    Name the final variable LEFTDT and add a LABEL statement to label the new value "Date & Time Left." Submit the code and verify that the correct changes have been made.

    ```
    I= FIND(TIMELEFT," ");
    P=FIND(TIMELEFT,"P");
    TIMELEFT=SUBSTR(TIMELEFT,1,I-1);
    TIMELEFTT=INPUT(TRIM(_____),TIME8.);
    IF P>0 THEN TIMELEFTT=_____+43200;
    LEFTDT=DHMS(DATELEFT2,0,0,TIMELEFTT)  ;
    ```

3. Once you have times for arrival and departure, you can calculate how long a subject stayed in the clinic. Enter this code, using the INTCK function with a "MINUTES" argument

    ```
    STAYMINUTES=INTCK( 'MIN', ARRIVEDT, LEFTDT );
    ```

    And then divide by 60 and round it off to get the number of hours stayed in the clinic:

    ```
    STAYHOURS=ROUND(STAYMINUTES/60,.1);
    ```

4. Add STAYHOURS to the PROC PRINT VAR list. Submit the code and observe the results in Table 6.14.

    Note that some STAYHOURS are negative and (if you look at more of the output) some values are very large – both positive and negative. It is obvious that some of the

TABLE 6.14 Calculated Values for Time Stayed in Clinic

Obs	Subject ID	Date Arrived	Time Arrived	DATE-ARRIVED2	TIME-ARRIVET	Date & Time Arrived	STAY-HOURS
1	00001	2/7/2005	11:18:00	16474	40680	07FEB05:11:18:00	3.0
2	00002	3/10/2005	1:23:00	16505	48180	10MAR05:13:23:00	23.3
3	00003	3/2/2005	1:08:00	16497	47280	02MAR05:13:08:00	4.0
4	00004	5/2/2005	11:46:00	16558	42360	02MAY05:11:46:00	10.0
5	00005	1/31/2005	12:09:00	16467	43740	31JAN05:12:09:00	13.4
6	00006	2/1/2005	2:27:00	16468	52020	01FEB05:14:27:00	0.2
7	00007	2/3/2005	11:50:00	16470	42600	03FEB05:11:50:00	9.3
8	00008	2/4/2005	10:47:00	16471	38820	04FEB05:10:47:00	
9	00009	2/5/2005	11:42:00	16472	42120	05FEB05:11:42:00	13.2
10	00010	1/30/2005	12:53:00	16466	46380	30JAN05:12:53:00	10.8
11	00011	2/6/2005	9:22:00	16473	33720	06FEB05:09:22:00	5.3
12	00012	1/30/2005	2:30:00	16466	9000	30JAN05:02:30:00	
13	00013	2/6/2005	7:10:00	16473	25800	06FEB05:07:10:00	8789.2

dates and times were entered incorrectly. For example, in observation 13, the `TIMELEFT` is more than a year after `TIMEARRIVE`.

If possible you could check the original records to reconcile the values. For this exercise, set any negative `STAYHOURS` or any values greater than 48-h to missing using this code:

```
IF STAYHOURS<0 or STAYHOURS>48 then STAYHOURS=.;
```

Resubmit the program and observe the results.

6.4.6 Looking for Duplicate Records

A final check for this dataset is to determine if there are duplicate records. Typically, this is accomplished by looking for duplicate `ID`s. A simple way to do this is with `PROC FREQ`. Hands-On Example 6.11 illustrates this process.

HANDS-ON EXAMPLE 6.11

This example shows how to find and remove duplicate records.

1. Open the program file `DISCOVER3.SAS` and observe the code. `PROC FREQ` is performed on the `ID` (in this case, the variable `SUBJECT`). The results are stored in a file named `FREQCNT` (using the statement `OUT=FREQCNT`). The `NOPRINT` tells SAS not to display the frequency table, which might be very large. Finally, the `PROC PRINT` displays only records where `COUNT` is 2 or more.

```
PROC FREQ data=MYSASLIB.cleaned NOPRINT;
    TABLES SUBJECT / OUT=FREQCNT;
RUN;
```

(Continued)

(*Continued*)

```
PROC PRINT DATA=FREQCNT;
WHERE COUNT>1;
RUN;
```

Submit this code and observe the results in Table 6.15. Subject ID=26 is used for two records.

TABLE 6.15 Discovering Duplicate IDs

Obs	SUBJECT	COUNT	PERCENT
26	00026	2	2.53165

2. Add this code to DISCOVER3.SAS to list only records were SUBJECT=26.

```
PROC PRINT DATA=MYSASLIB.CLEANED;
   WHERE FIND(SUBJECT,"00026");
RUN;
```

An alternative method that accomplishes the same thing is

```
PROC PRINT DATA=MYSASLIB.CLEANED;
   WHERE COMPRESS(SUBJECT)='00026';
RUN;
```

The first method uses the FIND function to locate the character string "0026" in the SUBJECT variable. The second method removes all blanks from SUBJECT using the COMPRESS() function and then looks for exact matches to "00026." Choose one of these options, enter it into the code, submit the updated code, and observe the results in Table 6.16. (Some variables are not shown.) Note that the SUBJECT records 26 are together. Using this Information you may be able to refer to original records to find that the second observation should be SUBJECT=27.

To correct the problem, add this code to the end of your edited MESSY5.SAS that changes the value of SUBJECT to 27 on record number _N_27.

```
IF _N_27 THEN SUBJECT="00027";
```

Run the code to complete the exercise.

TABLE 6.16 Discovering Observation Number for Supplicate IDs

Obs	SUBJECT	DateArrived	TimeArrive	DateLeft	TimeLeft	Single	Gender
26	26	1/25/2005	10:29:00 AM	1/25/2005	11:30:00 PM	0	M
27	26	1/25/2005	2:52:00 PM	1/26/2005	12:30:00 AM	0	F

> **HANDS-ON EXAMPLE 6.12**
>
> A final version of the code used to clean the dataset is found in the file `MESSY_ALL.SAS`.
>
> 1. Open program the file `MESSY_ALL.SAS` and run the program. This will create the final dataset named `CLEANED.SAS7BDAT` with all of the previous fixes completed.
> 2. Verify that the dataset named `CLEANED` is in your `MYSASLIB` library.
>
>   When working with a new dataset, this type of code is a typical example of what you will need to do before you use the data to create any reports or perform any statistical analysis.

6.4.7 Cleaning a Dataset: Summary

This case study illustrated typical steps for correcting problems in a messy dataset. (And almost every real-world dataset is messy.) Every dataset is different, and there are multiple ways to perform each of the tasks shown in this example, so there is no need to limit yourself to the following procedures only.

Although this process of cleaning a dataset can be tedious and time consuming, spending some time cleaning your data can increase your chances of getting more reliable and meaningful results. In summary, use these steps as a guideline whenever you get a dataset to analyze:

1. After importing or receiving a dataset, create an initial report and visually inspect the dataset for obvious problems. Correct any obvious problems using the procedures listed below (but they do not have to be in this order).
2. Rename variables that have strange or unclear names.
3. Label your variables. This will help prevent confusion and will result in more readable output.
4. Fix case problems by converting case in all character categories to all upper case (or all lower case).
5. Delete any unneeded records.
6. Use `PROC FREQ` to discover if there are inappropriate values present in categorical variables. Either fix or replace the inappropriate values or set these values as missing values.
7. Check for and correct any duplicated variables.
8. Use `PROC MEANS` (or `PROC UNIVARIATE`) to check for unusual minimum or maximum values in numeric variables. Either fix/replace any incorrect values or set the values to missing.
9. For all variables with known missing value codes, set those missing value codes to SAS missing dot (.) or blank.
10. Convert any variables that are not in the correct format to an appropriate format.

11. Separate information that is initially contained in a single variable (i.e., convert 120/80 blood pressure to two separate variables) or combine variables that need to be combined (i.e., date and time to datetime).
12. Search for and reconcile any duplicate records.

Although these steps do not cover all problems concerning messy datasets, they go a long way to cleaning up messy information to make your data more accurate and (hopefully) more reliable.

6.5 SUMMARY

This chapter provides additional information on common programming topics for the SAS language. The subjects covered are not exhaustive but were selected because they are often used for preparing data for analysis. Many more topics could have been covered, and readers are encouraged to refer to the SAS documentation for additional information.

EXERCISES

6.1 **Using PROC TRANSPOSE.**
   Use PROC TRANSPOSE to summarize multiple records.

   Using DTRANSPOSE2.SAS as an example, complete the program below (EX_6.1.SAS). The purpose of this program is to expand the AMOUNTOWED values onto one line for each ID, then to sum the rows and display the results. Use the following guidelines:

   - Name the output file created by PROC TRANSPOSE as EXPANDED.
   - Use AMT as the prefix to the expanded variable.
   - Expand on the variable AMOUNTOWED.
   - Expand by ID.

```
DATA OWED;
INPUT ID $3. AMOUNTOWED DOLLAR9.;
DATALINES;
001 $3,209
002 $29
002 $34.95
003 2,012
003 312.45
003 34.23
004 2,312
004 $3.92
005 .98
;
RUN;
PROC TRANSPOSE DATA=OWED OUT=_____ PREFIX=____;
```

TABLE 6.17 Listing of Transposed Data

Obs	ID	AMT1	AMT2	AMT3	TOTAL
1	001	3209.00			3209.00
2	002	29.00	34.95		63.95
3	003	2012.00	312.45	34.23	2358.68
4	004	2312.00	3.92		2315.92
5	005	0.98			0.98

```
    BY ___;
    VAR _____;
RUN;
DATA SUMMARIZE;SET EXPANDED;
TOTAL=SUM(of AMT1-AMT3);
DROP _NAME_;
RUN;
PROC PRINT DATA=SUMMARIZE;
RUN;
```

Your output should look like Table 6.17.

a. In the DATA SUMMARIZE data step, include the statement

```
FORMAT TOTAL DOLLAR9.2;
```

Rerun the program. How does that change the results?

b. Note that SUM (of a list of variables) summed the data and ignored missing values. Change the TOTAL= statement to

```
TOTAL = AMT1+AMT2+AMT3;
```

Rerun the program. What are the differences in the answers?

6.2 **Using SELECT.**

Using the WOUND dataset, recode the variable AGE to a new variable named AGEGP using the SELECT statement and the following criteria:

```
AGE LE 4 - Toddler
AGE GT 4 AND LE 12 - Child
AGE GT 12 AND LE 18 - Teen
AGE GT 18 - Adult
```

Here is a start to the SAS program to do this:

```
DATA NEW; SET MYSASLIB.WOUND;
    SELECT ;
*recode AGE to AGEGP in code here;
    END;
RUN;
```

6.3 **Cleaning a messy dataset.**

Using the SAS data file SOMEMISS, do the following to correct some problems and save the results in a SAS data file named FIXMISS:

a. Discover problems with GENDER and fix them. Replace any obvious bad data with missing data values.

b. Discover problems with AGE and replace any obvious bad data with missing values, and create a numeric variable named AGEN.

6.4 As a follow-up to Hands-On Example 6.4 this exercise will determine how many subjects had renal failure.

a. Using the SAS dataset named COMP_OUT created in Hands-On Example 6.4, create a dataset named RENAL, with the following code. Note the CATT function used in the statement

```
DATA RENAL;SET COMP_OUT;
CCAT=CATT(OF COMP1-COMP7);
RUN;
```

Look at Appendix B for a description of the CATT function. Briefly, CATT concatenates all of the complications for the subject (COMP1-COMP7) into a single variable named CCAT. For example, examine the dataset created and note that subject 2076 has CCAT= "PneumoniaHeart AttackRenal Failure" – three complications concatenated into the values of a single variable.

b. Add the following IF statement code (still in the DATA step – after the CCAT statement and before the RUN; statement) to categorize subjects with or without renal failure. The FIND function locates records where CCAT has the uppercase word "RENAL" within the data value. (Again, reference Appendix B for definitions of these functions.)

```
IF FIND(UPCASE(CCAT),"RENAL") NE 0 THEN
   RENALFAILURE="YES";
   ELSE RENALFAILURE="NO";
   RUN;
```

c. Use a PROC FREQ (frequencies) to display a table counting the number of "NO" and "YES"; values. The number of "YES"; values tells you how many of the subjects had RENALFAILURE as a complication. How many did you find?

```
PROC FREQ DATA=RENAL;
TABLES RENALFAILURE;RUN;
```

d. Using the same procedure, determine how many subjects had PNEUMONIA as a complication

e. Going Deeper: Change the code to create a table using an expanded IF statement to display how many subjects had either RENAL FAILURE or PNEUMONIA.

7

SAS® ADVANCED PROGRAMMING TOPICS PART 2

LEARNING OBJECTIVES

- To be able to create and use arrays
- To be able to use DO loops
- To be able to use the RETAIN statement
- To be able to create and use SAS macro variables
- To be able to create and review SAS macros

This chapter continues with the presentation of topics that are useful in preparing a dataset for analysis. The topics here are designed to provide you with more tools to use when constructing SAS programs. Using these techniques can simplify coding for tasks that involve repeated or related observations.

7.1 USING SAS ARRAYS

A SAS `ARRAY` statement provides a method for placing a series of observations into a single variable called an array. Often these variables are referenced as part of a `DO` loop,

---

SAS® Essentials: Mastering SAS for Data Analytics, Third Edition. Alan C. Elliott and Wayne A. Woodward.
© 2023 John Wiley & Sons, Inc. Published 2023 by John Wiley & Sons, Inc.
Companion website: https://www.alanelliott.com/SASED3/

which is a technique for iterating through a set of data. (DO loops are presented in Section 7.2.) A simplified syntax for SAS arrays is the following:

ARRAY array-name(subscript) <$> array-values

One-dimensional arrays are specified with an ARRAY statement, such as

ARRAY TIME(1:6) TIME1-TIME6;

This ARRAY statement creates a one-dimensional array called TIME1 with an index value indicated in parenthesis. So, TIME(1)=TIME1, TIME(2)=TIME2, and so on. In this case, six values are assigned to the values TIME(1) through TIME(6). Here are some more ways to create an ARRAY:

ARRAY TIME(6) TIME1-TIME6;

The result is the same as the previous example.

ARRAY TIME(0:5) A B C D E F;

TIME(0)=A, TIME(1)=B, and so on. Note that in this case, the index begins with a 0 instead of 1 because the initial definition of the array size in the parentheses (0:5) specified a beginning and end for the index values.

ARRAY ANSWER(*) Q1-Q50;

ANSWER(1)=Q1, ANSWER (2)=Q2, and so on. In this example, the size of the array is not specified. Instead, it is indicated as (*). SAS figures out how many items are being assigned to the array and creates the array with the required size.

ARRAY ANSWER(1:5) $ Q1-Q5;

where Q1–Q5 are character values. Arrays containing character variables are created in the same way as numeric arrays but with character values. The $ after the array definition indicates to SAS that the array is of character type.

ARRAY VALUE[5] (2 4 6 8 10);

Specifies actual values for the array where VALUE(1)=2, VALUE(2)=4, and so on. Note the difference here is that the values are within parentheses (you can use [], {}, or () for brackets in the ARRAY statement).

ARRAY ABC[5];

When no values or variables are given, SAS creates the variables based on the array name: ABC1, ABC2, and so on, and their initial values are missing values.

ARRAY FRUIT[3] $10 ("apple" "orange" "lemon");

Specifies actual character values of the FRUIT array as character values. The $10 specifies that the items in the array are character, and that the maximum number of characters for each item in the array is 10.

```
ARRAY DRINKSIZE[*] SMALL REGULAR LARGE MEGA (8, 16, 24, 32);
```
Or
```
ARRAY CUPSIZE[*] CUP1-CUP4 (8, 16, 24, 32);
```

Specifies initial values for the items in an array. In the first example, `DRINKSIZE(1)` equals `SMALL`, which equals 8. In the second example, `CUPSIZE(1)` is equal to `CUP1`, which is equal to 8.

7.1.1 Referring to Values in an Array

Once an array is created (such as `TIME[1:6]`), you can refer to a value in the array by specifying an index value, and use that value in any common SAS statement. For example,

```
NEWVAL =(TIME[5]/365);
```

This expression uses the value of `TIME` at index 5 in the calculation.
When using (*) in an array specification, as in

```
ARRAY TIME(*) TIME1-TIME6;
```

you may not know the value of the largest index. You can access these values using SAS functions. In this case, the upper bound is set to the value `DIM(arrayname)` or `DIM(TIME)`, which is 6. Moreover, the function `LBOUND(arrayname)` represents the lowest index in the specified array and `HBOUND(arrayname)` represents the highest index.

Once you create an array, you'll want to access the values in the array. It should be clear in the above example that to refer to `TIME1`, you could use `TIME(1)` and to refer to `TIME4`, you could use `TIME(4)`. In other words, you can use the original variable name or the corresponding indexed array value. The following example illustrates how to refer to values of an array when using a SAS function.

HANDS-ON EXAMPLE 7.1

1. Open the program file `DARRAY1.SAS`.

```
DATA A;
INPUT ONE TWO THREE FOUR;
ARRAY T(4) ONE TWO THREE FOUR;
TSUM1=SUM(OF ONE TWO THREE FOUR);
TSUM2=SUM(OF T(1) T(2) T(3) T(4));
TSUM3=SUM(OF T(*));
DATALINES;
1 2 3 4
5 6 7 8
;
RUN;
PROC PRINT DATA=A;RUN;
```

(*Continued*)

(*Continued*)

2. Submit this code and observe the results in Table 7.1.

    TABLE 7.1 Values in an Array

Example of using an ARRAY							
Obs	ONE	TWO	THREE	FOUR	TSUM1	TSUM2	TSUM3
1	1	2	3	4	10	10	10
2	5	6	7	8	26	26	26

    Note that all three versions of the SUM() function yield the same answer. They calculate the sum of the values in the T() array. Look at the three references to SUM. This code illustrates that when referring to values in an array, you can use the variable's original (ONE TWO THREE FOUR) or the indexed array names (T(1) T(2) T(3) T(4)) to mean the same thing. The T(*) reference is a shortcut way to represent all of the T() array elements.

3. Change the functions in the code from SUM() to MEDIAN() and the variable names from TSUM1-TSUM3 to MED1-MED3. Resubmit and observe the results. You should get the same answers for all three versions of the MEDIAN() function.

4. Another interesting function to try is the CAT() function, which concatenates values. (See Appendix B for more information about this function.) Change the function in the example from MEDIAN() to CAT() and the variable names to CAT1()-CAT3(). Resubmit and observe the results. Note that in this case, the resulting values CAT1()-CAT3() are character variables.

There are a number of tasks that are made easier by using arrays. One example is searching the contents of an array for a specific number or character value. This is accomplished with an IN statement. For example, the statement

```
IF 3 IN MYARRAY THEN some expression;
```

performs the expression when there is a 3 somewhere in the array named MYARRAY.

HANDS-ON EXAMPLE 7.2

As an example illustrating the use of the IN statement, do the following example.

1. Open the program file DARRAY2.SAS.

    ```
    DATA B;
    FORMAT ONE TWO THREE FOUR $10.;
    INPUT ONE $ TWO $ THREE $ FOUR $;
    ARRAY FRUIT(4) ONE TWO THREE FOUR;
    IF "ORANGE" IN FRUIT THEN ISORANGE=1;ELSE ISORANGE=0;
    DATALINES;
    ```

```
    APPLE ORANGE PINEAPPLE APRICOT
    LEMON APPLE KIWI STRAWBERRY
    ;
    RUN;
    PROC PRINT DATA=B;
    RUN;
```

2. Submit the code and observe the results in Table 7.2.
   Note that the array FRUIT contains the word ORANGE in observation 1, so the variable ISORANGE takes on the value 1. If there is no ORANGE fruit in the list, ISORANGE=0;
3. Change the IF statement to

   ```
   IF "ORA" IN FRUIT THEN ISORANGE1;ELSE ISORANGE=0;
   ```

   How does that change the answer? Does the IN statement find a partial word?

TABLE 7.2 Example Output Using the IN Statement with an Array

Example of using an ARRAY					
Obs	ONE	TWO	THREE	FOUR	ISORANGE
1	APPLE	ORANGE	PINEAPPLE	APRICOT	1
2	LEMON	APPLE	KIWI	STRAWBERRY	0

7.2 USING DO LOOPS

The DO statement is a way to control the flow of your program. These are the varieties of DO loops available in SAS:

1. **IF and DO:** Use a DO in an IF statement to cause a group of statements to be conditionally executed.
2. **Iterative DO loop:** Iterate through a series of statements based on some incremental index.
3. **DO UNTIL:** Execute a group of statements until some condition is met.
4. **DO WHILE:** Execute a group of statements as long as the specified condition is true.

The IF and DO combination statement has been briefly discussed in Chapter 4. The primary DO statement discussed in this chapter is the iterative DO loop. An *iterative* DO *loop* is a way of incrementing through a series of statements based on some index. A simplified syntax of an iterative DO loop is as follows:

```
DO indexvariable=initialvalue TO endvalue <BY incrementvalue>;
  . . . SAS statements . . .
END;
```

176 CHAPTER 7: SAS® ADVANCED PROGRAMMING TOPICS PART 2

For example, consider the SAS code that follows. In this case, the index variable is ILOOP (choose any legitimate SAS variable name that is not being used for another purpose within your SAS program), and the initial value of ILOOP is 1. The last referenced value is 4, and the implied increment value is 1. (You can optionally specify a different increment value.)

```
SUM=0;
DO ILOOP=1 to 4;
   SUM=SUM+ILOOP;
END;
```

Here is how to follow the action of this code: The value of SUM is initially set to 0. In the DO loop, the index ILOOP is first set to the initial value 1. The following code is executed:

```
SUM=0+1;
```

When the code encounters the END statement, ILOOP is incremented by 1 (the increment value) and the flow of the program goes back to the DO statement. As long as the index is not greater than the end value, another pass through the loop is performed. Once ILOOP is greater than 4, the loop stops. In that case, the process yields the following statements:

```
SUM=0;
SUM=0+1;
SUM=1+2;
SUM=3+3;
SUM=6+4;
```

The final value of SUM at the end of the loop is 10.

In a DO UNTIL loop, the increment is replaced with an UNTIL clause with (ILOOP GT 4) in parentheses. For example,

```
ILOOP=1;
SUM=0;
DO UNTIL (ILOOP GT 4);
     SUM=SUM+ILOOP;
ILOOP+1;
END;
```

This code is similar to the previous example except for the user now has direct control of the variable ILOOP. It is initially set to one and then incremented by 1 (ILOOP+1) during the loop. The loop continues until ILOOP is greater than 4. The results are the same as in the previous example.

In DO WHILE, the loop continues until the condition is no longer true. Once it is false, the loop stops. For example, the following code results in the same loop as the two previous examples.

```
ILOOP=1;
SUM=0;
```

```
DO WHILE(ILOOP LT 5);
     SUM=SUM+ILOOP;
ILOOP+1;
END;
```

HANDS-ON EXAMPLE 7.3

This program illustrates the uses of the DO WHILE loop.

1. Open the program file DWHILE.SAS.

   ```
   DATA TMP;
   N=0;
   SQUARESUM=0;
   DO WHILE(N<5);
      SQUARESUM=SQUARESUM+N**2; OUTPUT;
      N+1;
   END;
   RUN;
   PROC PRINT;
   RUN;
   ```

   The OUTPUT statement in the code causes SAS to save each SQUARESUM as a record in the dataset TMP so you'll see the values at each iteration.

2. Run the program and observe the output dataset given in Table 7.3. Why do the values of N range from 0 to 4?
3. Change the code to use a DO UNTIL to get the same results.
4. Change the code to use an iterative DO loop to get the same results.

TABLE 7.3 Example of Using the DO WHILE Statement

Obs	N	SQUARESUM
1	0	0
2	1	1
3	2	5
4	3	14
5	4	30

One of the most powerful uses of the DO loop is realized when it is used in conjunction with an array. For example, suppose you take five systolic blood pressure readings on each subject. Suppose that if any of these readings is above 140, you want to classify the subject with the value HIGHBP=1; otherwise the subject will have the value HIGHBP=0. The following code will perform that task:

```
DATA PRESSURE;SET MEDDATA;
ARRAY SBP(5) READING1-READING5;
HIGHBP=0;
DO I=1 TO 5;
    IF SBP(I) GT 140 THEN HIGHBP=1;
END;
```

- The ARRAY statement sets up an array named SBP that contains the values of the variables READING1 through READING5.
- In the DO statement, the variable I iteratively takes on the values from 1 to 5.
- Within the DO loop, if any of the five readings is greater than (GT) 140, the variable HIGHBP is set to 1. Otherwise, the value of HIGHBP for that subject remains at 0.

Once the value of HIGHB takes on the value 1, there is no way for it to be set back to 0 within the same loop. Therefore, for each record HIGHBP starts out at 0 and when changed within the loop, it doesn't get reset to 0 until the next record in the dataset it read.

For another example, suppose you have a questionnaire in which the variables (answers) are named Q1 to Q50 and the answers recorded are YES, NO. Missing answers were originally coded in the dataset as NA. Suppose you want to change the missing values to the value " " (two quote marks together representing a blank, which is typically a missing value for a character variable). Using an array and a DO loop, you can quickly perform this task within a DATA step.

```
DATA NEW;SET OLD;
ARRAY ANSWER(1:50) $ Q1-Q50;
DO I=1 TO 50;
    IF ANSWER(I)="NA" then ANSWER(I)="";
END;
```

- The ARRAY statement sets up an array named ANSWER that contains the text values of the variables Q1–Q50.
- The DO statement begins a loop in which the variable I iteratively takes on the values from 1 to 50. The DO loop ends with the END statement.
- Within the DO loop, the IF statement is performed 50 times, each time with a different value of ANSWER(I) representing the different values of Q1, Q2, and so on.
- Thus, for the first subject, each of the 50 questions is examined. If an answer is "NA" it is recoded with value " ".

```
IF ANSWER(1)="NA" THEN ANSWER(1)="";
```

which can be interpreted as

IF Q1 has an answer NA, then recode Q1 as " ".

In a similar way, if your answers are numeric, 1=Yes, 0=No, and −9=Missing, then you could use the following code to set all missing values (-9) to the standard SAS missing value code dot (.). In this case, note that the variables Q1-Q50 are numeric.

```
DATA NEW;SET OLD;
ARRAY ANSWER(1:50)Q1-Q50;
DO I=1 to 50;
    IF ANSWER(I)= -9 then ANSWER(I)=.;
END;
```

HANDS-ON EXAMPLE 7.4

This example illustrates how to use ARRAYs and DO loops in a SAS program. Suppose you have a record of a measure taken on subjects who repeatedly came to a clinic. However, the number of times they came was not the same for each patient. You want to know the value of the measurement (any non-missing value) observed the last time a subject was in the office.

1. Open the program file DDOLOOP.SAS.

```
DATA CLINIC;
INPUT @1 ID $3. @4 VISIT1 4. @7 VISIT2 4.
      @10 VISIT3 4. @13 VISIT4 4. @16 VISIT5 4.;
ARRAY TIME (5) VISIT1-VISIT5;
LAST=VISIT1;
DO I=1 TO 5;
   IF TIME(I) > 0 THEN LAST=TIME(I);
END;
DROP I;
DATALINES;
001 34 54 34 54 65
002 23 43 54  .
003 23 43  .  43
004 45 55 21 43 23
005 54
;
PROC PRINT DATA=CLINIC; RUN;
```

2. Run this program and observe the output in Table 7.4. Note that the variable LAST contains the value of the observation for the last "good" value from the list of visits.

3. If you also want to know the number of times a value was recorded for each subject, then put the following statement, using the N() function, in the DATA step to get that value. Resubmit the code. You should have a new column containing the number of visits.

```
NUMVISITS=N(OF TIME(*));
```

4. In the same way, use the MAX function to find the maximum value for each subject. Resubmit the code and observe the results.

```
MAXVISITS=MAX(OF TIME(*));
```

TABLE 7.4 Example of ARRAYs and DO Loops

Illustrates the use of DO LOOPS and ARRAYS							
Obs	ID	VISIT1	VISIT2	VISIT3	VISIT4	VISIT5	LAST
1	001	34	54	34	54	65	65
2	002	23	43	54	-	-	54
3	003	23	43		43	-	43
4	004	45	55	21	43	23	23
5	005	54	-	-	-	-	54

HANDS-ON EXAMPLE 7.5

This example illustrates the use of a DO loop when you don't know the lower and upper bounds of an array. The dataset used is a report of crimes in Washington, DC, between the years 1978 and 2007. The code converts the raw incident values into a percentage (incident/population $\times$ 100.)

1. Open the program file DARRAY3.SAS.

   ```
   DATA CRIME;SET MYSASLIB.DC_CRIME78_07;
   FORMAT TOTAL 6.;
   ARRAY INCIDENTS(*) VIOLENT-CARTHEFT;
   DO I= LBOUND(INCIDENTS) TO HBOUND(INCIDENTS);
        TOTAL=SUM(OF TOTAL,INCIDENTS(i));
   END;
   DROP I;
   RUN;
   PROC PRINT DATA=CRIME;
   VAR YEAR POP-CARTHEFT TOTAL;
   RUN;
   ```

   Note in this code that the array named INCIDENTS uses (*) instead of a fixed number of elements. That is, we don't specify in the code how many variables are included in the list VIOLENT-CARTHEFT. Even though we don't know the bounds of the INCIDENTS array, we can still increment over the entire list by using the LBOUND and HBOUND functions in the DO loop.

   Submit this code and observe the results. Notice in the displayed results that the large numbers are shown without commas.

2. To also find the total number of crimes per year, and the percentage of crimes by population:

   ```
   TOTALCRIMES=SUM(OF INCIDENTS(*));
   PCTTOTAL=TOTALCRIMES/POP*100;
   ```

   Add these statements to the code. The trick is where to put these statements. As they use the array INCIDENTS, they must come *after* the ARRAY statement. However, they do not need to be within the DO loop. Put these lines in the proper location in the code.

3. Change the TOTALCRIMES format so that it includes commas by including the statement

   ```
   FORMAT TOTALCRIMES COMMA10.;
   ```

   on a program line before you do the TOTALCRIMES calculation. Change the PROC PRINT code in the VAR statement to

   ```
   VAR YEAR POP--CARTHEFT TOTALCRIMES PCTTOTAL;
   ```

   Submit the code and observe the results in Table 7.5. You should see that the TOTALCRIMES column values are now formatted with commas.

TABLE 7.5 Results of Using LBOUND and HBOUND in a DO Loop

Obs	Year	POP	Index	Violent	Property	Murder	Rape	Robbery	Assault	Burglary	Larceny	CarTheft	TOTAL	TOTALCRIMES	PCTTOTAL
1	1978	674000	50950	9515	41435	189	447	6333	2546	12497	25744	3194	101900	101,900	15.1187
2	1979	656000	56430	10553	45877	180	489	6920	2964	13452	28819	3606	112860	112,860	17.2043
3	1980	635233	63668	12772	50896	200	439	8897	3236	16260	31068	3568	127336	127,336	20.0456
4	1981	636000	67910	14468	53442	223	414	10399	3432	16832	32845	3765	135820	135,820	21.3553
5	1982	631000	65692	13397	52295	194	421	9137	3645	14774	33435	4086	131384	131,384	20.8216
6	1983	623000	57776	11933	45843	183	406	7698	3646	12483	29405	3955	115552	115,552	18.5477
7	1984	623000	53524	10725	42799	175	366	6087	4097	10954	27471	4374	107048	107,048	17.1827
8	1985	626000	50075	10171	39904	147	337	5230	4457	10005	24874	5025	100150	100,150	15.9984
9	1986	626000	52204	9423	42781	194	328	4720	4181	10815	25861	6105	104408	104,408	16.6786
10	1987	622000	52569	10016	42553	225	245	4462	5084	11244	25012	6297	105138	105,138	16.9032
11	1988	620000	61471	11914	49557	369	165	5690	5690	12300	28624	8633	122942	122,942	19.8294
12	1989	604000	62172	12937	49235	434	186	6542	5775	11780	29164	8291	124344	124,344	20.5868

7.2.1 Quick Notes for Using SAS Arrays

The following Figures 7.1 and 7.2 provide Quick Tips for Using SAS Arrays. These sheets are also available to print out from www.alanelliott.com. Click on the Quick Tips link.

7.3 USING THE RETAIN STATEMENT

The RETAIN statement, used in a DATA step, allows you to retain values of a variable across iterations as data records are read into the dataset. The basic syntax for RETAIN is as follows:

```
RETAIN <var1 <initial-value(s) > var2 <initial-values(s)> and
so on;
```

For example, to instruct SAS to remember the value of the variable SCORE, you would use the statement

```
RETAIN SCORE;
```

A statement that retains the values of several variables could be

```
RETAIN TIME1-TIME10 SCORE VISITDATE;
```

By default, the initial value of retained variables is missing. A statement to retain the value of SCORE, with an initial value of 1, is

```
RETAIN SCORE 1;
```

In the following code, CATEGORY has an initial value of "NONE" and variables TIME1-TIME10 have initial values of 0.

SAS® Arrays Reference

SAS ESSENTIALS Ed. 3 by Alan Elliott and Wayne Woodward

SAS Arrays Overview

Arrays are used in the SAS DATA STEP to store a series of related values. The general format for an ARRAY statement is

```
ARRAY arrayname[] <$> opt list;
```

You can use [] {} or () for brackets in the ARRAY statement. Within the [] put the dimension of the array. Examples:

```
ARRAY TIME(6) TIME1-TIME6;

ARRAY LETTERS(0:5) A B C D E F;

ARRAY ANSWER(*) Q1-Q50;
```

Refer to values: In the above array TIME, the value TIME(2) refers to the value of TIME2. In LETTERS, the value LETTERS(1) refers to the value of B, etc.

Specify Actual Values

To **specify actual values** for the array

```
ARRAY VALUE[5] (1 2 3 4 5);
```

Creates VALUE1=1 VALUE2=2 etc.

```
ARRAY FRUIT[4] $10 ("apple",
"orange", "lemon",
"pomegranate");
```

Creates FRUIT1=apple, FRUIT2=orange, etc. Note: $10 is length of the character values (default is $8.)

If **no values or variables are given**, such as

```
ARRAY ABC[5];
```

SAS creates the variables ABC1, ABC2, etc. each set as a missing value.

Specify a Range of Variables

To **specify a range of variables** in a data set. Suppose you have variables Q1 through Q50 that are (numeric) answers in a questionnaire.

```
ARRAY ANSWERS[50] Q1-Q50;
```

Where the Q1 variable contains the Q1 value, Q2 contains the Q2 value, etc.

If the Q answers are of character type (say A, B, C, etc.) then use

```
ARRAY ANSWERS[50]  $ Q1-Q50;
```

You could also use the form

```
ARRAY ANSWERS[*]  $ Q1-Q50;
```

Assign Initial Values

To **assign initial values** to an array you can use:

```
ARRAY DRINKSIZE[*] SMALL REGULAR LARGE MEGA
(8, 16, 24, 32);
```

Where DRINKSIZE(1) equals SMALL equals 8. Or for

```
ARRAY CUPSIZE[*] CUP1-CUP4 (8, 16, 24, 32);
```

The value of CUPSIZE(1) is equal to CUP1 is equal to 8.

For all of the above arrays, the indexes for values start at 1. You can **set up a different index** by using [beginning:end]. For example.

```
ARRAY TIME[0:4] VISIT1-VISIT5;
```

Is set up so that TIME(0) = VISIT1, TIME(2)=VISIT 2, etc.

VNAME Function

The VNAME function allows you to discover the name of a variables associated with an index. For example

```
VAR_NAME=VNAME(T[2]);
```

Assigns the value TIME2 to VAR_NAME if the following ARRAY was used:

```
ARRAY T[4] TIME1-TIME4;
```

Figure 7.1 Quick Tips for Using SAS Arrays

SAS Arrays Page 2

Example Simple Array

Create an ARRAY and **use a DO LOOP** to SUM the data.

```
DATA TIMES;
INPUT TIME1 TIME2 TIME3 TIME4;
ARRAY T[4] TIME1-TIME4;
SUMTIME=0;
DO I=1 TO 4;
   SUMTIME=SUM(OF SUMTIME,T(I));
END;
DATALINES;
22.3     25.3     28.2     30.6
22.8     27.5     33.3     35.8
18.5     26.0     29.0     27.9
22.5     29.3     32.6     33.7
;
RUN; PROC PRINT DATA=TIMES;RUN;
```

Gives the following:

Obs	TIME1	TIME2	TIME3	TIME4	SUMTIME
1	22.3	25.3	28.2	30.6	106.4
2	22.8	27.5	33.3	35.8	119.4
3	18.5	26.0	29.0	27.9	101.4
4	22.5	29.3	32.6	33.7	118.1

The SUMTIME column is created by the DO LOOP. **The DIM function** allows you to create a more general DO loop:

```
SUMTIME=0;
DO I=1 TO DIM(T);
   SUMTIME=SUM(OF SUMTIME,T(I));
END;
```

produces the same results. If you want all variables that are numbers in the array, can use the statement

```
ARRAY T[*] _NUMERIC_;
```

Example: (using DIM for dimension)

```
DATA TIMES;
INPUT TIME1 TIME2 TIME3 TIME4;
ARRAY T[*] _NUMERIC_;
SUMTIME=0;
DO I=1 TO DIM(T);
   SUMTIME=SUM(OF SUMTIME,T(I));
END;
```

Also, to use all character variables use _CHARACTER_ or to use all variables of any type in the array use _ALL.

When the key word **_TEMPORARY_** is used in a ARRAY statement, data elements are created but are not stored in the data file. Example:

```
ARRAY VAR{6} _TEMPORARY_ (0.06 0.09 0.23 0.44);
```

Using the OF Operator

Use the **OF operator** and statistical functions to calculate values across the contents of an array. Examples:

```
*CALCULATE THE MEAN;
MEAN_TIME=MEAN(OF T(*));

*CALCULATE MINIMUM;
MIN_TIME=MIN(OF T(*));

*CONCATENATE VALUES;
ALL_TIME=CATX(" " ,OF T(*));
```

Results in the following new variables

mean_time	min_time	all_time
26.6	22.3	22.3 25.3 28.2 30.6
29.85	22.8	22.8 27.5 33.3 35.8
25.35	18.5	18.5 26 29 27.9
29.525	22.5	22.5 29.3 32.6 33.7

Note that all_time is a character variable. Other operators:

MAX finds maximum value in list.
MEDIAN find median of a list of values.
CALL MISSING sets all values in as missing.
Example: `CALL MISSING(OF T(*));`
CALL SORTN sorts values in the array
Example: `CALL SORTN(OF T(*));`

Using the IN Operator

Find an instance of a value in the contents of an array.

Find a number (using the TIMES dataset from above):

`if 22.3 in t then FOUND='Yes'; else FOUND="No";`

Find a character string:

Suppose the array complication contains a list of complications for subjects, the following code would find an instance of Pneumonia in the list

```
IF 'Pneumonia' IN COMPLICATION THEN
FOUND='Yes'; ELSE FOUND="No";
```

Note that a character search must match case.

© Alan Elliott and Wayne Woodward, 2022

SAS ESSENTIALS Ed. 3
by Alan Elliott and Wayne Woodward
www.alanellitt.com/sas

Figure 7.2   Quick Tips for Using SAS Arrays, page 2

```
RETAIN CATEGORY "NONE" TIME1-TIME10 0;
```

To set the values of a list, use this format:

```
RETAIN TIME1-TIME4  (1 2 3 4);
```

or

```
RETAIN TIME1-TIME4  (1, 2, 3, 4);
```

or

```
RETAIN TIME1-TIME4  (1:4);
```

To retain the values of all variables, use

```
RETAIN _ALL_;
```

You may also set a value of a variable before you initiate the RETAIN, in which case the initial value becomes the current value of that variable.

For example, suppose you have a sequence of observations that includes a date variable and you want to calculate the number of days after the first date. One way to do that is with the RETAIN statement. Assuming the data are sorted by date, consider the following code:

```
DATA DAYS;SET MYDATA;
IF _N_=1 THEN FIRST=VISIT_DATE;
RETAIN FIRST;
DAYS=VISIT_DATE-FIRST;
```

- The IF _N_=1 statement creates a new variable called FIRST and sets the initial value of FIRST to the value of VISIT_DATE in the first record.
- The RETAIN statement tells SAS to keep the value of FIRST as it was read from the previous record (otherwise it will be missing).
- The DAYS= statement uses the FIRST variable to calculate a difference between the current record's VISIT_DATE value and FIRST, giving you the number of days since the first visit.

HANDS-ON EXAMPLE 7.6

This example calculates the number of days since the first visit.

1. Open the program file DRETAIN1.SAS. Note the ANYDTDTE10 format interprets any input value as a date, meaning that you don't have to specify as the particular date format.

```
DATA MYDATA;
INPUT VISIT_DATE ANYDTDTE10.;
DATALINES;
01/04/2015
01/29/2015
03/07/2015
04/25/15
06Jul2015
08-30-2015
;
DATA DAYS;SET MYDATA;
IF _N_=1 THEN FIRST=VISIT_DATE;
RETAIN FIRST;
DAYS=VISIT_DATE-FIRST;
RUN;
PROC PRINT DATA=DAYS;
Title 'These are the contents of the dataset DAYS';
FORMAT VISIT_DATE FIRST DATE10.;
RUN;
```

2. Run this example and observe the output in Table 7.6.
3. Remove the RETAIN statement and rerun the program. Note that the variable FIRST loses its value after the first iteration, and the variable DAYS has a missing value for all records after the first record.

TABLE 7.6 Results of Calculation Using the RETAIN Statement

| These are the contents of the dataset DAYS ||||
Obs	VISIT_DATE	FIRST	DAYS
1	04JAN2015	04JAN2015	0
2	29JAN2015	04JAN2015	25
3	07MAR2015	04JAN2015	62
4	25APR2015	04JAN2015	111
5	06JUL2015	04JAN2015	183
6	30AUG2015	04JAN2015	238

As a second example of the use of the RETAIN statement, suppose you have a lemonade kiosk at an outdoor mall. You have data for the first 7 days of sales, and you want to create a report that provides the accumulated sales and also keeps track of the maximum daily sales you've had. The following example illustrates how you could do this using the RETAIN statement. The following code contains the sales data for the lemonade kiosk and a code to calculate the desired values.

```
DATA LEMON;
INPUT DAY SALES;
IF _N_=1 THEN DO; TOTAL=0; END;
RETAIN TOTAL MAX;
TOTAL=SALES+TOTAL;
IF SALES>MAX THEN MAX=SALES;
```

Examine this code and note the statement that accumulates the sum:

`TOTAL=SALES+TOTAL;`

and the statement that sets MAX to the largest observed values.

`IF SALES>MAX THEN MAX=SALES;`

To illustrate how this program is using the RETAIN statement, do Hands-On Example 7.7.

HANDS-ON EXAMPLE 7.7

This example illustrates the RETAIN statement for accumulating data and identifying minimum and maximum.

1. Open the program file DRETAIN2.SAS.

```
DATA LEMON;
INPUT DAY SALES;
IF _N_=1 THEN TOTAL=0;
RETAIN TOTAL MAX;
TOTAL=SALES+TOTAL;
IF SALES>MAX THEN MAX=SALES;
DATALINES;
1 23.40
2 32.50
3 19.80
4 55.55
5 34.90
6 65.30
7 33.40
;
RUN;
PROC PRINT DATA=LEMON; RUN;
```

Submit the code and verify that you get the output given in Table 7.7.

TABLE 7.7 Using RETAIN to Accumulate TOTAL and Find MAX

Obs	DAY	SALES	TOTAL	MAX
1	1	23.40	23.40	23.40
2	2	32.50	55.90	32.50
3	3	19.80	75.70	32.50
4	4	55.55	131.25	55.55
5	5	34.90	166.15	55.55
6	6	65.30	231.45	65.30
7	7	33.40	264.85	65.30

2. Comment out the `RETAIN` statement and resubmit the program. How does this change the output and why?
3. Take the comment off of the `RETAIN` statement, comment out the `IF _N_=1` statement

    ```
    /*IF _N_=1 then DO; TOTAL=0; END;*/
    ```

    Resubmit the program. How does this change the output and why? (Hint: Since an accumulation using + is set to missing if any variable is missing, then the first

    ```
    TOTAL=SALES+TOTAL;
    ```

    is set to missing, as are all the rest.) On the other hand, the `IF SALES>MAX` works correctly because a positive number is greater than a missing value.
4. Change `TOTAL=SALES+TOTAL` to

    ```
    TOTAL=SUM(SALES,TOTAL);
    ```

    Resubmit the code. Does this version work correctly?
5. The current code finds the `MAX` sales amount. Add code to find the `MIN` sales amount.

7.4 USING SAS MACROS

Macros, a powerful programming tool within SAS, allow you to

- set aside repetitive sections of code and to use them again and again when needed,
- create dynamic variables (macro variables) within the code that can take on specified values.

Two types of SAS macros types are described:

- **Macro Variable** – a symbolic name that stores information that can be used to *dynamically modify SAS code* through symbolic substitution.
- **Macro Process** – a set of SAS code that can be "called" to do repetitive tasks

The following brief and limited introduction to this topic illustrates some common ways to use SAS macros. For more information about macros, refer to the SAS documentation.

7.4.1 Creating and Using SAS Macro Variables

A SAS macro variable is a powerful programming concept that allows you to write code that can be easily changed to reflect new values for selected variables or code. This is accomplished using a SAS macro. A SAS macro variable is the symbolic name that stores information that can be used to dynamically modify SAS code through symbolic substitution.

You can think of this substitution in a way that is similar to a "replace" in a word processor. SAS can dynamically (during the process of running a program) replace the value of a specified macro variable with another value you specify throughout your program code.

For example, suppose you have written a SAS program that uses the year value 2016 repeatedly in your code. Later on, you want to update the program to use 2017 instead. But if you attempt to manually change the entire 2016 year references in the program, you might miss one. Moreover, in a program of hundreds of lines of code, this task could take a long time. If you need to make many code changes such as this in a long SAS program, the task becomes daunting.

A better programming technique is to use a macro variable rather than fixed code in places where you know you'll need to make changes in the future. This is just one setting in which macro variables come in handy.

For example, suppose you have a program that calculates values based on a given year and a comparison based on the value DEPT. You can create macro variables called YEAR and DEPT using the statements

```
%LET YEAR=2015;
%LET DEPT=ENGINEERING;
```

The SAS %LET statement allows you to assign a value to a SAS macro variable. Note the % prefix to the %LET statement. Also note that these are not typical assignment statements like you would use in a DATA step. In fact, %LET statements do not have to appear in a DATA step. This type of statement typically appears at the beginning of your program before any DATA step or PROC.

The general format for creating a macro variable is as follows:

```
%LET macrovariablename = macrotext;
```

The value assigned to the macro variable is always a character value, even if (as in the above example) the assigned value is 2015.

Once you have created the macro variable, you refer to it in your SAS program by placing an & (ampersand) as a prefix to the variable name. Thus, after you define YEAR as a macro variable, you use it in a statement in the following way:

```
IND_DAY=MDY(7,4,&YEAR);
```

In this example, &YEAR is replaced by the value of the macro variable, 2015, before SAS executes the code. Thus, SAS "sees" the replaced code

```
IND_DAY=MDY(7,4,2015)
```

and executes that statement. In the case of the macro variable DEPT, the statement

```
GROUP = "&DEPT";
```

is "seen" by SAS as the code

```
GROUP = "ENGINEERING";
```

The &DEPT macro variable is replaced with its value ENGINEERING. The quotations were required in the statement to make the completed statement resolve into acceptable SAS code.

The following code uses these two macro variables within a TITLE statement:

```
TITLE "Analysis for &DEPT for the year &YEAR";
```

If there is no text provided in a %LET statement, the contents of the macro variable is a null value (0 characters). For example,

```
%LET ISNULL=;
```

It is important to understand that any leading or trailing blanks in the *macrotext* value are ignored. For example,

```
%LET CITY=DALLAS;
```

produces the same result as code that has several blanks around the name:

```
%LET CITY  =  DALLAS;
```

The blanks in front of or after the text are ignored.

HANDS-ON EXAMPLE 7.8

This example illustrates the use of macro variables.

Open the program file DMACRO1.SAS. Note the two macro variables assigned in the %LET statement and their inclusion in the PROC MEANS statement. The DNAME macro variable is used for the name of the dataset to use, and the VARIABLES macro variable provides a list of variables to be analyzed.

```
%LET DNAME=C:\SASDATA\SUBJECTS;
%LET VARIABLES = AGE TIME_EXPOSED;
PROC MEANS DATA="&DNAME";
VAR &VARIABLES;
RUN;
```

(Continued)

(Continued)

TABLE 7.8 Output Obtained Using the %LET Statement to Create Macro Variables

Variable	Label	N	Mean	Std Dev	Minimum	Maximum
AGE	AGE	50	10.4600000	2.4261332	4.0000000	15.0000000
TIME_EXPOSED	TIME_EXPOSED	50	21.6744000	1.7464894	17.7900000	25.1200000

Submit the code and observe the results given in Table 7.8. The code produces descriptive statistics for the specified list of variables.

Change the dataset named from SUBJECTS to SOMEDATA and the list of variables to TIME1-TIME4. Resubmit the program and observe the results.

This exercise illustrates how to define macro variables at the beginning of your program and use them throughout your code. If you make any changes in the value of the text in the %LET statement, they are changed in the entire program at the time SAS compiles the SAS code.

7.4.2 Combining Macro Variables

Sometimes you want to combine macro variables together or with some other text. This may allow you to create code that is more generalizable and to reuse the code again and again by only changing the values of the macro variables. For example, consider the following macro variable assignments:

```
%LET PATH=C:\SASDATA\;
%LET DSN=SOMEDATA;
%LET CLASS=GP;
%LET SELECTVAR=AGE TIME1-TIME4;
```

The macro variable PATH is the Windows location of data files, the DSN is the dataset name, CLASS is a grouping variable, and SELECTVAR is a list of variables for some analysis. An example of how these could be used in a PROC MEANS statement is given here:

```
PROC MEANS DATA="&PATH&DSN" MAXDEC=2;
     CLASS &CLASS;
     VAR &SELECTVAR;
RUN;
```

Note how &PATH&DSN are placed together. When the names are resolved, the results are C:\SASDATA\SOMEDATA, which is the reference to the data file needed for PROC MEANS. The CLASS and SELECTVAR macro variables provide additional information for the procedure. Separating PATH and DSN makes it simpler to change the value of one or the other. If there is ambiguity with respect to how two items are concatenated, you can use a dot (.) as a separator. For example, suppose you have

```
%LET STATUS=PRE;
```

If you want to combine &PRE with the word PRODUCTION, and if you used "&PREPRODUCTION", SAS could not resolve the macro variable correctly. Using the code "&PRE.PRODUCTION" works.

HANDS-ON EXAMPLE 7.9

This example illustrates combining macro variables.

1. Open the program file DMACRO2.SAS. Note the four macro variable assignments and how they are used in PROC MEANS. In particular, note how PATH and DSN are concatenated.

```
%LET PATH=C:\SASDATA\;
%LET DSN=SOMEDATA;
%LET CLASS=GP;
%LET SELECTVAR=AGE TIME1-TIME4;
TITLE "Descriptive Statistics";
PROC MEANS DATA="&PATH&DSN" MAXDEC=2;
    CLASS &CLASS;
    VAR &SELECTVAR;
RUN;
```

2. Submit this code and observe the results given in Table 7.9.

TABLE 7.9 Output Created Using Macro Variables

Intervention Group	N Obs	Variable	Label	N	Mean	Std Dev	Minimum	Maximum
A	11	AGE	Age on Jan 1, 2000	11	10.36	2.87	4.00	13.00
		TIME1	Baseline	11	20.90	2.08	17.00	23.00
		TIME2	6 Months	11	27.00	2.93	21.30	31.30
		TIME3	12 Months	11	30.35	3.20	22.70	34.20
		TIME4	24 Months	11	30.44	3.96	21.20	35.80
B	29	AGE	Age on Jan 1, 2000	29	10.31	2.48	6.00	15.00
		TIME1	Baseline	29	21.19	1.59	18.50	24.20
		TIME2	6 Months	29	27.43	2.64	23.10	32.30
		TIME3	12 Months	29	30.33	2.87	25.30	34.20
		TIME4	24 Months	29	30.51	3.47	25.10	35.60

(Continued)

(Continued)

Intervention Group	N Obs	Variable	Label	N	Mean	Std Dev	Minimum	Maximum
C	10	AGE	Age on Jan 1, 2000	10	11.00	1.83	8.00	14.00
		TIME1	Baseline	10	21.91	1.66	19.00	23.90
		TIME2	6 Months	10	27.94	2.61	22.50	31.10
		TIME3	12 Months	10	31.12	3.52	24.90	35.90
		TIME4	24 Months	10	32.24	3.19	26.60	36.10

3. Add a new macro variable assignment statement

   ```
   %LET STATUS=Pre;
   ```

   and change the `TITLE` to

   ```
   TITLE "Descriptive Statistics for &STATUS.Production";
   ```

4. Submit the code and note the title in the output that reads "Descriptive Statistics for PreProduction." Take the dot out from between `&STATUS` and Production. Rerun the code and note how the concatenation is no longer correctly resolved.

The use of macro variables can become quite complex, and there are a number of capabilities and uses that are not covered in this discussion. However, the following section does include information on how to create and use macro variables as a part of a macro routine. For more information on macro variables refer to SAS Documentation.

7.4.3 Creating Callable Macro Routines

A SAS macro is a series of SAS statements that performs a general task, such as creating a report. Instead of "hard coding" this routine with fixed values, you write your code so that certain portions of the code can be altered using macro variables.

After a macro routine is defined properly, you can use it again and again in your program. This is referred to as "calling" the macro.

When you call a SAS macro routine, you typically "send" it a series of macro variables and these variables contain information that is substituted into portions of the macro routine code to customize it dynamically as the program is run. The simplified syntax of a SAS macro is

```
%MACRO macroname <(parameter-list)></ option-1 < ... option-n>>;
    SAS code that optionally contains macro variables;
%MEND macroname;
```

Here is a brief explanation of how this macro code works:

- The information between the `%MACRO` statement and the `%MEND` statement is a SAS macro routine consisting of SAS code.

- The *macroname* is the name you give to the macro (and it should obey standard SAS naming conventions).
- Use of the parameter list indicates a method for "sending" information into the macro routine (illustrated in Hands-On Example 7.10).
- Within the SAS macro (between %MACRO to %MEND), the variables referenced in the parameter list become macro variables within the macro and may be referred to with the standard macro variable syntax and variable names.

For example, the following program code, which contains a macro routine named REPORT, prints out a short report on a requested subject. The subject is specified by the macro variable SUBJECTS, from a dataset, specified by the macro variable name DSN. Because you may want to create these reports for a number of people, you can use a macro routine and call it as many times as needed. Within the macro report code, the dynamic variable that is replaced by a subject name is &SUBJ and the dynamic variable replaced by a dataset name is &DSN.

```
%MACRO REPORT(SUBJ, DSN);
    DATA REPORT;SET "&DSN";
    IF SUBJ=&SUBJ;
    TITLE "REPORT ON SUBJECT # &SUBJ";
     PROC PRINT NOOBS DATA=REPORT;
         VAR GENDER TIME_EXPOSED DIAGNOSED;
    RUN;
%MEND REPORT;
```

The first line of the macro routine

```
%MACRO REPORT(SUBJ, DSN);
```

specifies the name of the macro (REPORT) and the two macro variables that are used to send information into the macro routine (SUBJ and DSN).

The macro routine is typically placed at the beginning of your SAS code file so that when you submit the code, SAS sees and compiles the macro first. Once the macro is defined, you can "call" it using a macro call command. The call for any macro is a % followed by the macro routine name. In this case, the routine name is REPORT so an example call to the routine is

```
%REPORT(SUBJ=001, DSN=C:\SASDATA\SUBJECTS)
```

Note that for the call, certain values are assigned to the macro variables. For this example, SUBJ is assigned the value 001, and the value of DSN is a path to a SAS dataset named SUBJECTS. (There are no quotation marks around either value.)

Once the macro routine REPORT is compiled by SAS, a call to it using the example information does the following:

1. The values assigned to SUBJ and DSN are sent to the macro routine.
2. Within the routine wherever &SUBJ is found, it is substituted for the value of SUBJ. Wherever &DSN is found, it is substituted for the value of DSN.
3. For this example, the SAS code after the substitutions becomes

194 CHAPTER 7: SAS® ADVANCED PROGRAMMING TOPICS PART 2

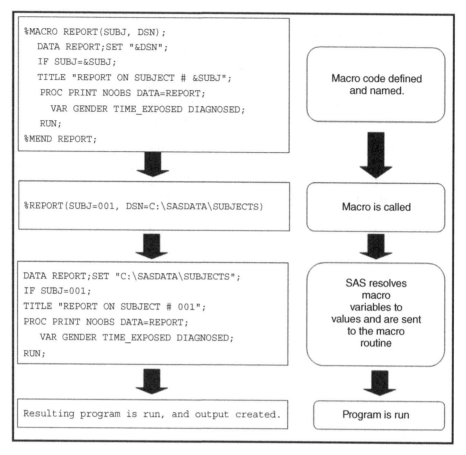

Figure 7.3 How a SAS macro works

```
DATA REPORT;SET "C:\SASDATA\SUBJECTS";
IF SUBJ=001;
TITLE "REPORT ON SUBJECT # 001";
 PROC PRINT NOOBS DATA=REPORT;
      VAR GENDER TIME_EXPOSED DIAGNOSED;
RUN;
```

The revised code is run by SAS, and a report is created by the PROC PRINT statement. This process is illustrated in Figure 7.3.

HANDS-ON EXAMPLE 7.10

This example illustrates how to use a SAS macro routine.

1. Open the program file DMACRO3.SAS. This is the same code illustrated previously in this section. Note that following the macro routine code is a series of macro calls

```
%REPORT(SUBJ=001, DSN=C:\SASDATA\SUBJECTS)
%REPORT(SUBJ=017, DSN=C:\SASDATA\SUBJECTS)
%REPORT(SUBJ=040, DSN=C:\SASDATA\SUBJECTS)
```

2. Submit the program and observe the output given in Table 7.10.
3. Change the SUBJ= values to 002, 010, and 020. Submit the program and observe the reports created for the requested subjects.
4. Change the %MACRO statement by adding the variable VARLIST after DSN.

```
%MACRO REPORT(SUBJ, DSN, VARLIST);
```

Replace the VAR statement in the PROC PRINT with this statement

```
VAR &VARLIST;
```

Change the macro calls by adding the third macro variable VARLIST and assign it the value GROUP--DIAGNOSED. (Note that it is a double dash between the variable names.) For example,

```
%REPORT(SUBJ=002, DSN=C:\SASDATA\SUBJECTS,
   VARLIST=GROUP--DIAGNOSED)
```

Comment out the other two calls. Resubmit the program to see how this has changed the reports.

TABLE 7.10   Output from Macro Program

REPORT ON SUBJECT # 001		
GENDER	TIME_EXPOSED	DIAGNOSED
M	23.31	1

REPORT ON SUBJECT # 017		
GENDER	TIME_EXPOSED	DIAGNOSED
F	20.95	1

REPORT ON SUBJECT # 040		
GENDER	TIME_EXPOSED	DIAGNOSED
F	18.1	0

7.4.4 Using a Macro with No Parameters

A particularly simple type of macro is one that contains no parameters. In this case, the macro routine is defined as

```
%MACRO macroname;
    SAS code;
%MEND macroname;
```

This type of macro could be used to call a set of standard code within a larger program so you don't have to rewrite the same code again and again. It could be used to display a title, run a series of interim reports that change when a dataset is updated, run a standard report, and so on. For example, suppose you have a standard disclaimer that you use when you create reports. The following macro routine contains code that outputs a message using PROC GSLIDE. (Options for the output, such as J for justification and H for height, are similar to the options previously discussed for titles and footnotes.) The following macro code is in DMACRODISCLAIM.SAS.

```
%MACRO DISCLAIM;
PROC GSLIDE;
NOTE J=C H=1
"Data provided by ACME Company exchanges may be delayed";
NOTE J=C H=1
"as specified by financial exchanges or our data providers. ";
NOTE J=C H=1
"ACME does not verify any data and disclaims any obligation to
    do so.";
RUN;
QUIT;
%MEND DISCLAIM;
```

After this macro is defined, it can be called whenever you want that disclaimer to appear in your output.

```
%DISCLAIM
```

This is illustrated in Hands-On Example 7.11.

7.4.5 Including SAS Macro Code

You may create a number of SAS macro routines that you want to use again and again. Instead of including the macro code in the program where you are going to call it, you may want to include the macro code dynamically in your program. For example, the file that contains the DISCLAIM macro is DMACRODISCLAIM.SAS. By placing a %INCLUDE command, given here, at the top of your code, you can make this macro available for use in a program.

```
%INCLUDE "C:\SASDATA\DMACRODISCLAIM.SAS";
```

> **HANDS-ON EXAMPLE 7.11**
>
> This example illustrates how to include external SAS code in a program.
>
> 1. Open the program file FIRST.SAS. This was the first SAS program we used. On the line before the DATA statement, enter
>
>    ```
>    %INCLUDE "C:\SASDATA\DMACRODISCLAIM.SAS";
>    ```
>
>   And on the line after the last RUN statement, enter

```
%DISCLAIM;
```

(You are not required to put a semicolon after a macro call, but sometimes it helps if you have other code following.)

When you run this program, SAS includes the code from the file `DMACRODISCLAIM.SAS` in the program stream (because of the `%INCLUDE` statement) causing that macro code to be compiled at the beginning of the `FIRST` program. Therefore, it is ready to be called when the code sees the `%DISCLAIM` statement at the end of the program.

2. Submit the code and observe the output given in Table 7.11. It contains the disclaimer from the macro routine. This illustrates that you can keep certain macros in files, include them in any program you want to use them in with a `%INCLUDE` statement, and call them at locations in your program where you want the macro to run.

TABLE 7.11 Output Showing GSLIDE Results

	Analysis Variable: AGE					
GENDER	N Obs	N	Mean	Std Dev	Minimum	Maximum
FEMALE	9	9	40.8888889	14.2954927	24.0000000	72.0000000
MALE	9	9	38.7777778	15.1226467	18.0000000	64.0000000

Data provided by ACME Company. Exchanges may be delayed as specified by financial exchanges or our data providers. ACME does not verify any data and disclaims any obligation to do so.

7.4.6 Using The SAS Macro %DO Loop

The SAS `%DO` loop statement is similar to the previously discussed `DO` loop, but in this macro version (note the percentage sign before the `DO`), you are able to define increment macro variables that take on dynamic values. The `%DO` loop statement must appear inside a SAS macro routine, but unlike the standard `DO` loop, it does not have to appear within a `DATA` step. A simplified syntax is as follows:

```
%DO macrovariable=start %TO stop <%BY increment>;
  SAS statements
%END;
```

Note that one of the big differences in this `%DO` loop is the variable being incremented, called *macrovariable* here. That is, this variable is like a macro variable created in a `%LET` statement, but you have the ability to make it increment over a selected range in the loop. For example, the following `%DO` loop iterates from the variable `I` (which we refer to as `&I`, a macro variable) to 5 and assigns the name of a `DATA` file based on the value of `&I`.

```
%DO I=1 %TO 5;
   DATA GP
     SET MYSASLIB.SOMEDATA;;
     WHERE STATUS=&1;
   RUN;
%END;
```

The code GP&I resolves to the dataset names GP1, GP2, GP3, and so on within the loop as I increments from 1 to 5.

HANDS-ON EXAMPLE 7.12

This example illustrates how to use a %DO loop in a SAS macro.

1. Open the program file DMACRO4.SAS. This routine contains a macro named GROUPS that creates SAS datasets from the SOMEDATA dataset based on the value of the variable STATUS. Submit this code and examine the log. Verify that five datasets GP1, GP2, GP3, GP4, and GP5 were created.

   ```
   %MACRO GROUPS(COUNT);
      %DO I=1 %TO &COUNT;
         DATA GP&I;
            SET MYSASLIB.SOMEDATA;
            WHERE STATUS=&I;
         RUN;
      %END;
   %MEND GROUPS;
   %GROUPS(COUNT=5);
   TITLE "LISTING FOR SOCIOECONOMIC STATUS=1";
   PROC PRINT DATA=GP1;RUN;
   TITLE "LISTING FOR SOCIOECONOMIC STATUS=2";
   PROC PRINT DATA=GP2;RUN;
   ```

   Note that the call to the macro program (%GROUPS(COUNT=5);) sends the value 5 to the macro variable named COUNT, which was used in the DO loop at the %TO value. The dataset names are created by concatenating GP with the increment variable &I and they are GP1, GP2, and so on. The PROC PRINT statements list the contents of the first two datasets created. Submit this code and observe the results in Table 7.12.

2. Change the macro call to

   ```
   %GROUPS(5)
   ```

   Resubmit and see that the call still works. As long as there is no ambiguity in the call, you do not have to include a specific assignment such as COUNT=5.

TABLE 7.12 Output Showing Results of %DO Loop

Listing for SOCIOECONOMIC STATUS=1									
Obs	ID	GP	AGE	TIME1	TIME2	TIME3	TIME4	STATUS	SEX	GENDER
1	343	B	11	19.0	25.2	27.7	25.1	1	1	Male
2	410	B	13	22.2	30.4	32.1	35.2	1	0	Female
3	307	C	11	22.6	30.1	34.4	35.2	1	0	Female

Listing for SOCIOECONOMIC STATUS=2									
Obs	ID	GP	AGE	TIME1	TIME2	TIME3	TIME4	STATUS	SEX	GENDER
1	187	A	11	22.5	29.3	32.6	33.7	2	0	Female
2	162	B	7	20.9	28.9	29.7	25.9	2	1	Male
3	198	B	12	23.4	29.2	30.4	35.1	2	0	Female
4	282	B	13	21.3	28.7	32.3	30.3	2	0	Female
5	575	B	8	21.2	25.7	28.9	31.6	2	0	Female
6	589	B	7	22.0	25.5	28.2	30.9	2	0	Female
7	593	B	11	18.9	23.1	27.7	25.6	2	1	Male

7.4.7 USING CALL SYMPUT TO CREATE A MACRO VARIABLE

The SAS CALL SYMPUT routine assigns a value to a macro variable within a DATA step. The call does not have to be within a macro routine, although it can be. The syntax of the call is as follows:

CALL SYMPUT('macrovariablename', value);

The macrovariablename is the name you want to assign a value (similar to how the %LET statement assigns a value). However, in this case, the value in the routine can be an expression, and the name itself can be created using SAS text operators. This is illustrated in the following example:

HANDS-ON EXAMPLE 7.13

This example illustrates how to use CALL SYMPUT.

1. Open the program file DMACRO5.SAS. The dataset CARS contains two MPG variables CITYMPG and HYWMPG. This macro allows you to specify which version of MPG to use in a procedure (PROC MEANS in this case). Submit this code and observe the output in Table 7.13. The DATA _NULL_ statement is used to allow you to run procedures that must take place in a DATA step, but you do not want to save any data values to a SAS dataset.

(Continued)

(*Continued*)

TABLE 7.13 Output Showing Results of Call Symput

The MEANS Procedure				
Analysis Variable: CityMPG				
N	Mean	Std Dev	Minimum	Maximum
1081	19.29	4.60	10.00	60.00

```
%MACRO CARINFO(WHICHPLACE);
DATA _NULL_;
CALL SYMPUT('TIMEVAR', CATT('&WHICHPLACE.','MPG'));
RUN;
TITLE "THE VALUE OF THE MACRO VARIABLE TIMEVAR IS &TIMEVAR ";
PROC MEANS MAXDEC=2 DATA=MYSASLIB.CARS;
VAR &TIMEVAR;
RUN;
%MEND CARINFO;
%CARINFO(WHICHPLACE=CITY);
```

Note that the macro variable sent to the macro, WHICHPLACE has the value CITY. The SYMPUT call concatenates '&WHICHPLACE.' with 'MPG' to create the string CITYMPG. The dot (.) after &WHICHPLACE is a macro variable delimiter required to avoid ambiguity when there is some character following the macro variable. In this case, the quote character follows the macro variable name, so the dot separates the two.

2. Take out the dot after the 'WHICHPLACE' in the CALL SYMPUT statement and resubmit the code. You get an output message

```
WARNING: Apparent symbolic reference WHICHPLACEMPG not
resolved.
WARNING: Apparent symbolic reference WHICHPLACEMPG not
resolved.
ERROR: Variable WHICHPLACEMPG not found.
```

Thus, the value of the macro variable is &WHICHPLACEMPG, is not what you want. Put the dot back, rerun, and verify that it works again.

3. An important SAS concept is that a macro variable created in a DATA step cannot be used in that same data step. That is why a RUN ends the DATA step before the TITLE statement is given. To test this, put the RUN statement after the

```
TITLE "THE VALUE OF THE MACRO VARIABLE TIMEVAR IS &TIMEVAR ";
```

statement instead of before it. Resubmit the code and verify that you do not get the desired output. Fix these lines and verify that the code now works again. (This might seem to be a trivial situation until you spend hours trying to figure out why your code doesn't work.)

7.5 SUMMARY

This chapter provides information on two topics that allow you to write SAS code that can simplify complex procedures: mainly arrays and macros. Arrays can simplify creating and using large data constructs that must be manipulated by code. Macro variables allow you to create generalized programs, and macro routines allow you to create code that can be used over and over again.

EXERCISES

7.1 **Creating and using array and a DO loop.**
   You want to create a program that calculates an average grade based on the best 4 of 5 recorded grades. Open the program file EX_7.1.SAS, a partial SAS program to perform this task:

```
DATA GRADES;
INPUT IDNO GRADE1 GRADE2 GRADE3 GRADE4 GRADE5;
ARRAY G(5); *FINISH THE STATEMENT;
* FIND LOWEST GRADE;
LOW=100;
DO; * FINISH THE STATEMENT;
   IF G(I)<LOW then LOW=G(I); END;
TOTAL=; * FINISH THE STATEMENT;
AVERAGE=TOTAL/4;
DATALINES;
001 90 34 88 79 88
002 99 69 87 86 98
003 91 75 85 94 100
004 88 57 68 74 89
;
RUN;
PROC PRINT DATA=GRADES;RUN;
```

a. Complete the missing statements.
   - For the ARRAY statement, specify the range GRADE1-GRADE5.
   - For the DO statement, increment from 1 to 5.
   - For the TOTAL statement, use the SUM(of variables) statement and subtract the value LOW from the result.
   - For the AVERAGE statement, divide the TOTAL result by 4.
   - Submit the code and verify that the result in AVERAGE is the average grade for each student, leaving out the lowest grade for each average.

b. Examine how LOW is calculated as the lowest grade for each student. Answer the following questions:
   Why is LOW initially set to 100?
   In the DO loop, when is LOW assigned a new value?
   At the end of the DO loop, what is the value of LOW and why?

Why subtract LOW from the SUM(Of GRADE1-GRADE5) in the TOTAL= statement?

7.2 Use a SAS macro.

Using the SAS code in DMACRO3.SAS as an example, complete the program below to create a report using the CARS dataset. This code is in EX_7.2.SAS. Use this information:

- The name of the macro is GETCARS.
- The name of the SAS dataset is "C:\SASDATA\CARS". Use this in the SET statement.
- For the second and third reports, use FORD and TOYOTA. Note that the name must be in all caps.
- Place the macro variable named &CHOICE in the title.

```
%MACRO _____(CHOICE=);
  DATA TEMP; SET "C:_____";
  IF BRAND="&CHOICE";
  TITLE "CAR SELECTION FOR _____";
  PROC PRINT; VAR BRAND MODEL CITYMPG;
  RUN;
%MEND GETCARS;
ODS HTML;
%GETCARS(CHOICE=SCION)
%GETCARS(CHOICE=_____)
%GETCARS(CHOICE=_____)
```

a. Run the program and observe the results. It should create three reports, one for each brand of car.

b. Create a new macro call using VOLKSWAGEN.

c. To verify that case matters, perform the following macro call:

```
%GETCARS(CHOICE=Honda)
```

Why didn't it produce a report?

d. Change the IF statement to

```
IF BRAND=UPCASE("&CHOICE");
```

Perform a macro call using

```
%GETCARS(CHOICE=Honda)
```

Why does it now work?

7.3 Using RETAIN.

You are a football fan and are keeping up with the number of yards gained by your favorite running back. The data are in EX_7.3.SAS. Using DRETAIN2.SAS as an example, write a SAS program that calculates the accumulated yards gained and finds the maximum and minimum yards gained in any one game.

Game	Yards Gained
1	55
2	105
3	110
4	153
5	99
6	69
7	34
8	101
9	110
10	120

7.4 CREATING A MACRO REPORT

The SAS code in EX_7.4.SAS displays a report on MPG for cars for a selected company.

```
%MACRO CARMPG(B);
  DATA &B;
     SET MYSASLIB.AUTOMPG;
     WHERE BRAND="&B";
  RUN;
  TITLE "Miles Per Gallon Averages for &B ";
  PROC MEANS DATA=&B;
  VAR CITYMPG HWYMPG;
  RUN;
%MEND CARMPG;
%CARMPG(FORD);
```

a. Run the program to see how it works. Notice that it creates a data file named FORD in the WORK library. Why?

b. Change FORD to TOYOTA and rerun this program. Notice the name of the file it now creates.

c. Redo the macro to display MPG by the number of cylinders for a car. For example, we want a macro call CARMPG2(4) to display a report for all cars with 4 cylinders.

d. This illustrates several items about using character and numeric values in macro calls. You can keep the first line similar (Change CARMPG to CARMPG2.)

```
%MACRO CARMPG2(4);
DATA CYLINDERS&B;
     SET MYSASLIB.AUTOMPG;
     WHERE CYLINDERS=&B;
  RUN;
  TITLE " Miles Per Gallon Averages for CYLINDERS = &B ";
  PROC MEANS DATA=CYLINDERS&B;
  VAR CITYMPG HWYMPG;
  RUN;
```

Notice that instead of the call value (4) being a text value (i.e., FORD) it will be a numeric value (i.e., 4) Therefore, you have to change the statement

```
DATA &B;
```

which creates the working dataset. Note that if &B is 4, the macro creates a SAS file named 4 which is not an allowable SAS file name. (i.e., DATA 4 will not work.) For this new macro, simply change the filename to CYLINDERS&B

```
DATA CYLINDERS&B;
```

which will create a dataset named CYLINDERS4   (if 4 is the variable value of &B.)

e. In the previous WHERE statement

```
WHERE BRAND="&B";
```

was resolved to WHERE BRAND="FORD"; That is, we needed the quotes in the statement to make sure the statement resolved to a statement that was viable.

But now we don't want the numeric value in quotes, (i.e., we don't; want CYLINDERS="4") so this statement must be changed to

```
WHERE CYLINDERS=&B;
```

(without the quotes around &B.)

f. Change the PROC MEANS report statements to

```
TITLE "Miles Per Gallon Averages for CYLINDERS = &B ";
PROC MEANS DATA=CYLINDERS&B;
  VAR CITYMPG HWYMPG;
  RUN;
```

g. Call the macro with the statement

```
%CARMPG2(4);
```

h. Call the macro to create a report for 6 and 8 cylinder cars. What happens if you put 20 as the number of cylinders?

8

CONTROLLING OUTPUT USING ODS

LEARNING OBJECTIVES

- To be able to specify ODS output format and destination
- To be able to specify ODS style
- To be able to select specific tables for output
- To be able to capture information from ODS tables
- To be able to use ODS graphics from SAS® procedures
- To be able to use Traffic Lighting to highlight selected values

ODS, or Output Delivery System, is a method for controlling the output from SAS procedures. ODS began with version 8 and continues with added enhancements in more recent versions.

8.1 SPECIFYING THE ODS OUTPUT FORMAT AND DESTINATION

The SAS ODS is set up so that you "turn on" or initiate output into a designated output format. Once the output format has been initiated, SAS procedures send information to that output format. You can send output from one or more procedures to the output stream.

SAS® Essentials: Mastering SAS for Data Analytics, Third Edition. Alan C. Elliott and Wayne A. Woodward.
© 2023 John Wiley & Sons, Inc. Published 2023 by John Wiley & Sons, Inc.
Companion website: https://www.alanelliott.com/SASED3/

Following the procedures for which you want the ODS output, you "turn off" or end the output. To initiate output using ODS, use the following statement:

ODS *output-format* <*options*>;

To end the ODS output, use the CLOSE statement:

ODS *output-format* CLOSE;

Here are several ODS output formats:

```
ODS LISTING <FILE='file-specification'>;
ODS HTML <BODY='html-file-pathname.html'>;
ODS PDF <FILE='pdf-file-pathname.pdf'>;
ODS RTF <FILE='rtf-file-pathname.rtf'>;
ODS PS <FILE='ps-file-pathname.ps'>;
ODS PLC <FILE='pcl-file-pathname.pcl'>;
ODS EXCEL <FILE='Excel-file-pathname.xlsx'>;
ODS POWERPOINT <FILE='ppoint-file-pathname.pptx'>;
ODS WORD <FILE='Word-file-pathname.docx'>;
```

There are many more ODS types not discussed here such as EPUB, DOCUMENT, MARKUP and others. Refer to the SAS documentation for more information on these ODS types.

Note that most of the output formats include the option to specify a filename with a FILE= option. The HTML format uses BODY= rather than FILE= since that is the terminology used by HTML to describe a displayed page of information. These output formats are described more fully as follows:

- **HyperText Markup Language (HTML):** HTML is the current standard (default) output format used in Windows SAS versions. This HTML format is the same used for displaying Web pages on the Internet and is also recognized by most word processors such as Microsoft Word. To reset SAS to this default (if it becomes turned off, use the statement ODS PREFERENCES;)
- **LISTING:** This "old-style" SAS output format was the default output location used in SAS Windows versions prior to Version 9.1. It is sometimes called "typewriter output" and is based on a monospaced font.
- **Portable Document Format (PDF):** PDF is widely used to distribute documents by e-mail or via Internet downloads. Use of a PDF document requires that the (free) Adobe Reader or a similar program be installed on your computer. One of the benefits of the PDF format is that the contents of your document will appear the same when viewed under many types of computer operating systems, such as Windows, Mac OS X, and UNIX and its Linux derivations.
- **Rich Text Format (RTF):** RTF is a common language for word processors. When you capture information in this format, it can be opened and edited by most word processors such as Microsoft Word or even WordPad.
- **Postscript (PS):** PS is a printer language, and the output captured in this format is usually designed to be printed on a PS printer.
- **Print Command Language (PCL):** PCL is a printer language, and the output captured in this format is usually designed to be printed on a PCL printer.

EXCEL (XLSX): A Microsoft Excel spreadsheet is a file type created through ODS.
POWERPOINT (PPTX): Microsoft PowerPoint presentation files can be produced using ODS.
WORD (DOCX): ODS can be used to create a Microsoft Word document. Extensible Markup Language (XML): XML is a plain text data interchangeable format that can be read using a normal text editor but is in a standard machine-readable format.
Tagsets.ExcelXP: The **Tagsets.ExcelXP** option allows you to output XML tables that can be read by Microsoft Excel version 2002 or later. Other specialty output formats are DOCUMENT, MARKUP, and WML (Wireless Markup Language).

The file names following the ODS output format are optional, but SAS may prompt you for a file name if it is omitted. Note that when you specify a file name in Windows, it must include the correct extension if Windows is to recognize the output file type. For example, to send data to a PDF file, you could use the following syntax:

```
ODS PDF file='windows-path\filename.pdf';
```

or, as a specific example:

```
ODS PDF FILE='C:\SASDATA\MYOUTPUT.PDF';
```

If you leave off the destination, as in

```
ODS PDF;
... some procedures
ODS PDF CLOSE;
```

SAS will send the output to the SAS **Results Viewer** window without prompting you for a file name. For output formats such as ODS PDF and ODS RTF, if you do not indicate a file name, you may be prompted to enter one (or you may choose to have the results go into the SAS Results Viewer). When you use the default ODS HTML output format, you are not prompted for a file name.

It is important to remember that after you specify an ODS output, you also need to close it to tell SAS to finish sending the output to that location. If you do not close the location, you may not be able to see the results. For example, if you use ODS RTF to open the RTF (Word) output, then you must at some point close with the statement ODS RTF CLOSE. Thus, the statements would be

```
ODS RTF;
SAS code that produces output ...;
ODS RTF CLOSE;
```

Beginning with SAS 9.2, the default output location became HTML. (Previously, it was LISTING.) Beginning with SAS Version 9.4, the output defaults are as follows:

- Default output is HTML and the default style is HTMLBlue
- ODS Graphics is on
- Listing is closed

TABLE 8.1   Current ODS Styles for SAS 9.4

Analysis	Harvest	MonochromePrinter	Rtf
BarrettsBlue	HighContrast	Monospace	Sapphire
Dtree	HighContrastLarge	Moonflower	SasDocPrinter
Daisy	Ignite	Netdraw	SasWeb
Default	Illuminate	NoFontDefault	Seaside
Dove	Journal	Normal	SeasidePrinter
EGDefault	Journal1a	NormalPrinter	Snow
Excel	Journal2	Ocean	StatDoc
FancyPrinter	Journal2a	Pearl	Statistical
Festival	Journal3	PearlJ	Word
FestivalPrinter	Journal3a	Plateau	vaDark
Gantt	Listing	PowerPointDark	vaHighContrast
GrayscalePrinter	Meadow	PowerPointLight	vaLight
HTMLBlue	MeadowPrinter	Printer	
HTMLEncore	Minimal	Raven	

However, if you use the ODS statement to close the HTML location (ODS HTML CLOSE), SAS may not produce any output. This will cause an error message in the **Log Window** that indicates that there is no output destination open. If this happens, you should reset the HTML default option in SAS. Use this statement:

```
ODS PREFERENCES;
```

to restore the default HTML output settings.

8.2   SPECIFYING ODS OUTPUT STYLE

When you output information using a SAS ODS format, the tables, graphs, and text are defined with default colors and fonts. You can select from several built-in ODS styles. Each style is based on some theme or a format appropriate for a specific purpose. For example, the JOURNAL style formats your output in black and white that is appropriate for inclusion in a journal article. The option to specify a style is

```
STYLE=styletype;
```

This option appears in the ODS statement. For example,

```
ODS RTF STYLE=JOURNAL;
... some procedures
ODS RTF CLOSE;
```

To see a listing of the available SAS styles, use the code:

```
PROC TEMPLATE; LIST STYLES; RUN;
```

The default style is STYLES.DEFAULT. The SAS format styles for version 9.4 as reported by PROC TEMPLATE are given in Table 8.1 (with the prefix "STYLES." removed to make it easier to read the table).

HANDS-ON EXAMPLE 8.1

This example illustrates outputting information to a specific WORD file. (You'll need to have Microsoft Word installed on your computer for this to work. If you don't have Word installed, use ODS1A.SAS instead – with similar results except an HTML file is created.)

1. Open the program file ODS1.SAS.

   ```
   DATA TEST; SET "C:\SASDATA\SOMEDATA";
   * DEFINE WHERE WORD OUTPUT WILL GO;
   TITLE 'ODS WORD Example';
   ODS WORD FILE='c:\SASDATA\ODS.docx';
   PROC MEANS MAXDEC=2; VAR AGE TIME1-TIME4;
   RUN;
   * CLOSE THE Word OUTPUT;
   ODS WORD CLOSE;
   RUN;
   ODS PREFERENCES;
   ```

   The ODS WORD statement tells SAS where to save the resulting Word file.

2. Submit the program. Observe the output in Microsoft Word (or in the Viewer if you used ODS1.SAS). (see Table 8.2). Go to Windows and open Windows Explorer (or click on the **My Computer** Icon on your desktop). Drill down to the C:\SASDATA folder. You should see a file named ODS.DOCX. Thus, the SAS output from this program has been saved locally in WORD format.

   TABLE 8.2 Microsoft Word Output Using SAS ODS

ODS WORD Example						
The MEANS Procedure						
Variable	Label	N	Mean	Std Dev	Minimum	Maximum
AGE	Age on Jan 1, 2000	50	10.46	2.43	4.00	15.00
TIME1	Baseline	50	21.27	1.72	17.00	24.20
TIME2	6 Months	50	27.44	2.66	21.30	32.30
TIME3	12 Months	50	30.49	3.03	22.70	35.90
TIME4	24 Months	50	30.84	3.53	21.20	36.10

3. Return to SAS and note that the output in the Word document is also displayed in the SAS **Results Viewer**.

   (Continued)

210 CHAPTER 8: CONTROLLING OUTPUT USING ODS

(Continued)

4. Close the Word document created by SAS. Then change the ODS statement to

   ```
   ODS WORD FILE='c:\SASDATA\ODS.docx' STYLE=FESTIVAL;
   ```

   Change the title to

   ```
   TITLE "Word Output Example Using Festival Style";
   ```

5. Resubmit the program and observe the output. Note that the new program produces output with different color combinations (see Table 8.3).

TABLE 8.3 ODS HTML Output Using Festival Style

\multicolumn{7}{c}{Word Output Example Using Festival Style}
The MEANS Procedure						
Variable	Label	N	Mean	Std. Dev.	Minimum	Maximum
AGE	Age on Jan 1, 2000	50	10.46	2.43	4.00	15.00
TIME1	Baseline	50	21.27	1.72	17.00	24.20
TIME2	6 Months	50	27.44	2.66	21.30	32.30
TIME3	12 Months	50	30.49	3.03	22.70	35.90
TIME4	24 Months	50	30.84	3.53	21.20	36.10

6. To cause the output to be sent to a PDF file, change the ODS statement to

   ```
   ODS PDF FILE='C:\SASDATA\ODS.PDF' STYLE=STATISTICAL;
   ```

   Change the ODS WORD CLOSE statement to

   ```
   ODS PDF CLOSE;
   ```

   Run the program and observe the output and the file saved, named ODS.PDF. (See Table 8.4) File ODS2.SAS contains the code after the PDF changes.

TABLE 8.4 Output in ODS PDF Format Using Statistical Style

						09:42 Thursday, March 31, 2022
ODS PDF Example						
The MEANS Procedure						
Variable	Label	N	Mean	Std. Dev.	Minimum	Maximum
AGE	Age on Jan 1, 2000	50	10.46	2.43	4.00	15.00
TIME1	Baseline	50	21.27	1.72	17.00	24.20
TIME2	6 Months	50	27.44	2.66	21.30	32.30
TIME3	12 Months	50	30.49	3.03	22.70	35.90
TIME4	24 Months	50	30.84	3.53	21.20	36.10

> The results will appear in the Results Viewer. Depending on your particular computer setup, the Adobe Reader program may start up and display the ODS.PDF file. The results are given in Table 8.4. This output is created in the STATISTICAL style format as specified in the ODS statement FILE= *statement*. Similar results can be observed by outputting information in other ODS formats.

8.3 USING ODS TO SELECT SPECIFIC OUTPUT TABLES FOR SAS PROCEDURES

SAS procedures often output a lot of information you don't want or need. In ODS output, each part of the output is contained in a table. Using ODS options, you can customize which tables you want SAS to output to the ODS destination. For example, suppose you are calculating a number of 2×2 cross-tabulations in a chi-square analysis. SAS automatically includes the Fisher's Exact test results in the output. If you do not want that part of the output displayed (maybe you want to save paper when you print results), you can use an ODS statement to instruct SAS to exclude the Fisher table (or any table) from the output.

To include or exclude a table from the output, you first need to know the table's name. You can discover this information by using the ODS TRACE command in the following way:

```
ODS TRACE ON;
    procedure specifications;
ODS TRACE OFF;
```

This ODS option outputs information in the **Log Window** that indicates the name of each output table. Once you've discovered the name of one or more tables you want to include (in this example) in the output, you can use the following code to make that request:

```
ODS SELECT tables-to-include;
PROC FREQ CODE;
```

When you run the program, only the tables requested in the SELECT statement are included in the output. Similarly, you can use an ODS EXCLUDE statement to exclude tables from the output. Hands-on Example 8.2 illustrates the procedure for selecting or excluding part of the computer output.

> **HANDS-ON EXAMPLE 8.2**
>
> This example illustrates how to use ODS to limit output from a SAS analysis.
>
> 1. Open the program file ODS3.SAS. This program performs a chi-square analysis on a 2×2 contingency table.
>
>    ```
>    DATA TABLE;
>    INPUT A B COUNT;
>    DATALINES;
>    ```
>
> *(Continued)*

(*Continued*)

```
0 0 12
0 1 15
1 0 18
1 1 3
;
ODS TRACE ON;
PROC FREQ;WEIGHT COUNT;
   TABLES A*B /CHISQ;
   TITLE 'CHI-SQUARE ANALYSIS FOR A 2X2 TABLE';
RUN;
ODS TRACE OFF;
```

2. Run the program and observe the results. Note that the results consist of the following tables (not shown here):
   (a) The A by B cross-tabulation table
   (b) Statistics for the A by B analysis
   (c) Fisher's Exact test.

   Examine the information in the **Log Window**. The portion of the **Log Window** shown in Figure 8.1 is the information created from the ODS TRACE ON and ODS TRACE OFF statements.

```
Output Added:
-------------
Name:       CrossTabFreqs
Label:      Cross-Tabular Freq Table
Template:   Base.Freq.CrossTabFreqs
Path:       Freq.Table1.CrossTabFreqs
-------------

Output Added:
-------------
Name:       ChiSq
Label:      Chi-Square Tests
Template:   Base.Freq.ChiSq
Path:       Freq.Table1.ChiSq
-------------

Output Added:
-------------
Name:       FishersExact
Label:      Fisher's Exact Test
Template:   Base.Freq.ChisqExactFactoid

Path:       Freq.Table1.FishersExact
```

Figure 8.1 ODS TRACE output

Note the names of the tables:
- CrossTabFreqs
- ChiSq
- FishersExact

Knowing these names (they are not case sensitive), you can, for example, limit your output to only the CrossTabFreqs and ChiSq tables.

3. Open the program file ODS4.SAS. Note the statement beginning with ODS SELECT. It is the statement that specifies what tables to include in the output.

```
DATA TABLE;
INPUT A B COUNT;
DATALINES;
0 0 12
0 1 15
1 0 18
1 1 3
;
ODS SELECT CROSSTABFREQS CHISQ;
PROC FREQ;WEIGHT COUNT;
    TABLES A*B /CHISQ;
    TITLE 'CHI-SQUARE ANALYSIS FOR A 2X2 TABLE';
RUN;
```

4. Submit this program. Note that it produces (almost) the same output as the previous code, but without the Fisher's table. You would get the same results by using the EXCLUDE option as in the following code:

```
ODS EXCLUDE FISHERSEXACT;
```

8.4 GOING DEEPER: CAPTURING INFORMATION FROM ODS TABLES

Another way to use the information from ODS TRACE is to capture output from procedures and save that information in a SAS output file. Once you know the name of the output table, you can use ODS to output the table as a SAS data file using the command

```
ODS OUTPUT nameoftable=outputdataset;
```

After outputting the information to a new dataset, you can merge that information into the current SAS dataset using the following code:

```
DATA NEW; SET originaldataset;
IF _N_=1 THEN SET outputdataset;
```

The _N_ is a special variable whose value is the sequential number of a record as a SAS dataset is read into the memory buffer. Thus, the statement tells SAS to merge the information from record 1 of OUTPUTDATASET into the NEW dataset (along with the information from ORIGINAL). This is illustrated in Hands-on Example 8.3.

```
IF _N_=1 then SET outputdataset;
```

HANDS-ON EXAMPLE 8.3

This example illustrates how to capture a specific statistic from the output of a procedure and save that information in a SAS dataset.

1. Open the program file ODS5.SAS.

```
DATA WT;
INPUT WEIGHT @@;
DATALINES;
64 71 53 67 55 58
77 57 56 51 76 68
;
ODS TRACE ON;
PROC MEANS DATA=WT;
RUN;
ODS TRACE OFF;RUN;
QUIT;
```

2. Run this program. It uses the ODS TRACE option to display the names of the output tables for PROC MEANS in the **Log Window**, as shown in Figure 8.2. Note that the name of the output table is "Summary."

```
Output Added:
-------------
Name:        Summary
Label:       Summary statistics
Template:    base.summary

Path:        Means.Summary
```

Figure 8.2 Trace output from a PROC MEANS

3. Close the **Results Viewer** and clear the **Output** and **Log Windows**. Open the program ODS6.SAS. Note that the SAS code has been modified by the removal of the TRACE ON and TRACE OFF commands and the addition of the following statement just before the PROC MEANS statement, as shown here:

```
ODS OUTPUT SUMMARY=STATS;
PROC MEANS;
```

SUMMARY in the ODS OUTPUT statement is the name of the table, and STATS is the name of the output dataset you are creating.

4. Run the SAS program that includes the above changes (ODS6.SAS). Look in the **Log Window** and see the following statement, which tells you that SAS created the requested dataset.

```
NOTE: The dataset WORK.STATS has 1 observation and 5 variables.
```

5. Examine the **Results Viewer** window to see information about the new dataset created, as given in Table 8.5. This information can now be used in other calculations, as illustrated in the next step.

TABLE 8.5 The SAS Dataset Created Using ODS Output

Output statistics to a data set					
Obs	WEIGHT_N	WEIGHT_Mean	WEIGHT_StdDev	WEIGHT_Min	WEIGHT_Max
1	12	62.75	8.9861003778	51	77

The WEIGHT_MEAN and WEIGHT_STDDEV values calculated and stored in the STATS dataset can be used to calculate a Z-score for the weights of subjects in the WT dataset. The following two lines of code can be used to calculate the Z-score for each WEIGHT value:

```
DIFF=WEIGHT-WEIGHT_MEAN;
Z=DIFF/WEIGHT_STDDEV;
```

However, in order to do this calculation, you must merge the WT dataset with the statistics in the STAT dataset. The next step shows how this is accomplished.

6. Open the program file ODS7.SAS. The purpose of this code is to use the information created by PROC MEANS and saved in the STATS dataset as information used to calculate the DIFF and Z values in a new dataset.

```
DATA WT;
INPUT WEIGHT @@;
DATALINES;
64 71 53 67 55 58
77 57 56 51 76 68
;
DATA WTDIFF;SET WT;
IF _N_=1 THEN SET STATS;
DIFF=WEIGHT-WEIGHT_MEAN;
Z=DIFF/WEIGHT_STDDEV; * Creates standardized scoreE (Z-score);
RUN;
PROC PRINT DATA= WTDIFF;VAR WEIGHT DIFF Z;
RUN;
```

This program creates a new dataset called WTDIFF that consists of the original dataset WT merged with the first (and only) row of the new STATS dataset. This merge is accomplished with the statement

```
IF _N_=1 THEN SET STATS;
```

This statement merges in the needed variables WEIGHT_MEAN and WEIGHT_STDDEV, from the STAT dataset, making them available for the Z-score calculation. A calculation is then made to create the Z-score variable.

(Continued)

(*Continued*)
7. Run this program and observe the output, which contains Z-scores for each record, as given in Table 8.6. Using the technique illustrated here, you can capture any statistic from a SAS output table, merge it into another dataset, and use that information to calculate new variables.

TABLE 8.6 Results of Calculating Z-Score

Obs	WEIGHT	DIFF	Z
1	64	1.25	0.13910
2	71	8.25	0.91808
3	53	−9.75	−1.08501
4	67	4.25	0.47295
5	55	−7.75	−0.86244
6	58	−4.75	−0.52859
7	77	14.25	1.58578
8	57	−5.75	−0.63988
9	56	−6.75	−0.75116
10	51	−11.75	−1.30757
11	76	13.25	1.47450
12	68	5.25	0.58424

8.4.1 Using ODS Graphics from SAS Procedures

ODS graphics refers to graphs that are created from SAS procedures. Although ODS graphics have been around for several SAS versions, the graphics only recently began appearing automatically when you run a SAS PROC (at least in the Windows edition of SAS).

SAS offers ways to customize the automatically created graphs using PROC TEMPLATE and the Graph Template Language (GTL). However, this feature is not covered in this book. Refer to SAS documentation for more about the GTL programming features. The creation of custom graphs using SAS/GRAPH is covered in Chapter 19.

Sometimes, the automatic creation of ODS graphics takes a while, and if you are doing a lot of analyses and only want the tabled output, you might be interested in shutting off ODS graphics for a while. You can do this (temporarily) using the command

```
ODS GRAPHICS CLOSE;
```

To turn ODS graphics back on use

```
ODS GRAPHICS ON;
```

You can also turn off ODS graphics by selecting (in SAS) the menu options **Tools → Options → Preferences**, and select the **Results Tab**. The dialog box shown in Figure 8.3 is displayed. Note the option **Use ODS Graphics**. If you want to turn off graphics, uncheck this box.

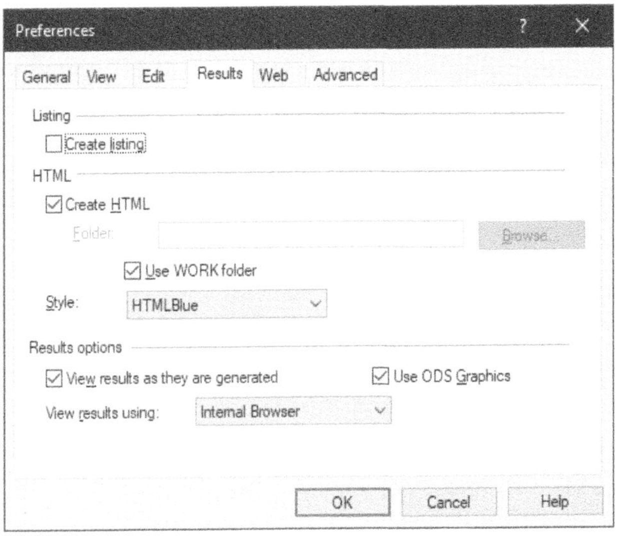

Figure 8.3 Preferences/results dialog box

Although SAS automatically produces plots for numerous procedures, there may be additional plots you want to display that are not automatically created. In this case, many procedures have added `PLOT` options that allow you to make these types of requests. As an example, consider Hands-on Example 8.4.

HANDS-ON EXAMPLE 8.4

1. Open the program file `ODS8.SAS`. This example creates a table of frequencies for the variable `STATUS` in the dataset `SOMEDATA`. Run this code and observe the output. There is no graph created with the tabled output (see Table 8.7).

```
PROC FREQ DATA="C:\SASDATA\SOMEDATA";
    TABLES STATUS;
TITLE 'Simple Example of PROC FREQ';
RUN;
```

TABLE 8.7 Frequency Output from PROC FREQ

	The FREQ Procedure			
	Socioeconomic Status			
STATUS	Frequency	Percent	Cumulative Frequency	Cumulative Percent
1	3	6.00	3	6.00
2	7	14.00	10	20.00
3	6	12.00	16	32.00
4	8	16.00	24	48.00
5	26	52.00	50	100.00

(*Continued*)

(*Continued*)

2. Make a small change in the code by adding an option to the TABLES statement

    TABLES STATUS/ PLOTS=FREQPLOT;

    Submit this code and observe the results. A graph is included in the output, as shown in Figure 8.4.

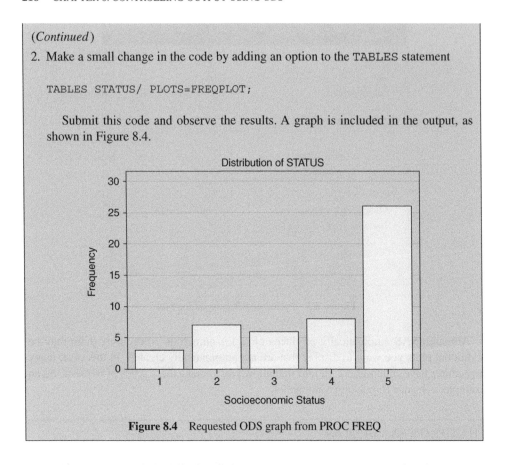

Figure 8.4 Requested ODS graph from PROC FREQ

As procedures are discussed in future chapters, ODS PLOT options will be introduced to show how to create graphs for procedures when they are not automatically generated.

8.5 GOING DEEPER: USING "TRAFFIC LIGHTING" TO HIGHLIGHT SELECTED VALUES

There may be an instance when you are creating a report using PROC PRINT in which you want to highlight certain values according to some criteria. (This feature only works in PROC PRINT, TABULATE, and REPORT. This example only discusses its use in PROC PRINT.) For example, if a value is importantly high or low, you might want to highlight the value with a color. "Traffic Lighting" is a way to do that. Using Traffic lighting is a two-step procedure:

1. Create a format that indicates colors for certain values. For example, suppose you want to highlight values of ISS, the Injury Severity Score according to severity. A format to do that could be created using this code:

    PROC FORMAT;
    VALUE FMTSEVERE

GOING DEEPER: USING "TRAFFIC LIGHTING" TO HIGHLIGHT SELECTED VALUES

```
        LOW-<10="GREEN"
        10-<20="BLUE"
        20-HIGH="RED";
RUN;
```

2. In PROC PRINT, you tell SAS which variable is to be assigned this format in the VAR statement. For example,

```
VAR ISS/STYLE={FOREGROUND=FMTSEVERE.};
```

The following example illustrates Traffic Lighting.

HANDS-ON EXAMPLE 8.5

1. Open the program file `ODS9.SAS`. This example lists the first 10 records in the TRAUMA dataset. Note the three VAR statements. They specify the order of variables for the listing. For the ISS variable, the STYLE is set to the format FMTSEVERE. Submit the code and observe the results, which are given here; see Table 8.8.

TABLE 8.8 Traffic Light Example

Trauma Data Report					
Subject ID	Age in 2014	Gender	Injury Severity Score	Injury Type	Discharge Status
468879	9.1	Female	1	Blunt	Alive
468942	17.7	Female	20	Blunt	Alive
468961	16.6	Female	5	Blunt	Alive
468971	6.5	Female	5	Blunt	Alive
469030	6.7	Male	−81		Alive
469055	10.8	Male	1		Alive
487580	17.7	Male	25	Penetrating	Alive
597075	14.6	Male	9	Blunt	Alive
603091	5.1	Female	17	Blunt	Alive
603104	8.3	Male	5	Blunt	Alive

```
PROC FORMAT;
VALUE FMTSEVERE
        LOW-<0="BLACK"
        0-<10="GREEN"
        10-<20="BLUE"
        20-HIGH="RED";
RUN;
TITLE "Trauma Data Report";
PROC PRINT LABEL DATA=MYSASLIB.TRAUMA (FIRSTOBS=1 OBS=20);
ID INC_KEY;
```

(Continued)

(*Continued*)
```
VAR AGE GENDER INJTYPE DISSTATUS;
VAR ISS/STYLE={FOREGROUND=FMTSEVERE.};
LABEL   INC_KEY='Subject ID '
        AGE='Age in 2014 '
        GENDER='Gender '
        ISS='Injury Severity Score
        'INJTYPE='Injury Type '
        DISSTATUS='Discharge Status';
RUN;
```

Note the on-screen color coding in the `ISS` column. Table 8.8 shows the output but does not show the color distinctions.

2. Create a second format for `AGE` where <13 is displayed as `RED` and 13 or older is displayed as black.
3. Add the following format to the SAS code:

```
VALUE FMTISS
      LOW-<0="MISSING"
      0-<10="OKAY"
      10-<20="CONCERN"
      20-HIGH="SEVERE";
```

Apply this format to the `ISS` variable. Submit the code and observe that the `ISS` column is now displayed as different color words for each category of injury.

4. Add the following attributes by changing the `VAR` statement for `ISS` to

```
VAR ISS/STYLE={FOREGROUND=FMTSEVERE.
               BACKGROUND=LIGHTYELLOW
               FONT_WEIGHT=BOLD
               FONT_FACE=SCRIPT
               FLYOVER='Severity Rating'};
```

Run this new code and observe how the additional attributes change the output. Put your cursor over one of the cells and see that a pop-up appears that says "Severity Rating." This was set up with the `FLYOVER` attribute. Other `ODS` style attributes, colors, and fonts are listed in Appendix A.

8.6 CREATING A POWERPOINT SLIDE USING GMAPS

This final example of the use of `ODS` output uses `PROC GMAP` to create a US map by population, and places the results in a PowerPoint slide. We've selected GMAPS to illustrate that almost any PROC can create output for `ODS POWERPOINT`. The code to open the `ODS POWERPOINT` destination is

```
ODS POWERPOINT FILE="filename/path";
```

CREATING A POWERPOINT SLIDE USING GMAPS

Run a SAS program that creates output. Then close the ODS output. In this case, we'll use code to close all (_ALL_) ODS destinations (except ODS HTML.)

```
ODS _ALL_ CLOSE;
```

This is illustrated in Hands-On Example 8.6.

HANDS-ON EXAMPLE 8.6

1. Open the program file ODS10.SAS. This example uses a file in your MYSASLIB library named POP2019. It contains population data from the US Census. The STATE variable is the official state code for each state, and is needed for GMAPS to work correctly. Here is the code

   ```
   TITLE1 "PowerPoint Slide Example";
   FOOTNOTE "US Map By Population";
   ODS POWERPOINT FILE="C:\SASDATA\POP.PPTX";
   PROC GMAP DATA = MYSASLIB.POP2019 MAP=MAPS.US;
        ID STATE;
        CHORO POP_2019;
   RUN;
   QUIT;
   ODS _ALL_ CLOSE;
   ```

2. Note that PROC GMAP uses the POP2019 data and assigns it to a SAS map called MAPS.US. The map uses the ID named STATE to identify what data goes with each state. The STATE variable is the official census state code. The CHORO statement creates a choropleth map, which displays results as levels of magnitude to the corresponding CHORO variable. SAS fills in the map areas with different colors and patterns. More details may be specified in SAS GMAPS documentation, but that technique is beyond the scope of this book. Refer to SAS documentation. Run the program and observe the results.

3. Note that the map appears both in the Results Viewer in SAS, and (if you have it installed on your computer) the map is displayed in PowerPoint. Verify that the file POP.PPTX is also saved in your C:\SASDATA folder. See Figure 8.5

4. Add the following code between the run and quit statements.

   ```
   PROC MEANS DATA=MYSASLIB.POP2019;
   RUN;
   ```

5. Run the revised code and note that this creates a PowerPoint presentation with 2 slides. You can create many slides by designing a program that outputs from a number of SAS procedures.

(Continued)

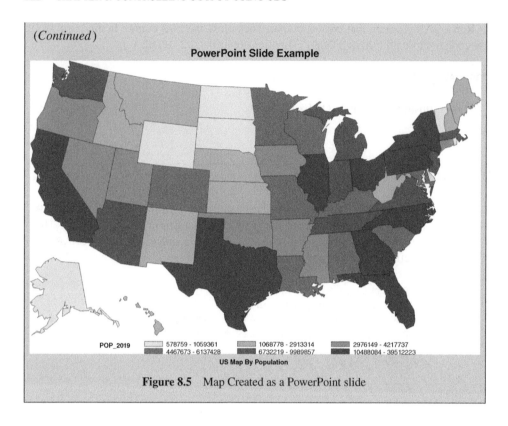

Figure 8.5 Map Created as a PowerPoint slide

8.7 EXTENDED ODS FEATURES

This chapter has introduced a number of features available in SAS ODS. However, there are many features not covered here. For example, every ODS table output by a SAS procedure uses a template to control the way it is created. Using PROC TEMPLATE, you can modify these templates to customize the content and look of each output table.

You can store the components of any ODS report in order to modify and replay them using PROC DOCUMENT. These features are beyond the scope of this book, and we refer you to the SAS documentation if you want to learn more about these topics.

8.8 SUMMARY

SAS ODS provides several ways to manipulate output from SAS procedures. Many examples in the book have used ODS to output results from procedures into HTML and other formats. This chapter introduced other ways to use ODS, including additional options for outputting results from SAS procedures, outputting SAS datasets, and using ODS to enhance graphic output.

EXERCISES

8.1 **Using ODS PDF.**
Using the FIRST.SAS program file, include output to the ODS PDF format using the STATISTICAL style.

8.2 Using ODS TRACE.

Open the program file EX_8.2.SAS. Note the TRACE ON and TRACE OFF commands. Run this program to discover the names of the output tables (in the **Log Window**). They are as follows:

a. Statistics

b. Conflimits

c. _____

d. _____

e. _____

f. _____

Use this information in an ODS SELECT statement to limit the output of the procedure to the tables for statistics and *t*-tests only.

8.3 Use ODS SELECT.

Open the program file EX_8.3.SAS

```
DATA TABLE;
INPUT EXPOSURE DISEASE COUNT;
DATALINES;
0 0 24
0 1 8
1 0 9
1 1 19
;
PROC FREQ DATA=TABLE; WEIGHT COUNT;
TABLES EXPOSURE*DISEASE/CHISQ;
RUN;
```

This file contains the code for analyzing the 2×2 information in this table.

		DISEASED	
		YES	NO
EXPOSED	YES	24	8
	NO	9	19

The output from this program consists of three tables: a cross-tabulation, chi-square results, and Fisher's Exact test results. Using TRACE and SELECT, modify this code so that it will display only the cross-tabulation and the Fisher's Exact test (and not display the chi-square table).

8.4 Use ODS to extract table statistics.

Using information from Exercise 8.3, write a program that will save the contents of the Fisher's Exact test to a SAS data file. Use the command

```
ODS OUTPUT nameoftable=STATSOUT;
```

where `nameoftable` is the name of the code for Fisher's Exact test you discovered using the TRACE commands in Exercise 8.3.

a. Use **Viewtable** to examine the output SAS dataset named STATSOUT (or use a PROC PRINT).

b. Note the name of the variable that contains the value of XP2_FISH. The value is _____

c. Note the name of the variable that contains the *p*-value for XP2_FISH. The value is _____

8.5 **Use ODS to select a specific value in a table.**
In order to minimize the amount of output, suppose you want to see only the *p*-value for the two-sided Fisher's Exact test. You can do this by capturing specific values from the STATSOUT dataset.

a. Use the following code to print out the *p*-value from the Fisher's Exact test. (This code uses several features that were discussed in previous chapters.)

```
PROC PRINT LABEL DATA=STATSOUT;
LABEL CVALUE1='P-VALUE' NAME1='TEST';
WHERE NAME1="XP2_FISH";
VAR NAME1 CVALUE1;
ID TABLE;
RUN;
```

b. Why is the LABEL option included in the PROC PRINT statement? (See Section 4.1.)

c. What does the WHERE statement do?

d. What does the ID statement do?

Hint: To figure out what each statement does, comment out the statement and rerun the program to see what happens (or doesn't happen).

8.6 **SAS LISTING format**
Occasionally, you need to create output in the older SAS LISTING format. This was the output for SAS program used until the HTML type output became standard.

a. Open the file SECOND.SAS and run the program. This produces a simple table of statistics on the variable AGE.

b. Before the PROC MEANS statement, add

```
ODS LISTING;
```

and after then RUN statement add

```
ODS LISTING CLOSE;
```

c. Rerun the program and observe the output. Probably, the output looks the same because SAS defaults to the **Results Viewer**. Older style output appears in the **Output** tab. Click on that tab to view the old-style listing.

9

INTRODUCTION TO PROC SQL

LEARNING OBJECTIVES

- What is SQL, and How Does it Fit in SAS?
  - How is SQL useful in SAS?
  - SQL Nomenclature
  - The Structure of a SQL Statement
- To be able to use the SELECT Statement for variable selection in a PROC SQL Statement
- To be able to use the FROM Statement
- To be able to create SAS Datasets
- To be able to use a WHERE Clause
- To be able to use a GROUP BY Clause
- To be able to use a HAVING Clause
- To be able to use an ORDER BY Clause
- GOING DEEPER: Doing Calculations using Calculated Variables
- GOING DEEPER: Unions and Joins

SAS® Essentials: Mastering SAS for Data Analytics, Third Edition. Alan C. Elliott and Wayne A. Woodward.
© 2023 John Wiley & Sons, Inc. Published 2023 by John Wiley & Sons, Inc.
Companion website: https://www.alanelliott.com/SASED3/

9.1 WHAT IS SQL, AND HOW DOES IT FIT IN SAS?

SQL (pronounced "see-qwell") stands for Structured Query Language. It is designed to allow you to communicate with databases. It's also the standard language for relational database management systems. SAS PROC SQL statements are used to perform tasks like updating a database/dataset and/or retrieving data from a database/dataset.

So, why would a SAS programmer need to know SQL? Doesn't the SAS DATA step manage data? It's true that for many users and many applications, there is no need to know SQL to use SAS effectively. However, there are also compelling reasons to learn SQL. For many programmers, SQL is simpler and more straightforward than the DATA step, plus some programmers already know SQL because they used it in other database applications. Unlike the DATA step, SQL does not require data to be sorted for many functions such as merging datasets. Using SQL in this case saves time and programming. SQL has simple methods for combing data, finding averages across records, etc., that are easier to program than in the DATA step. SQL also has easier ways to name, rename, and manage variables. Of course, there are some things that are also easier to accomplish in the DATA step. . . . Because SQL is a PROC, you can use it in conjunction with the DATA step – using it a little or as your complete data manipulation tool. So, a good SAS programmer should have both data manipulation tools in their programming toolbox.

SQL was initially conceived in 1970 and developed more fully at IBM by Donald D. Chamberlin and Raymond F. Boyce. Although there are vendor differences, it was standardized in 1986 by American National Standards Institute. So the SAS version of SQL, while very similar to other versions, is slightly different. In June 1979, Oracle introduced the first commercial implementation of SQL. IBM also introduced its version in 1979. Other versions followed, including Microsoft SQL, MySQL, PostgresSQL, and SAP. SAS introduced PROC SQL in Version 6.06 (1990). Changes were made over time, and the current PROC SQL was introduced with SAS Version 8 (1999).

Because SQL is a language on its own, it has a different syntax than other SAS statements. However, knowing PROC SQL will provide you with additional data handling methods you can use within SAS and provides additional techniques for accessing corporate or organizational databases. It may even give you additional skills beneficial in some future job search.

SQL is important because it is implemented in so many different software programs. This means that there is portability between software programs. Thus, once you learn SAS PROC SQL, you'll know 90% or more of what you'd need to know to use SQL on some other platform. SAS also provides powerful benefits for using a standardized relational language. Once you know SQL, you'll understand ways that it can enhance your productivity.

9.1.1 How Is SQL Useful in SAS?

Just like the DATA step, PROC SQL manipulates SAS tables. It's important to realize that the tables accessed and created by SQL are the same as those used by the DATA step. In this chapter, we'll concentrate on the following uses of PROC SQL:

- Selecting or subsetting data from a larger dataset using robust selection criteria (similar to but different than SAS statements: SET, IF, WHERE, etc.)
- Ordering and sorting data, again using different and sometimes more powerful techniques not found in PROC SORT.

- SQL provides powerful ways of restructuring a dataset not easily performed in standard SAS routines.
- SQL also provides ways of combining dataset that are not easily accomplished in the DATA step.

In general, using PROC SQL provides you with techniques for retrieving data from databases in ways that may be faster than those in other SAS routines. For example, many SQL features do not require the presorting often required in SAS procedures. SQL is also less particular about data types, so it makes it easy to use common variable names for merging datasets.

It's important to also note that SQL is not used in creating datasets from raw numbers, which is a major function of the DATA step. We'll create new datasets by manipulating current data, but in this chapter, we'll always begin with SAS datasets that currently exist.

Here are some of the terms and concepts we'll use when discussing SQL that are somewhat different from the terms used in typical SAS usage:

- PROC SQL will primarily be used to *create, modify, and combine tables* (datasets) and is *not used to analyze data*.
- The term "**Quer**" is used to describe the process of requesting data from a dataset or database. Query is a term not usually used for the DATA step, but is part of the common language for SQL.
- The term "**View**" is also a common SQL term, which is used to describe the default results of SQL queries.

9.1.2 SQL Nomenclature

A simple PROC SQL program needs only 3 statements (assuming you already have a library defined.) (See Hands-On Example 9.1)

```
PROC SQL;
      SELECT * from MYSASLIB.SOMEDATA;
QUIT;
```

This program selects all variables (indicated by the *) from the SOMEDATA dataset that's in the MYSASLIB library. It's important to note that the result of a SQL query (the **View**) in these early examples does not create any new dataset. How to save the **View** as a new dataset will be discussed later in the chapter.

HANDS-ON EXAMPLE 9.1

This example illustrates the use of PROC SQL for retrieving data from a current dataset and putting it into a SQL **View**.

1. Open the program file SQL1.SAS.
2. Run the program. See the (partial) results in Table 9.1. Note that there is no PROC PRINT statement, but the results of this program (its **View**) are displayed in the SAS

(Continued)

(Continued)

TABLE 9.1 A SQL View Resulting from the PROC SAS Program (Results not Ordered by Age)

My First SQL Program								
ID Number	Intervention Group	Age on Jan 1, 2000	Baseline	6 Months	12 Months	24 Months	Socioeconomic Status	SEX	GENDER
101	A	12	22.3	25.3	28.2	30.6	5	0	Female
102	A	11	22.8	27.5	33.3	35.8	5	0	Female
110	A	12	18.5	26	29	27.9	5	1	Male
187	A	11	22.5	29.3	32.6	33.7	2	0	Female
312	A	4	17	21.3	22.7	21.2	5	1	Male
390	A	6	21.3	30.5	30.7	30.1	4	0	Female
445	A	12	22.1	31.3	31.6	31.6	3	1	Male
543	A	10	20.8	28.6	34.2	30	5	1	Male
544	A	10	18.1	23.8	29	27.8	5	1	Male
550	A	13	23	26.4	29.9	31.8	3	0	Female

Results Viewer. Unless you tell SQL otherwise, it will always show the results of a SELECT statement (the results of the query) in this way.

3. Change the SELECT statement to

   ```
   SELECT * from MYSASLIB.SOMEDATA;
   ORDER BY AGE;
   ```

   (We'll learn more about the ORDER clause later.)

4. Rerun the program and note that the **View** is now ordered by AGE.

9.1.3 The Structure of a SQL Statement

Notice in the SQL1.SAS example, that the main statement between PROC SQL and QUIT is a SELECT statement.

```
PROC SQL;
      SELECT * FROM MYSASLIB.SOMEDATA;
QUIT;
```

The SELECT statement is the primary and most important statement in PROC SQL. Much of this chapter is dedicated to describing how to use the SELECT statement. Here are the typical components of a SQL program:

1. Begin with a PROC SQL statement.
2. The SELECT statement tells SAS what variables to display and where to get the information.
3. The asterisk (*) means to display all variables (we'll learn more options in the following sections).

4. The FROM clause indicates the dataset name that contains the data to use for this SELECT.
5. End the PROC SQL with a QUIT statement.

HANDS-ON EXAMPLE 9.2

This example illustrates retrieving data from a current dataset into a SQL **View**.

1. Open the program file SQL2.SAS. This example includes some additional clauses in the SELECT statement (WHERE and ORDER BY) that will be covered in detail later in the chapter.

   ```
   PROC SQL;
   SELECT ID, GP, AGE
               from MYSASLIB.SOMEDATA
   WHERE GENDER CONTAINS "Female"
   ORDER BY ID;
   QUIT;
   ```

2. Run the program to see the results.
   a. There are three columns in the **View**: ID, GP, and AGE. Take a moment, and see if you know why the SELECT statement produced a table with three columns.
   b. The data are in the dataset (table) MYSASLIB.SOMEDATA.
   c. The **View** only shows female subjects but that variable is not shown in the **View** because GENDER is not one of the variables listed in the SELECT statement (before the FROM statement).
   d. The data are ordered by ID.
3. Change the list of variables to

   ```
   SELECT ID, GP, AGE, GENDER
   ```

   Rerun the program and verify that now GENDER is displayed for each subject and that only FEMALE subjects are in the **View**.

The order of the clauses in the SELECT statement is important. If you create a SQL statement with these items in the wrong order, it will not work. This chapter will cover each of these components of the PROC SQL statement with examples. It would be helpful to memorize (or at least become very familiar with this order.) Once you know the order, and how to use each of these clauses, you'll be able to utilize the power of SAS PROC SQL.

```
PROC SQL;
   SELECT column(s)
   FROM table-name | view-name
   WHERE expression
   GROUP BY column(s)
   HAVING expression
   ORDER BY column(s);
QUIT;
```

9.2 USING THE SELECT STATEMENT: VARIABLE SELECTION IN A PROC SQL STATEMENT

One of the main purposes of a query is to "select" certain variables from a database. The SELECT statement is the primary tool for getting data (querying) from an existing dataset. A simple SELECT statement in PROC SQL might read

SELECT * from MYSASLIB.SOMEDATA;

The asterisk (*) in this statement tells SQL to retrieve all of the variables in the SOMEDATA dataset. However, there are other ways to tell SQL which variables to select. Another simple way is to list the variables of interest. For example

SELECT ID, GENDER, AGE from MYSASLIB.SOMEDATA;

would select only the three listed variables (ID, GENDER, and AGE) from the dataset MYSASLIB.SOMEDATA. Note that variables are separated by commas. The resulting **View** from running the query would show a table with three columns in the order specified in the SELECT statement. As in any SAS statement, you can also indicate a range of variables:

SELECT ID--TIME4 from MYSASLIB.SOMEDATA;

using a double dash to select all variables from ID to TIME4 that are stored in the dataset SOMEDATA or

SELECT TIME1-TIME4

to indicate variables that have consecutive numbers appended. You can also combine individual variables and lists of variables. For example:

SELECT ID, TIME1-TIME4, GENDER

You may also signify a variable name (column name) using "AS alias" to take a variable in the FROM source (FAMILY) dataset) and rename the column as specified in the AS clause. For example:

SELECT FOODEXP, INCOME, AGE AS AGE2020 FROM MYSASLIB.FAMILY;

Note that AGE will be renamed AGE2020, resulting in a **View** that is a table with a column named AGE2020 that contains the AGE variable data.

To select variables based on some logical selection criteria, use a CASE clause. For example:

```
SELECT ID, CASE GP
    WHEN "A" then "TRT 1"
    WHEN "B" then "TRT 2"
        ELSE "CONTROL"
    End AS TREATMENT,
AGE from MYSASLIB.SOMEDATA;
```

In this situation, the ID variable will be the first column in the **View** table. The second column will be called TREATMENT (specified by the AS TREATMENT clause). The CASE GP clause says, if GP is "A" then assign TREATMENT = "TRT 1" (a character type variable.) If GP is "B" then assign TREATMENT = "TRT 2". Thus, in the column named TREATMENT, the values will be TRT 1 or TRT 2. Note the comma after TREATMENT. This signals that another variable/column is specified. The AGE variable will be a third column in the **View**. Thus, **View** contains three columns: ID, TREATMENT, and AGE.

You may also label and format variables as they appear in the table using a FORMAT= or LABEL = option after a column (variable) name.

```
COLUMNNAME FORMAT=somesasformat.
COLUMNNAME LABEL = "some label"
```

For example:

```
PROC SQL;
SELECT FOODEXP, INCOME FORMAT=DOLLAR7.2,
       AGE LABEL" Age in 2020",
           OWNRENT
FROM MYSASLIB.FAMILY;
QUIT;
```

In this program, the INCOME column is displayed using the DOLLAR7.2 (notice the specification is not in quotes), and AGE2020 will have a column name "Age in 2020" (notice that this specification does use quotes.)

HANDS-ON EXAMPLE 9.3

1. Open the program file SQL3.SAS. Examine the code.

   ```
   PROC SQL;
   SELECT ID,GP,AGE,GENDER,
   CASE GP
      WHEN "A" THEN "TRT 1"
           WHEN "B" THEN "TRT 2"
           ELSE "CONTROL"
   END AS TREATMENT,
   STATUS
       FROM MYSASLIB.SOMEDATA;
   QUIT;
   ```

2. Notice the CASE clause and determine what will be the contents of the column it creates. How many total columns will be created by this program?
4. Run the program to see the resulting SQL **View**.
5. Change END AS TREATMENT, to

   ```
   END AS TREATMENT LABEL='RX TREATMENT',
   ```

   and add a new variable (TIME1) to the end of the variable list with a format specified as FORMAT=5.2.
6. Rerun the program and observe the results.

232 CHAPTER 9: INTRODUCTION TO PROC SQL

Functions within SAS can also create custom values from data in the dataset. For example, use the SUBSTR () function to display the brand names of cars in a table using only the first four letters of BRAND. (See Appendix B for an explanation of how SUBSTR() works.) The code to do this is:

```
PROC SQL;
SELECT SUBSTR(BRAND,1,4), MODEL, CITYMPG
FROM MYSASLIB.CARS;
QUIT;
```

Or suppose you want to display FOODEXP to the nearest dollar (whole number) from the FAMILY dataset. In this case, using the ROUND() function.

```
PROC SQL;
SELECT ROUND(FOODEXP), INCOME, FAMILYSIZE
FROM MYSASLIB.FAMILY;
QUIT;
```

HANDS-ON EXAMPLE 9.4

1. Open the program file SQL4.SAS. Examine the code. This program creates a dataset using the standard DATA statement in SAS, and then displays it as a report using PROC SQL. This illustrates how mixing the two data manipulation features of SAS is easily accomplished. Run the program.

```
DATA ROOMSIZE;
INPUT ROOM $ WIDTH LENGTH @@;
DATALINES;
LIVING    14 22    DINING   14 12    BREAKFAST 10 12
KITCHEN   12 16    BEDROOM1 18 12    BEDROOM2  12 14
BEDROOM3  13 16    BATH1     8 12    BATH2      7 10
BATH3      6  8    GARAGE   23 24
;
RUN;

PROC SQL;
SELECT *,
ROUND(WIDTH*LENGTH, 0.01) FORMAT=6.2 AS AREA
FROM ROOMSIZE;
QUIT;
```

2. Note that in the SELECT statement, there is an asterisk (*), which tells SQL to select all variables, then a comma (,) and then a calculated variable that is rounded ROUND(), formatted (FORMAT=), and named AREA.
7. Put a comma after AREA to signal that you're adding another variable to the list. Add this code (which assumes the home has 8-foot ceilings)

```
ROUND(8*WIDTH*LENGTH, 0.01) FORMAT=6.2 AS VOLUME
```

8. Run the program and observe that you now have a new column named VOLUME containing the results of the calculation. The results are shown in Table 9.2.

TABLE 9.2 Results of Calculating Values in SQL

ROOM	WIDTH	LENGTH	AREA	VOLUME
LIVING	14	22	308.00	2464.0
DINING	14	12	168.00	1344.0
BREAKFAST	10	12	120.00	960.00
KITCHEN	12	16	192.00	1536.0
BEDROOM1	18	12	216.00	1728.0
BEDROOM2	12	14	168.00	1344.0
BEDROOM3	13	16	208.00	1664.0
BATH1	8	12	96.00	768.00
BATH2	7	10	70.00	560.00
BATH3	6	8	48.00	384.00
GARAGE	23	24	552.00	4416.0

9.3 USING THE FROM STATEMENT IN PROC SQL

The FROM clause in PROC SQL tells SAS the location of the original data source. The reference will almost always refer to a SAS dataset using a specification such as MYSASLIB.FAMILY (using a libname) or "C:\SASDATA\SOMEDATA" (using the Windows filename/path. For example,

```
PROC SQL;
    SELECT * from MYSASLIB.SOMEDATA;
QUIT;
```

displays data from the SOMEDATA dataset and

```
PROC SQL;
    SELECT * from "C:/SASDATA/FAMILY";
QUIT;
```

displays data from the FAMILY dataset. If your dataset (i.e., MYDATA) was in the WORK library, you could simply use

```
PROC SQL;
    SELECT * from MYDATA;
QUIT;
```

There may be instances in a corporate or organizational setting in which a dataset, table, or database is located on a server or network that has a different filepath than these examples. You'll have to determine if this is the case in your own location, and now that might change how you'd specify the dataset name in the FROM clause.

9.4 CREATING SAS DATASETS

Before examining the other clauses in the SELECT statement, we'll take a look at a how to create SAS datasets in PROC SQL. So far, all of the output we've seen has been in SQL **Views**. These **Views** are displayed only on the screen. These **Views** let you know how your query worked, but the table/**View** displayed is not stored anywhere on disk, and when you exit SAS, those **Views** vanish. However, there is a way to save the resulting table in the **View** to a SAS dataset. This is done with the CREATE TABLE statement.

To save the result of a query to a SAS dataset, use a CREATE TABLE *sasdatasetname* AS statement before the SELECT statement. For example

```
PROC SQL;
     CREATE TABLE MYQUERY AS
     SELECT * FROM MYSASLIB.SOMEDATA;
QUIT;
```

When you use a CREATE TABLE statement, no **View** is displayed. However, if you examine the **Log Window**, you'll find the message

```
NOTE: Table WORK.MYQUERY created, with 50 rows and 10 columns.
```

To display the contents of the resulting dataset named MYQUERY, add the statement

```
PROC SQL; SELECT * FROM MYQUERY;QUIT;
```

At this point, the program displays the results of the query. Of course, you can also use the standard PROC PRINT SAS code:

```
PROC PRINT DATA=MYQUERY; RUN;
```

to display the contents of the MYQUERY dataset.

HANDS-ON EXAMPLE 9.5

1. Open SQL1.SAS and change the code to create a SAS dataset named NEW1 by adding the following statement:

   ```
   CREATE TABLE NEW1 AS
   ```

   before the SELECT statement. Also, add this code after the QUIT statement.

   ```
   PROC SQL; SELECT * FROM NEW1; QUIT;
   ```

   Run the new program and observe the **View**. Did it work?

2. In this example, a temporary dataset is created (named NEW1 or WORK.NEW1). Change NEW1 in the CREATE statement to MYSASLIB.NEW2. Also, change the name to MYSASLIB.NEW2 in the second PROC SQL statement that displays the results. Run the program and observe the results. Also, verify that the NEW2 table is in the MYSASLIB library by clicking on the **Explorer** tab.

You could easily add a CREATE TABLE AS statement in any of the following exercises to save the results to a permanent dataset.

9.5 USING A WHERE CLAUSE

A WHERE clause allows you to select a portion of a dataset based on a logical specification. The WHERE clause follows the FROM clause. Note that there is no semicolon separating the FROM clause and the WHERE clause – they are all parts of the SELECT statement. For example

```
PROC SQL;
      SELECT * FROM MYSASLIB.SOMEDATA
      WHERE AGE LE 12;
QUIT;
```

This code produces a table from the SOMEDATA dataset containing only subjects less than or equal to 12 years of age.

Also consider

```
PROC SQL;
      SELECT * from MYSASLIB.COMPLICATIONS
      WHERE COMPLICATION CONTAINS "Pneumonia";
QUIT;
```

This code produces a table from the COMPLICATIONS dataset that only includes records containing the word Pneumonia (case matters) in the variable COMPLICATION. The resulting **View** shows subjects whose COMPLICATION is Pneumonia or Aspiration Pneumonia because the word Pneumonia appears in each. Care must be taken since uppercase and lowercase are important. A better statement for this selection might be

```
WHERE UPCASE(COMPLICATION) CONTAINS "PNEUMONIA";
```

Note also that this statement produces different results from

```
WHERE UPCASE(COMPLICATION) = "PNEUMONIA";
```

In the first instance, the resulting **View** contains Pneumonia and Aspiration Pneumonia records. In contrast, only records with Pneumonia as the COMPLICATION are displayed in the second case. Thus, be careful regarding the logical SAS expression in the WHERE clause within the SELECT statement.

HANDS-ON EXAMPLE 9.6

1. Open the program file SQL2.SAS and examine the code. Note that this code creates a dataset called FEMALES in the MYSASLIB library.

   ```
   PROC SQL;
   SELECT ID, GP, AGE
               FROM MYSASLIB.SOMEDATA
   ```

   (Continued)

> (*Continued*)
> ```
> WHERE GENDER CONTAINS "Female"
> ORDER BY ID;
> QUIT;
> ```
>
> 2. Count the number of columns that will be in the **View**. How many are there? Which columns are they? Observe the output and note that GENDER/FEMALE is nowhere in **View**. Add the necessary code to include GENDER as a column.
> Create a similar table containing only MALE subjects.

9.6 USING A GROUP BY CLAUSE IN PROC SQL

The GROUP clause in the SELECT statement allows you to display output by some grouping variable. Here is an example code using the GROUP clause (SQL5.SAS):

```
PROC SQL;
SELECT GP, COUNT(GP) AS N,
 TIME1, TIME2,
 MAX(TIME1) FORMAT=6.2 AS TIME1MAX
 FROM MYSASLIB.SOMEDATA
 GROUP BY GP;
QUIT;
```

When you run this code, you'll see the output, part of which is shown in Table 9.3. The **View** is displayed alphabetically by GP (A, B, or C). Note that there are also a couple of variables whose values are calculated.

TABLE 9.3 Grouped Output from PROC SQ

Intervention Group	N	Baseline	6 Months	TIME1MAX
A	11	22.3	25.3	23.00
A	11	22.8	27.5	23.00
A	11	18.5	26	23.00
A	11	22.5	29.3	23.00
A	11	17	21.3	23.00
A	11	21.3	30.5	23.00
A	11	22.1	31.3	23.00
A	11	20.8	28.6	23.00
A	11	18.1	23.8	23.00
A	11	23	26.4	23.00
A	11	21.5	27	23.00
B	29	22.8	30	24.20
B	29	19.5	25	24.20

The code runs as follows:
  a. First, the variable N is created using the COUNT() function, (COUNT(GP) AS N), which gives you the count within the group. Because there are 11 subjects in Group A, the value 11 is repeated in the N column for every subject in Group A.
  b. The variable TIME1MAX is also calculated by GP. It is the maximum value for TIME1 in each group. For example, in Group A, the maximum value for TIME1 is 23. So, 23 is repeated for each subject in Group A.
  c. Finally, the entire **View** is displayed by group, with the A group displayed first, then B, and then C.

HANDS-ON EXAMPLE 9.7

1. Open the program file SQL6.SAS. Beginning with this snippet complete the code by filling in the blank lines to display a column for the minimum CITYMPG BY BRAND. Also save the results in a table named CARSMPG (in the WORK library.)

   ```
   PROC SQL;
      CREATE TABLE _____ AS
      SELECT BRAND, CITYMPG, HWYMPG,
      MAX(CITYMPG) FORMAT=6.2 AS CITYMAX,
      MIN(_____) FORMAT=6.2 AS CITYMIN
      FROM MYSASLIB.CARS
      GROUP BY _____;
   QUIT;
   PROC SQL; SELECT * FROM CARSMPG;QUIT;
   ```

2. How many columns does the **View** have? Why?
3. Add similar MAX and MIN statements to calculate HWYMAX and HWYMIN columns. Run the program to verify that your code works.

9.7 USING A HAVING CLAUSE

The HAVING clause in SQL is typically used in conjunction with a GROUP BY clause. The HAVING clause allows you to specify conditions for selecting which group results appear in your results. This is different from the WHERE clause which places conditions on the selected columns. The HAVING clause, on the other hand, places conditions on groups created by the GROUP BY clause.

For example: The file SQL8.SAS contains the following code:

```
PROC SQL;
SELECT * from MYSASLIB.ACCIDENTS
GROUP BY GENDER
HAVING ISS GT 5;
QUIT;
```

238 CHAPTER 9: INTRODUCTION TO PROC SQL

Run this code and you will see that it produces a **View** that is ordered by GENDER, and only contains values of ISS (Injury Severity Score) greater than 5.

> You may think you could use a WHERE clause instead of a HAVING clause because they behave similarly. But remember, HAVING places conditions on a *group* created by GROUP BY, whereas WHERE places conditions on *columns*.

HANDS-ON EXAMPLE 9.8

1. Open the file SQL8.SAS and run the code. Note that the **View** contains a listing ordered by GENDER. In the ISS column, only subjects with values greater than 5 are displayed. If you examine the original dataset, you'll see many subjects with values of 5 or less.
2. Experiment by changing HAVING to WHERE and rerun the code. Did it work? Why not?
3. To illustrate that the HAVING statement can contain a more complex SAS expression, change it to

   ```
   HAVING ISS GT 5 AND AGE GE 10;
   ```

   and rerun the program. How did this change the output?

Technically, you could have a HAVING statement without a GROUP BY statement. In that case, it treats the output as if it only had one group.

9.8 USING AN ORDER BY CLAUSE

The ORDER BY clause sorts a dataset in ASCENDING or DESCENDING order. The default order, if you don't specify, is ASCENDING. For example, the following code (SQL8.SAS) creates a **View** of a list of phone numbers ordered (in ASCENDING order) by FIRST name within LAST name.

```
PROC SQL;
SELECT LAST, FIRST, PHONE
     from MYSASLIB.PHONE
     ORDER by LAST, FIRST;
QUIT;
```

HANDS-ON EXAMPLE 9.9

1. Open SQL8.SAS. Note the ORDER BY statement in the program. Run the program. The resulting **View** displays the phone directory by LAST name, then within same LAST names (Smith) FIRST names are sorted.
2. Insert DESC after LAST in the ORDER BY clause (and before the comma.) Run the program again. How does this change the **View**?

A **Quick Tips SAS PROC SQL Reference** summary is shown in Figures 9.1 and 9.2. These tips sheets are also available on the www.alanelliott.com/sas website. You may find it useful to obtain a PDF copy of the sheet by clicking on the **Quick Tips** button.

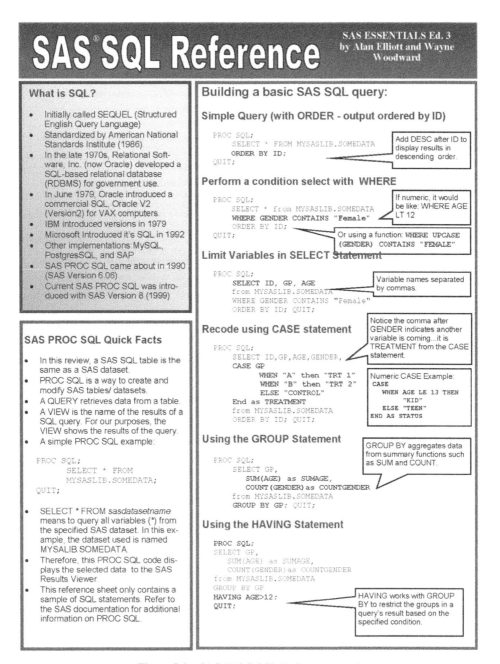

Figure 9.1   SAS PROC SQL Reference, part 1

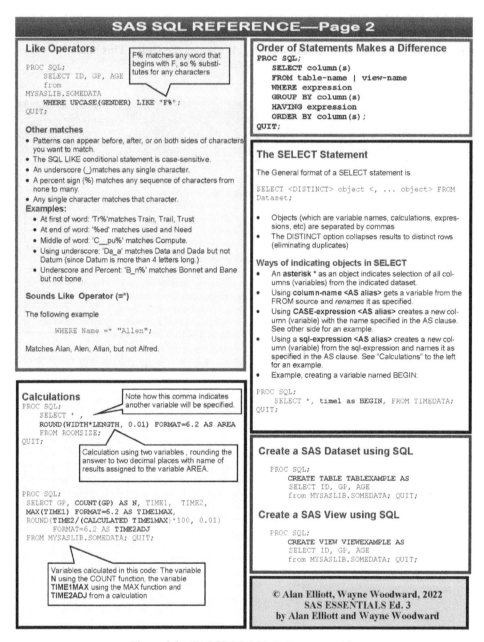

Figure 9.2 SAS PROC SQL Reference, part 2

9.9 GOING DEEPER: DOING CALCULATIONS USING CALCULATED VARIABLES

In a previous example, we showed how to create a variable by calculation. However, if you create a new variable from a calculated variable, you must indicate that the variable you're referring to was just calculated within the current SELECT statement. Hands-On

Example 9.10 using `SQL10.SAS` illustrates this. Note that in `SQL10.SAS`, the variable `TIME1MAX` is calculated by the function `MAX(TIME1)`. Then, in a subsequent statement, `TIME1MAX` is used in a calculation for the new variable named `TIME2ADJ`.

Because `TIME1MAX` was not in the original dataset (it was calculated), you must inform SQL that `TIME1MAX` was a calculated variable using a `CALCULATED` keyword before the name of the calculated variable you're using. Notice the expression `CALCULATED TIME1MAX` in the following code (`SQL10.SAS`).

```
PROC SQL;
 SELECT GP, COUNT(GP) AS N,
 TIME1,   TIME2,
 MAX(TIME1) FORMAT=6.2 AS TIME1MAX,
 ROUND(TIME2/(CALCULATED TIME1MAX)*100, 0.01) FORMAT=6.2 AS
TIME2ADJ
 FROM MYSASLIB.SOMEDATA
 GROUP BY GP;
QUIT;
```

HANDS-ON EXAMPLE 9.10

1. Open `SQL9.SAS` and notice the keyword `CALCULATED` before `TIME1MAX`. Run the program. Notice the two calculated columns, `TIME1MAX` and `TIME2ADJ` included in the **View**.
2. Take out `CALCULATED` before `TIME1MAX`. Go to **Explorer**, to the **Work** directory, right-click on the dataset named `TIMEADJ` and choose **Delete**. Now run the new program (without `CALCULATED`). No **View** is created. Why?
3. Go to the **Log** and note the message

   ```
   ERROR: The following columns were not found in the contributing
       tables: TIME1MAX.
   ```

   This error occurs because SQL couldn't calculate the value `TIME2ADJ` using `TIME1MAX` because it didn't know it had been calculated within the `SELECT` paragraph.

9.10 GOING DEEPER: UNIONS AND JOINS

There are many features of the SQL language that are beyond the scope of this introductory chapter. However, one final topic that will be briefly discussed is the use of SQL to append and merge files (`UNION` and `JOIN` in SQL terminology.) SQL unions and joins often provide a quicker (and easier to code) way of combining datasets than using the SAS DATA step.

9.10.1 Using SQL UNIONS

The following program (`SQL10.SAS`) appends two datasets (named `OLD1` and `OLD2`). That is, it adds new records to `OLD1` from the dataset `OLD2`. The statement that appends these two sets is called a `UNION`. Note that there are two `SELECT` statements with a `UNION` between them in the given `PROC SQL` code.

```
DATA OLD1;
INPUT SUBJ $ AGE YRS_SMOKE;
DATALINES;
001 34 12
003 44 14
004 55 35
006 21 3
;
DATA OLD2;
INPUT SUBJ $ AGE YRS_SMOKE MARRIED;
DATALINES;
006 21 3 .
011 33 11 1
012 25 19 0
023 65 45 1
032 71 55 1
;
RUN;
PROC SQL;
      SELECT * FROM OLD1
      UNION
      SELECT * FROM OLD2;
QUIT;
```

> **HANDS-ON EXAMPLE 9.11**
>
> 1. Open SQL10.SAS and run the program. The resulting **View** is shown in Table 9.4. The OLD2 dataset has been appended to the OLD1 dataset. Note that because OLD1 doesn't have a MARRIED variable when the datasets are appended, the values for MARRIED in the original OLD1 dataset are set to missing in the **View**. Also, because the record for SUBJ 006 is the same in both OLD1 and OLD2, it is only listed once in the **View**.
> 2. To experiment with other types of SQL unions, change UNION to UNION ALL, and run the program. What is the difference?
> 3. Change UNION ALL to EXCEPT and run the program. How does the output change?
> 4. Change EXCEPT to INTERSECT and run the program. How does the output change?
> 5. Change INTERSECT to OUTER UNION and run the program. How does the output change?
>
> **TABLE 9.4 Results of a SQL Append**
>
> | SUBJ | AGE | YRS_SMOKE | MARRIED |
> |---|---|---|---|
> | 001 | 34 | 12 | . |
> | 003 | 44 | 14 | . |
> | 004 | 55 | 35 | . |
> | 006 | 21 | 3 | . |
> | 011 | 33 | 11 | 1 |
> | 012 | 25 | 19 | 0 |
> | 023 | 65 | 45 | 1 |
> | 032 | 71 | 55 | 1 |

In general, a SQL UNION appends datasets vertically (adding records to the **View**) in a similar manner as dataset concatenation in the SAS DATA step. A brief description of the types of UNION statements is shown in Table 9.5.

TABLE 9.5 Types of SQL UNIONS

UNION	Produces all unique rows from both queries.
EXCEPT	Produces rows that are part of the first query only.
INTERSECT	Produces rows that are common to both query results.
UNION ALL	Produces rows from both queries.
OUTER UNION	Concatenates the query results.

9.10.2 Using SQL JOINS

Another method of combining data from multiple datasets in SQL uses the JOIN command. Joins merge datasets horizontally (typically adding variables to the **View**) similar to a SAS DATA step MERGE. Various types of JOIN statements are described in Table 9.6.

TABLE 9.6 Types of SQL JOINS

INNER JOIN	Selects records where observations in two datasets have matching key values.
RIGHT OUTER JOIN	Similar to an INNER JOIN with rows from the right (or second) table that do not match any row from the other (left) table included in the join.
LEFT OUTER JOIN	Same as RIGHT OUTER JOIN except variables in the left (or first) table that do not match any row from the other (right) table are included in the join.
FULL OUTER JOIN	Returns all rows from the left table and from the right table. That is, it combines the results of the RIGHT and LEFT OUTER JOINS.

For example, an INNER JOIN selects records where observations in two datasets have matching key values. Look at the following code. (SQL11.SAS) Notice the WHERE statement equating SUBJ from the first dataset (OLD1.SUBJ) with SUBJECT in the second dataset (OLD2.SUBJECT.) Placing the dataset name as a prefix to a variable allows you to specify variables from different datasets without ambiguity (i.e., OLD1.SUBJ and OLD2.SUBJECT).

```
DATA OLD1;
INPUT SUBJ $ AGE YRS_SMOKE;
datalines;
001 34 12
003 44 14
004 55 35
006 21 3
;
```

```
DATA OLD2;
INPUT SUBJECT $ SBP MARRIED;
datalines;
003 110 1
001 120 .
004 90 0
006 100 1
011 110 1
;
RUN;
Title "Results of INNER Join (1-to-1 merge)";
PROC SQL;
     SELECT *
     FROM OLD1 INNER JOIN OLD2
     ON OLD1.SUBJ=OLD2.SUBJECT;
QUIT;
```

HANDS-ON EXAMPLE 9.12

1. Open SQL11.SAS and run the program. Notice the output in Table 9.7. The JOIN statement in PROC SQL merges records based on the key variables OLD1.SUBJ and OLD2.SUBJECT, combining them where there was a match. Note that SUBJECT 11 in OLD2 was not a match. So, it does not appear in the results.
2. Change JOIN to RIGHT OUTER JOIN. Run the program. What is the difference in the results? Note that the RIGHT dataset (OLD2) contains the extra SUBJECT11. So, it is included in the output.
3. Change RIGHT OUTER JOIN to LEFT OUTER JOIN and run the program. Why does this yield the same results as JOIN? (In step 1.)
4. Change LEFT OUTER JOIN to FULL OUTER JOIN and run the program. Why are these results the same (in this case) as those obtained using RIGHT JOIN?

A Quick Tips SAS PROC SQL UNIONs and JOINs summary is shown in Figures 9.3 and 9.4. These tips sheets are also available on the www.alanelliott.com/sas website. You can obtain a PDF copy of the sheet by clicking on the **Quick Tips** button.

TABLE 9.7 Output Illustrating the Use of the JOIN Command

SUBJ	AGE	YRS_SMOKE	SUBJECT	SBP	MARRIED
003	44	14	003	110	1
001	34	12	001	120	–
004	55	35	004	90	0
006	21	3	006	100	1

JOINs merge datasets horizontally (typically adding variables to the **View**) similar to a SAS DATA step MERGE.

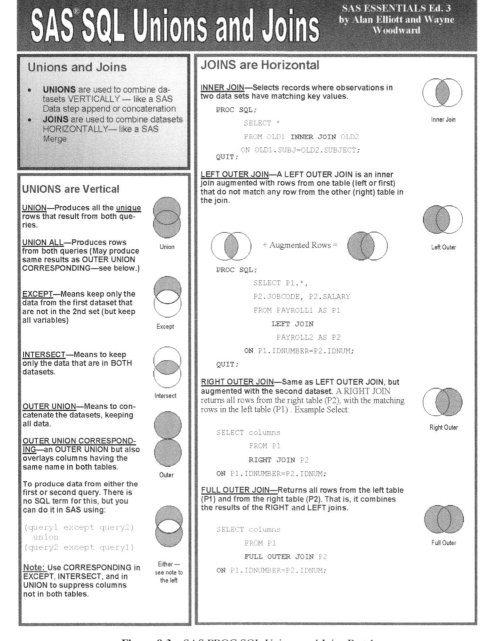

Figure 9.3 SAS PROC SQL Unions and Joins Part 1

SAS SQL JOINS, UNIONS — Page 2

COALESCE STATEMENT

Use the **COALESCE** keyword to overlay like-named columns.

```
COALESCE(P2.SALARY,P1.SALARY) LABEL='CURRENT
SALARY' FORMAT=DOLLAR8.
```

Returns the first nonmissing value of either P2.JOBCODE or P1 JOBCODE. Because P2.JOBCODE is the first argument, if there is a nonmissing value for P2.JOBCODE, COALESCE returns that value.

DISTINCT STATEMENT

Tables sometimes contain duplicate values. If you only want the results to contain different (distinct) values, use the **DISTINCT** statement:

```
SELECT DISTINCT COLUMN_NAME, COLUMN_NAME
FROM TABLE_NAME;
```

ALTER TABLE Statement

ALTER TABLE can ADD columns to or DROP columns from an existing table. For example:

```
PROC SQL ;
    ALTER TABLE PAYROLL1
        ADD NEWVAR1 NUM FORMAT=6.
        ADD NEWVAR2 CHAR(10) FORMAT=$10.
        DROP OLDVAR;
QUIT;
```

DESCRIBE TABLE

Displays a CREATE TABLE statement in SAS log for the specified table. For example, this program

```
PROC SQL ;
    DESCRIBE TABLE PAYROLL2;
QUIT;
```

writes the following information to the SAS LOG:

```
create table WORK.PAYROLL2( bufsize=65536 )
  (
    birth num format=DATE7.,
    hired num format=DATE7.,
    Idnum char(8),
    Gender char(8),
    Jobcode char(8),
    Salary num
  );
```

The UPDATE Statement

The UPDATE Statement modifies the column values of an existing row in a data table. For example, the following code updates the value of the SALARY variable:

```
PROC SQL;
UPDATE PAYROLL2
    SET SALARY=SALARY*
    CASE WHEN JOBCODE EQ "TA3" THEN 1.04
         WHEN JOBCODE EQ "ME3" THEN 1.05
         ELSE 1.06
    END;
QUIT;
```

WHERE BETWEEN Statement

Use the BETWEEN statement in a WHERE clause to select records between two values.

```
PROC SQL; SELECT * FROM PAYROLL1
    WHERE SALARY BETWEEN 30000 AND 40000;
QUIT;
```

WHERE CONTAINS Statement

To select records according to a string within a string, use the CONTAINS Statement:

```
PROC SQL; SELECT * FROM PAYROLL2
    WHERE JOBCODE CONTAINS "3";
QUIT;
```

The CALCULATED Statement

Use the CALCULATED statement in a SELECT clause when you use a variable in an expression that has been previously calculated (created) within the SQL statement. Example:

```
PROC SQL;
SELECT
    CASE
        WHEN AGE LT 10 AND DISSTATUS IN ("DEAD")
THEN "CHILD DEATH"
        WHEN AGE GE 10 AND DISSTATUS IN ("DEAD")
THEN "OTHER DEATH"
        ELSE "ALIVE"
    END AS TYPEDEATH,    Note new variable, TYPEDEATH
    COUNT(CALCULATED TYPEDEATH) AS COUNTDEATH
FROM TRAUMA
QUIT;
```

CALCULATED used here since TYPEDEATH was previously created above.

© Alan Elliott and Wayne Woodward, 2022

SAS ESSENTIALS Ed. 3
by Alan Elliott and Wayne Woodward

Figure 9.4 SAS PROC SQL Unions and Joins, part 2

9.11 SUMMARY

This brief introduction to SAS PROC SQL provides an overview of the SQL language available in SAS. There are many aspects of the SQL language that are not covered here. We encourage you to refer to SAS Documentation for more details. Nevertheless, this introduction provides valuable beginning concepts that will allow you to use the basic SQL language in SAS.

EXERCISES

9.1 Creating new variables in SQL

1. Open the SAS file EX_9.1.SAS. This contains a partial SQL program.

```
PROC SQL;
  CREATE _____ MEANTIME ___
  SELECT COUNT(*) AS __,
         MEAN(_____) FORMAT=____ AS _____,
         _____(TIME2)_____=6.2 ___  TIME2MEAN
  FROM MYSASLIB.SOMEDATA;
QUIT;

PROC SQL; SELECT * FROM MEANTIME;QUIT;
```

Using data from MYSASLIB.SOMEDATA, create a column named N using the COUNT() function that contains the number of records in the dataset in the SQL View.

2. Create a column named TIME1MEAN that contains the mean of TIME1 variable, using the MEAN() function.
3. Report TIME1MEAN using FORMAT7.2.
4. Create TIME2 mean in the same way.
5. The resulting **View** should be as shown here:

N	TIME1MEAN	TIME2MEAN
50	21.27	27.44

9.2 Using the CASE statement

1. The dataset TRAUMA_MISS contains several missing values. Use the following code to **View** the contents of this dataset:

```
PROC SQL; SELECT * FROM MYSASLIB.TRAUMA_SM; QUIT;
```

The first few records are shown below. Note how the columns AGE, EDGCSTOTAL, and ISS contain some missing values coded as −81. If you do an analysis with these numbers in the dataset, your analysis will all be wrong. Therefore, you need to recode the missing values as the standard SAS missing value code, a dot (.), before running any analyses.

INC_KEY	AGE	GENDER	RACE	INJTYPE	FSBP	EDTEMP	HEADCT	EDGCSTOTAL	ISS	LOS	DISSTATUS
468879	9.1	Female	Black	Blunt	120	36.9		-81	1	1	Alive
468942	-81.0	Female	Black	Blunt	102	36.4	Positive	-81	20	4	Alive
468961	16.6	Female	White, not of Hispanic Origin	Blunt	118	35.8	Positive	-81	5	2	Alive
468971	6.5	Female	Black	Blunt	124	36.3		15	5	1	Alive
469030	6.7	Male	Black		128	36.6		-81	-81	1	Alive

2. Open the file EX_9.2.SAS. It contains the following SELECT statement that recodes missing values for ISS (-81) to a dot (.):

```
SELECT
    INC_KEY,GENDER, RACE, INJTYPE,
    CASE
       WHEN ISS=-81 THEN .
       ELSE ISS
    END AS ISS
```

Also note that the results are stored in a SAS dataset named TRAUMA_FIXED using a CREATE AS clause.

3. Run the program to see that the resulting ISS column no longer contains any −81 values (they have all been changed to.)
4. Use the CASE statement to create code to recode EDGCSTOTAL and AGE values that are −81 to a dot(.).
5. After fixing the missing values, run the following SAS code:

```
PROC MEANS DATA=TRAUMA_FIXED; RUN;
```

Note that the PROC MEANS reported how many non-missing values were used to calculate statistics. For AGE, EDGCSTOTAL, and ISS that number is less than 100 because the missing values were not used in the calculation.

6. The variable INJTYPE contains the type of injury observed: Burn, Penetrating, or Blunt. You're interested only in an analysis that *excludes* Burn subjects. Add this CASE statement to recode INJTYPE to " " (a null value) to create a new variable called BLUNTORPEN (Blunt or Penetrating.)

```
CASE
   WHEN UPCASE(INJTYPE)="BURN" then ""
   ELSE INJTYPE
END AS BLUNTORPEN
```

Note that the recoded variable, in this case, is BLUNTORPEN. Also, note that the CASE statement used the UPCASE function in the comparison in case the coded value "Burn" didn't always have the letter case correct.

7. Enter this SAS code to display counts of BURN and PENETRATE for the new variable BLUNTORPEN:

```
PROC FREQ DATA=TRAUMA_FIXED; TABLE BLUNTORPEN; RUN;
```

Note that only BLUNT and PENETRATE are counted in this table. How many missing values are reported? That is the number of recoded "Burn" values.

9.3 Using JOIN to combine datasets

1. Open the file EX_9.3.SAS. This contains two datasets, MOMS (containing mother's information) and KIDS (containing children's information.) You want to combine these two datasets into one that includes both mother and children's data.
2. Use an INNER JOIN in the SELECT * statement

   FROM MOMS INNER JOIN KIDS

   matching on the MOMS.FID variable from the first dataset and on KIDS.FAMID on the second dataset.
3. Enter the INNER JOIN code and run the results. Does the resulting **View** show a table containing both mother and children's data in one table?

PART II

STATISTICAL ANALYSIS USING SAS PROCEDURES

10

EVALUATING QUANTITATIVE DATA

LEARNING OBJECTIVES

- To be able to calculate basic descriptive statistics using PROC MEANS
- To be able to produce graphs using PROC MEANS
- To be able to create output from PROC MEANS
- To be able to calculate basic and advanced descriptive statistics and analyze the characteristics of a set of data using PROC UNIVARIATE
- To be able to create graphs using PROC UNIVARIATE

The ancient Greek philosopher Socrates had a saying: "Know thyself." In statistics, it's important to know thy data. You need to have knowledge of the distribution and characteristics of your data before you can successfully perform other analyses. If your data are quantitative in nature (e.g., a measure such as height or the volume of liquid in a bottle), the SAS® procedures in this chapter can help you glean a lot of information from the numbers.

10.1 USING PROC MEANS

PROC MEANS is useful for evaluating quantitative data in your dataset and for creating a simple summary report. Using this procedure, you can calculate simple statistics that

---

SAS® Essentials: Mastering SAS for Data Analytics, Third Edition. Alan C. Elliott and Wayne A. Woodward.
© 2023 John Wiley & Sons, Inc. Published 2023 by John Wiley & Sons, Inc.
Companion website: https://www.alanelliott.com/SASED3/

include the mean and standard deviation, plus the minimum and maximum, all of which allow you to see quickly whether data values fall within reasonable limits for each variable. If a data value is smaller or larger than expected, check your data to determine if the value has been miscoded. PROC MEANS can be used for the following:

- Describing quantitative data (numerical data on which it makes sense to perform arithmetic calculations)
- Describing means by group
- Searching for possible outliers (unusually small or large values) or incorrectly coded values
- Creating a simple report showing summary statistics.

The following section includes a list of commonly used options and statements illustrated in this chapter along with a few options that might be helpful when you use this procedure. For more information, refer to the SAS documentation.

10.1.1 PROC MEANS Statement Syntax and Options

The syntax of the PROC MEANS statement is as follows:

PROC MEANS <options> <statistics keywords>; <statements>;

The items listed in angle brackets "<>" are optional. The most commonly used options are listed in Table 10.1.

TABLE 10.1 Common Options for PROC MEANS

Option	Meaning
DATA=datasetname	Specifies dataset to use
MAXDEC=n	Uses n decimal places to print output
NOPRINT	Suppresses output of descriptive statistics
ALPHA=p	Sets the level for confidence limits (default 0.05)

Statistics keywords for PROC MEANS are listed in Table 10.2. In the table, the options in bold are the default statistics reported by SAS if no statistics keywords are listed. On the other hand, you can select your own list of statistics to report. For example, the statement

PROC MEANS DATA=RESEARCH MEAN MEDIAN STD;RUN;

instructs SAS to calculate the mean, median, and standard deviation on the quantitative data in the RESEARCH dataset. The confidence interval settings are based on a default alpha = 0.05. This can be changed using the ALPHA= option as shown here:

PROC MEANS DATA=RESEARCH ALPHA=.10 MEAN MEDIAN STD CLM;RUN;

TABLE 10.2 Statistics Keywords for PROC MEANS

Option	Meaning
CLM	Two-sided confidence interval
CSS	Corrected sum of squares
CV	Coefficient of variation
KURTOSIS	Kurtosis
LCLM	Lower confidence limit
MAX	Maximum (largest)
MEAN	Arithmetic average
MEDIAN (P50)	50th percentile
MIN	Minimum (smallest)
MODE	Mode
N	Number of observations
NMISS	Number of missing observations
P1	1st percentile
P5	5th percentile
P10	10th percentile
P90	90th percentile
P95	95th percentile
P99	99th percentile
PROBT	p-Value associated with t-test requested by the option T
Q1 (P25)	1st quartile
Q3 (P75)	3rd quartile
QRANGE	Quartile range
RANGE	Range
SKEWNESS (SKEW)	Skewness
STDDEV (STD)	Standard deviation
STDERR	Standard error
SUM	Sum of observations
SUMWGT	Sum of the WEIGHT variable values
T	Student's t-value
UCLM	Upper confidence limit
USS	Uncorrected sum of squares
VAR	Variance

10.1.2 Commonly Used Statements for PROC MEANS

Statements related to PROC MEANS appear after the semicolon in the initial PROC MEANS statement. Table 10.3 lists commonly used statements for PROC MEANS. For more information, consult the SAS documentation. The following examples illustrate the use of PROC MEANS options and statements:

```
* Simplest invocation - on all numeric variables *;
PROC MEANS;

*Specified statistics and variables *;
PROC MEANS N MEAN STD;
VAR SODIUM CARBO;
```

```
* Subgroup descriptive statistics in separate table
  using the BY statement*;
PROC SORT; BY GENDER;RUN;
PROC MEANS; BY GENDER;
  VAR FAT PROTEIN SODIUM;
* Subgroup descriptive statistics in the same table
  using the CLASS statement*;
PROC MEANS; CLASS GENDER;
  VAR FAT PROTEIN SODIUM;
```

TABLE 10.3 Common Statements for PROC MEANS

Statement	Meaning
VAR variable(s);	Identifies one or more variables to analyze. This can be any standard SAS variable list such as AGE SBP GENDER, or a range such as TIME1-TIME10
CLASS variable(s);	Specifies variables that the procedure uses to group the data into classification levels
OUTPUT OUT=dataname;	Specifies a SAS output dataset that will contain statistics calculated in PROC MEANS. For example, OUTPUT OUT=SAS-data-set; To specify the names of specific variables' output, use the following format: OUTPUT <OUT=SAS-data-set> statistic-keyword-1=name(s) <...statistic-keyword-n-name(s)> <percentiles-specification>;
FREQ weight variable;	Specifies a variable that represents a count of observations
BY, FORMAT, LABEL, WHERE	These statements are common to most procedures and may be used here

HANDS-ON EXAMPLE 10.1

In this example, you will learn how to use some of the basic options in PROC MEANS. Suppose you have a dataset of measurements (weight, height, and age) from several children. You will perform the following tasks using PROC MEANS:

- Report simple descriptive statistics.
- Limit the output to two decimal places.
- Request a selected list of statistics.

1. Open the program file AMEANS1.SAS.

   ```
   DATA CHILDREN;
   INPUT WEIGHT HEIGHT AGE;
   DATALINES;
   64 57 8
   71 59 10
   53 49 6
   67 62 11
   ```

```
55 51 8
58 50 8
77 55 10
57 48 9
56 42 10
51 42 6
76 61 12
68 57 9
;
PROC MEANS DATA=CHILDREN;
TITLE 'PROC MEANS, simplest use';
RUN;
PROC MEANS MAXDEC=2; VAR WEIGHT HEIGHT;
TITLE 'PROC MEANS, limit decimals, specify variables';
RUN;
PROC MEANS MAXDEC=2 N MEAN STDERR MEDIAN; VAR WEIGHT HEIGHT;
TITLE 'PROC MEANS, specify statistics to report';
RUN;
```

2. Run the program and observe the output in Table 10.4. Note the following:
   - In the first output table, statistics are reported to seven decimal points in the first PROC MEANS output and are limited to two decimal points for the other two calls to PROC MEANS, as controlled by the MAXDEC=2 option.
   - In the second table, descriptive statistics are reported only for the two variables specified in the VAR statement.
   - In the third table, only the statistics selected in the PROC statement are reported.

TABLE 10.4   Output for PROC MEANS

PROC MEANS, simplest use					
The MEANS Procedure					
Variable	N	Mean	Std. Dev.	Minimum	Maximum
WEIGHT	12	62.7500000	8.9861004	51.0000000	77.0000000
HEIGHT	12	52.7500000	6.8240884	42.0000000	62.0000000
AGE	12	8.9166667	1.8319554	6.0000000	12.0000000

PROC MEANS, limit decimals, specify variables					
The MEANS Procedure					
Variable	N	Mean	Std. Dev.	Minimum	Maximum
WEIGHT	12	62.75	8.99	51.00	77.00
HEIGHT	12	52.75	6.82	42.00	62.00

PROC MEANS, specify statistics to report				
The MEANS Procedure				
Variable	N	Mean	Std. Error	Median
WEIGHT	12	62.75	2.59	61.00
HEIGHT	12	52.75	1.97	53.00

10.1.3 Using PROC MEANS with BY Group and CLASS Statements

It is common to compare means across the levels of some grouping factor that, for example, specifies treatment times. Hands-On Example 10.2 shows how to use PROC MEANS to calculate weight gain in chickens by the type of feed. This example uses two methods for producing the desired calculations:

- Results reported for each group in separate tables using the BY group option (following a call to PROC SORT).
- Results reported for each group in a single table using the CLASS option.

HANDS-ON EXAMPLE 10.2

In this example, you will learn how to display basic statistics in groups using two techniques.

1. Open the program file AMEANS2.SAS.

   ```
   DATA FERTILIZER;
   INPUT FEEDTYPE WEIGHTGAIN;
   DATALINES;
   1 46.20
   1 55.60
   1 53.30
   1 44.80
   1 55.40
   1 56.00
   1 48.90
   2 51.30
   2 52.40
   2 54.60
   2 52.20
   2 64.30
   2 55.00
   ;
   PROC SORT DATA=FERTILIZER;BY FEEDTYPE;
   PROC MEANS; VAR WEIGHTGAIN; BY FEEDTYPE;
   TITLE 'Summary statistics by group';
   RUN;
   PROC MEANS; VAR WEIGHTGAIN; CLASS FEEDTYPE;
   TITLE 'Summary statistics USING CLASS';
   ```

2. Examine the SAS code. Note that the data must be sorted before you can use a BY group option in a PROC MEANS statement. In the second PROC MEANS, where you are using the CLASS option, sorting is not necessary (even if the data are not already sorted).

3. Run the program. Examine the output from the first PROC MEANS, which uses the BY option (Table 10.5). The output contains two separate tables, one for each value of FEEDTYPE.

4. Examine the output given in Table 10.6 which is the output from the second PROC MEANS. In this instance, which uses the CLASS option, the output for both variables appears in a single table, broken down by FEEDTYPE. (Note the N Obs and N columns. If there are missing values, the N column will be less than the N Obs column.)
5. Modify the second call to PROC MEANS (using the CLASS statement) to output the following statistics:

N MEAN MEDIAN MIN MAX

and use MAXDEC=2 to limit the number of decimals in output. Run the edited program and observe the output.

TABLE 10.5 PROC MEANS Output Using the BY Option

Summary statistics by group				
The MEANS Procedure FEEDTYPE=1				
Analysis Variable: WEIGHTGAIN				
N	Mean	Std. Dev.	Minimum	Maximum
7	51.4571429	4.7475808	44.8000000	56.0000000
FEEDTYPE=2				
Analysis Variable: WEIGHTGAIN				
N	Mean	Std. Dev.	Minimum	Maximum
6	54.9666667	4.7944412	51.3000000	64.3000000

TABLE 10.6 PROC MEANS Output Using the CLASS Option

Summary statistics USING CLASS						
The MEANS Procedure						
Analysis Variable: WEIGHTGAIN						
FEEDTYPE	N Obs	N	Mean	Std. Dev.	Minimum	Maximum
1	7	7	51.4571429	4.7475808	44.8000000	56.0000000
2	6	6	54.9666667	4.7944412	51.3000000	64.3000000

10.1.4 Output Statistics from PROC MEANS

PROC MEANS allows you to capture statistics into a new SAS Data file that you could then use in another DATA step or procedure. The OUTPUT statement is used to specify the name of the output dataset. The syntax for this statement is

OUTPUT OUT=*filename* statisticname=<*varname*>;

260 CHAPTER 10: EVALUATING QUANTITATIVE DATA

You can specify a number of statistics and names, or leave off the name such as MEAN=. The output will include all of the variables in the VAR list, or all numeric variables if no VAR list is specified. For example,

```
PROC MEANS DATA="C:\SASDATA\SOMEDATA";
OUTPUT OUT=TIME1MEAN MEAN=T1;
VAR TIME1;
RUN;
```

outputs the means of TIME1 to a data SAS set named TIME1MEAN and gives it the SAS variable name T1.

HANDS-ON EXAMPLE 10.3

In this example, you will learn how to output statistics into a new SAS data file using PROC MEANS.

1. Open the program file AMEANS3.SAS. Note the OUTPUT statement including OUT=TIME1MEAN and MEANS=T1. This OUTPUT statement names the output file TIME1MEAN and gives the mean of TIME1 the variable name T1. Submit this program and observe the output given in Table 10.7.

```
PROC MEANS DATA="C:\SASDATA\SOMEDATA";
OUTPUT OUT=TIME1MEAN MEAN=T1;
VAR TIME1;
RUN;
PROC PRINT DATA=TIME1MEAN;RUN;
```

TABLE 10.7 PROC MEANS Using OUTPUT Statement

The MEANS Procedure				
Analysis Variable: TIME1 Baseline				
N	Mean	Std. Dev.	Minimum	Maximum
50	21.2680000	1.7169551	17.0000000	24.2000000

Obs	_TYPE_	_FREQ_	T1
1	0	50	21.268

2. Take off the T1 from the MEAN= statement (keep it as MEAN=) and resubmit. The only change is that the variable name reported in the PROC PRINT is now TIME1 instead of T1 since the name was no longer in the OUTPUT statement.
3. Change the OUTPUT statement to

```
OUTPUT OUT=TIME1MEAN MEAN=T1MEAN STD=T1STD;
```

Resubmit and observe the resulting variables listed in the TIME1MEAN dataset.

4. To use the output from PROC MEANS to calculate a z-score for TIME1, add the following DATA step at the end of the code:

```
DATA ZSCORE;SET "C:\SASDATA\SOMEDATA";
IF _N_1 then SET TIME1MEAN;
ZSCORE=(TIME1-T1MEAN)/T1STD;
DROP _TYPE_ _FREQ_;
RUN;
PROC PRINT DATA=ZSCORE;RUN;
```

The IF statement where _N_=1 includes the first observation from the dataset TIME1MEAN in the new ZSCORE dataset, making the information shown in Table 10.7 available for the subsequent ZSCORE calculation.

Resubmit and observe the output. A portion of the output is given in Table 10.8. Note the columns containing the mean and standard deviation of TIME1 plus the calculated column named ZSCORE. This illustrates how you can use calculated values from PROC MEANS output as an import into another dataset calculation.

TABLE 10.8 Output Includes Calculated ZSCORE

Obs	ID	GP	AGE	TIME1	TIME2	TIME3	TIME4	STATUS	SEX	GENDER	T1MEAN	T1STD	ZSCORE
1	101	A	12	22.3	25.3	28.2	30.6	5	0	Female	21.268	1.71696	0.60106
2	102	A	11	22.8	27.5	33.3	35.8	5	0	Female	21.268	1.71696	0.89228
3	110	A	12	18.5	26.0	29.0	27.9	5	1	Male	21.268	1.71696	−1.61216
4	187	A	11	22.5	29.3	32.6	33.7	2	0	Female	21.268	1.71696	0.71755
5	312	A	4	17.0	21.3	22.7	21.2	5	1	Male	21.268	1.71696	−2.48580

10.2 USING PROC UNIVARIATE

Section 10.1 described how to use the PROC MEANS statement to calculate basic statistics for a quantitative variable. PROC UNIVARIATE provides a wider variety of statistics and graphs and is better suited to helping you discover important information about the distribution of each variable, such as whether

- the data are approximately normally distributed.
- there are outliers in the data (if so, where?)
- data distributions for variables differ by group.

With this type of information in hand, you can make informed decisions about how best to present and analyze your data. The syntax of the PROC UNIVARIATE statement is as follows:

PROC UNIVARIATE <options>; <statements>;

Common options and statements for PROC UNIVARIATE are listed in Tables 10.9 and 10.10, respectively.

TABLE 10.9 Common Options for PROC UNIVARIATE

Option	Meaning
DATA=dataname	Specifies dataset to use
CIBASIC	Requests confidence limits for the mean, standard deviation, and variance based on normally distributed data
ALPHA=p	Specifies the level for the confidence limits
MU0=n	Specifies the value of the location parameter for the one-sample t-test, sign test, and signed rank test (MU stands for the Greek letter µ, which symbolizes the population mean)
NORMAL	Requests tests for normality
PLOTS	Requests stem-and-leaf, box, and probability plots
	Request a series of explanatory plots to be created. These include a horizontal histogram, a box plot, and a normal probability plot.
NOPRINT	Suppresses the output of descriptive statistics (primarily used in conjunction with the OUTPUT statement)
TRIMMED=value	Trims the top and bottom proportion of the data before calculating the mean; this can be used to eliminate outliers from the calculation (e.g., TRIMMED=0.1)

TABLE 10.10 Common Statements for PROC UNIVARIATE

Option	Meaning
VAR variable(s);	Identifies one or more variables to analyze
OUTPUT	Identifies output dataset and statistics to include in the dataset; the syntax for this statement is
	OUTPUT <OUT=SAS-data-set> statistic-keyword-1=name(s)
	<... statistic-keyword-n=name(s)> <percentiles-specification>;
HISTOGRAM	Produces a histogram of one or more specified variables; the syntax is HISTOGRAM <variable(s)> </ option(s)>;
	A simple program that creates a histogram using the variable AGE in the SOMEDATA dataset is
	PROC UNIVARIATE DATA="C:\SASDATA\SOMEDATA";
	VAR AGE;
	HISTOGRAM AGE;
	RUN;
	See options for this statement below
INSET	Puts a box containing summary statistics in a graph associated with CDFPLOT, HISTOGRAM, PPPLOT, PROBPLOT, or QQPLOT statements
BY, CLASS, WHERE	Statements common to most PROCs.

Histogram Options: Table 10.11 lists options for the HISTOGRAM statement. These provide a number of ways you can enhance the basic histogram. These options appear after a slash / in the HISTOGRAM option. For example,

    HISTOGRAM variablename/options;

TABLE 10.11 Common OPTIONS for the PROC UNIVARIATE HISTOGRAM Statement

Option	Meaning
Density curves such as NORMAL	Overlays a density curve on the histogram. Common options would be NORMAL or KERNEL. NORMAL displays a Gaussian (bell-shaped) curve over the histogram. KERNEL displays a smoothed nonparametric density curve. Specifies characteristics of the curve in parentheses. For example, HISTOGRAM AGE/NORMAL (COLOR=GREEN W=5); specifies a green normal curve with width 5 (1 is default). Another example: HISTOGRAM AGE/NORMAL (COLOR=green W=5) CFILL=YELLOW; fills under the curve with the color specified by the CFILL option. You could also use PFILL to indicate a pattern. Note: Some color options may not work in some versions of SAS unless you turn off ODS graphics using code such as ODS GRAPHICS OFF;
MIDPOINTS	This option allows you to specify the intervals for the histogram bars MIDPOINTS = min to max by units For example, MIDPOINTS =0 to 40 by 2 specifies that the horizontal axis will have a minimum of 0, a maximum of 40, and increment of two units. The midpoints of the intervals would be 0, 2, 4, . . ., 40
VSCALE	This option specifies the scale of the vertical axis VSCALE=`type of scale` where `type of scale` can be COUNT, PROPORTION, or PERCENT For example, VSCALE=COUNT specifies that the scale of the vertical axis will be the number of observations per bin (count). You can alternately request that the scale be proportion or percent
WAXIS=`width`	This option specifies the width of the histogram bars For example, WAXIS= 3 specifies that the axis width be three units wide (1 is the default)
WBARLINE=`width`	This option specifies the width of the line around the bar WBARLINE= `width-scale` (1 is the default) For example, WBARLINE= 3 specifies that the lines around the histogram bars are to be three units wide
NROWS=`n` NCOLS=`n`	Specifies the number of rows or columns in a matrix of histograms specified when a CLASS statement is used to produce multiple histograms. For example, PROC UNIVARIATE; CLASS GENDER; VAR AGE; HISTOGRAM AGE/NROWS=2; produces two histograms one over the other (two rows)

or more specifically

```
HISTOGRAM AGE/NORMAL;
```

Color Options: Color options for `PROC UNIVARIATE` are shown in Table 10.12. They specify color for various parts of the histogram. These options appear after a slash / in the `HISTOGRAM` option and typically relate to a displayed `NORMAL` curve. For example,

```
HISTOGRAM AGE/ NORMAL (COLOR=RED);
```

(In addition to the `HISTOGRAM`, these `OPTION` statements also work with `CDFPLOT`, `PPPLOT`, `PROBPLOT`, and `QQPLOT`.)

A listing of SAS colors can be found in Appendix A. Note: For recent versions of SAS, you may have to "turn off" ODS graphics for these options to work. Do this with the code

```
ODS GRAPHICS OFF:
```

You can subsequently turn ODS graphics back on with

```
ODS GRAPHCIS ON;
```

Bar Characteristics: Use the statements in Table 10.13 to specify bar characteristics in the /*options* section of the `HISTOGRAM` statement. A listing of SAS pattern codes can be found in Appendix A.

TABLE 10.12 Common OPTIONS for the COLOR Statement (in a Univariate Graph)

Option	Meaning
COLOR=color)	Specifies the color of the density line. Put the COLOR option in parentheses. For example, HISTOGRAM AGE/NORMAL (COLOR=RED);
CAXIS=color	Specifies the color of axes. After a density curve request (such as HISTOGRAM), this option does not appear in parentheses. For example, HISTOGRAM AGE/NORMAL (COLOR=BLUE) CAXIS=RED; Color choices are described in Appendix A
CBARLINE=color	Specifies the color of the line around the bars. For example, HISTOGRAM AGE/NORMAL CAXIS=RED CBARLINES=GREEN;
CFILL=color	Specifies the color of the bars. For example, HISTOGRAM AGE/NORMAL CAXIS=RED CFILL=GREEN;
CFRAME=color	Specifies the background color within the frame. For example, HISTOGRAM AGE/NORMAL CAXIS=RED CFRAME=YELLOW;
CTEXT=color	Specifies the color of tick mark values and labels For example, HISTOGRAM AGE/CTEXT=GREEN; specifies that the borders around the histogram bars will be green

TABLE 10.13 Common OPTIONS for the PROC UNIVARIATE BAR Statement (Related to a Histogram)

Option	Meaning
BARWIDTH=n	Specifies bar width where n is the percent of the screen. For example, HISTOGRAM AGE/BARWIDTH=10;
PFILL=patterncode	Indicates which fill pattern to use in the histogram bars. For example, HISTOGRAM AGE/PFILL=R4; specifies that right 45° stripes will fill the boxes. The 4 specifies the width of the stripes. See also CBARLINE and WBARLINE above. Fill options are described in Appendix A.
BARLABEL	Places a label above the bar. Options include COUNT, PERCENT, and PROPORTION. For example, HISTOGRAM AGE/NORMAL KERNEL BARLABEL=COUNT;

Inset Statement: The INSET statement puts a box containing summary statistics in a graph associated with CDFPLOT, HISTOGRAM, PPPLOT, PROBPLOT, or QQPLOT statements. Commonly used INSET options are listed in Table 10.14. For example, the code

```
HISTOGRAM AGE/NORMAL;
INSET MEAN STD;
```

creates a plot with mean and standard deviation listed in an inset box within the graph frame. For additional options see the SAS documentation.

PROC MEANS is useful for calculating descriptive statistics for many variables at a time, whereas PROC UNIVARIATE is best for the case in which you want to analyze only one or a few variables in depth and create detailed histograms. Hands-On Example 10.4 illustrates some of these commonly used options and statements.

TABLE 10.14 Common OPTIONS for the PROC UNIVARIATE INSET Statement

Option	Meaning
Display statistics	Indicates which statistics to report in the inset box. The list is the same as the statistics used in PROC MEANS. Commonly used statistics include N, NOBS, MEAN, STD, SUM, MIN, MAX, MEDIAN, and so on. For example, PROC UNIVARIATE; VAR AGE HISTOGRAM AGE/NORMAL; INSET MEAN STD MIN MAX;
Format values	Allows you to specify labels and formats for displayed statistics. For example, PROC UNIVARIATE; VAR AGE HISTOGRAM AGE/NORMAL; INSET MEAN="Mean" (5.2) STD ='St. Dev." (5.2); where the item in parentheses is a format applied to the displayed statistics

(*Continued*)

266 CHAPTER 10: EVALUATING QUANTITATIVE DATA

TABLE 10.14 (*Continued*)

Option	Meaning
NORMALTEST	Displays a test for normality. To use, place a NORMALTEST option in the PROC UNIVARIATE statement and NORMALTEST and PNORMAL INSET option.
PNORMAL	PROC UNIVARIATE NORMALTEST; VAR AGE; HISTOGRAM AGE; INSET NORMALTEST PNORMAL; displays results of Shapiro–Wilk test for normality
Inset position	Positions the inset box inside the graph frame using map-type specifications such as NE (northeast – upper right), NW, SE, SW, N (top middle), S (bottom middle). You may also position the box outside the frame using TM (top middle), BM, RM, or LM
Inset box title	Includes a title for the inset box by indicating it after a slash in the INSET statement using HEADER=, FORMAT=, and POSITION=. For example, HISTOGRAM AGE; INSET NORMALTEST PNORMAL/HEADER="Key Title" POSITION=RM FORMAT=5.2; places the box outside the frame at the right middle (RM) and formats all statistics using 5.2
NOFRAME	Does not display a box around the inset
HEIGHT=	Height of characters in the inset box. The default is 1

HANDS-ON EXAMPLE 10.4

In this example, you will learn how to use PROC UNIVARIATE.

1. Open the program file AUNI1.SAS.

```
DATA EXAMPLE;
INPUT AGE @@;
DATALINES;
12 11 12 12 9 11 8 8 7 11 12 14 9 10 7 13
6 11 12 4 11 9 13 6 9 7 13 9 13 12 10 13
11 8 11 15 12 14 10 10 13 13 10 8 12 7 13
11 9 12
;
PROC UNIVARIATE DATA=EXAMPLE;
    VAR AGE;
RUN;
```

2. Run this program and observe that this procedure produces several tables of output, and that the output in the Results Viewer is in HTML format. Below is a description of the statistics reported in these tables.

10.2.1 Understanding PROC UNIVARIATE Output

The output from Hands-On Example 10.4 is given in several tables, and we describe the contents in each table in the following:

Moments Output: The first table (Moments; Table 10.15) provides a list of descriptive statistics for the variable AGE.

N is the sample size.

Sum Weights is the same as the sample size unless a WEIGHT statement is used to identify a separate variable that contains counts for each observation.

Mean is the arithmetic mean (also known as the average).

Sum Observations is the total (sum) of all the data values.

Std Deviation is the standard deviation.

Variance is a measure of the spread of the distribution (and is the square of the standard deviation).

Skewness is a measure of the symmetry of the data. For normally distributed data, skewness should be close to 0. A positive value indicates that the skew is to the right (long right tail), and a negative value indicates a skew to the left (long left tail).

Kurtosis measures the shape of the distribution (0 indicates normality), where a positive value indicates a distribution that is more peaked with heavier tails than a normal distribution and a negative value indicates a flatter distribution with lighter tails.

Uncorrected SS is the sum of the squared data values.

Corrected SS is the sum of the squared deviations from the mean, a quantity that is used in the calculation of the standard deviation and other statistics.

Coef Variation is the coefficient of variation, which is a unitless measure of variability. It is usually expressed as a percentage and is helpful in comparing the variability between two measures that may have differing units of measurement.

Std. Error Mean is the standard error of the mean. This is calculated as the standard deviation divided by $\sqrt{N}$, and it provides a measure of the variability of the sample mean. That is, if we were to take similar samples over and over again, the standard error of the mean would approximate the standard deviation of the sample means.

Basic Statistical Measures: The second table from PROC UNIVARIATE (Table 10.16) provides several measures of the central tendency and spread of the data. Some values are repeated from the previous table.

TABLE 10.15 Moments Output for PROC UNIVARIATE

Moments			
N	50	Sum Weights	50
Mean	10.46	Sum Observations	523
Std Deviation	2.42613323	Variance	5.88612245
Skewness	−0.5119219	Kurtosis	−0.2610615
Uncorrected SS	5759	Corrected SS	288.42
Coeff Variation	23.1943903	Std. Error Mean	0.34310705

TABLE 10.16 Basic Statistical Measures from PROC UNIVARIATE

Basic Statistical Measures			
Location		Variability	
Mean	10.46000	Std Deviation	2.42613
Median	11.00000	Variance	5.88612
Mode	12.00000	Range	11.00000
		Interquartile Range	3.00000

Median is the centermost value of the ranked data.

Mode is the most frequent value in the data.

Range is the maximum value minus the minimum value, which is a measure of the spread of the data.

Interquartile Range (IQR) is the difference between the 25th and 75th percentiles of the data and is a measure of spread of the data.

Tests for Location: The tests for location given in Table 10.17 are used to decide whether the mean (or central tendency) of the data is significantly different from 0 (or another hypothesized value). These tests are discussed more thoroughly in upcoming chapters and are only briefly described here.

Student's *t*-Test is a single sample *t*-test of the null hypothesis that the mean of the data is equal to the hypothesized value (in this case 0 by default). When $p < 0.05$, you would typically reject the null hypothesis.

Sign Test is a test of the null hypothesis that the probability of obtaining a positive value (often these values are differences) is the same as that for a negative value.

Signed Rank Test is a nonparametric test often used instead of the Student's *t*-test when the data are not normally distributed and the sample sizes are small.

Quantiles: Table 10.18 provides a listing of commonly used quantiles of the data including the median (listed as the 50th quantile). To interpret the values in the quantile table, note that the *k*th quantile is the value below which *k*% of the values in the data fall. ("Definition 5" in the table refers to the specific method used to calculate the quantiles using an empirical distribution with averaging; see Frigge et al. 1989).

Extreme Observations: The Extreme Observations table (Table 10.19) provides a listing of the largest and smallest values in the dataset. This information is useful for locating outliers in your data. Note that with each extreme value listed in the table, the observation number is also provided. This helps you to be able to go back to locate these extreme observations in your dataset.

TABLE 10.17 Tests for Location from PROC UNIVARIATE

Tests for Location: Mu0=0				
Test		Statistic	p Value	
Student's t	t	30.48611	Pr > |t|	<.0001
Sign	M	25	Pr >= |M|	<.0001
Signed Rank	S	637.5	Pr >= |S|	<.0001

TABLE 10.18 Quantiles from PROC UNIVARIATE

Quantiles (Definition 5)	
Level	Quantile
100% Max	15
99%	15
95%	14
90%	13
75% Q3	12
50% Median	11
25% Q1	9
10%	7
5%	6
1%	4
0% Min	4

TABLE 10.19 Extreme Values Output from PROC UNIVARIATE

Extreme Observations			
Lowest		Highest	
Value	Obs	Value	Obs
4	20	13	42
6	24	13	47
6	17	14	12
7	46	14	38
7	26	15	36

10.2.2 Using PROC UNIVARIATE to Assess the Normality of the Data

When you select certain options in the PROC UNIVARIATE statement, SAS produces information that is helpful in assessing the normality of your data. This is important information because a number of statistical tests are based on an assumption of normality. For example, consider the following SAS code that can be used to assess normality:

```
PROC UNIVARIATE NORMAL PLOT DATA=EXAMPLE; VAR AGE;
    HISTOGRAM AGE/NORMAL;
```

The NORMAL and PLOT options produce the following output:

- Tests for normality
- Stem-and-leaf plot (for some SAS versions)
- Box plot
- Normal probability plot.

The HISTOGRAM statement along with the NORMAL option in addition produces a

- histogram
- superimposed normal distribution curve.

HANDS-ON EXAMPLE 10.5

In this example, you will learn how to request options to assess normality using PROC UNIVARIATE.

1. Open the program file AUNI2.SAS.

   ```
   DATA EXAMPLE;
   INPUT AGE @@;
   DATALINES;
   12 11 12 12 9 11 8 8 7 11 12 14 9 10 7 13
   6 11 12 4 11 9 13 6 9 7 13 9 13 12 10 13
   11 8 11 15 12 14 10 10 13 13 10 8 12 7 13
   11 9 12
   ;
   PROC UNIVARIATE NORMAL PLOT DATA=EXAMPLE; VAR AGE;
   TITLE 'PROC UNIVARIATE EXAMPLE';
   RUN;
   ```

2. Run this program and observe the output, which is described below. PROC UNIVARIATE provides several tests and graphical methods useful in assessing normality. Observe the graphs shown in Figure 10.1. These are the default graphs produced by PROC UNIVARIATE.

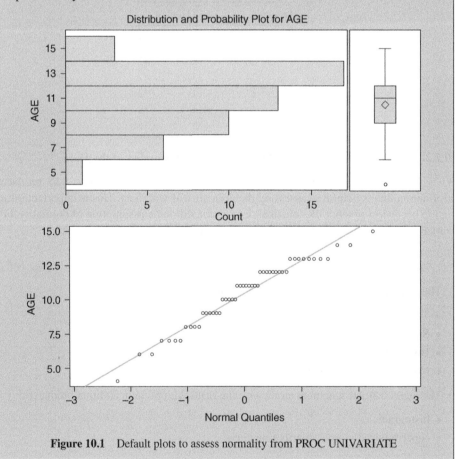

Figure 10.1 Default plots to assess normality from PROC UNIVARIATE

Using Graphs as a Visual Check for Normality: The graphics shown in Figure 10.1 may differ among versions of SAS. The output includes a distribution plot (horizontal histogram) along with a box-and-whiskers plot and a normal probability plot. Here are brief explanations of each plot:

- The **horizontal histogram** (top left) is a visual representation of the distribution of the values of AGE. To assess normality, you want the histogram to have a peak in the middle with equal tails trailing off on either side. (In some versions of SAS, this plot is a stem-and-leaf plot.)
- The **box-and-whiskers plot** or boxplot (top right) is a graphical representation of the quartiles of the data with 50% (the middle) of the data represented by the box and the whiskers representing 25% of the data on each side. The center line represents the median (50th percentile) and the diamond (◊) indicates the mean. The "o" that stands alone at the bottom of the box-and-whiskers plot indicates an extreme value or outlier. If the data are normally distributed, the box-and-whiskers plot is approximately symmetric with similar values for the mean and median.
- The **normal probability plot** (bottom) provides a graphic that (if the data are normally distributed) is a plot of points (shown as dots) that lie in a tight random scatter around the reference (diagonal) line. (In this case, the fact that AGE is recorded in whole numbers causes a sequence on the horizontal line that may make the plot confusing to interpret.)

3. Because the NORMAL option is included in PROC UNIVARIATE, tests for normality are displayed. These are shown in Table 10.20, labeled Tests for Normality.

TABLE 10.20 Test for Normality from PROC UNIVARIATE

Tests for Normality				
Test	Statistic		p Value	
Shapiro-Wilk	W	0.958283	Pr < W	0.0753
Kolmogorov-Smirnov	D	0.148067	Pr > D	<0.0100
Cramer-von Mises	W-Sq	0.145762	Pr > W-Sq	0.0259
Anderson-Darling	A-Sq	0.834989	Pr > A-Sq	0.0301

Tests for Normality: Tests for normality given in Table 10.20 are one way to assess whether the data appear normally distributed. Three tests for normality are provided: Kolmogorov–Smirnov, Cramer–von Mises, and Anderson–Darling. Some versions of SAS also include a Shapiro–Wilk test. The p-values in this table are for testing the null hypothesis that the AGE variable is normally distributed. In this case, all three tests indicate an issue with normality since the p-values for all three tests are all <0.05.

The three tests for normality represent differing statistical approaches for answering the same question. Frankly, most statisticians believe that these tests are helpful tools but should not be used as the only basis for assessing normality. Moreover, some references indicate that these tests are helpful only when the size of the dataset is less than 50.

10.2.3 Creating a Histogram Using PROC UNIVARIATE

A histogram is a commonly used plot for visually examining the distribution of a set of data. You can produce a histogram in PROC UNIVARIATE with the following statement:

HISTOGRAM AGE;

The statement HISTOGRAM AGE produces a histogram for the variable AGE. If you use the code

HISTOGRAM AGE/NORMAL;

then the NORMAL option produces a superimposed normal curve. The command including the NORMAL option produces the plot shown in Figure 10.2.

HANDS-ON EXAMPLE 10.6

1. As a continuation of the previous example using the program file AUNI2.SAS, add the statement

   HISTOGRAM AGE/NORMAL;

   after the PROC UNIVARIATE statement and before the TITLE statement. Run the revised program and observe the new histogram with superimposed normal curve included in the output, as shown in Figure 10.2.

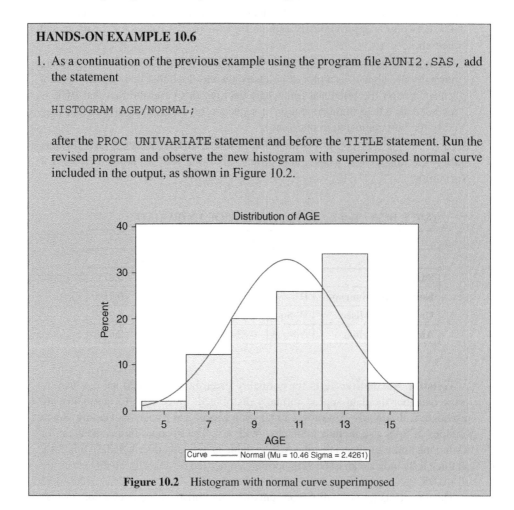

Figure 10.2 Histogram with normal curve superimposed

The superimposed normal curve on the histogram allows you not only to assess whether the data are approximately normally distributed but also to visualize the apparent departure from normality. If the data are normally distributed, you will expect the peak of the histogram bars to approximately coincide with the peak of the normal curve. In this case, the data again appear to be skewed to the left.

The NORMAL option provides several methods for you to assess the normality of your data. No plot or test by itself is definitive. They should be taken as a whole.

10.3 GOING DEEPER: ADVANCED PROC UNIVARIATE OPTIONS

With a few additional commands using PROC UNIVARIATE, it is relatively easy to create a series of histograms that allow you to compare subgroups of your data. In this Going Deeper section, we show how to display the distributions of a variable at each of several levels of a grouping variable using data in a simulated dataset. Hands-On Example 10.7 illustrates how you could create histograms for systolic blood pressure by RACE using PROC UNIVARIATE.

HANDS-ON EXAMPLE 10.7

In this example, you will learn how to create multiple histograms using a grouping (CLASS) variable.

1. Open the program file AUNI3.SAS. The variables in the WOUND dataset include AGE, GENDER, RACE_CAT, ISS (Injury Severity Score), DISSTATUS (Discharge Status), WOUND (Wound type), SBP, and TEMP_C (Temperature in Celsius).

    ```
    PROC UNIVARIATE DATA="C:\SASDATA\WOUND" NOPRINT;
      CLASS RACE_CAT;
      LABEL RACE_CAT="RACE";
      VAR SBP;
      HISTOGRAM /NORMAL (COLOR=GREEN W=5) NROWS=3;
    RUN;
    ```

    The new statements used in this example include the following:
    - The NOPRINT statement suppresses most output except the graph because we're interested only in producing the graph.
    - The CLASS RACE statement indicates that the data are to be examined for each category (classification) of the RACE variable.
    - LABEL creates an output label for RACE_CAT as RACE.
    - NROWS=3 indicates that three graphs (three rows) will appear on each output page. Because there are three RACE groups, all graphs will appear together. Similarly, you can use NCOLS=3 to place all three graphs on the same page in three columns.
    - The COLOR=GREEN and W=5 statements in parentheses refer to the NORMAL option and tell SAS to display the fitted normal curve with a green line and with a width of 5.

    (Continued)

(*Continued*)

- The DATA= statement accesses the data from the file WOUND.SAS7BDAT. You could also use the statement DATA=MYSASLIB.WOUND to access the data if the MYSASLIB library is defined.

2. Run the program and observe the output in the **Results Viewer** (see Figure 10.3). Note that the three histograms are for the variable SBP separately for three values of RACE_CAT. In this case, there is a visual agreement that SBP is similarly distributed for all races, and the assumption of normality for each race appears plausible. (Note: If the above commands do not produce the plot in Figure 10.3, place the command ODS GRAPHICS OFF; at the beginning of your code. This is needed for some versions of SAS.)

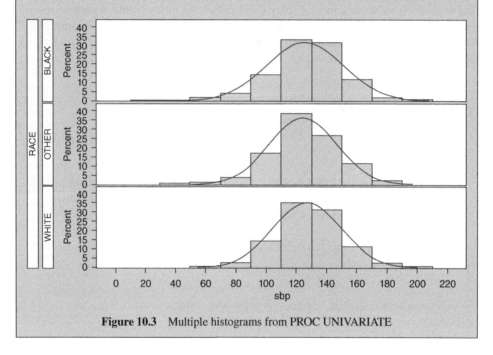

Figure 10.3 Multiple histograms from PROC UNIVARIATE

There are a number of formatting options you can apply to histograms to customize their appearance. Briefly, here are common ways to enhance a histogram:

Use PROC FORMAT: Prepare your categorical variables with formats that allow you to display the names of categories in the histogram. The following PROC FORMAT code could be used with the WOUND dataset to assign names to the two categories of WOUND. Using this format allows you to display names of the group categories causing PENETRATE and NONPENETRATE to appear in the graph rather than with the 0, 1 data codes.

```
PROC FORMAT;
VALUE FMTWOUND 0="NONPENETRATE"
               1="PENETRATE";
RUN;
```

Hands-On Example 10.8 further illustrates these comments.

Include a CLASS statement: Including a CLASS statement allows you to display comparative histograms by grouping factors such as (in the case of the WOUND dataset) by WOUND type and GENDER.

Include HISTOGRAM Options: The options within the HISTOGRAM statement define how the graph will appear. The statements NROWS=2 NCOLS=2 produce two histograms per row (for WOUND, the first item in the CLASS statement) and two histograms per column (for GENDER or the second item in the CLASS statement).

Use CFILL to specify color fill: The bar colors are specified by the CFILL (color fill) statement: CFILL=BLUE. Some of the colors available in SAS (there are thousands to choose from) are listed in Appendix A.

Use PFILL to specify patterns: The pattern for the bars is specified by the PFILL (pattern fill) statement: PFILL=M3N45. You can select from a number of available patterns. The default pattern is solid. Additional patterns are listed in Appendix A.

Use INSET to create a key: This option defines an inset or a key to the graph. An example INSET statement is as follows:

```
INSET N='N:' (4.0) MIN='MIN:' (4.1) MAX='MAX:' (4.1)
            / HEADER="My Key" NOFRAME POSITION=NE HEIGHT=2;
```

In this code

```
N='N:' (4.0) MIN='MIN:' (4.1) MAX='MAX:' (4.1)
```

defines which statistics will be included in the inset. In this case, N (the sample size) will be designated with N: and will be displayed using the SAS output format 4.0 (designated in the code in parentheses). The MIN and MAX are similarly defined. The remaining options

```
/ HEADER="My Key" NOFRAME POSITION=NE HEIGHT=2;
```

specify the following:

- A header titled MY KEY will be included for the box.
- No frame will be printed around the inset.
- The inset box position will be in the NE (northeast) corner of the graph.
- The height of the characters will be set at 2 units.

Hands-On Example 10.8 makes use of these PROC UNIVARIATE options when creating a custom HISTOGRAM.

HANDS-ON EXAMPLE 10.8

1. Open the program file AUNI4.SAS.

```
PROC FORMAT;
VALUE FMTWOUND 0="NONPENETRATE"
              1="PENETRATE";
RUN;
TITLE 'HISTOGRAMS of SBP by GENDER and WOUND TYPE';
```

(Continued)

276 CHAPTER 10: EVALUATING QUANTITATIVE DATA

(*Continued*)
```
PROC UNIVARIATE DATA="C:\SASDATA\WOUND" NOPRINT;
  CLASS WOUND GENDER;
  VAR SBP;
  HISTOGRAM / NROWS=2 NCOLS=2 CFILL=BLUE PFILL=M3N45;
  INSET N='N:' (4.0) MIN='MIN:' (4.1) MAX='MAX:' (4.1)
            / NOFRAME POSITION=NE HEIGHT=2;
  FORMAT WOUND FMTWOUND.;
RUN;
```

2. Observe the statements in this program. In this example, you are telling SAS to display histograms by two grouping variables, WOUND and GENDER:
3. Run this SAS code and observe the output, as shown in Figure 10.4.

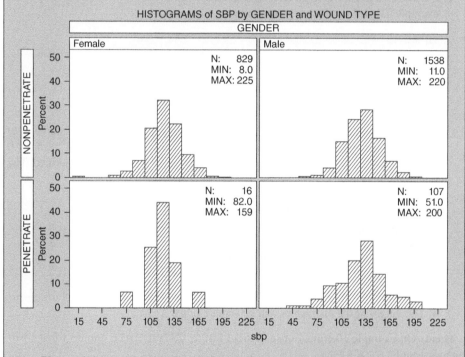

Figure 10.4 PROC UNIVARIATE histogram output using two grouping variables

4. Change the program using the following items:
   (a) Make the histogram color green.
   (b) Add the option MEAN='MEAN:' (4.1) to the INSET option. For example,

   ```
   N='N:' (4.0) MIN='MIN:' (4.1) MAX='MAX:' (4.1)
   MEAN='MEAN:' (4.1)
   ```

   (c) After HISTOGRAM/, add the option

   ```
   NORMAL (COLOR=BROWN W=3)
   ```

   to superimpose a normal plot on each histogram.
   (d) Run the revised program and observe the results.

10.4 SUMMARY

In this chapter, you learned how to use `PROC MEANS` and `PROC UNIVARIATE` to calculate basic statistics and graphics for describing the characteristics of a set of data.

EXERCISES

10.1 **Create histograms**.
   Use the `WOUND` dataset (`C:\SASDATA\WOUND`) as in Hands-On Examples 10.7 and 10.8 and create the following histograms:
   a. Create a matrix of histograms with `RACE_CAT` (three categories) using the pattern `M3XO` and `CFILL=RED`.
   b. Place the key on the upper left corner (`NW`).
   c. Add mean (`MEAN='MEAN:' (4.1)`) to the list of statistics reported and standard deviation.
   d. Put your name in a `TITLE` statement.
   e. Redo the plot using solid blue bars.
   f. For this exercise, capture the output using `ODS RTF`.
   (Note: If color options do not work in this exercise, place the command `ODS GRAPHICS OFF;` at the beginning of your code. This is needed for some versions of SAS.)

10.2 **Produce normal probability plots using `PROC UNIVARIATE`**.
   Using the `WOUND` dataset from Exercise 10.1, use `PROC UNIVARIATE` with the statement

   `PROBPLOT SBP/NORMAL (COLOR=BLUE W=5) NROWS=3;`

   to produce a normal probability plot by `RACE_CATEGORY`.

10.3 **Assess normality using a Q–Q plot**.
   Open the file `AUNI1.SAS`. After `VAR AGE;` and before the `RUN;` command, enter the statement

   `QQPLOT AGE;`

   to request a Q–Q plot, which can be used to assess normality. Run the program. If the data are normal, the plot will show a tight cluster of points around a straight line. Do you think this plot suggests that the data are normally distributed?

10.4 **Display multiple histograms.**
   Open the program file `AUNI5.SAS` and complete the code to create two columns and two rows of histograms from the `CARS` dataset showing a comparison of city MPG for vehicles that are SUVs or not and for vehicles that have automatic transmissions or not. Missing information in the code is indicated as ???.
   a. Complete the `PROC FORMAT` statement to define values for the 0/1 codes for the `SUB` and `AUTOMATIC` variables where 1 means "SUV" for the `SUV` variables and 0 means "Standard" for the `AUTOMATIC` variable.

278 CHAPTER 10: EVALUATING QUANTITATIVE DATA

b. Because there are two values for each CLASS variable, indicate NROWS=2 and NCOLS=2.
c. Indicate that the format for the MEAN should be (4.1).
d. Place the inset in the northeast corner of the graph.
e. In the FORMAT statement, indicate the name of the defined format for the AUTOMATIC variable as FMTAUTO. (Don't forget the dot at the end of the definition.)

```
PROC FORMAT;
VALUE FMTSUV 0="NOT SUV"
             1="???";
VALUE FMTAUTO 0="???"
              1="Automatic";
RUN;
TITLE 'HISTOGRAMS of CITY MPG by SUV and AUTOMATIC';
PROC UNIVARIATE DATA="C:\SASDATA\CARS" NOPRINT;
   CLASS AUTOMATIC SUV;
   VAR CITYMPG;
   HISTOGRAM / NROWS=??? NCOLS=??? CFILL=BLUE
   PFILL=M3N45;
   INSET N='N:' (4.0) MIN='MIN:' (4.1) MAX='MAX:' (4.1)
MEAN='MEAN' (???)
               /NOFRAME POSITION=??? HEIGHT=2;
   FORMAT AUTOMATIC ??? SUV FMTSUV.;
RUN;
```

Run the program. Your results should look like the histograms in Figure 10.5.

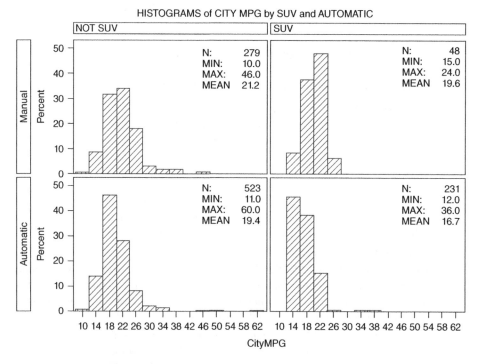

Figure 10.5 Multiple histograms from the CAR dataset

(Note: If color options do not work in this exercise, place the command ODS GRAPHICS OFF; at the beginning of your code. This is needed for some versions of SAS.)

REFERENCE

Frigge, M., Hoaglin, D.C., and Iglewicz, B. (1989). Some implementations of the boxplot. *The American Statistician* 43 (1): 50–54.

11

ANALYZING COUNTS AND TABLES

LEARNING OBJECTIVES

- To be able to use PROC FREQ to create one-way frequency tables
- To be able to use PROC FREQ to create two-way (cross-tabulation) tables
- To be able to use two-by-two contingency tables to calculate relative risk measures
- To be able to use Cohen's kappa to calculate inter-rater reliability

Data that are collected as counts require a specific type of treatment. It doesn't make sense to calculate means and standard deviations on this type of data. Instead, categorical data (also called qualitative data) are often analyzed using frequency and cross-tabulation tables. The primary procedure within SAS® for this type of analysis is PROC FREQ.

11.1 USING PROC FREQ

PROC FREQ is a multipurpose SAS procedure for analyzing count data. It can be used to obtain frequency counts for one or more individual variables or to create two-way tables (cross-tabulations) from two variables. (Tables using three variables are discussed later in the chapter.)

SAS® Essentials: Mastering SAS for Data Analytics, Third Edition. Alan C. Elliott and Wayne A. Woodward.
© 2023 John Wiley & Sons, Inc. Published 2023 by John Wiley & Sons, Inc.
Companion website: https://www.alanelliott.com/SASED3/

PROC FREQ can also be used to perform statistical tests on count data. The syntax for PROC FREQ and commonly used options and statements that are discussed in this chapter is as follows:

PROC FREQ <options(s)>; <statements> TABLES requests</options>;

Some common options for PROC FREQ are listed in Table 11.1.
Some common statements for PROC FREQ are listed in Table 11.2.

TABLE 11.1 Common Options for PROC FREQ

Option	Meaning
DATA=dataname	Specifies which dataset to use
ORDER=option	Specifies the order in which results are listed in the output table Options are DATA, FORMATTED, FREQ, and INTERNAL. This is illustrated in an upcoming example
PAGE	Specifies that only one table will appear per page (not applicable to HTML output)
ALPHA=a	Sets the level for confidence limits (default 0.05)
COMPRESS	Begins the next table on the same page when possible (not applicable to HTML output)
NOPRINT	Used when you want to capture output but not display tables

TABLE 11.2 Common Statements for PROC FREQ

Option	Meaning
EXACT	Produces exact p-values for tests. Fisher's Exact test automatically calculated for a 2×2 table
OUTPUT=dataname	Creates an output dataset containing statistics from an analysis
WEIGHT variable	Identifies a weight variable that contains summarized counts
TABLES <variable-combinations/options>;	Specifies which tables will be displayed. More information about this statement is given below (Required)
TEST	Specifies which statistical tests will be performed (requires a TABLES statement)
BY, FORMAT, LABEL, WHERE	These statements are common to most procedures and may be used with PROC FREQ

11.1.1 The TABLES Statement

The TABLES statement is required for all of the examples in this chapter. Its format is as follows:

TABLES <variable-combinations/options>;

where *variable-combinations* specifies frequency or cross-tabulation tables.
For examples, see Table 11.3.
Options for the TABLE statement follow a slash (/). For example,

TABLES A*B / CHISQ;

requests the chi-square and related statistics to be reported for the cross-tabulation A*B. Commonly used TABLE options following a / are listed in Table 11.4.

For example, in a simple use of PROC FREQ, to obtain counts of the number of subjects observed in each category of group (GP), use the following:

PROC FREQ; TABLES GP; RUN;

To produce a cross-tabulation table of GENDER by treatment GP, use

PROC FREQ; TABLES GENDER*GP;RUN;

The variables specified in the TABLES statement can be either categorical/character or numeric (where numeric values represent discrete categories). To request chi-square statistics for a table, include the option /CHISQ at the end of the TABLES statement. For example,

PROC FREQ; TABLES GENDER*GP/CHISQ;

The following sections describe in more detail how to use PROC FREQ to create and analyze tables of counts.

TABLE 11.3 Sample TABLES Statements

Table Specification	Description
TABLES A;	Specifies frequencies for a single variable
TABLES A*B;	Specifies a cross-tabulation between two variables
TABLES A*B B*C X*Y; Also, TABLES A*(B C D);is the same as TABLES A*B A*C A*D;	Specifies several cross-tabulation tables
TABLES (A -- C)*X; is the same as TABLES A*X B*X C*X;	Uses a range of variables in a TABLES statement

TABLE 11.4 Options for the TABLE Statement

Option	Description
AGREE	Requests kappa statistic (inter-rater reliability). To include significance tests for kappa, add a TEST statement. For example, TABLES A*B /AGREE; TEST KAPPA;
RELRISK	Requests relative risk calculations
FISHER	Requests Fisher's Exact test for tables greater than 2×2
SPARSE	Requests all possible combinations of variable levels even when a particular combination never occurs

TABLE 11.4 (*Continued*)

Option	Description
MISSING	Requests that missing values be treated as nonmissing
CELLCHI2	Displays the contribution to chi-square for each cell
NOCOL	Suppresses column percentages for each cell
NOCUM	Suppresses cumulative frequencies and cumulative percentages in one-way frequency tables and in list format
NOFREQ	Suppresses frequency count for each cell
NOPERCENT	Suppresses row percentage and column percentage in cross-tabulation tables, or percentages and cumulative percentages in one-way frequency tables and in list format
NOPRINT	Suppresses tables but displays statistics
NOROW	Suppresses row percentage for each cell
TESTP=(*list*)	Specifies a test based on percentages for the chi-square test (see the goodness-of-fit example)

11.2 ANALYZING ONE-WAY FREQUENCY TABLES

When count data are collected, you can use PROC FREQ to produce tables of the counts by category as well as to perform statistical analyses on the counts. This section describes how to create tables of counts by category and how to perform a goodness-of-fit test.

The number and type of tables produced by the PROC FREQ procedure are specified with a TABLES statement. When individual variables are used in the TABLES statement, PROC FREQ produces a frequency table. Hands-On Example 11.1 illustrates how to create a frequency table for a single variable.

> Recall that sample datasets are in the C:\SASDATA folder and the SOMEDATA dataset contains a sample dataset of 50 records. In Hands-On Example 11.1, note that the DATA= statement could also be DATA=MYSASLIB.SOMEDATA if you have the MYSASLIB library defined.

HANDS-ON EXAMPLE 11.1

In this example, you will learn how to use PROC FREQ to display frequency information for a qualitative (categorical) data variable.

1. Open the program file AFREQ1.SAS.

```
PROC FREQ DATA="C:\SASDATA\SOMEDATA";
    TABLES STATUS;
TITLE 'Simple Example of PROC FREQ';
RUN;
PROC FREQ DATA="C:\SASDATA\SOMEDATA" ORDER=FREQ;
    TABLES STATUS;
TITLE 'Example of PROC FREQ Using ORDER= Option';
RUN;
```

(*Continued*)

(*Continued*)

This SAS code includes two `PROC FREQ` statements. In the first example, a table is requested for the variable `STATUS`. The second example is similar, but an additional option, `ORDER=FREQ`, is used to request that the table be sorted by the descending order of the frequency count.

Following are some of the `PROC FREQ ORDER=` options.

DATA	Orders values by their order in the dataset.
FORMATTED	Orders values by ascending values determined by a FORMAT statement (see PROC FORMAT in Chapter 4).
FREQ	Orders values by descending frequency count.
ORDER	Orders values by order in dataset.

2. Run this program and observe the output (Table 11.5) from the first `PROC FREQ`.

   The "Frequency" column gives the count of the number of times the `STATUS` variable takes on each value in the `STATUS` column. The "Percent" column is the percentage of the total (50). The "Cumulative Frequency" and "Percent" columns report an increasing count and percent for values of `STATUS`. You use this type of analysis to learn about the distribution of the categories in your dataset. For example, in these data, more than half of the subjects are in the `STATUS=5` category. If you had planned to have approximately the same number of observations in each category, this table would indicate that that criterion had not been met. Note that the order of categories in the table is based on ascending values of `STATUS`.

TABLE 11.5   Simple Output for PROC FREQ

Simple Example of PROC FREQ				
The FREQ Procedure				
Socioeconomic Status				
STATUS	Frequency	Percent	Cumulative Frequency	Cumulative Percent
1	3	6.00	3	6.00
2	7	14.00	10	20.00
3	6	12.00	16	32.00
4	8	16.00	24	48.00
5	26	52.00		

3. The second `PROC FREQ` in the code includes the `ORDER=FREQ` option. The output from this version of `PROC FREQ` is given in Table 11.6. In this table, the order of categories is based on frequencies (as requested by the `ORDER=FREQ` option). Using the `ORDER=FREQ` option helps you quickly identify which categories have the most and fewest counts.

TABLE 11.6 PROC FREQ Output Using the ORDER=FREQ Option

Example of PROC FREQ Using ORDER= Option				
The FREQ Procedure				
Socioeconomic Status				
STATUS	Frequency	Percent	Cumulative Frequency	Cumulative Percent
5	26	52.00	26	52.00
4	8	16.00	34	68.00
2	7	14.00	41	82.00
3	6	12.00	47	94.00
1	3	6.00	50	100.00

Hands-On Example 11.2 illustrates another way to change the order in which the categories are displayed. This example uses the ORDER=FORMATTED option for PROC FREQ. In this case, you must first create a custom format using a PROC FORMAT command to define the order that you want to be used in your output table.

HANDS-ON EXAMPLE 11.2

1. Open the program file AFREQ2.SAS.

```
PROC FORMAT;
VALUE $FMTRACE "AA"="African American"
               "H"="Hispanic"
               "OTH"="Other"
               "C"="White";
RUN;
PROC FREQ ORDER=FORMATTED DATA="C:\SASDATA\SURVEY";
     TABLES RACE;
     TITLE 'Example of PROC FREQ using OPTION=Formatted';
     FORMAT RACE $FMTRACE.;
RUN;
```

This program uses the ORDER=FORMATTED option to control the order in which the categories will be displayed in the table. To use this option, you must define a format which specifies that the labels are in the ascending order you want to use. In this program, a format named $FMTRACE is created in a PROC FORMAT statement. The code

ORDER=FORMATTED

(Continued)

(*Continued*)

specifies that the table is to be sorted in ascending order by the formatted values for RACE based on the $FMTRACE format. To let SAS know to apply the format $FMTRACE. to the RACE variables, include the statement

```
FORMAT RACE $FMTRACE.;
```

which applies the $FMTRACE. format to the RACE variable.

2. Submit this program and observe the results given in Table 11.7. Note that it is the assigned label that specifies the order, and not the original values of RACE.

TABLE 11.7 Output from PROC FREQ Using the ORDER= FORMATTED Option

Example of PROC FREQ using OPTION=Formatted				
The FREQ Procedure				
RACE				
RACE	Frequency	Percent	Cumulative Frequency	Cumulative Percent
African American	37	46.84	37	46.84
Hispanic	30	37.97	67	84.81
Other	4	5.06	71	89.87
White	8	10.13	79	100.00

3. To emphasize the effect of this format statement, place an asterisk at the beginning of the SAS statement containing the FORMAT statement:

```
* FORMAT RACE $FMTRACE.;
```

This turns the statement into a comment. Rerun the program and observe how the table has changed. Because no format is associated with RACE, the frequencies are displayed in the table based on the alphabetical order of the category values AA, C, H, and OTH.

11.3 CREATING ONE-WAY FREQUENCY TABLES FROM SUMMARIZED DATA

Suppose you broke into your piggy bank and made stacks of the coins you found. In this case, your data are already summarized into counts. To analyze this type of data using SAS, use the WEIGHT statement to indicate the variable that represents the counts.

HANDS-ON EXAMPLE 11.3

This example illustrates how to summarize counts from a dataset into a frequency table.

1. Open the program file AFREQ3.SAS.

   ```
   DATA COINS;
           INPUT @1 CATEGORY $9. @11 NUMBER 3.;
   DATALINES;
   CENTS     152
   CENTS     100
   NICKELS    49
   DIMES      59
   QUARTERS   21
   HALF       44
   DOLLARS    21
   ;
   PROC FREQ; WEIGHT NUMBER;
     TITLE 'Reading Summarized Count data';
     TABLES CATEGORY;
   RUN;
   ```

   Note that the data are already in counts. The statement

   `WEIGHT NUMBER;`

   tells PROC FREQ that the data for the variable NUMBER are counts. Even though there are two records for CENTS, the program is able to combine the WEIGHT (counts) into a single CENTS category (252 CENTS).

2. Run the program and observe the output, as given in Table 11.8.

   TABLE 11.8 Simple Output for PROC FREQ

Reading Summarized Count data				
The FREQ Procedure				
CATEGORY	Frequency	Percent	Cumulative Frequency	Cumulative Percent
CENTS	252	56.50	252	56.50
DIMES	59	13.23	311	69.73
DOLLARS	21	4.71	332	74.44
HALF	44	9.87	376	84.30
NICKELS	49	10.99	425	95.29
QUARTERS	21	4.71	446	100.00

3. To tell SAS to reorder this table in the order in which you specified the categories in the dataset (which makes more sense in this example), use the ORDER=DATA option in the statement:

   `PROC FREQ ORDER=DATA;`

   Rerun the analysis and observe the new frequency table.

11.3.1 Testing Goodness of Fit in a One-Way Table

A goodness-of-fit test of a single population is a test to determine if the distribution of observed frequencies in the sample data closely matches the expected number of occurrences under a hypothetical distribution for the population. The observations are assumed to be independent, and each data value can be counted in one and only one category. It is also assumed that the number of observations is fixed. The hypotheses being tested are as follows:

H_0: The population follows the hypothesized distribution.
H_a: The population does not follow the hypothesized distribution.

A chi-square statistic is calculated, and a decision can be made based on the *p*-value associated with that statistic. A low *p*-value indicates that the data do not follow the hypothesized, or theoretical, distribution. If the *p*-value is sufficiently low (usually <0.05), you will reject the null hypothesis. The syntax to perform a goodness-of-fit test is as follows:

```
PROC FREQ; TABLES variable / CHISQ TESTP=(list of ratios);
```

As an example, we will use data from an experiment conducted by the nineteenth-century monk Gregor Mendel. According to a genetic theory, crossbred pea plants show a 9:3:3:1 ratio of yellow smooth, yellow wrinkled, green smooth, and green wrinkled offspring. Out of the 556 plants, under the theoretical ratio (distribution) of 9:3:3:1, you would expect about

(9/16) × 556 = 312.75 yellow smooth peas (56.25%)
(3/16) × 556 = 104.25 yellow wrinkled peas (18.75%)
(3/16) × 556 = 104.25 green smooth peas (18.75%)
(1/16) × 556 = 34.75 green wrinkled peas (6.25%)

After growing 556 of these pea plants, Mendel observed the following:

315 have yellow smooth peas
108 have yellow wrinkled peas
101 have green smooth peas
32 have green wrinkled peas

Do these offspring support the hypothesized ratios? Hands-On Example 11.4 illustrates the use of `PROC FREQ` to run a goodness-of-fit test to assess whether the observed frequencies seem to support the theory.

HANDS-ON EXAMPLE 11.4

This example illustrates how to perform a goodness-of-fit test.

1. Open the program file `AFREQ4.SAS`.

   ```
   DATA GENE;
        INPUT @1 COLORTYPE $13. @15 NUMBER 3.;
   DATALINES;
   YELLOWSMOOTH  315
   YELLOWWRINKLE 108
   ```

```
GREENSMOOTH 101
GREENWRINKLE 32
;
* HYPOTHESIZING A 9:3:3:1 RATIO;
PROC FREQ ORDER=DATA; WEIGHT NUMBER;
  TITLE 'GOODNESS OF FIT ANALYSIS';
  TABLES COLORTYPE / NOCUM CHISQ
       TESTP=(0.5625 0.1875 0.1875 0.0625);
RUN;
```

Note the following components of this SAS program:

- The data are summarized as indicated by the WEIGHT NUMBER statement in PROC FREQ.
- The ORDER=DATA option causes the output frequencies to be ordered as they were input into the dataset.
- Frequencies are based on the variable COLORTYPE.
- The /NOCUM CHISQ and TESTP= statements request the goodness-of-fit test. The test ratios are based on the percent progeny expected from each of the four categories. The NOCUM option requests a table without the cumulative column.
- The NOCUM option instructs SAS not to include the "Cumulative Frequency" column in the table.

Note that you must use the ORDER=DATA option to ensure that the hypothesized ratios listed in the TESTP= statement match up correctly with the categories in the table.

2. Run the program and observe the output, as given in Table 11.9. A graph showing deviations is also included in the output (in the latest versions of SAS), but not shown here.

It is a good idea to make sure the "Test Percent" column matches the hypothesized percentages for the categories of the tested variable.

Note that in this case, the p-value for the chi-square test is >0.05 ($p = 0.9254$), which leads us to not reject the null hypothesis and conclude that the evidence indicates support for the theory that the population exhibits the 9:3:3:1 phenotypic ratios.

TABLE 11.9 Goodness-of-Fit Test Using **PROC FREQ**

GOODNESS OF FIT ANALYSIS			
The FREQ Procedure			
COLORTYPE	Frequency	Percent	Test Percent
YELLOWSMOOTH	315	56.65	56.25
YELLOWWRINKLE	108	19.42	18.75
GREENSMOOTH	101	18.17	18.75
GREENWRINKLE	32	5.76	6.25

Chi-Square Test for Specified Proportions	
Chi-Square	0.4700
DF	3
Pr > ChiSq	0.9254

11.4 ANALYZING TWO-WAY TABLES

To create a cross-tabulation table using PROC FREQ for relating two variables, use the TABLES statement with both variables listed and separated by an asterisk (*), (e.g., A*B). A cross-tabulation table is formed by counting the number of occurrences in a sample across two grouping variables. The number of columns in a table is usually denoted by c and the number of rows by r. Thus, a table is said to be an $r \times c$ table, that is, it has $r \times c$ cells. For example, in a dominant-hand (left–right) by hair color table (with five hair colors used), the table would be referred to as a 2×5 table.

The hypotheses associated with a test of independence are as follows:

H_0: The variables are independent (no association between them).
H_a: The variables are not independent.

Thus, in the "dominant-hand/hair color" example, the null hypothesis is that there is no association between the dominant hand and hair color (each hand-dominance category has the same distribution of hair color). The alternative hypothesis is that left- and right-handed people have different distributions of hair color – perhaps left-handed people are more likely to be brown-haired.

Another test that can be performed on a cross-tabulation table is a test of homogeneity. In this case, the table is built of independent samples from two or more populations, and the null hypothesis is that the populations have the same distribution (they are homogeneous). In this case, the hypotheses are as follows:

H_0: The populations are homogeneous.
H_a: The populations are not homogeneous.

Rows (or columns) represent data from different populations (e.g., treated and not treated), and the other variable represents categorized data observed on the population.

The chi-square test of independence or homogeneity is reported by PROC FREQ (the tests are mathematically equivalent) by the use of the /CHISQ option in the TABLES statement. For example,

```
PROC FREQ; TABLES GENDER*GP/CHISQ;
```

will create a two-way cross-tabulation table and statistics associated with the table. Also included in the output are the likelihood ratio chi-square, Mantel–Haenszel chi-square, phi, contingency coefficient, and Cramer's V. For a 2×2 table, a Fisher's Exact test is also performed.

HANDS-ON EXAMPLE 11.5

Data for this example come from a study performed by Karl Pearson in 1909 involving the relationship between criminal behaviors and drinking alcoholic beverages. The category "Coining" refers to counterfeiting. For the DRINKER variable, 1 means yes and 0 means no.

1. In the Editor window, open the program AFREQ5.SAS.

```
DATA DRINKERS;
INPUT CRIME $ DRINKER COUNT;
DATALINES;
```

ANALYZING TWO-WAY TABLES

```
Arson     1   50
Arson     0   43
Rape      1   88
Rape      0   62
Violence  1  155
Violence  0  110
Stealing  1  379
Stealing  0  300
Coining   1   18
Coining   0   14
Fraud     1   63
Fraud     0  144
;
PROC FREQ DATA=DRINKERS;WEIGHT COUNT;
    TABLES CRIME*DRINKER/CHISQ;
TITLE 'Chi Square Analysis of a Contingency Table';
RUN;
```

2. Run the program and observe the cross-tabulation table given in Table 11.10. The four numbers in each cell are the overall frequency, the overall percent, the row

TABLE 11.10 Cross-Tabulation from PROC FREQ

Frequency Percent Row Pct Col Pct	Table of CRIME by DRINKER		
		DRINKER	
CRIME	0	1	Total
Arson	43 3.02 46.24 6.39	50 3.51 53.76 6.64	93 6.52
Coining	14 0.98 43.75 2.08	18 1.26 56.25 2.39	32 2.24
Fraud	144 10.10 69.57 21.40	63 4.42 30.43 8.37	207 14.52
Rape	62 4.35 41.33 9.21	88 6.17 58.67 11.69	150 10.52
Stealing	300 21.04 44.18 44.58	379 26.58 55.82 50.33	679 47.62
Violence	110 7.71 41.51 16.34	155 10.87 58.49 20.58	265 18.58
Total	673 47.19	753 52.81	1426 100.00

(Continued)

(*Continued*)

percent, and the column percent. The cells in the margins show total frequencies and percentages for columns and rows.

3. Observe the statistics table given in Table 11.11. The chi-square value is 49.7 and $p<0.0001$. Thus, you reject the null hypothesis of no association (independence) and conclude that there is evidence of a relationship between drinking status and type of crime committed.

TABLE 11.11 Statistics for Drinking and Crime Cross-Tabulation

Statistics for Table of CRIME by DRINKER			
Statistic	DF	Value	Prob
Chi-Square	5	49.7306	<.0001
Likelihood Ratio Chi-Square	5	50.5173	<.0001
Mantel-Haenszel Chi-Square	1	13.0253	0.0003
Phi Coefficient		0.1867	
Contingency Coefficient		0.1836	
Cramer's V		0.1867	
Sample Size = 1426			

Note that the likelihood ratio statistic given in Table 11.9 is an alternative to the chi-square. While the results are usually similar, many practitioners prefer to report the chi-square statistic. The other statistics reported in this table are for special settings and are not discussed here.

4. To discover the nature of the relationship suggested by the significant chi-square test, alter the program by changing the TABLES statement to the following:

```
TABLES CRIME*DRINKER/CHISQ EXPECTED NOROW NOCOL NOPERCENT;
```

EXPECTED specifies that expected values under the null hypothesis of no association are to be included in the table, and NOROW, NOCOL, and NOPERCENT tell SAS to exclude row, column, and overall percentages, respectively, from the table. The results are given in Table 11.12.

Run the revised program and note that while most of the expected values are close to the observed values, those for Fraud are very different from what was expected. This information leads to the conclusion that those involved in Fraud as a crime are less likely to drink alcoholic beverages than those involved in other crimes.

TABLE 11.12 Expected Values for Drinking and Crime Data

Frequency Expected	Table of CRIME by DRINKER		
		DRINKER	
CRIME	0	1	Total
Arson	43	50	93
	43.891	49.109	
Coining	14	18	32
	15.102	16.898	
Fraud	144	63	207
	97.694	109.31	
Rape	62	88	150
	70.792	79.208	
Stealing	300	379	679
	320.45	358.55	
Violence	110	155	265
	125.07	139.93	
Total	673	753	1426

11.4.1 Creating a Contingency Table from Raw Data, the 2 × 2 Case

If your data are stored as individual observations in a SAS dataset, you can use the PROC FREQ procedure to create a table of counts and analyze the results. The following Hands-On Example illustrates this.

HANDS-ON EXAMPLE 11.6

In this example, we consider a dataset collected for the purpose of studying the relationship between two commercial floor cleaners and the presence or absence of the appearance of a skin rash on the hands of the users. The data for 40 subjects are in a SAS dataset named RASH.SAS7BDAT. A few of the records are shown in Figure 11.1. The floor cleaner brands are coded 1 or 2, and the presence of a rash is coded Y or N.

1. Open the program file AFREQ6.SAS.

```
PROC FREQ DATA="C:\SASDATA\RASH";
   TABLES CLEANER*RASH /CHISQ;
   TITLE 'CHI-SQUARE ANALYSIS FOR A 2X2 TABLE';
RUN;
```

(Continued)

(*Continued*)

	Subject	Cleaner	Rash
1	001	1	N
2	002	1	Y
3	003	1	Y
4	004	1	Y
5	005	1	N
6	006	1	Y
7	007	1	Y
8	008	1	Y
9	009	1	Y
10	010	1	N
11	011	1	Y
12	012	1	Y
13	013	1	Y
14	014	1	Y

Figure 11.1 Partial viewtable for the RASH dataset

2. Run the program and observe the output. The cross-tabulation table is given in Table 11.13, which contains the cell count along with overall, row, and column percents for each combination of cleaner type and observed rash.

TABLE 11.13 Cross-Tabulation of Cleaner By Rash

Frequency Percent Row Pct Col Pct	Table of Cleaner by Rash			
	Cleaner (Cleaner)	Rash (Rash)		
		N	Y	Total
	1	7 17.50 35.00 30.43	13 32.50 65.00 76.47	20 50.00
	2	16 40.00 80.00 69.57	4 10.00 20.00 23.53	20 50.00
	Total	23 57.50	17 42.50	40 100.00

3. Observe the statistics tables as given in Table 11.14. Statistics for a 2×2 cross-tabulation are reported in two tables. The first table is similar to the one reported for the crime data, while the second reports the results of a Fisher's Exact test. In this

TABLE 11.14 Statistics Tables for a 2×2 Analysis

Statistic	DF	Value	Prob
Chi-Square	1	8.2864	0.0040
Likelihood Ratio Chi-Square	1	8.6344	0.0033
Continuity Adj. Chi-Square	1	6.5473	0.0105
Mantel-Haenszel Chi-Square	1	8.0793	0.0045
Phi Coefficient		−0.4551	
Contingency Coefficient		0.4143	
Cramer's V		−0.4551	

Fisher's Exact Test	
Cell (1,1) Frequency (F)	7
Left-sided Pr <= F	0.0048
Right-sided Pr >= F	0.9995
Table Probability (P)	0.0042
Two-sided Pr <= P	0.0095
Sample Size = 40	

case, the chi-square statistic, 8.29, $p = 0.004$, indicates an association between CLEANER and RASH (rejects the null hypothesis that the proportion of users who experience a rash is the same for the two cleaners). The Continuity Adj. Chi-Square (sometimes called Yates's chi-square) is an adjustment that some statisticians use to improve the chi-square approximation in the case of a 2×2 table. The p-value for the adjusted chi-square is also <0.05. The second table of statistics reports the Fisher's Exact test. This test is based on all possible 2×2 tables that have the same marginal counts as those observed. This test is often reported instead of the chi-square when counts in the table are small. Typically, the two-sided p-value $\Pr \leq P$ value is the correct value to report except in specialized cases. For this example, Fisher's p-value is $p < 0.0095$.

11.4.2 Tables with Small Counts in Cells

When you summarize counts in tables and there are small numbers in one or more cells, a typical chi-square statistical analysis may not be valid. For example, suppose you have sampled freshmen and sophomores at a college to see if they took advantage of the Learning Enhancement Center (LEC) tutorials and you want to know if the proportion using the center differs by class. Hands-On Example 11.7 illustrates how a small sample size can lead to interpretation problems.

HANDS-ON EXAMPLE 11.7

This example creates a 2×2 table from count data, but the sample size is small enough that there are interpretation problems.

1. Open the program file AFREQ7.SAS.

```
DATA LEARN;
INPUT CLASS $ LEC $ COUNT;
DATALINES;
S Y 4
S N 11
F Y 5
F N 3
;
PROC FREQ DATA=LEARN;WEIGHT COUNT;
    TABLES CLASS*LEC/CHISQ;
TITLE 'Chi Square Analysis of a Contingency Table';
RUN;
```

This program analyzes a 2×2 table from summary data. Submit the program and observe the results. Partial results are given in Table 11.15.

In the chi-square table, observe the warning message "WARNING: 50% of the cells have expected counts <5. Chi-square may not be a valid test."

TABLE 11.15 Cross-Tabulation of a 2×2 Table with Small Cell Sizes

Statistic	DF	Value	Prob
Chi-Square	1	2.8126	0.0935
Likelihood Ratio Chi-Square	1	2.8066	0.0939
Continuity Adj. Chi-Square	1	1.5094	0.2192
Mantel-Haenszel Chi-Square	1	2.6903	0.1010
Phi Coefficient		−0.3497	
Contingency Coefficient		0.3301	
Cramer's V		−0.3497	
WARNING: 50% of the cells have expected counts less than 5. Chi-Square may not be a valid test.			

Fisher's Exact Test	
Cell (1,1) Frequency (F)	3
Left-sided Pr <= F	0.1102
Right-sided Pr >= F	0.9834
Table Probability (P)	0.0935
Two-sided Pr <= P	0.1793
Sample Size = 23	

2. Change the TABLE statement to

```
TABLES CLASS*LEC/CHISQ EXPECTED NOPERCENT NOCOL NOROW;
```

Resubmit the program and observe the cross-tabulation table, given in Table 11.16. Because this code contains the EXPECTED option the output displays expected frequencies in each cell. Observe that the expected number of Females for both No and Yes (LEC) are <5. This indicates that the chi-square test may not be valid. In this case, the Fisher's Exact test (given in Table 11.13) is more reliable. Usually, the two-sided p-value ($p = 0.1793$) would be the correct value to report for this analysis.

TABLE 11.16 Cross-Tabulation of a 2×2 Table Showing Expected Values

Frequency Expected	Table of CLASS by LEC		
		LEC	
	CLASS	N	Y	Total
	F	3	5	8
		4.8696	3.1304	
	S	11	4	15
		9.1304	5.8696	
	Total	14	9	23

11.5 GOING DEEPER: CALCULATING RELATIVE RISK MEASURES

Two-by-two contingency tables are often used when examining a measure of risk. In a medical setting, these tables are often constructed when one variable represents the presence or absence of an outcome (e.g. a disease) and the other is some risk factor. A measure of this risk in a retrospective (case–control) study is called the odds ratio (OR). In a case–control study, a researcher takes a sample of subjects and looks back in time for exposure (or nonexposure). If the data are collected prospectively, where subjects are selected by presence or absence of risk and then observed over time to see if they develop an outcome, the measure of risk is called relative risk (RR).

In either case, a risk measure (OR or RR) equal to 1 indicates no risk. A risk measure different from 1 represents a risk. Assuming the outcome studied is undesirable, a risk measure >1 indicates that exposure is harmful and a risk measure <1 implies that exposure is a benefit.

In PROC FREQ, the option to calculate the values for OR or RR is RELRISK and appears as an option to the TABLES statement as shown here:

```
TABLES CLEANER*RASH /RELRISK;
```

Hands-On Example 11.8 illustrates how to calculate the OR for the cleaner/rash data described previously.

HANDS-ON EXAMPLE 11.8

This example uses the RASH data in Hands-On Example 11.6. Here we assume that the data were collected in a retrospective (case–control) study.

1. Open the program file AFREQ6.SAS (used in a previous example) and change the TABLES statement to read

    TABLES CLEANER*RASH /RELRISK;

2. Run the revised program and observe the output as given in Table 11.17. In this table, the OR=0.1346 specifies the odds of Row1/Row2 – that is, for cleaner 1 versus cleaner 2. Because OR is <1, the interpretation is that the odds of a person having a rash who is using cleaner 1 is less than the odds when the person is using cleaner 2.

    It is helpful to note that the inverse of this OR (1/0.1346=7.49) indicates that the odds of a person using cleaner 2 getting a rash is 7.49 times greater than that of a person using cleaner 1. A method for altering this program to report the odds of cleaner 2 versus cleaner 1 is given in the chapter exercises.

TABLE 11.17 Relative Risk Measures for the Floor Cleaner Data

Statistics for Table of Cleaner by Rash		
Odds Ratio and Relative Risks		
Statistic	Value	95% Confidence Limits
Odds Ratio	0.1346	0.0322 0.5625
Relative Risk (Column 1)	0.4375	0.2316 0.8265
Relative Risk (Column 2)	3.2500	1.2776 8.2673

11.6 GOING DEEPER: INTER-RATER RELIABILITY (KAPPA)

A method for assessing the degree of agreement between two raters is Cohen's kappa coefficient. For example, kappa is useful for analyzing the consistency of two raters who evaluate subjects on the basis of a categorical measurement.

Using an example from Fleiss (1981, p. 213), suppose you have 100 subjects rated by two raters on a psychological scale that consists of three categories. The data are given in Table 11.18.

TABLE 11.18 Data for Inter-Rater Reliability Analysis

		RATER A			
		Psyc.	Neuro.	Organic	
Rater B	Psych.	75	1	4	80
	Neuro.	5	4	1	10
	Organic.	0	0	10	10
		80	5	15	100

Source: Fliess (1981, table 13.1, p. 213). Reproduced with permission from Wiley.

HANDS-ON EXAMPLE 11.9

This example illustrates how to calculate the kappa statistic in inter-rater reliability analysis.

1. Open the program file AKAPPA1.SAS.

```
DATA KAPPA;
INPUT RATER1 RATER2 WT;
DATALINES;
1   1   75
1   2   1
1   3   4
2   1   5
2   2   4
2   3   1
3   1   0
3   2   0
3   3   10
;
PROC FREQ;
   WEIGHT WT;
   TABLE RATER1*RATER2 / AGREE; TEST KAPPA;
   TITLE 'KAPPA EXAMPLE FROM FLEISS';
RUN;
```

This SAS command is similar to that used for a chi-square analysis except with a /AGREE option to request the kappa statistic and a TEST KAPPA statement to request a kappa analysis.

2. Run this program and observe the tables of output. Partial results are given in Table 11.19.

TABLE 11.19 Results for Kappa Analysis

Symmetry Test		
Chi-Square	DF	Pr > ChiSq
7.6667	3	0.0534

Kappa Statistics				
Statistic	Estimate	Standard Error	95% Confidence Limits	
Simple Kappa	0.6765	0.0877	0.5046	0.8484
Weighted Kappa	0.7222	0.0843	0.5570	0.8874

Test of H0: Kappa = 0				
Estimate	H0 Std Err	Z	Pr > Z	Pr > |Z|
0.6765	0.0762	8.8791	<.0001	<.0001

The "Symmetry Test" table provides Bowker's test of marginal homogeneity. A nonsignificant result (which is usually what you want) indicates that there is no evidence that the two raters have differing tendencies to select categories. In this case, the results are marginally nonsignificant ($p = 0.053$).

(Continued)

(*Continued*)

The "Kappa Statistics" table reports the kappa statistic and related measures. In this case, kappa = 0.6765. ASE is the asymptotic standard error. A 95% confidence interval is also reported. A large value of kappa (many would say 0.61 or higher) indicates a moderate to substantial level of agreement. This is based on a widely referenced interpretation of kappa suggested by Landis and Koch (1977) and given in Table 11.20.

The test of hypothesis "Test of H_0:" table in Table 11.19 provides a test of the null hypothesis that kappa = 0. In this case, the test indicates you would reject the null hypothesis ($p < 0.0001$).

TABLE 11.20 Interpretation of Kappa Statistic

Kappa Value	Interpretation
< 0	No agreement
0.0-0.20	Poor agreement
0.21-0.40	Fair agreement
0.41-0.60	Moderate agreement
0.61-0.80	Substantial agreement
0.81-1.00	Almost perfect agreement

3. Observe the graphical output as shown in Figure 11.2. (This may not appear if you are using an older version of SAS.) This graph provides a visual inspection of how well raters agreed on the three categories. The larger squares indicate larger sample sizes in the cell. The inner square indicates exact agreement and the outer square is partial agreement. The larger the area represented with the dark, the larger the kappa will be.

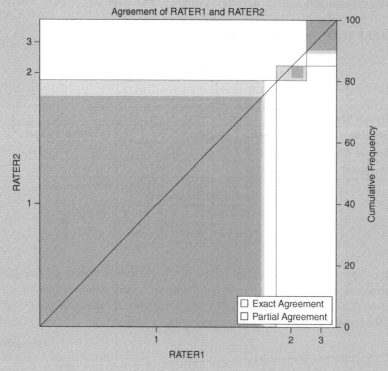

Figure 11.2 Example graphical output from a kappa test

11.6.1 Calculating Weighted Kappa

For the case in which rated categories are ordinal, it is appropriate to use the weighted kappa statistic because it is designed to give partial credit to ratings that are close to but not on the diagonal.

For example, in a test of recognition of potentially dangerous airline passengers, suppose a procedure is devised that classifies passengers into three categories: 1 = No threat/Pass, 2 = Concern/Recheck, and 3 = Potential threat/Detain. To assess the reliability of this measure, suppose two security officers are trained in the procedure, and they rate 99 passengers. A partial listing of the data is given in Table 11.21. Note that the categories are ordinal, so weighted kappa is an appropriate analysis.

TABLE 11.21 Partial SECURITY Data

RATER1	RATER2
1	1
1	1
1	1
1	1
1	3
3	2
2	2
etc.	etc.

HANDS-ON EXAMPLE 11.10

This example uses the hypothetical security office training dataset to illustrate how to calculate the kappa statistic in an inter-rater reliability analysis when categories are ordinal. Unlike the previous example, in which the data are summarized in the dataset, in this example, the data are listed by subject (see Table 11.21), and the PROC FREQ procedure creates the appropriate cross-tabulation of counts for the analysis.

1. Open the program file AKAPPA2.SAS.

```
PROC FREQ data="C:\SASDATA\SECURITY";
   TABLE RATER1*RATER2 / AGREE; TEST WTKAP;
   TITLE 'Security Data';
RUN;
```

2. Run the program and observe the cross-tabulation table given in Table 11.22.
3. The AGREE option specifies the kappa statistic, and the TEST WTKAP produces tests for the weighted statistic. The kappa and weighted kappa results are given in Table 11.23.

The test of symmetry is nonsignificant ($p = 0.84$) and suggests there is no evidence that the two raters have differing tendencies to select categories. Note that both the simple and weighted kappa statistics and confidence limits are calculated by SAS. Because the data are ordinal, the appropriate statistic for this analysis is the

(Continued)

(*Continued*)

weighted kappa. The test of hypothesis "Test of H_0:" table indicates that you would reject the hypothesis that kappa = 0 ($p < 0.0001$) indicates agreement.

Weighted kappa = 0.74 in this example statistic can be interpreted as suggesting a "substantial agreement" between the two raters using the criteria of Landis and Koch. The researcher involved in this experiment must determine, in relation to the importance of the decision involving the lives of passengers, if the result indicates sufficient agreement to adopt the classification method.

TABLE 11.22   Cross-Tabulation of Security Data

Frequency Percent Row Pct Col Pct	Table of RATER1 by RATER2			
	RATER1 (RATER1)	RATER2 (RATER2)		
		1	2	3	Total
	1	58	3	1	62
		58.59	3.03	1.01	62.63
		93.55	4.84	1.61	
		90.63	10.71	14.29	
	2	5	24	2	31
		5.05	24.24	2.02	31.31
		16.13	77.42	6.45	
		7.81	85.71	28.57	
	3	1	1	4	6
		1.01	1.01	4.04	6.06
		16.67	16.67	66.67	
		1.56	3.57	57.14	
	Total	64	28	7	99
		64.65	28.28	7.07	100.00

TABLE 11.23   Kappa Results for Security Data

Symmetry Test		
Chi-Square	DF	Pr > ChiSq
0.8333	3	0.8415

Kappa Statistics				
Statistic	Estimate	Standard Error	95% Confidence Limits	
Simple Kappa	0.7386	0.0654	0.6104	0.8667
Weighted Kappa	0.7413	0.0674	0.6092	0.8735

Test of H0: Weighted Kappa = 0				
Estimate	H0 Std Err	Z	Pr > Z	Pr > |Z|
0.7413	0.0845	8.7682	<.0001	<.0001

11.7 SUMMARY

This chapter discusses the capabilities of PROC FREQ for creating one- and two-way frequency tables, analyzing contingency tables, calculating measures of risk, and measuring inter-rater reliability (using KAPPA).

EXERCISES

11.1 **Perform a goodness-of-fit analysis.**

Suppose you conduct a marketing survey in a city where you hypothesize that people frequent restaurants at lunchtime in the following proportions: Mexican food (40%), home cooking (20%), Italian food (20%), and Chinese food (20%). The results of a random sample from the population are given here in which subjects were asked to specify the type of restaurant at which they most recently ate lunch.

Mexican: 66

Home cooking: 25

Italian: 33

Chinese: 38

The DATA statement to read in this information is in EX_11.1.SAS:

```
DATA FOOD;
INPUT @1 CATEGORY $13. @14 NUMBER 3.;
DATALINES;
Mexican 66
Home cooking 25
Italian 33
Chinese 38
;
RUN;
```

Using this as a starting point, perform a goodness-of-fit test using PROC FREQ. Hint: Use Hands-On Example 11.3 as a template for the analysis. Use the statements:

`PROC FREQ ORDER=DATA; WEIGHT NUMBER;`

and

`TESTP=(.4,.2,.2,.2);`

in your SAS program to perform the appropriate goodness-of-fit test.

11.2 **Perform a contingency table analysis.**

Historical data used to research the relationship between smoking and cancer yields the following data:

Smoking Habit/Cancer	Cancer	No Cancer	Total
None to slight	56	956	1012
Moderate to excessive	269	1646	1915
Total	325	2602	2927

a. Using the cleaner/rash SAS program example as a template for your SAS program code (AFREQ6.SAS), perform a chi-square test on these data using PROC FREQ and state your conclusions.

b. What is the hypothesis tested in this example?

11.3 **Going deeper: calculate risk.**

a. Using the code in AFREQ6.SAS, change the TABLES statement to

```
TABLES CLEANER*RASH /RELRISK;
```

Run the program and observe the OR = 0.1346 representing the risk of getting a rash when using Cleaner 1 versus Cleaner 2.

b. Change the table to calculate the risk of Cleaner 2 versus Cleaner 1. To do this, you must change the order of the cleaner categories. By default, SAS orders them in numerical order (1 and 2). To change this order, create a formatted value for CLEANER and use the ORDER=FORMATTED option to put the categories in the desired order. Enter this code at the top of the program:

```
PROC FORMAT;
VALUE $FMTYN "Y"="1 YES"
  "N"="2 NO";
RUN;
```

Moreover, change the PROC FREQ statement to read

```
PROC FREQ DATA="C:\SASDATA\RASH" ORDER=FORMATTED;
```

and add a FORMAT statement after the TITLE statement and before RUN.

```
FORMAT RASH $FMTYN.;
```

Run the revised program. The order of CLEANER should be changed according to the formatted values, and the OR reported is the reciprocal of 0.1346, or 7.429.

c. What does OR = 7.429 mean?

11.4 **Going deeper: calculate kappa.**

Suppose a researcher is examining the reliability of a method for interpreting X-rays displayed on photographic film. Two raters examine a series of images and classify the severity of a bone break using a 1, 2, 3 system with 1 indicating a minor break and 3 indicating a severe break. The data are as follows:

		\multicolumn{3}{c}{Rater 1}		
			1	2	3
		1	43	7	1
Rater 2		2	9	24	5
		3	4	8	19

a. Which is the appropriate kappa statistic to use for this analysis (kappa or weighted kappa)? Do the rates seem to be in agreement? (Assess the agreement visually before calculating any statistics.)
b. Write a SAS program to read the data and calculate the appropriate kappa statistic.
c. What is the value of the kappa statistic?
   According to the Landis and Koch criteria, how would you characterize the strength of this result?

REFERENCES

Fleiss, J.L. (1981). *Statistical Methods for Rates and Proportions*, 2e. New York: Wiley.
Landis, J.R. and Koch, G.G. (1977). The measurement of observer agreement for categorical data. *Biometrics* 33: 159–174.

12

COMPARING MEANS USING *T*-TESTS

LEARNING OBJECTIVES

- To understand the use of the one-sample *t*-test using PROC UNIVARIATE and PROC TTEST
- To understand the use of the two-sample *t*-test using PROC TTEST
- To understand the use of the paired *t*-test using PROC TTEST

Experiments whose outcome measures are quantitative variables are often analyzed by comparing means. The Student's *t*-test is the most commonly used statistical test for comparing two means or for comparing an observed mean with a known value. If more than two groups are observed, an analysis of variance (ANOVA) is used to compare means across groups (discussed in Chapter 14).

12.1 PERFORMING A ONE-SAMPLE *T*-TEST

A one-sample *t*-test is often used to compare an observed mean with a known or "gold standard" value. For example, in a quality control setting, you may be interested in comparing a sample of data to an expected outcome, such as the actual number of calories in cans of baby formula against the claim on the label. The purpose of the one-sample *t*-test, in this

---

SAS® Essentials: Mastering SAS for Data Analytics, Third Edition. Alan C. Elliott and Wayne A. Woodward.
© 2023 John Wiley & Sons, Inc. Published 2023 by John Wiley & Sons, Inc.
Companion website: https://www.alanelliott.com/SASED3/

case, is to determine if there is enough evidence to dispute the claim. In general, for a one-sample *t*-test, you obtain a random sample from some population and then compare the observed sample mean to some fixed value. The typical hypotheses for a one-sample *t*-test are as follows:

$H_0: \mu = \mu_0$: The population mean is equal to a hypothesized value, μ_0.
$H_a: \mu \neq \mu_0$: The population mean is not equal to μ_0.

The key assumption underlying the one-sample *t*-test is that the population from which the random sample is selected is normal. If the data are non-normal, then nonparametric tests such as the sign test and the signed rank test are available (see Chapter 16). However, because of the central limit theorem, whenever the sample size is sufficiently large, the distribution of the sample mean is approximately normal even when the population is non-normal. A variety of rules of thumb have been recommended to help you determine whether to go ahead and trust the results of a one-sample *t*-test even when your data are non-normal. The following are general guidelines (see Moore et al. 2014).

- Small sample size ($N < 15$): You should not use the one-sample *t*-test if the data are clearly skewed or if outliers are present.
- Moderate sample size ($N > 15$): The one-sample *t*-test can be safely used except when there are severe outliers.
- Large sample size ($N > 40$): The one-sample *t*-test can be safely used without regard to skewness or outliers.

12.1.1 Running the One-Sample *t*-Test in SAS®

In SAS there are (at least) two ways to perform a one-sample *t*-test. Using PROC UNIVARIATE, you can specify the value of μ_0 for the test reported in the "Tests for Location" table given in Table 12.1 using the option MU0=*value*. For example,

```
PROC UNIVARIATE MU0=4;VAR LENGTH; RUN;
```

would request a one-sample *t*-test of the null hypothesis that $\mu_0 = 4$.

A second method in SAS for performing this one-sample *t*-test is to use the PROC TTEST procedure. For this procedure, the corresponding SAS code is

```
PROC TTEST H0=4;VAR LENGTH; RUN;
```

Hands-On Example 12.1 illustrates these two techniques.

TABLE 12.1 *t*-Test Results Using PROC UNIVARIATE

Tests for Location: Mu0 = 4				
Test		Statistic		p Value
Student's *t*	t	−1.40593	Pr > |t|	0.1759
Sign	M	−1.5	Pr >= |M|	0.6291
Signed Rank	S	−26.5	Pr >= |S|	0.2240

HANDS-ON EXAMPLE 12.1

In this example, you will learn how to perform a one-sample *t*-test using SAS commands. A certain medical implant component is reported to be 4 cm in length by its manufacturer. The precision of the component length is of the utmost importance. To test the accuracy of the manufacturer's claim, a random sample of 20 components is collected, and the following SAS code is used to test $\mu_0 = 4$.

1. Open the program file `ATTEST1.SAS`.

```
DATA ONESAMPLE;
INPUT LENGTH @@;
DATALINES;
4       3.95    4.01    3.95    4.00
3.98    3.97    3.97    4.01    3.98
3.99    4.01    4.02    4.02    3.98
4.01    3.99    4.03    4.00    3.99
;
Title 'Single sample t-test, using PROC UNIVARIATE';
PROC UNIVARIATE DATA=ONESAMPLE MU0=4;VAR LENGTH; RUN;
Title 'Single sample t-test using PROC TTEST';
PROC TTEST DATA=ONESAMPLE H0=4;var LENGTH;
RUN;
```

Note that the *t*-test is performed twice: once using `PROC UNIVARIATE` and again using `PROC TTEST`. If you are using older versions of SAS, you may need to add the statement

```
ODS GRAPHICS ON;
```

at the first of your code to obtain the graphs mentioned in this example.

2. Run this program and observe the "Test for Location" table from `PROC UNIVARIATE` given in Table 12.1. The row titled "Student's *t*" contains the test statistic ($t = -1.40$) and the *p*-value ($p = 0.1759$). This *p*-value indicates that you would not reject $H_0: \mu = 4$ at the $\alpha = 0.05$ level of significance. `PROC UNIVARIATE` also reports that the sample mean is 3.993 and the sample standard deviation is 0.022 (not shown here). Two nonparametric tests are also performed: the sign and signed rank tests. These tests are discussed in Chapter 16. Because the sample size is small, normality is an issue that needs to be addressed. The results of these tests show that we make the same decision (i.e., to not reject the null hypothesis) whether or not normality is assumed. A more direct assessment of the normality of the data is discussed in step 5.

3. Observe the results from `PROC TTEST`, given in Table 12.2. Note that the *t*-statistic is the same as that given by `PROC UNIVARIATE`, and the sample mean and standard deviation are 3.993 and 0.022, respectively, as before. Table 12.2 also reports the degrees of freedom (`DF`). The results of this analysis are that the *t*-statistic (which has 19 DF) is equal to −1.41 and the *p*-value is 0.1759. The "*t*(19)" shows the number of degrees of freedom (19 in this case), which should be included when you report these results. The `TTEST` output also gives you a 95% confidence interval for the mean

under the "95% CL Mean" heading (3.9826, 4.0034). Both the *p*-value and the fact that the confidence interval contains the hypothesized value 4 indicate that $H_0: \mu = 4$ would not be rejected at the 0.05 level of significance.

TABLE 12.2 *t*-Test Results Using PROC TTEST

N	Mean	Std Dev	Std Err	Minimum	Maximum
20	3.9930	0.0223	0.00498	3.9500	4.0300

Mean	95% CL Mean		Std Dev	95% CL Std Dev	
3.9930	3.9826	4.0034	0.0223	0.0169	0.0325

DF	*t* Value	Pr > |*t*|
19	−1.41	0.1759

4. Although the statistical test did not reject the null hypothesis, the researcher should also consider the minimum and maximum values in the table to determine if, in a clinical setting, either of these lengths would present a detrimental clinical issue or whether they might indicate incorrectly coded data.
5. Observe the plots in Figure 12.1. The blue (or solid curve in some SAS versions) is a normal curve based on the mean (3.993) and standard deviation (0.022) estimated from the data. The red (or dashed in some SAS versions) curve is a kernel

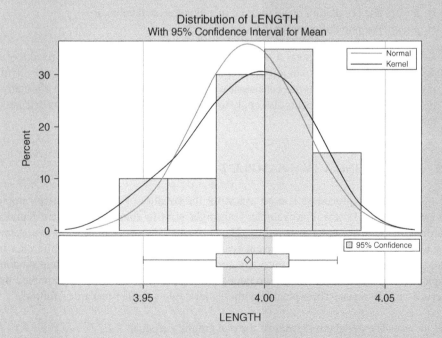

Figure 12.1 Histogram, kernel density estimate, and normal curve output for single sample *t*-test

(*Continued*)

(*Continued*)

density estimator, which is a smoothed version of the histogram. If dramatic skewness were evident in the data, then the skewness would also be seen in the kernel density estimator. Another plot (a Q–Q plot) that can help you assess normality is also displayed in the output (although not shown here.) Q–Q plots were discussed earlier in Section 10.2.

These plots provide information about the normality assumption. There does not appear to be a dramatic departure from normality because the kernel density estimate is fairly bell-shaped. At the bottom of the histogram is a boxplot that plots the minimum, 25th, 50th, and 75th percentiles along with the maximum of the data. The boxplot is fairly symmetrical in shape.

One-tailed tests: SAS always reports two-tailed *p*-values for the *t*-test. If you are interested only in rejecting the null hypothesis if the population mean differs from the hypothesized value in a particular direction of interest, you may want to use a one-tailed (sometimes called a one-sided) test. For example, if you want to reject the null hypothesis only if there is sufficient evidence that the mean is smaller than the hypothesized value, the hypotheses become as follows:

$H_0: \mu = \mu_0$: The population mean is equal to a hypothesized value, μ_0.
$H_a: \mu < \mu_0$: The population mean is less than μ_0.

Because SAS always reports a two-tailed *p*-value, in order to report the results of a one-tailed test you need to modify the reported *p*-value to fit a one-tailed test by dividing it by 2 (if your results are consistent with the direction specified in the alternative hypothesis). These comments also apply to the two-sample and paired *t*-tests discussed later in this chapter.

12.2 PERFORMING A TWO-SAMPLE *T*-TEST

The `PROC TTEST` procedure is used to test for the equality of means for a two-sample (independent group) *t*-test. For example, you might want to compare males and females regarding their reactions to a certain drug. The purpose of the two-sample *t*-test is to determine whether your data provide sufficient evidence to conclude that there is a difference in mean reaction levels. In general, for a two-sample *t*-test you obtain independent random samples of size N_1 and N_2 from the two populations of interest, and then you compare the observed sample means. The typical hypotheses for a two-sample *t*-test are as follows:

$H_0: \mu_1 = \mu_2$: The population means of the two groups are equal.
$H_a: \mu_1 \neq \mu_2$: The population means are not equal.

As in the case of the one-sample *t*-test, SAS provides *p*-values for a two-tailed test. If you have a one-tailed alternative, the *p*-values will need to be modified as mentioned in the section on one-sample *t*-tests (i.e., divide the *p*-value by 2 if the results are consistent with the alternative hypothesis). Key assumptions underlying the two-sample *t*-test are that the random samples are independent and that the populations are normally distributed with equal variances. If the data are non-normal, then nonparametric tests such as the Mann–Whitney U are available (see Chapter 16).The following are guidelines regarding normality and equal variance assumptions:

Normality: As in the one-sample case, rules of thumb are available to help you determine whether to go ahead and trust the results of a two-sample *t*-test even when your data are non-normal. The sample size guidelines given earlier in this chapter for the one-sample test can be used in the two-sample case by replacing N with $N_1 + N_2$ (see Moore et al. 2014).

Equal variances: There are two *t*-tests reported by SAS in this setting: one based on the assumption that the variances of the two groups are equal (and thus using a pooled estimate of the common variance) and one (Satterthwaite) not making that assumption. Both methods assume normality. There is no universal agreement concerning which *t*-test to use.

A conservative approach suggested by some statisticians (see Moore, McCabe, and Craig, 2014) is to *always use the version of the t-test (Satterthwaite) that does not assume equal variances*. The following is the classical approach to deciding which version of the *t*-test to use:

Formally test the equal variance assumption using an *F*-test that SAS gives in the *t*-test output. The *F*-test is a test of the null hypothesis that the variances are equal.

- If the *p*-value for the *F*-test test is less than α (say 0.05), then conclude that the variances are unequal and use the Satterthwaite *t*-test.
- If the *p*-value for the *F*-test is greater than α, use the *t*-test based on a pooled estimate of the variances.

At least one of the reasons for the recommendation to always use the Satterthwaite test is that studies have shown that the *F*-test for assessing whether the variances are equal is unreliable.

> If your observations are related across "groups" as paired or repeated measurements, the two-sample *t*-test is an *incorrect* version of the *t*-test. For that case, see the discussion of the paired *t*-test in Section 12.3.

12.2.1 Running the Two-Sample *t*-Test in SAS

Two sample *t*-tests can be obtained using the PROC TTEST procedure, which was previously introduced in the context of a one-sample test. In this section, we'll describe PROC TTEST more completely. The syntax for the TTEST procedure is as follows:

PROC TTEST <*options*>; CLASS variable; <*statements*>;

Table 12.3 lists common options and statements used with PROC TTEST.

Other options and statements are available for PROC TTEST concerning analyses not discussed in this text. Consult the SAS documentation for more information.

TABLE 12.3 Common Options and Statements for PROC TTEST

Option	Explanation
DATA = datasetname	Specifies which dataset to use
COCHRAN	Specifies that the Cochran and Cox probability approximation is to be used for unequal variances. (The Satterthwaite test is another approximation used for unequal variances. It is the default.)
H0=c	Specifies the hypothesized value under H_0 (null hypothesis) in a one-sample *t*-test
WEIGHT var;	Specifies that an observation is to be counted a number of times according to the value of the WEIGHT variable. This is the same usage as discussed for PROC FREQ in Chapter 11
Common statements for PROC TTEST	
CLASS variables;	The CLASS statement is required for a two-sample *t*-test, and it specifies the grouping variable for the analysis. The data for this grouping variable must contain two and only two values. An example PROC TTEST command is PROC TTEST;CLASS GROUP; VAR SCORE; In this example, where GROUP contains two values, say 1 or 2, a *t*-test will be performed on the variable SCORE
VAR variables;	Specifies which variables will be used in the analysis. For example, PROC TTEST; CLASS GROUP; VAR SCORE WEIGHT HEIGHT; would produce three *t*-tests, one for each variable in the VAR statement
PAIRED x*y;	Specifies that a paired *t*-test is to be performed and which variables to use. For example, PROC TTEST; PAIRED BEFORE*AFTER;
BY, FORMAT, LABEL, WHERE	These statements are common to most procedures, and may be used with PROC TTEST

HANDS-ON EXAMPLE 12.2

In this example, a biologist experimenting with plant growth designs an experiment in which 15 seeds are randomly assigned to one of two fertilizers and the height of the resulting plant is measured after 2 weeks. She wants to know if one of the fertilizers provides more vertical growth than the other.

1. Open the program file ATTEST2.SAS.

```
DATA GROW;
INPUT BRAND $ HEIGHT;
DATALINES;
A    20.00
A    23.00
A    32.00
A    24.00
A    25.00
A    28.00
A    27.50
B    25.00
B    46.00
```

```
B   56.00
B   45.00
B   46.00
B   51.00
B   34.00
B   47.50
;
PROC TTEST;
   CLASS BRAND;
   VAR HEIGHT;
   Title 'Independent Group t-Test Example';
RUN;
```

2. Run this program and observe the output for PROC TTEST, given in Table 12.4.

TABLE 12.4 Two-Sample *t*-Test Output from PROC TTEST

BRAND	Method	N	Mean	Std Dev	Std Err	Minimum	Maximum
A		7	25.6429	3.9021	1.4748	20.0000	32.0000
B		8	43.8125	9.8196	3.4717	25.0000	56.0000
Diff (1–2)	Pooled		−18.1696	7.6778	3.9736		
Diff (1–2)	Satterthwaite		−18.1696		3.7720		

BRAND	Method	Mean	95% CL Mean		Std Dev	95% CL Std Dev	
A		25.6429	22.0340	29.2517	3.9021	2.5145	8.5926
B		43.8125	35.6031	52.0219	9.8196	6.4925	19.9855
Diff (1–2)	Pooled	−18.1696	−26.7541	−9.5852	7.6778	5.5660	12.3692
Diff (1–2)	Satterthwaite	−18.1696	−26.6479	−9.6914			

Method	Variances	DF	*t* Value	Pr > |*t*|
Pooled	Equal	13	−4.57	0.0005
Satterthwaite	Unequal	9.3974	−4.82	0.0008

Equality of Variances				
Method	Num DF	Den DF	F Value	Pr > F
Folded F	7	6	6.33	0.0388

- There is a lot of output, and you have to be careful about what you are reading. Note that there are actually four tables. The first table gives you the sample mean, standard deviation, standard error of the mean, and the minimum and maximum for each value of BRAND (A and B). The last line includes similar information for the mean difference.
- The second table shows the mean and standard deviations again along with 95% confidence intervals for the means and standard deviations in each group along with two 95% confidence intervals for the difference in the group means: (1) using a pooled estimate of the variance (if you believe the variances are equal) and

(*Continued*)

314 CHAPTER 12: COMPARING MEANS USING *T*-TESTS

(*Continued*)

    (2) based on the Satterthwaite method that does not make an assumption of equal variances.

- The third table gives the results of the two *t*-tests. The *t*-test based on a pooled estimate of the variance has a *p*-value of 0.0005 while for the Satterthwaite version, $p = 0.0008$. Note that the Satterthwaite version generally reports fractional DF. In this case, that value is 9.4. Note also that the decision regarding which *t*-test to use is not crucial here because both suggest rejecting the null hypothesis. These two tests will typically agree.
- The fourth table gives the results of the *F*-test for deciding whether the variances can be considered equal. The *p*-value for this test is 0.0388. So, at the 0.05 level, we would conclude that the variances are not equal, and it appears that the variance for BRAND B is larger than that for BRAND A. Thus, the classical approach would lead us to use the Satterthwaite *t*-test.

3. Observe the graphical information shown in Figure 12.2. This figure is similar to Figure 12.1, except that it shows histograms, normal curves, kernel density estimators, and boxplots for HEIGHT separately for BRAND A and BRAND B. In both cases, normality looks like a fairly good assumption, but there seems to be evidence that the variance for BRAND B is larger than that for A. Q–Q plots are also included in the output (not shown here). The information in the plots is consistent with the conclusions of the *F*-test given in Table 12.3.

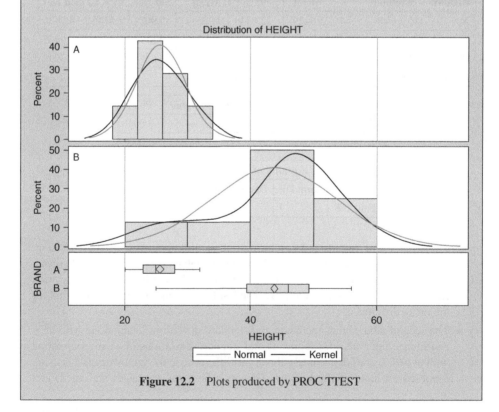

Figure 12.2 Plots produced by PROC TTEST

12.2.2 Using PROC BOXPLOT

A complementary graphical representation that can be used to compare means visually can be obtained using PROC BOXPLOT. Although boxplots are produced by default (as shown in the preceding example), you might want to create a plot that includes only the side-by-side boxplots. You can use PROC BOXPLOT to create this type of graph (see Chapter 19).

12.3 PERFORMING A PAIRED T-TEST

To perform a paired *t*-test to compare two repeated measures (such as in a before–after situation), where both observations are taken from the same or matched subjects, use PROC TTEST with the PAIRED statement. Suppose your data contain the variables WBEFORE and WAFTER (before and after weight on a diet) for eight subjects, the hypotheses for this test are as follows:

$H_0: \mu_{\text{Loss}} = 0$: The population average weight loss is zero.
$H_a: \mu_{\text{Loss}} \neq 0$: The population average weight loss is not zero.

HANDS-ON EXAMPLE 12.3

In this example, you will learn how to perform a paired *t*-test using PROC TTEST.

1. Open the program file ATTEST3.SAS.

   ```
   DATA WEIGHT;
   INPUT WBEFORE WAFTER;
   DATALINES;
   200 185
   175 154
   188 176
   198 193
   197 198
   310 275
   245 224
   202 188
   ;
   PROC TTEST;
   PAIRED WBEFORE*WAFTER;
   TITLE 'Paired t-test example';
   RUN;
   ```

2. Run the program and observe the output in Table 12.5. The mean of the difference is 15.25. The *t*-statistic used to test the null hypothesis is 3.94, and the *p*-value for this paired *t*-test is $p = 0.0056$, which provides evidence to reject the null hypothesis. The output also gives a 95% confidence interval on the difference (7.23, 22.26), which indicates that the population's mean weight loss could plausibly have been

   (*Continued*)

(*Continued*)

as little as 7.23 lb and as much as 22.26 lb. Consequently, it appears that the mean weight loss is greater than zero. A one-sided test might have been used here because the goal of the study (before any data were collected) was to produce a positive weight loss.

TABLE 12.5 Paired *t*-Test Output

N	Mean	Std Dev	Std Err	Minimum	Maximum
8	15.2500	10.9381	3.8672	−1.0000	35.0000

Mean	95% CL Mean		Std Dev	95% CL Std Dev	
15.2500	6.1055	24.3945	10.9381	7.2320	22.2621

DF	*t* Value	Pr > |t|
7	3.94	0.0056

3. Observe the graphic illustrating the distribution of the difference in Figure 12.3. In this graph, the histogram, the kernel density estimator, and the boxplot all suggest that the differences are reasonably normally distributed. Thus, even though the sample size is quite small, it seems reasonable to use the paired *t*-test. The `PAIRED` option produces several other plots that are not shown here.

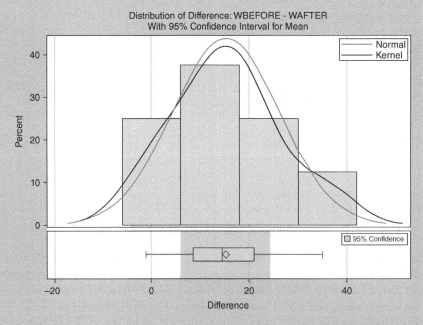

Figure 12.3 Plots produced by PROC TTEST using the PAIRED option

The paired *t*-test is actually a one-sample *t*-test computed on the differenced data. That is, if we computed a new variable DIF=WBEFORE-WAFTER and ran a one-sample *t*-test on the variable DIF testing $H_0: \mu_{DIF} = 0$, we would get results identical to those obtained using the PAIRED statement. The normality assumption in this case applies to the differenced data.

12.4 SUMMARY

This chapter illustrated how to perform tests on one or two means using PROC UNIVARIATE and PROC TTEST. In particular, the chapter emphasizes the fact that it is important to keep in mind the distinction between an independent group's comparison and a repeated (paired) observation's comparison.

EXERCISES

12.1 **Perform a *t*-test.**

Suppose you are interested in knowing if comprehension of certain medical instructions is dependent on the time of day when the instructions are given. In a school setting, you randomize members of a 12th-grade class into two groups. In the morning at 8:30, Group A is shown a video describing how to use an infant forehead thermometer. The same video is shown to Group B at 3:00 on the same day. The next day all students are given a test over the material. The following scores were observed.

Group A Subjects	Test Score	Group B Subjects	Test Score
1	88	1	87
2	89	2	69
3	79	3	78
4	100	4	79
5	98	5	83
6	89	6	90
7	94	7	85
8	95		

a. Is this analysis an independent or a paired comparison?
b. Enter these data into a SAS dataset and perform the appropriate *t*-test. Here is some code to get you started:

```
DATA TEMP;
INPUT GROUP $ SCORE;
DATALINES;
A 88
A 89
... etc ...
;
```

```
PROC TTEST DATA=TEMP;
CLASS _____ ; VAR _____ ;
... etc ...
RUN;
```

c. What is your conclusion?

12.2 Perform a *t*-test on pre- and postdata.

To test if a memory technique was effective, a researcher provided five people with a list of 10 objects for 45 seconds and then tested them to see how many they remembered (pretest). He then taught them a memory technique and then presented them with a list of 10 different (although similar) objects and administered a similar test (posttest). The results follow:

Subject	Pretest	Posttest
AD	6	10
GE	7	7
KL	9	10
MM	4	7
OU	7	9

a. Assuming that the differenced data are approximately normally distributed, what analysis is appropriate to test the hypothesis that training subjects on a memory technique would improve their ability to remember a list of 10 objects?

b. Using code similar to that in Hands-On Example 12.3, write a SAS program to perform the appropriate test for this analysis. Here is some code to get you started:

```
DATA MEMORY;
INPUT PRETEST _____ ;
DATALINES;
6 10
... etc ...
;
PROC TTEST DATA=MEMORY;
... etc ...
RUN;
```

c. What are your conclusions?

REFERENCE

Moore, D., McCabe, G., and Craig, B. (2014). *Introduction to the Practice of Statistics*, 8e. New York: Freeman.

13

CORRELATION AND REGRESSION

LEARNING OBJECTIVES

- To be able to use SAS® procedures to calculate Pearson and Spearman correlations
- To be able to use SAS procedures to produce a matrix of scatterplots
- To be able to use SAS procedures to perform simple linear regression
- To be able to use SAS procedures to perform multiple linear regression
- To be able to use SAS procedures to calculate predictions using a linear regression model
- To be able to use SAS procedures to perform residual analysis

Correlation measures the association between two quantitative variables, and the closely related regression analysis uses the relationship between independent and dependent variables for predicting the dependent (response or outcome) variable using one or more independent (predictor or explanatory) variables. In simple linear regression, there is a single dependent variable and a single independent variable, while in multiple linear regression (MLR), there are two or more independent variables. In this chapter, we discuss the use of SAS to perform correlation and regression analyses.

13.1 CORRELATION ANALYSIS USING PROC CORR

Before proceeding to the use of SAS to perform correlation analysis, we first provide a brief discussion of the basics of correlation analysis.

13.1.1 Correlation Analysis Basics

The correlation coefficient is a measure of the linear relationship between two quantitative variables measured on the same subject (or entity). For example, you might want to study the relationship between height and weight for a sample of teenage boys. For two variables of interest, say X and Y, the correlation, ρ, measures the extent to which a scatterplot of data from the bivariate distribution of X and Y tends to fall along a line. The correlation ρ is a unitless quantity (i.e., it does not depend on the units of measurement) that ranges from -1 to $+1$ where $\rho = -1$ and $\rho = +1$ correspond to perfect negative and positive linear relationships, respectively, and $\rho = 0$ indicates no linear relationship.

In practice, it is often of interest to test the hypotheses:

$H_0: \rho = 0$: There is no linear relationship between the two variables.

$H_a: \rho \neq 0$: There is a linear relationship between the two variables.

The correlation coefficient ρ is typically estimated from data using the Pearson correlation coefficient, usually denoted as r. `PROC CORR` provides a test of the above hypotheses designed to determine whether the estimated correlation coefficient, r, is significantly different from zero. This test assumes that the data represent a random sample from some bivariate normal population. If normality is not a good assumption, nonparametric correlation estimates are available, the most popular of which is Spearman's ρ, which `PROC CORR` also provides.

> Scatterplots: To examine the nature of the relationship between two variables, it is always a good practice to look at scatterplots of the variables.

13.1.2 Using SAS PROC CORR for Correlation Analysis

The SAS procedure most often used to calculate correlations is `PROC CORR`. The syntax for this procedure is as follows:

`PROC CORR <options>; <statements>;`

Table 13.1 lists common options and statements used with `PROC CORR`.

TABLE 13.1 Common Options for PROC CORR

Option	Explanation
DATA = datasetname	Specifies which SAS dataset to use
PEARSON	Requests Pearson correlation (default).
SPEARMAN	Requests Spearman rank correlations
NOSIMPLE	Suppresses display of descriptive statistics
NOPROB	Suppresses the display of p-values
PLOTS=	PLOTS=MATRIX requests a scatterplot matrix and PLOTS=SCATTER requests individual scatterplots. The PLOTS=MATRIX(HISTOGRAM) statement requests a scatterplot matrix with embedded histograms
OUTP=	Specifies an output dataset containing Pearson correlations

(Continued)

TABLE 13.1 (*Continued*)

Option	Explanation
Common statements for PROC CORR	
`VAR variable list`	All possible pairwise correlations are calculated for the variables listed and displayed in a table
`WITH variable(s);`	All possible correlations are obtained between the variables in the `VAR` list and variables in the `WITH` list
`MODEL`	Specifies the dependent and independent variables for the analysis. More specifically, it takes the form `MODEL dependentvar=independentvar(s);` More explanation follows
`BY, FORMAT, LABEL, WHERE`	These statements are common to most procedures and can be used with `PROC CORR`

HANDS-ON EXAMPLE 13.1

This example illustrates how to calculate Pearson correlations on several variables in a dataset.

1. Open the program file `ACORR1.SAS`.

   ```
   PROC CORR DATA="C:\SASDATA\SOMEDATA";
        VAR AGE TIME1 TIME2;
   TITLE "Example correlations using PROC CORR";
   RUN;
   ```

2. Run the program and observe the results given in Table 13.2. The output also includes descriptive statistics not shown here. Table 13.1 contains pairwise Pearson correlations between each of the three variables listed in the `VAR` statement. For example, the correlation between `AGE` and `TIME1` is $r = 0.50088$, and the p-value associated with the test of $H_0: \rho = 0$ is $p = 0.0002$.

TABLE 13.2 Output from PROC CORR

Pearson Correlation Coefficients, N = 50 Prob > |r| under H0: Rho = 0			
	AGE	TIME1	TIME2
AGE Age on Jan 1, 2000	1.00000	0.50088 0.0002	0.38082 0.0064
TIME1 Baseline	0.50088 0.0002	1.00000	0.76396 <.0001
TIME2 6 Months	0.38082 0.0064	0.76396 <.0001	1.00000

(*Continued*)

> *(Continued)*
> 3. If normality is questionable, you may want to request the rank correlations (Spearman) to be calculated (either by themselves or in combination with the Pearson correlations). Use the following PROC CORR statement to request both Pearson and Spearman correlations:
>
> ```
> PROC CORR DATA="C:\SASDATA\SOMEDATA" PEARSON SPEARMAN;
> ```
>
> Rerun the program and observe that a second table contains the Spearman correlations and the associated *p*-values.
>
> To simplify the output by suppressing the *p*-values and descriptive univariate statistics, change the PROC CORR statement to
>
> ```
> PROC CORR DATA="C:\SASDATA\SOMEDATA"
> PEARSON SPEARMAN NOPROB NOSIMPLE;
> ```
>
> Rerun the program and observe the output.

13.1.3 Producing a Matrix of Scatterplots

As indicated earlier, it is important to examine a scatterplot of the relationship between two variables to determine the nature of the relationship (e.g., is it linear?). The use of the ODS GRAPHICS mode provides a method for displaying a matrix of the scatterplots associated with correlation estimates. This is illustrated in Hands-On Example 13.2.

> **HANDS-ON EXAMPLE 13.2**
>
> This example illustrates how to create a matrix of scatterplots.
>
> 1. Open the program file ACORR2.SAS.
>
> ```
> PROC CORR DATA="C:\SASDATA\SOMEDATA" PLOTS=MATRIX;
>     VAR AGE TIME1 TIME2;
> TITLE 'Example correlations using PROC CORR';
> RUN;
> ```
>
> Note the option PLOTS=MATRIX. This option tells PROC CORR to include a graph that is a matrix of correlations in the output.
> 2. Run the program and observe the matrix of scatterplots shown in Figure 13.1. Note that two scatterplots are produced for each pair of variables, so you need to examine only the upper or lower half of the table.
> 3. To add more information to the plot, change the PLOTS statement to
>
> ```
> PLOTS=MATRIX(HISTOGRAM);
> ```

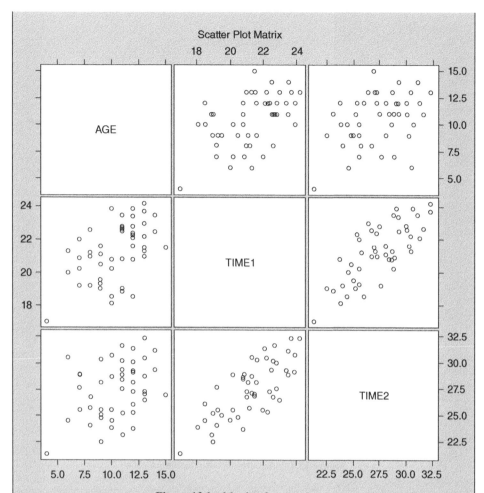

Figure 13.1 Matrix of scatterplots

Resubmit the code and observe the results (see Figure 13.2). This time, a histogram is displayed for each variable on the horizontal axis to show the distribution of the data for that variable.

4. To look at scatterplots in more detail, change the PLOTS statement to

PLOTS=SCATTER;

Resubmit the code and observe the results (see Figure 13.3). This option produces a scatterplot for each pair of variables. It also displays a 95% prediction ellipse. A number of other sub-options are available when using the PLOTS= option (see SAS documentation for more information).

(*Continued*)

324 CHAPTER 13: CORRELATION AND REGRESSION

(*Continued*)

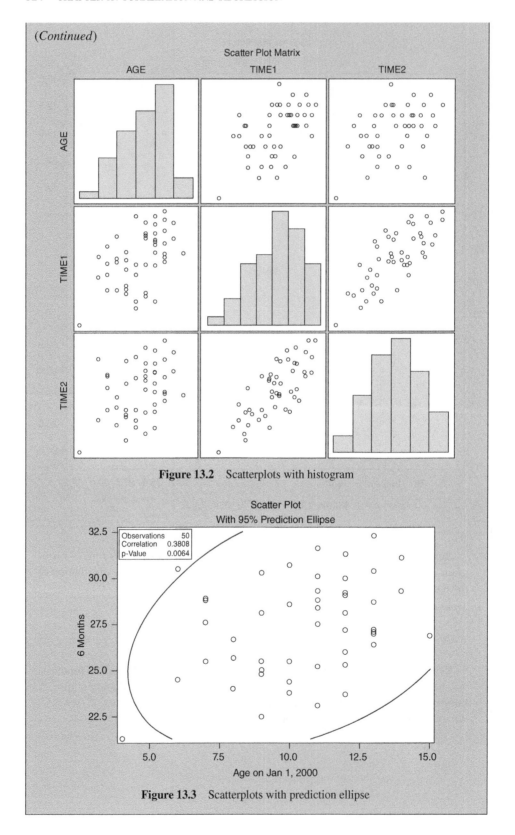

Figure 13.2 Scatterplots with histogram

Figure 13.3 Scatterplots with prediction ellipse

13.1.4 Calculating Correlations Using the WITH Statement

In Hands-On Example 13.2, all pairwise correlations were calculated from the list of variables. At times you may want to produce a list of correlations of one or more variables (possibly outcome variables) with several other variables. Use the `WITH` statement to produce the abbreviated list of correlations, as illustrated in Hands-On Example 13.3.

HANDS-ON EXAMPLE 13.3

This example illustrates how to calculate correlations using a `WITH` statement.

1. Open the program file `ACORR3.SAS`.

   ```
   PROC CORR DATA="C:\SASDATA\SOMEDATA";
       VAR TIME1-TIME4;
       WITH AGE;
   TITLE "Example correlation calculations using a WITH statement";
   RUN;
   ```

   Note that the `VAR` statement contains `TIME1-TIME4`, and a `WITH` statement is included that requests correlations between `AGE` and the four variables in the `VAR` statement.

2. Run this code and observe the output given in Table 13.3. Note that the output includes only correlations between `AGE` and the other variables.
3. Note that a table of simple statistics was included in the output. Put the option `NOSIMPLE` as an option in the `PROC CORR` statement (before the semicolon). Resubmit the program and observe the difference.
4. Add the option `PLOTS=MATRIX` to the `PROC CORR` statement. Resubmit the program and observe that a single row of scatterplot matrices is displayed.

TABLE 13.3 Correlations Using a WITH Statement

Pearson Correlation Coefficients, N = 50 Prob > |r| under H0: Rho=0				
	TIME1	TIME2	TIME3	TIME4
AGE	0.50088	0.38082	0.44952	0.48846
Age on Jan 1, 2000	0.0002	0.0064	0.0011	0.0003

13.2 SIMPLE LINEAR REGRESSION

Simple linear regression is used to predict the value of a dependent variable from the value of an independent variable. For example, in a study of factory workers, you could use simple linear regression to predict a pulmonary measure, forced vital capacity (`FVC`), from asbestos exposure (`ABS`). That is, you could determine whether increased exposure to asbestos is predictive of diminished FVC. The following PROC REG code produces the simple linear regression equation for this analysis:

```
PROC REG;
MODEL FVC=ASB;
RUN;
```

Note that the MODEL statement is used to tell SAS which variables to use in the analysis. As in the ANOVA procedure, which will be discussed in Chapter 14, the MODEL statement has the following form:

```
MODEL dependentvar = independentvar;
```

where the dependent variable (`dependentvar`) is the measure you are trying to predict and the independent variable (`independentvar`) is your predictor.

13.2.1 The Simple Linear Regression Model

The regression line that SAS calculates from the data is an estimate of a theoretical line describing the relationship between the independent variable (X) and the dependent variable (Y). The theoretical line is

$$Y = \alpha + \beta X + \varepsilon$$

where α is the y-intercept, β the slope, and ε an error term that is normally distributed with zero mean and constant variance. It should be noted that $\beta = 0$ indicates that there is no linear relationship between X and Y. A simple linear regression analysis is used to develop an equation (a linear regression line) for predicting the dependent variable given a value (x) of the independent variable. The regression line calculated by SAS is given by

$$\hat{Y} = a + bx$$

where a and b are the least squares estimates of α and β.

The null hypothesis states that there is no predictive linear relationship between the two variables. Because $\beta = 0$ indicates that there is no linear relationship between X and Y, the null hypothesis of no linear relationship is tested using the hypotheses

$$H_0 : \beta = 0$$
$$H_a : \beta \neq 0$$

A low p-value for this test (say, <0.05) indicates there is significant evidence to conclude that the slope of the line is not 0 (zero). That is, the knowledge of X would be useful in predicting Y.

> The t-test for slope is mathematically equivalent to the t-test of $H_0: \rho = 0$ in a correlation analysis.

13.2.2 Using PROC REG for Simple Linear Regression

The general syntax for `PROC REG` is as follows:

`PROC REG <options>; <statements>;`

Table 13.4 lists common options and statements used with `PROC REG`.

TABLE 13.4 Common Options for PROC REG

Option	Explanation
`DATA = datasetname`	Specifies which dataset to use
`SIMPLE`	Displays descriptive statistics
`CORR`	Displays a correlation matrix for variables listed in the `MODEL` and `VAR` statements
`PLOTS=option`	`PLOTS =NONE` suppresses graphs. Otherwise, several diagnostic graphs are produced by default
`NOPRINT`	Suppresses output when you want to capture results but not display them
`ALPHA=p`	Sets significance levels for confidence and prediction intervals
Common statements for **PROC REG**	
`MODEL dependentvar = independentvar </ options >;`	Specifies the variable to be predicted (`dependentvar`) and the variable that is the predictor (`independentvar`)
`OUTPUT OUT=dataname`	Specifies output dataset information. For example, `MODEL Y=A1 B1;` `OUTPUT OUT=OUTREG P=YHAT R=YRESID;` creates the variables `YHAT` for predicted values (`P`) and `YRESID` for residual values (`R`). Other handy variables include `LCL` and `UCL` (confidence limits on individual values) and `LCLM` and `UCLM` (confidence limits on the mean)
`PLOTS=option(s)`	Requests plots. Some options include `COOKD`, `LCL`, `UCLM`, `UCL`, `UCLM`, and `RESIDUALS`. See SAS documentation for others
`BY, FORMAT, LABEL, WHERE`	These statements are common to most procedures and may be used with `PROC REG`

The `MODEL` statement in `PROC REG` is used to specify which variable is to be used to predict the outcome. For example,

`MODEL TASK=CREATE;`

indicates that you want to predict `TASK` (called the dependent variable or outcome variable) from the value of `CREATE` (called the independent variable or predictor). Similar model statements are used for a number of SAS procedures, so it is helpful to understand fully how this `MODEL` statement works. You can think of it this way; the variable(s) on the left side of the equal sign is/are what you want to predict, and the variable(s) on the right side of the equal sign are the predictors. (In more complicated cases, there may be more than one variable on either side of the equal sign.) The `MODEL` statement is illustrated in Figure 13.4.

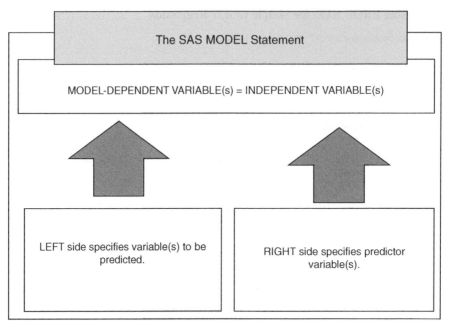

Figure 13.4 The SAS model statement

HANDS-ON EXAMPLE 13.4

In this example, a random sample of 14 elementary school students is selected from a school, and each student is measured on a creativity score (X) using a new testing instrument and on a task score (Y) using a standard instrument. The task score is the mean time taken to perform several hand–eye coordination tasks. Because administering the creativity test is much cheaper, the researcher wants to know if the CREATE score is a good substitute for the more expensive TASK score (i.e., whether the TASK score can be well predicted from the CREATE score). The data are shown in the code below where the CREATE column is a list of the scores on the CREATE test, and the TASK scores on the same data lines are the scores for the same individual for the TASK test.

1. Open the program file AREG1.SAS.

```
DATA ART;
INPUT SUBJECT $ CREATE TASK;
DATALINES;
AE    28    4.5
FR    35    3.9
HT    37    3.9
IO    50    6.1
DP    69    4.3
YR    84    8.8
QD    40    2.1
SW    65    5.5
DF    29    5.7
ER    42    3.0
```

```
RR    51    7.1
TG    45    7.3
EF    31    3.3
TJ    40    5.2
;
PROC REG;
MODEL TASK=CREATE;
TITLE "Example simple linear regression using PROC REG";
RUN;
QUIT;
```

2. Examine the code for PROC REG. The MODEL statement indicates that you want to predict the TASK score (the old test) from the CREATE score (the new test). Run the program and observe the output, some of which are given in Table 13.5.

   TABLE 13.5 Output from PROG REG for a Simple Linear Regression

Root MSE	1.60348	R-Square	0.3075
Dependent Mean	5.05000	Adj R-Sq	0.2498
Coeff Var	31.75213		

Parameter Estimates					
Variable	DF	Parameter Estimate	Standard Error	t Value	Pr > $|t|$
Intercept	1	2.16452	1.32141	1.64	0.1273
CREATE	1	0.06253	0.02709	2.31	0.0396

   Information of particular interest in these two tables includes the following:

   - **R-Square:** This value is given in the first table. It is a measure of the strength of the association between the dependent and independent variables. (It is the square of the Pearson correlation coefficient.) The closer this value is to 1, the stronger the association. In this case, R-Square = 0.31 indicates that 31% of the variability in TASK is explained by the regression with CREATE.
   - **Slope:** In the Parameter Estimates table, the statistical test on the "CREATE" row is for a test of $H_0: \beta = 0$, and $p = 0.0396$ provides evidence to reject the null hypothesis and conclude that the slope is not zero. (The statistical test on the intercept is generally of little importance.)
   - **Estimates:** The column labeled "Parameter Estimate" gives the least squares estimates of the slope and intercept (A and B) of the regression equation

   ```
   TASK = A + B * CREATE;
   ```

   Which, in this case, is

   ```
   TASK = 2.16452 + 0.06235 * CREATE;
   ```

   Because the slope is significantly different from zero, you can gain some information from the CREATE score for predicting the TASK score.

   (*Continued*)

(Continued)

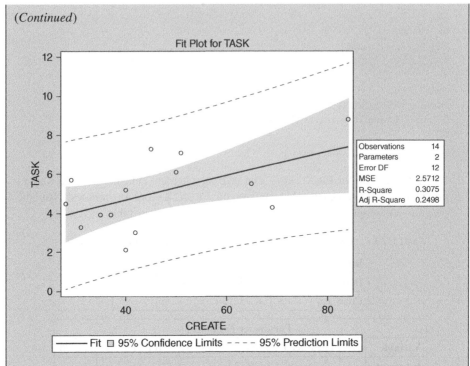

Figure 13.5 A scatterplot of CREATE * TASK with a regression fit

However, before making any predictions using this equation, you should analyze the linear relationship further. Some additional output provides visual information on how well this linear relationship might be predictive of TASK. Examine the plots in Figures 13.5 and 13.6, which are included in the output for this analysis.

The scatterplot in Figure 13.5 for CREATE * TASK provides a visual picture of the relationship between the two variables. In it you can see that as CREATE increases, TASK tends to increase. The dotted bands show prediction limits for individual values. That is, for a given value of CREATE say CREATE=40, about 95% of the corresponding values for TASK will fall between these two dotted lines. For CREATE=40, these values are approximately 1.0 and 8.3, that is, about 95% of the TASK scores will fall between 1.0 and 8.3 when the CREATE score is 40. (Not too reassuring remembering that the goal was to predict TASK from CREATE.) The shaded area represents a 95% confidence interval for the average TASK score for a given CREATE score. Again, when CREATE=40, we are 95% confident that the average value of TASK is between 3.5 and 5.5.

The diagnostic plots in Figure 13.6 provide additional information related to the regression analysis for predicting CREATE scores from TASK scores. Here are brief explanations of these plots:

- In the **Residual by Predicted Value** plot (upper left), we want to see a random scatter of points above and below the 0 line, which is the case here. A nonrandom pattern of dots could indicate an inadequate model.
- The **RStudent by Predicted Value** plot indicates whether any Studentized residuals fall beyond two standard deviations, which would indicate unusual values. In this case, none fall outside the ±2 standard deviation limits.

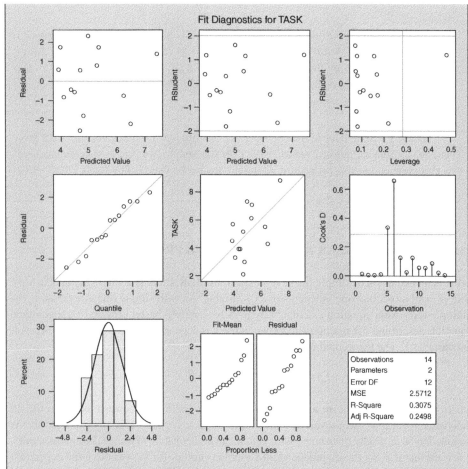

Figure 13.6 Diagnostic plots for a simple linear regression

- The **RStudent by Leverage** plot attempts to locate observations that might have unusual influence (leverage) on the calculation of the regression coefficients. In this case, there is possibly one observation that has undue influence. We'll identify this observation later.
- In the **Residual by Quartile** plot, a tight and random scatter along the diagonal line indicates an adequate fit to the model.
- The **Dependent Variable (TASK) by Predicted Value** plot visualizes variability in the prediction. For example, if there is a pattern of increasing variability as the predicted value increases, it indicates a nonconstant variance of the error.
- The **Cook's D** plot is designed to identify outliers or leverage points. In this case, it appears that observations 5 and 6 are suspect.
- **Residuals by Percent** plot assesses the normality of the residuals.
- The **Proportion Less** (Spread plot) plots the proportion of the data by the rank for two or more categories. If the vertical spread (based on ranked data) is about the same, it means that there is about the same variance in both the fitted and residual values.

(*Continued*)

> (*Continued*)
>
> For this model, you might conclude that there is a moderate linear fit between CREATE and TASK, but it is not impressive ($R^2 = 30.75$) or about 31% of the variation is accounted for by the regression using CREATE. Using the information in the regression equation, you could predict a value of TASK from CREATE=40.
>
> ```
> 4.67 = 2.16452 + 0.06235 * 40;
> ```
>
> The QUIT Statement in the PROC REG Examples is used because the REG procedure continues running even after the RUN statement. The QUIT; statement causes a complete end to this procedure. It should always be used following a PROC REG procedure.

13.3 MULTIPLE LINEAR REGRESSION USING PROC REG

MLR is an extension of simple linear regression. In MLR, there is a single dependent variable (Y) and more than one independent (X_i) variable. As with simple linear regression, the multiple regression equation calculated by SAS is a sample-based version of a theoretical equation describing the relationship between the k independent variables and the dependent variable Y. The theoretical equation is of the form

$$Y = \alpha + \beta_1 x_1 + \beta_2 x_2 + \cdots + \beta_k x_k + \varepsilon$$

where α is the intercept term and β_i is the regression coefficient corresponding to the ith independent variable. Moreover, as with simple linear regression, ε is an error term that is assumed to be normally distributed with zero mean and constant variance. From this model, it is clear that if $\beta_i = 0$ the ith independent variable is not useful in predicting the dependent variable. The multiple regression equation calculated by SAS for predicting the dependent variable from the k independent variables is

$$\hat{Y} = a + b_1 x_1 + b_2 x_2 + \cdots + b_k x_k$$

where, as with simple linear regression, the coefficients a, b_1, b_2, . . ., and b_k are least squares estimates of the corresponding coefficients in the theoretical model.

As part of the analysis, the statistical significance of each of the coefficients is tested using a Student's t-test to determine whether it contributes significant information to the predictor. These are tests of the hypotheses:

$$H_0 : \beta_i = 0$$
$$H_a : \beta_i \neq 0$$

For these tests, if the p-value is low (say, <0.05), the conclusion is that the ith independent variable contributes significant information to the equation. Care must be taken because variables in the equation may be related to other variables in the equation, so decisions about the inclusion or exclusion of a particular variable must take these interrelationships into consideration.

As in the simple linear regression model, the R^2 (R-squared) statistic is used to measure the strength of the relationship between the set of independent variables and the dependent variable. The overall significance of the regression is tested using an analysis of variance approach that tests the null hypothesis that all β_i's are equal to zero. The test statistic and the p-value are reported by SAS. If the p-value is low (say, <0.05), the conclusion is that at least one of the β_i's is not equal to zero, and thus the equation will have some predictive value for Y.

When there are several possible independent variables, you may want to find the subset of them that provides the best model. A number of model selection techniques, some manual and some automated (e.g., forward selection, backward elimination) can help you arrive at a parsimonious set of predictors (i.e., those that provide good prediction with few predictors).

13.3.1 Using SAS PROC REG for Multiple Linear Regression

As mentioned in the section on simple linear regression, the general syntax for PROC REG is

PROC REG <options>; <statements>;

Table 13.6 lists common options and statements used when analyzing a multiple regression model using PROC REG. Some of these are repeats of options and statements available in a simple linear regression. They are mentioned here for completeness.

When using PROC REG for multiple regression, there are a number of options associated with the MODEL statements. These follow the slash in the statement

MODEL dependentvar = independentvar <\ options >;

Typical MODEL statement options are given in Table 13.7.

TABLE 13.6 Common Options for PROC REG for Multiple Regression

Option	Explanation
DATA = datasetname	Specifies which dataset to use
SIMPLE	Displays descriptive statistics
ID variable	The variable specified here is displayed beside each observation in certain output tables to identify observations
CORR	Displays a correlation matrix for variables listed in the MODEL and VAR statements
PLOTS=option	PLOTS =NONE suppresses graphs. Otherwise, several diagnostic graphs are produced by default
NOPRINT	Suppresses output when you want to capture results but not to display them
ALPHA=p	Sets significance levels for confidence and prediction intervals
MODEL dependentvar = independentvar<\ options >;	Specifies the variable to be predicted (dependentvar) and the variables that are the predictors (independentvar). Model options are listed below
BY variable list;	Produces separate regression analyses for each value of the BY variable

(*Continued*)

TABLE 13.6 (*Continued*)

Option	Explanation
Common statements for PROC REG for multiple regression	

OUTPUT OUT= *datasetname*	Specifies output dataset information. For example,
	    MODEL Y=A1 B1;
	    OUTPUT OUT=OUTREG P=YHAT R=YRESID;
	 creates the variables YHAT for predicted values (P) and YRESID for residual values. Other handy variables include LCL and UCL (confidence limits on individual values) and LCLM and UCLM (confidence limits on the mean)

PLOTS=*options*	Requests plots. Some options include COOKD, LCL, UCLM, UCL, UCLM, and RESIDUALS (see SAS documentation for others)

TABLE 13.7 Common Statement Options for the PROC REG MODEL Statement (Options Follow /)

Option	Explanation
P	Requests a table containing predicted values from the model
R	Requests that the residuals be analyzed
CLM	Prints the 95% upper and lower confidence limits for the expected value of the dependent variable (mean) for each observation
CLI	Requests the 95% upper and lower confidence limits for an individual value of the dependent variable
INCLUDE=k	Includes the first k variables in the variable list in the model (for automated selection procedures)
SELECTION= *option*	Specifies an automated variable selection procedure. Options include BACKWARD, FORWARD, and STEPWISE (see below)
SLSTAY=p	Specifies the maximum p-value for a variable to stay in a model during automated model selection. Default values are 0.10 for BACKWARD and 0.15 for STEPWISE
SLENTRY=p	Minimum p-value for a variable to enter a model for forward or stepwise selection. Default values are 0.50 for FORWARD and 0.15 for STEPWISE

In particular, the SELECTION= options specify how variables will be considered for inclusion in the model. The BACKWARD method considers all predictor variables and eliminates the ones that do not meet the minimal SLSTAY criterion until only those meeting the criterion remain. The FORWARD method brings in the most significant variable that meets the SLENTRY criterion and continues entering variables until none meets the criterion. STEPWISE is a mixture of the two; it begins like the FORWARD method but re-evaluates variables at each step and may eliminate a variable if it does not meet the SLSTAY criterion. Additional model selection criteria are also available in SAS.

HANDS-ON EXAMPLE 13.5

In this example, an employer wants to be able to predict how well applicants will do on the job once they are hired. He devises four tests that he thinks will measure the skills required for the job. Ten prospects are selected at random from a group of applicants and given the four tests. Then they are given an on-the-job proficiency score (JOBSCORE) by a supervisor who observes their work.

1. Open the program file AREG2.SAS.

```
DATA JOB;
INPUT SUBJECT $ TEST1 TEST2 TEST3 TEST4 JOBSCORE;
CARDS;
1      75     100     90     88     78
2      51      85     88     89     71
3      99      96     94     93     85
4      92     106     84     84     67
5      90      89     83     77     69
6      67      77     83     73     65
7     109      67     71     65     50
8      94     112    105     91    107
9     105     110     99     95     96
10     74     102     88     69     63
;
PROC REG;
MODEL JOBSCORE=TEST1 TEST2 TEST3 TEST4;
TITLE 'Job Score Analysis using PROC REG';
RUN;
QUIT;
```

Note that the dependent variable in the MODEL statement is JOBSCORE and that the TEST variables, listed on the right side of the equal sign, are the independent variables.

2. Run the program and observe the analysis of variance given in Table 13.8. This table includes an overall test of the significance of the model. Because $p = 0.0003$, you would reject the null hypothesis that all $\beta_i = 0$ and conclude that the model has predictive value. In fact, because R-Square $= 0.9754$, you have evidence that the model is a good fit and should provide a good prediction of JOBSCORE.

TABLE 13.8 Analysis of Variance Output for PROC REG

Analysis of Variance					
Source	DF	Sum of Squares	Mean Square	F Value	Pr > F
Model	4	2495.96648	623.99162	49.58	0.0003
Error	5	62.93352	12.58670		
Corrected Total	9	2558.90000			

Root MSE	3.54777	R-Square	0.9754
Dependent Mean	75.10000	Adj R-Sq	0.9557
Coeff Var	4.72407		

(Continued)

(*Continued*)
3. To continue examining the model, look at the parameter estimates in Table 13.9. These estimates show results for tests of the significance (Pr > |*t*|) of the intercept and each coefficient. In this case, TEST1 ($p = 0.0446$) and TEST3 ($p = 0.0015$) seem to be the best predictors.

TABLE 13.9 Parameter Estimates

Parameter Estimates					
Variable	DF	Parameter Estimate	Standard Error	*t* Value	Pr > |*t*|
Intercept	1	−95.55939	12.82483	−7.45	0.0007
TEST1	1	0.17631	0.06616	2.66	0.0446
TEST2	1	−0.22344	0.14354	−1.56	0.1803
TEST3	1	1.74602	0.27770	6.29	0.0015
TEST4	1	0.26865	0.18424	1.46	0.2046

Examine the diagnostic plots for this analysis in Figure 13.7. General explanations for these plots were given in the previous example using simple linear regression. In

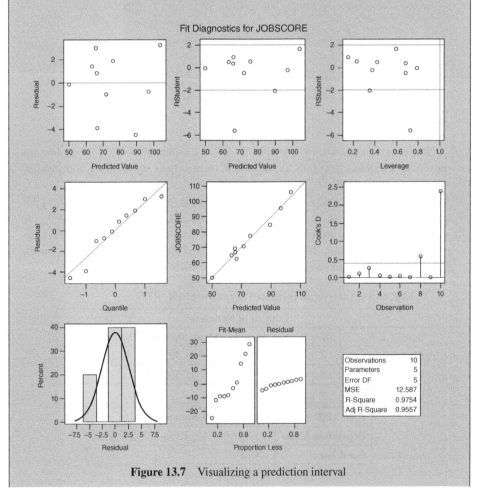

Figure 13.7 Visualizing a prediction interval

this instance, note that issues with the model include a possible outlier identified by Cook's plot (observation 10). Otherwise, the residual plots look reasonable. There are also some possible issues related to variability identified in the bottom plots. For now, we'll ignore these possible issues.

4. As in this case, the p-values for TEST2 and TEST4 are not significant ($p > 0.05$), a typical strategy is dropping them from the model and rerunning the program to see how the reduced model fits the data. Take TEST2 and TEST4 out of the model statement and resubmit the code.

   (a) How does this change the model? Does the R-Square still indicate a good fit?

   (b) Are the same issues present in terms of the residual plots?

   (c) If each test costs $20 to administer, and you intend to give it to 100 applicants, do you think the reduced model is still an adequate predictor of JOBSCORE?

13.3.2 Automated Model Selection

It is often the case in multiple regression analysis that one of your goals is to obtain a model that gives you an optimal regression equation with the fewest parameters. In Hands-On Example 13.5, the predictors TEST2 and TEST4 are not significant. As mentioned, following the listing of MODEL options in Table 13.7, a variety of automated model selection procedures are available in SAS. You can choose to select variables using manual or automated methods, or a combination of both. The various model selection techniques will not always result in the same final model, and the decision concerning which variables to include in the final model should not be based entirely on the results of any automated procedure. The researcher's knowledge of the data should always be used to guide the model selection process even when automated procedures are used.

Hands-On Example 13.6 illustrates the BACKWARD elimination technique for model selection. In this technique, the full model is examined, and in successive steps, the least predictive variable is eliminated from the model until all remaining variables show at least a $p < 0.10$ significance level for entry.

HANDS-ON EXAMPLE 13.6

This example illustrates how to use automated model selection techniques.

1. Open the program file AREG3.SAS. This code is similar to the previous example, but includes a SELECTION Statement:

```
MODEL JOBSCORE=TEST1 TEST2 TEST3 TEST4
    /SELECTION=BACKWARD;
```

(*Continued*)

(Continued)

2. Submit this program. Because the model contains BACKWARD selection criteria, it examines the coefficients in a full model first, and then systematically drops independent variables that are not predictive. In the output, not shown here, you will see that at Step 1, variable TEST4 is removed from the model and at Step 3, TEST2 is removed. The remaining two variables (TEST1 and TEST3) are not removed from the model based on the SLSTAY=.1 criterion. Examine the parameter estimates in Table 13.10. Note that the variables TEST1 and TEST3 remain in the model, while TEST2 and TEST4 were removed. In this table, TEST3 is highly significant ($p < 0.0001$) and TEST1 is marginally significant ($p = 0.0779$). Moreover, note that $R^2 = 0.9528$, which indicates a good predictive model.

TABLE 13.10 Parameter Estimates for Revised Model

Variable	Parameter Estimate	Standard Error	Type II SS	F Value	Pr > F
Intercept	−89.08684	14.33880	666.44871	38.60	0.0004
TEST1	0.15607	0.07562	73.55164	4.26	0.0779
TEST3	1.70426	0.14680	2327.04675	134.78	<.0001

3. Because TEST1 is only marginally significant (and not significant at the 0.05 level) and if each test is expensive, then you might want to use a more stringent rule to retain variables (e.g., use a smaller value for SLSTAY). Change the MODEL statement to read

```
MODEL JOBSCORE=TEST1 TEST2 TEST3 TEST4/
    SELECTION=BACKWARD
    SLSTAY=0.05;
```

SLSTAY=0.05 tells SAS to remove variables whose significance level is 0.05 or larger (the default is 0.10). Run the revised program and observe the parameter estimates in Table 13.11, which now includes only the predictor variable TEST3. Also note that $R^2 = 0.924$. Thus, this model is still highly significant and appears to be strongly predictive of JOBSCORE.

TABLE 13.11 Parameter Estimates for Final Model

Variable	Parameter Estimate	Standard Error	Type II SS	F Value	Pr > F
Intercept	−76.81121	15.47905	598.38485	24.62	0.0011
TEST3	1.71651	0.17402	2364.49377	97.30	<.0001

4. Examine the diagnostic plots for the final model. Note that there continue to be a few points (identified by Cook's *D*) that may be outliers, but the Fit Plot indicates a fairly tight fit for predicting JOBSCORE from TEST3 (see Figure 13.8).

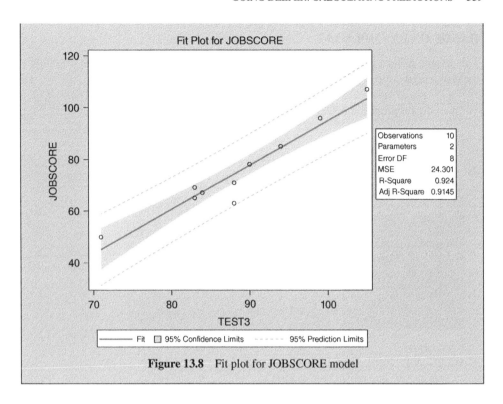

Figure 13.8 Fit plot for JOBSCORE model

13.4 GOING DEEPER: CALCULATING PREDICTIONS

After you decide on a "final" model, you may want to use the prediction equation to predict the values of the dependent variable for new subjects. In the JOBSCORE example, you could use the model given in Table 13.10 to predict how well a new job prospect will do on the job. The prediction equation is based on the parameter estimates shown in that table and is given by

```
JOBSCORE = -76.81121 +1.71651*TEST3;
```

Using this equation, you can manually calculate a JOBSCORE for new applicants. However, SAS can be used to do the calculations for you. This becomes particularly helpful if your final model contains several predictor variables or if you want to make predictions for a lot of new applicants. The following procedures can be used to predict new values:

- Create a new dataset containing new values for the independent variable(s).
- Merge (append) the new dataset with the old dataset.
- Calculate the regression equation and request predictions.
- Use the ID option to display the new values in the output.

Hands-On Example 13.7 shows how to use SAS to predict JOBSCORE for 10 applicants who received the following scores on TEST3: 79, 87, 98, 100, 49, 88, 91, 79, 84, and 87.

HANDS-ON EXAMPLE 13.7

This example illustrates how to calculate predictions for data following the development of a regression model.

1. Open the program file `AREG4.SAS`.

```
DATA JOB;
    INPUT SUBJECT $ TEST1 TEST2 TEST3 TEST4 JOBSCORE;
    DATALINES;
1    75    100   90   88   78
2    51    85    88   89   71
3    99    96    94   93   85
4    92    106   84   84   67
5    90    89    83   77   69
6    67    77    83   73   65
7    109   67    71   65   50
8    94    112   105  91   107
9    105   110   99   95   96
10   74    102   88   69   63
;
PROC REG;
MODEL JOBSCORE=TEST3;
TITLE 'Job Score Final Model';
RUN;
DATA NEWAPPS;
    INPUT SUBJECT $ TEST3;
DATALINES;
101 79
102 87
103 98
104 100
105 49
106 88
107 91
108 79
109 84
110 87
;
DATA REPORT; SET JOB NEWAPPS;
PREDICT_ID=CATS(SUBJECT,": ",TEST3);
RUN;
PROC REG DATA=REPORT;
    ID PREDICT_ID;
    MODEL JOBSCORE=TEST3 /P CLI;
RUN;
QUIT;
```

In the first DATA step, we recreate the dataset JOB that was used in Hands-On Example 13.6. The code then uses PROC REG to recalculate the final regression model that selected only TEST3 to predict JOBSCORE. In the second DATA step, a dataset NEWAPPS is created that contains the TEST3 scores for the 10 new applicants, which have been given SUBJECT numbers 101–110. In the third DATA step, the dataset NEWAPPS is appended to the JOB dataset to create a dataset named REPORT. In addition, a new variable is created named PREDICT_ID using a CATS function that concatenates the two variables SUBJECT and TEST3 (see Appendix D). This variable is used as the ID variable in the output so that the predicted scores can be matched to a particular subject. The /P CLI options tell SAS to output a table containing the predicted values and confidence limits on the actual JOBSCORE values. (Actually, simply using CLI will also produce both predictions and confidence limits.)

2. Run this example, and observe the (abbreviated) output in Table 13.12. The first 10 rows of the output table report information about the 10 subjects in the original dataset from which the regression model is obtained. Starting with observation 11, the information relates to the new data that did not include a dependent variable (JOBSCORE). The new data were not used to calculate the regression equation, but their predicted values were calculated. Thus, for observation 11 (i.e., subject 101), you can see that this subject had a score on TEST3 of 79, which is used to calculate a predicted JOBSCORE of 58.79.

The "95% CL Predict" column indicates confidence limits for the actual value of JOBSCORE for the TEST3 values. For example, JOBSCORE for subject 101 could plausibly be as low as 46.2760 or as high as 71.3103.

TABLE 13.12 Predictions Using Final Model

Output Statistics							
Obs	PREDICT_ID	Dependent Variable	Predicted Value	Std Error Mean Predict	95% CL Predict		Residual
1	1:90	78	77.6748	1.5806	65.7371	89.6124	0.3252
2	2:88	71	74.2417	1.5613	62.3176	86.1659	−3.2417
3	3:94	85	84.5408	1.8292	72.4158	96.6658	0.4592
4	4:84	67	67.3757	1.7445	55.3172	79.4342	−0.3757
5	5:83	69	65.6592	1.8292	53.5342	77.7842	3.3408
6	6:83	65	65.6592	1.8292	53.5342	77.7842	−0.6592
7	7:71	50	45.0611	3.4211	31.2242	58.8980	4.9389
8	8:105	107	103.4224	3.2671	89.7848	117.0601	3.5776
9	9:99	96	93.1234	2.4018	80.4783	105.7685	2.8766
10	10:88	63	74.2417	1.5613	62.3176	86.1659	−11.2417
11	101:79		58.7931	2.2722	46.2760	71.3103	
12	102:87		72.5252	1.5806	60.5876	84.4629	
13	103:98		91.4069	2.2722	78.8897	103.9240	
14	104:100		94.8399	2.5367	82.0555	107.6243	
15	105:49		7.2978	7.0482	−12.5361	27.1318	
16	106:88		74.2417	1.5613	62.3176	86.1659	
17	107:91		79.3913	1.6184	67.4267	91.3559	
18	108:79		58.7931	2.2722	46.2760	71.3103	
19	109:84		67.3757	1.7445	55.3172	79.4342	
20	110:87		72.5252	1.5806	60.5876	84.4629	

13.5 GOING DEEPER: RESIDUAL ANALYSIS

In the case of the simple linear regression model, scatterplots such as the scatterplot matrix shown in Figures 13.1 and 13.2 are useful graphs for visually inspecting the nature of the association. Hands-On Example 13.8 provides related residual analysis techniques for assessing the appropriateness of a linear regression fit to a set of data. These techniques are appropriate for simple and MLR analyses.

HANDS-ON EXAMPLE 13.8

This example illustrates how to examine model residuals as a part of model verification.

1. Open the program file AREG5.SAS. A portion of the code is shown here:

```
DATA JOB;
INPUT SUBJECT $ TEST1 TEST2 TEST3 TEST4 JOBSCORE;
CARDS;
1      75    100    90    88    78
2      51     85    88    89    71
3      99     96    94    93    85
4      92    106    84    84    67
5      90     89    83    77    69
6      67     77    83    73    65
7     109     67    71    65    50
8      94    112   105    91   107
9     105    110    99    95    96
10     74    102    88    69    63
;
RUN;
TITLE 'Residual Analysis';
PROC REG DATA=JOB;
MODEL JOBSCORE=TEST3/R;
RUN;
QUIT;
```

This code is based on data from the JOBSCORE example. The following statements are used for the model using TEST3 to predict JOBSCORE. The /R option requests graphs and diagnostics that assist with a residual analysis for this model.

2. Run the program and observe the Output Statistics in Table 13.13. The "Residual" column gives the difference between the observed dependent variable and the predicted value obtained using the regression equation. You can inspect this column to find out whether there are certain subjects whose scores were not as well predicted as others'. For example, subject 10's residual score is −11.24, which is the difference between the observed score from the JOBSCORE (63.0) and the predicted score (74.24). If a residual is much larger for one subject than for the others, examine the data for miscoding or to otherwise understand why the actual score for that subject is so different from its predicted value.

TABLE 13.13 Output Statistics for PROC REG

			Output Statistics				
Obs	Dependent Variable	Predicted Value	Std Error Mean Predict	Residual	Std Error Residual	Student Residual	Cook's D
1	78	77.6748	1.5806	0.3252	4.669	0.070	0.000
2	71	74.2417	1.5613	−3.2417	4.676	−0.693	0.027
3	85	84.5408	1.8292	0.4592	4.578	0.100	0.001
4	67	67.3757	1.7445	−0.3757	4.611	−0.081	0.000
5	69	65.6592	1.8292	3.3408	4.578	0.730	0.043
6	65	65.6592	1.8292	−0.6592	4.578	−0.144	0.002
7	50	45.0611	3.4211	4.9389	3.549	1.392	0.900
8	107	103.4224	3.2671	3.5776	3.691	0.969	0.368
9	96	93.1234	2.4018	2.8766	4.305	0.668	0.069
10	63	74.2417	1.5613	−11.2417	4.676	−2.404	0.322

The "Student Residual" column contains z-scores for residuals (called Studentized residuals) that provide a measure of the magnitude of the difference. Generally, a residual >2 or <−2 can be seen as statistically significant and may need further investigation. Because the Studentized residual for subject 10 is −2.404, this indicates that the prediction for this subject's JOBSCORE was substantially smaller than expected. The columns labeled "−2 −1 0 1 2" are a crude graph of the Studentized residuals. The "****" in that column for subject 10 provides a visual indication that the prediction is substantially smaller than expected.

The Cook's D statistic gives an indication of the "influence" of a particular data point. A value close to 0 indicates no influence; and the higher the value, the greater the influence. Note that subject 7 had a larger influence on the estimates of the regression coefficients than did most other subjects.

Based on this residual analysis, you should examine why subject 7 had such a high Cook's D score and why the prediction for subject 10 was so much less than expected. Sometimes, such an examination of the data turns up miscoded values or may indicate something different about some subjects that affect whether they should be included in the analysis. This analysis can also help you understand the visual output in the residuals plots also contained in the output (described previously but not shown here).

In particular, examine the plot of the regression line with mean and individual prediction limits (see Figure 13.8). The smaller, dark bands are 95% confidence limits for the mean of JOBSCORE given values of TEST3, and the larger dashed bands are 95% prediction limits for actual JOBSCORE values for given values of TEST3. The point that is farthest from the line is ID=10 (TEST3=88, JOBSCORE=63) and the lowest point, and one that is far away from the other point is ID=7 (TEST3=71, JOBSCORE=50). Understanding the nature of these two subjects and their scores may provide greater insight into the model.

13.6 SUMMARY

This chapter shows you how to measure the association between two quantitative variables (correlation analysis). It also discusses how to obtain and analyze the performance of a prediction equation using simple or MLR.

EXERCISES

13.1 **Run a regression analysis involving US crime data.**

This exercise uses crime statistics for Washington, DC from the *United States Uniform Crime Report* to examine the rise in assaults in the years 1978–1993 (Figure 13.9). In the early 1990s, there was a national concern about a greater-than-expected rise in crime. The media predicted that crime was spiraling up out of control and that cities would soon become war zones of violent crimes. They had the statistics to back them up.

a. Open the program file EX_13.1.SAS, which performs a simple linear regression model. Run the program and observe the results, which are summarized in Figure 13.9.

b. If your job was to predict future crimes in order to budget for police personnel, how would these results influence your decision?

c. Based on the results, fill in the blanks in the prediction equation below (Table 13.14):
ASSAULTS = _____ + _____ * YEAR

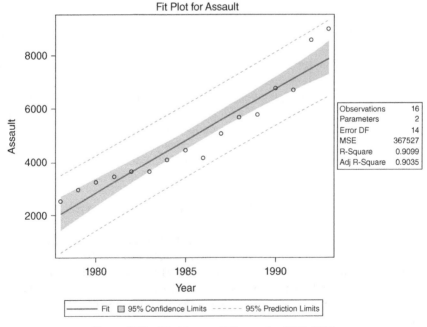

Figure 13.9 Washington, DC, assaults, 1978–1993

TABLE 13.14 Regression Estimates for Washington, DC, Crime Data 1978–1993

			Parameter Estimates			
Variable	Label	DF	Parameter Estimate	Standard Error	t Value	$\Pr > |t|$
Intercept	Intercept	1	−771153	65279	−11.81	<.0001
Year	Year	1	390.90441	32.87797	11.89	<.0001

 d. Using your prediction equation, predict the number of assaults for the years 1990, 1995, 2000, and 2007.

 e. How good do you think predictions will be, given the fact that the result shows an impressive fit ($R^2 = 0.91$)?

 f. Fill in the following table:

Year	Actual Number of Assaults	Estimated Number of Assaults
1990	7365	
1995	7228	
2000	4582	
2007	3686	

 g. Open the program file EX_13.1B.SAS. This contains the same regression model but with data through 2007. Run the model and observe the results.

  I. What is R^2?

  II. Examine the scatterplot of the year by the number of assaults. How does it differ from the earlier version (Figure 13.9)?

  III. Using the data from 1978 to 1993 to predict the number of assaults for 2000 and 2007 is called extrapolation. Based on these results, do you think extrapolation is a wise practice?

13.2 Perform variable selection in a MLR.

In this problem, we use SAS to predict CITYMPG using the four variables AUTOMATIC, ENGINESIZE, CYLINDERS, and HYBRID. In this exercise, you will find a "best" model.

 a. Open the program file EX_13.2.SAS.

 b. Run the program and note which variables are selected for the BACKWARD selection procedure.

 c. Change the selection option to FORWARD. Run the program and note which variables are in the final model.

 d. Change the selection option to STEPWISE. Run the program and note which variables are in the final model.

 e. Are the results the same? What criteria would you use to select the "best" model?

CHAPTER 13: CORRELATION AND REGRESSION

13.3 **Using the WITH statement.**

a. Open the program file ACORR4.SAS.

```
PROC CORR DATA="C:\SASDATA\SOMEDATA" NOSIMPLE;
VAR _____ ; WITH _____ ;
TITLE "Example correlation calculations using a WITH statement";
RUN;
```

Fill in the missing code so the VAR statement contains TIME1-TIME4, and a WITH statement is included that requests correlations between AGE and the four variables in the VAR statement.

b. Run this code and verify that the correlations listed are only between AGE and the TIME variables.

c. Add code that will print a series of scatterplots for AGE by the TIME variables.

14

ANALYSIS OF VARIANCE

LEARNING OBJECTIVES

- To be able to compare three or more means using one-way ANOVA with multiple comparisons
- To be able to perform a repeated measures (dependent samples) analysis of variance with multiple comparisons
- To be able to graph mean comparisons

This chapter illustrates how to use SAS to perform an analysis of variance (ANOVA) for several common designs. These procedures are used to compare means across groups or to compare three or more repeated measures (dependent samples). SAS provides three major PROCS for performing ANOVA: PROC ANOVA, PROC GLM, and PROC MIXED. We will discuss PROC MIXED in Chapter 15.

In this chapter, we use both PROC ANOVA and PROC GLM in our analysis of a one-way ANOVA. PROC ANOVA is a basic procedure that is useful for one-way ANOVA or for multiway factorial designs with fixed factors and an equal number of observations per cell.

PROC GLM is a SAS procedure that is similar to but more advanced than PROC ANOVA. (GLM stands for General Linear Model.) We use PROC GLM for the one-way repeated measures analysis because it involves techniques not supported by PROC ANOVA. In Chapter 15, we illustrate the use of `PROC MIXED` to analyze a model with both fixed and random factors and for a repeated measures design with a grouping factor. Section 14.3 describes how to perform *posthoc* comparisons for a one-way ANOVA.

14.1 COMPARING THREE OR MORE MEANS USING ONE-WAY ANALYSIS OF VARIANCE

A one-way ANOVA is an extension of the independent group *t*-test where there can be more than two groups. Assumptions for this test are similar to those for the *t*-test:

- Data within groups are normally distributed with equal variances across groups.
- Independent samples. That is, not only do the observations within a group represent a random sample but also there is no matching or pairing of observations among groups. This is analogous to the requirement of independent samples in the two-sample *t*-test.

As with the *t*-test, the ANOVA is robust against moderate departures from the assumptions of normality and equal variance (especially for larger sample sizes). However, the assumption of independence is critical. The hypotheses for the comparison of independent groups are as follows (*k* is the number of groups):

$H_0: \mu_1 = \mu_2 = \cdots = \mu_k$: Means of all the groups are equal.
$H_a: \mu_i \neq \mu_j$ for some $i \neq j$: At least two means are not equal.

The test statistic used to test these hypotheses is an *F* with $k-1$ and $N-k$ degrees of freedom, where *N* is the total number of subjects. A low *p*-value for the *F*-test is evidence for rejecting the null hypothesis. In other words, there is evidence that at least one pair of means are not equal. These tests can be performed using `PROC ANOVA`. The syntax for PROC ANOVA is as follows:

```
PROC ANOVA <options>;
    CLASS variable;
    MODEL dependentvar = independentvars;
    MEANS independentvars / typecomparison <meansoptions>;
```

Table 14.1 lists common options and statements used with `PROC ANOVA` and `PROC GLM`. Some options for GLM only are listed as (GLM only). Not every option listed here is illustrated in the chapter's examples.

The model statement is a key part of the `PROC ANOVA` procedures. It is similar to the `MODEL` statement described in Chapter 13 for `PROC REG`. The form of the statement is

```
MODEL dependentvariable=independentvariable;
```

The dependent variable is the quantitative variable of interest (your outcome variable), and the independent variable (there is one independent variable in the one-way ANOVA case) is the grouping variable (GROUP) for the analysis (the variable listed in the `CLASS` statement). For example, if you have measured FEV (pulmonary forced expiratory volume) and want to determine if there is a weight difference among groups, your model statement would be

```
MODEL FEV=GROUP;
```

TABLE 14.1 Common Options for PROC ANOVA and PROC GLM for Performing a One-Way ANOVA or Simple Repeated Measures

Option	Explanation
`DATA = datasetname`	Specifies which dataset to use
`NOPRINT`	Suppresses output. This is used when you want to extract information from ANOVA results but don't want SAS to produce output in the **Results Viewer**
`OUTSTAT=datasetname`	Names an output dataset that saves a number of the results from the ANOVA calculation
`PLOTS=options`	Specifies PLOTS=NONE to suppress plots that are generated by default
`ORDER=option`	Specifies the order in which to display the CLASS variable (similar to what was covered in Chapter 11). Options are DATA, FORMATTED, FREQ, or INTERNAL
`ALPHA=p`	Specifies alpha level for a confidence interval (GLM only)
Common statements for PROC ANOVA and PROC GLM (for one-way analyses)	
`CLASS variable list;`	This statement is required and specifies the grouping variable(s) for the analysis
`MODEL specificaion`	Specifies the dependent and independent variables for the analysis. More specifically, it takes the form `MODEL dependentvariable=independent variable(s);` More explanation follows
`FREQ var`	Specifies that a variable represents the count of values for an observation. Similar to the WEIGHT statement for PROC FREQ
`MEANS vars`	Calculates means for dependent variables and may include comparisons. This is illustrated in Hands-On Example 14.1
`LSMEANS vars`	Calculates least square means for a dependent variable and can be used to request comparisons (GLM only). This is illustrated in an example below
`REPEATED vars`	Used to specify repeated measure variables. This is described in Hands-On Example 14.2
`TEST specificaion`	Used to specify a hypothesis test value
`CONTRAST specification`	Allows you to create customized *posthoc* comparisons (GLM only)
`BY, FORMAT, LABEL, WHERE`	These statements are common to most procedures and may be used in PROC ANOVA and PROC GLM

That is, if you know to which group a subject belongs, you know something about the value of FEV. If they are related, you have some way to "predict" FEV by knowing GROUP.

You will see the MODEL statement in several SAS procedures. In general, the information on the left side of the equal sign is the dependent variable (the variable you are trying to predict), and the variable or variables on the right side of the equal sign are independent (predictor) variables.

14.1.1 Using the MEANS or LSMEANS Statement

A one-way ANOVA involves a two-step procedure:

1. Test $H_0: \mu_1 = \mu_2 = \cdots = \mu_k$ to determine whether any significant differences exist
2. If H_0 is rejected, then run subsequent multiple comparison tests to determine which differences are significantly different.

Pairwise comparison of means can be performed using one of several multiple comparison tests specified using the MEANS statement. The MEANS statement has the following format (where *independantvar* is a CLASS variable):

MEANS independentvar / typecomparison <meansoptions>;

For PROC GLM, use the LSMEANS statement:

LSMEANS independentvar / typecomparison <meansoptions>;

Table 14.2 lists multiple comparison options for both ANOVA and GLM.

For example, suppose you are comparing the time to relief of three headache medicines – brands 1, 2, and 3. The time-to-relief data are reported in minutes. For this experiment, 15 subjects were randomly placed on one of the three medications. Which medicine (if any) is the most effective? The data for this example are as follows:

Brand 1	Brand 2	Brand 3
24.5	28.4	26.1
23.5	34.2	28.3
26.4	29.5	24.3
27.1	32.2	26.2
29.9	30.1	27.8

TABLE 14.2 Common *type comparison* Options for the PROC ANOVA or GLM MEANS Statement (Options Following the Slash /)

Option	Explanation
BON	Bonferroni *t*-tests of difference
DUNCAN	Duncan's multiple range test
SCHEFFE	Scheffe multiple comparison procedure
SNK	Student Newman–Keuls multiple range test
LSD	Fisher's least significant difference test
TUKEY	Tukey's Studentized range test
DUNNETT ('x')	Dunnett's test – compares to a single control, where 'x' is the category value of the control group MEANS
ALPHA=pvalue	Specifies the significance level for comparisons (default: 0.05)
CLDIFF	Requests that confidence limits be included in the output

Common `typecomparison` options for the `PROC GLM LSMEANS` Statement (options following the slash /)

Option	Explanation
ADJUST=option	Specifies type of multiple comparisons. Examples are BON, DUNCAN, SCHFEE, SNK, LSD, and DUNNETT
PDIFF=	Calculates *p*-values base (default is T). You can also specify TUKEY or DUNNETT options

HANDS-ON EXAMPLE 14.1

This example illustrates how to compare the means of three or more independent groups using `PROC ANOVA`. We also illustrate here a technique for performing multiple comparisons.

1. Open the program file `AANOVA1.SAS`.

```
DATA ACHE;
INPUT BRAND RELIEF;
CARDS;
1  24.5
1  23.5
1  26.4
1  27.1
1  29.9
2  28.4
2  34.2
2  29.5
2  32.2
2  30.1
3  26.1
3  28.3
3  24.3
3  26.2
3  27.8
;
PROC ANOVA DATA=ACHE;
    CLASS BRAND;
    MODEL RELIEF=BRAND;
    MEANS BRAND/TUKEY;
TITLE 'COMPARE RELIEF ACROSS MEDICINES - ANOVA EXAMPLE';
RUN;
```

Examine the `PROC ANOVA` statement:

- BRAND is the CLASS or grouping variable (containing three levels). The CLASS variable can be either a numeric or character type.
- The MODEL statement indicates that RELIEF is the dependent variable, whose means across groups are to be compared. The grouping factor is BRAND.

(Continued)

(*Continued*)
- The MEANS statement requests a multiple comparison test for BRAND using the TUKEY method.
- Note how the data are set up in the DATA step, with one subject per line, where the grouping variable BRAND indicates the group membership for each subject.

2. Run this analysis and observe the results. The ANOVA (partial) results are given in Table 14.3.

    The first ANOVA table shows the results of a test of the full model, which in this case is the same as the test of the BRAND effect shown in the second table because only one grouping factor is in the model. This is a test of the null hypothesis shown earlier that the means of all groups are equal. The small p-value ($p = 0.009$) provides evidence for rejecting the null hypothesis that the means are equal.

    If the p-value for the model is not significant ($p > 0.05$), end your analysis here and conclude that you cannot reject the null hypothesis that all means are equal. If, as in this case, the p-value is small, you can perform multiple comparisons to determine which means are different.

TABLE 14.3 ANOVA Results

Source	DF	Sum of Squares	Mean Square	F Value	Pr > F
Model	2	66.7720000	33.3860000	7.14	0.0091
Error	12	56.1280000	4.6773333		
Corrected Total	14	122.9000000			

Source	DF	Anova SS	Mean Square	F Value	Pr > F
BRAND	2	66.77200000	33.38600000	7.14	0.0091

3. Observe the multiple comparison results given in Figure 14.1. This graph displays significant mean differences using Tukey's multiple comparison test. Groups that *are not* significantly different from each other are included in the same Tukey grouping. From Figure 14.1, we see that there are two groupings, shown (indicated by the red and blue bars) Note that BRAND 2 is in a group by itself (a blue bar). This indicates that the mean for BRAND 2 (30.88) is significantly higher than (different from) the means of BRAND 1 (26.28) and BRAND 3 (26.54). Because BRAND 1 and BRAND 3 are in the same grouping (a red bar), there is no significant difference between these two brands. Because a shorter time to relief is desirable, the conclusion would be that BRANDs 1 and 3 are preferable to BRAND 2.

    A comparative boxplot is also included in the output as shown in Figure 14.2. These comparative box plots help you visualize the distribution of observations in the three groups. Fifty percent of the data is in the highlighted box, and the mean is represented by a diamond. You can see that groups 1 and 3 have a lot of overlap, whereas group 2 has a substantially larger mean than both of them.

Figure 14.1 Tukey Multiple Comparisons Results

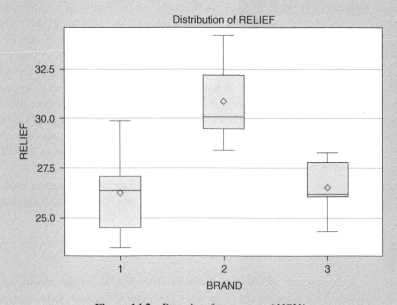

Figure 14.2 Box plots for one-way ANOVA

4. Another method for displaying Tukey's results is provided by including the option CLDIFF in the MEANS statement. Change the MEANS statement to read:

MEANS BRAND/TUKEY CLDIFF;

Rerun the program and observe the output in Table 14.4, showing simultaneous confidence limits on the difference between means.

(Continued)

(*Continued*)

TABLE 14.4 Simultaneous Confidence Limits

BRAND Comparison	Difference Between Means	Simultaneous 95% Confidence Limits		
2–3	4.340	0.691	7.989	***
2–1	4.600	0.951	8.249	***
3–2	–4.340	–7.989	–0.691	***
3–1	0.260	–3.389	3.909	
1–2	–4.600	–8.249	–0.951	***
1–3	–0.260	–3.909	3.389	

Comparisons Significant at the 0.05 Level Are Indicated By ***.

The asterisks (***) in the multiple comparisons table indicate paired comparisons that are significant at the 0.05 level. In this case, all comparisons except means 1 versus 3 are different. This indicates that the mean time to relief for BRAND 3 is not significantly different from that of BRAND 1, but that the mean time to relief for BRAND 2 is significantly different from those for BRAND 1 and BRAND 3. The simultaneous 95% confidence limits provide an estimate of how small or large the difference between the means is likely to be. You can use this information to assess the (clinical) importance of the difference. For example, the difference between BRANDs 2 and 3 could plausibly be as small as 0.69 minutes to relief and as large as 7.99 minutes to relief. If such differences are determined to be of clinical significance, then the conclusions are the same as those obtained from Figure 14.1 – that is, BRANDs 1 and 3 are preferable to BRAND 2.

5. Change PROC ANOVA to PROC GLM, the GLM procedure. This procedure produces the same output (in this case). It is more generalized and has more features, but for very large datasets, it takes more computer time. Add the statement

LSMEANS BRAND/PDIFF;

Resubmit the program and observe the additional table "Least Squares Means for effect BRAND." This is another way to show pairwise differences. (This technique works in PROC GLM but not PROC ANOVA.) The LSMEANS statement requests a comparison of least square means, which calculates an adjusted probability for each paired difference. This output is given in Table 14.5. These results also indicate no difference between groups 1 and 3 means ($p = 0.8523$) while all other paired comparisons are statistically significant at $p < 0.05$.

LSMEANS will be discussed in more detail later. For this one-way example, the least square means are no different than the unadjusted means. However, you must use the LSMEANS statement rather than a MEANS statement to use the PDIFF option and get the pairwise *p*-values in GLM.

TABLE 14.5 Analysis of Variance Results

	Least Squares Means for effect BRAND Pr > |t| for H0: LSMean(i)=LSMean(j) Dependent Variable: RELIEF		
i/j	1	2	3
1		0.0056	0.8524
2	0.0056		0.0080
3	0.8524	0.0080	

Some people choose to always use GLM in cases that either GLM or ANOVA would have applied. In the GLM output, you will see results in tables labeled TYPE I and TYPE III sums of squares. In this example, the results in the two tables will be the same. In some more complex settings – for example, multiway ANOVA designs with an unequal number of observations per cell – the TYPE I and TYPE III sums of squares will differ. When this occurs, the typical recommendation is to use the TYPE III sums of squares (see Elliott and Woodward 1986).

14.2 COMPARING THREE OR MORE REPEATED MEASURES

Repeated measures are observations taken from the same or related subjects over time or in differing circumstances. Examples include weight loss or reaction to a drug over time. When there are two repeated measures, the analysis of the data becomes a paired *t*-test (as discussed in Chapter 12). When there are three or more repeated measures, the corresponding analysis is a repeated measure ANOVA.

Assumptions for the repeated measures ANOVA are that the dependent variable is normally distributed and that the variances across the repeated measures are equal. Moreover, as in the one-way ANOVA case, the test is robust against moderate departures from the normality and equal variance assumptions. As in the one-way ANOVA, you will usually perform the analysis in two steps. First, an ANOVA will determine if there is a difference in means across time. If a difference is found, then multiple comparisons can be performed to determine where the differences lie.

The hypotheses being tested with repeated measures ANOVA are as follows:

H_0: There is no difference among the group means (repeated measures).
H_a: There is a difference among the group means.

For this analysis, the PROC GLM procedure will be used because the repeated measures ANOVA is not supported in PROC ANOVA. The abbreviated syntax for PROC GLM is similar to that for PROC ANOVA:

356 CHAPTER 14: ANALYSIS OF VARIANCE

```
PROC GLM <options>;
    CLASS variable;
    MODEL dependentvar = independentvars/options;
    MEANS independentvars / typecomparison
    <meansoptions>;
```

The CLASS, MODEL, and MEANS statements are essentially the same as for PROC ANOVA. These are not all of the options available in PROC GLM, but these three statements are sufficient for performing the analyses in this section. Also note that GLM output contains TYPE I and TYPE III sums of squares. In Hands-On Example 14.1, these two sums-of-squares will be the same. However, in more complex settings they may differ, in which case the typical advice is to use the TYPE III sums of squares.

> The repeated measures ANOVA is also called a within-subjects or treatment-by-subject design. Some call it a "Single-factor" experiment having repeated measures on the same element.
>
> The data in Hands-On Example 14.2 are repeated measures of reaction times (OBS) of five persons after being treated with four drugs in randomized order. (These types of data may come from a crossover experimental design.) The data are as follows where it is important to understand that, for example, the first row (i.e., 31, 29, 17, and 35) consists of results observed on Subject 1. The data must be entered into SAS in such a way that this relationship is identified. You will note that in the SAS code to follow, each reading on the dependent variable (RESULT) is identified with respect to its corresponding SUBJ and DRUG.

Subj	Drug1	Drug2	Drug3	Drug4
1	31	29	17	35
2	15	17	11	23
3	25	21	19	31
4	35	35	21	45
5	27	27	15	31

HANDS-ON EXAMPLE 14.2

This example illustrates how to compare three or more repeated measures (dependent samples) and perform pairwise comparisons using the DUNCAN procedure.

1. Open the program file AGLM1.SAS.

```
DATA STUDY;
INPUT SUBJ DRUG RESULT;
DATALINES;
1   1   31
1   2   29
1   3   17
```

```
1    4    35
2    1    15
... etc
5    3    15
5    4    31
;
RUN;
PROC GLM DATA=STUDY;
    CLASS SUBJ DRUG;
    MODEL RESULT= SUBJ DRUG;
    MEANS DRUG/DUNCAN;
    TITLE 'Repeated Measures ANOVA';
RUN;
```

2. Run the program and observe the results. Several tables are included in the output. Table 14.6 shows the overall ANOVA table and the "Type III SS" table. The test in the first table is an overall test to determine whether there are any significant differences across subjects or drugs. If this test is not significant, you can end your analysis and conclude that there is insufficient evidence to show a difference among subjects or drugs. In this case, $p < 0.0001$, so you continue to the Type III results table. In the Type III results table, the DRUG row reports a p-value of $p < 0.0001$. This is the test of the null hypothesis of interest, which is that there is no difference between the drugs. Because $p < 0.05$, you reject the null hypothesis and conclude that there is a difference between the drugs. Although SUBJ is included as a factor in the model statement, you will generally not be interested in a subject (SUBJ) effect.

TABLE 14.6 Analysis of Variance Results

Source	DF	Sum of Squares	Mean Square	F Value	Pr > F
Model	7	1331.800000	190.257143	25.03	<.0001
Error	12	91.200000	7.600000		
Corrected Total	19	1423.000000			

Source	DF	Type III SS	Mean Square	F Value	Pr > F
SUBJ	4	648.0000000	162.0000000	21.32	<.0001
DRUG	3	683.8000000	227.9333333	29.99	<.0001

3. The multiple comparison test results are given in Figure 14.3. This table is similar to the multiple comparisons discussed in the one-way ANOVA example given in Figure 14.1 except that in this example we have used the Duncan multiple range test rather than the Tukey test for multiple comparisons. The Duncan multiple range test for DRUG indicates that the time to relief for drug 3 is significantly lower than that for all other drugs. There is no statistical difference between drugs 2 and 1; drug 4 has the highest time to relief for all drugs tested. Thus, on this basis, drug 3 would be the preferred drug.

(Continued)

(*Continued*)

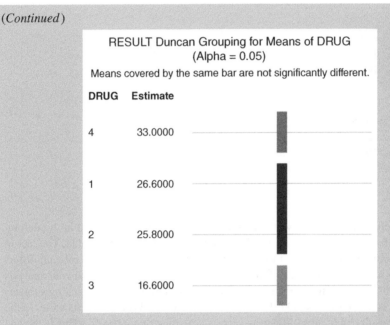

Figure 14.3 Duncan's Multiple Comparison Results

The interaction plot shown in Figure 14.4 visually shows that lines 1 and 2 are very close, but all other lines are far apart. Several other plots are also included in the output that can help you visualize the results. (The plots are by default in color, but we have created them here using an ODS STYLE=JOURNAL statement so the lines can be distinguished in black and white. For ODS STYLE information, see Chapter 8.)

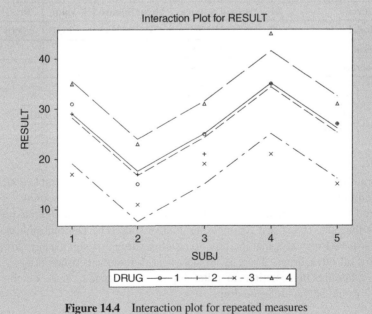

Figure 14.4 Interaction plot for repeated measures

4. Change the MEANS DRUG/DUNCAN line to

```
LSMEANS DRUG/PDIFF ADJUST=TUKEY
```

The LSMEANS statement requests a comparison of least square means with PDIFF ADJUST=TUKEY, which calculates Tukey-type adjusted probability for each paired difference. (Other common ADJUST= options are BON, DUNNETT, and SCHEFFE.) Resubmit the program and observe the results, particularly the LSMEANS results, as given in Table 14.7. The results from this analysis are similar to the DUNCAN comparison. DRUGs 1 and 2 are not different ($p = 0.966$) and all other comparisons are significant at the 0.05 level.

TABLE 14.7 LSMEANS Multiple Comparison Results

Least Squares Means for effect DRUG
Pr > |t| for H0: LSMean(i)=LSMean(j)
Dependent Variable: RESULT

i/j	1	2	3	4
1		0.9666	0.0005	0.0147
2	0.9666		0.0010	0.0066
3	0.0005	0.0010		<.0001
4	0.0147	0.0066	<.0001	

14.3 GOING DEEPER: CONTRASTS

At times, when you are comparing means across groups in a one-way ANOVA, you may be interested in specific posthoc comparisons. For example, suppose you have a dataset consisting of four groups. For some hypothesized reason, you wonder if the average of means 1 and 2 is different from mean 4. Using a CONTRAST statement, you can specify such a comparison. A CONTRAST statement has the following syntax:

```
CONTRAST 'label' indvar effectvalues
```

where the label is a display label for the contrast, *indvar* is an independent variable from the MODEL statement and *effectvalues* are a set of coefficients that specify the contrast. The tricky part of this statement is the *effectvalues*. The *effectvalues* are a series of numbers in the positional order of the classification variable that corresponds to the effect to be tested. For example, to create a CONTRAST statement to compare GROUP 1 versus the combined mean of GROUP 3 and 4, you could use this statement:

```
CONTRAST '1 vs 3+4' GROUP -1 0 .5 .5;
```

Note that the *effectvalues* (−1, 0, 0.5, and 0.5) sum up to zero. You can read them in the following way. The −1 represents the mean of GROUP 1. GROUP 2 is not included in

360 CHAPTER 14: ANALYSIS OF VARIANCE

the contrast so its coefficient is 0. GROUPS 3 and 4 both have coefficients of 0.5, which indicate that their means are combined equally, each contributing a half (0.5) to the value. The signs indicate the comparison. In this case, the mean represented by the negative (−1) is contrasted with the positive combination of the means represented by 0.5 and 0.5. The contrast statement

```
CONTRAST '1 vs 3+4' GROUP -2 0 1 1;
```

produces the same result although it compares twice the mean of GROUP 1 to the sum of the means of GROUPS 3 and 4. Hands-On Example 14.3 illustrates how this works in SAS.

HANDS-ON EXAMPLE 14.3

1. Open the program file AGLM CONTRAST.SAS. A portion of this code is given below. In this dataset, there are four groups and an observation for each record. The PROC GLM performs a one-way ANOVA analysis. The LSMEANS statement produces a pairwise comparison for all pairs.

```
DATA CONTRAST;
INPUT GROUP OBSERVATION;
DATALINES;
1 9.81
1 8.92
... Data continues ...
4 5.41
4 3.94
;
PROC GLM DATA=CONTRAST;
CLASS GROUP;
MODEL OBSERVATION=GROUP;
CONTRAST '1 vs 3+4' GROUP -1 0.5 .5;
RUN;
QUIT;
```

The CONTRAST statement requests a specific comparison of group 1 versus the mean of groups 3 and 4. Submit this code and observe the results. Of particular interest are the results given in Table 14.8. The result indicates that there is a statistically significant difference between the mean of DRUG1 and the average of the means of DRUG3 and DRUG4.

TABLE 14.8 CONTRAST Results from PROC GLM

Contrast	DF	Contrast SS	Mean Square	F Value	Pr > F
Groups 1 vs. 3&4	1	156.3497143	156.3497143	55.96	<.0001

2. Add the following CONTRAST statements and resubmit the code.

```
CONTRAST 'Drugs 1 vs 3&4 Again' GROUP -2 0 1 1;
CONTRAST 'Drugs 1&2 vs 3&4' GROUP -.5 -.5 .5 .5;
```

The first statement is an example of another way to specify the original CONTRAST statement and gives the same results. The second statement compares the average of DRUGS 1 and 2 to the average of GROUPS 3 and 4. The results are given in Table 14.9. Note that the results for the first two contrasts are identical. Moreover, there is a significant difference between the average of the means of DRUG1 and DRUG2 versus DRUG3 and DRUG4.

TABLE 14.9 Additional CONTRAST Results

Contrast	DF	Contrast SS	Mean Square	F Value	Pr > F
Groups 1 vs. 3&4	1	156.3497143	156.3497143	55.96	<.0001
Drugs 1 vs. 3&4 Again	1	156.3497143	156.3497143	55.96	<.0001
Drugs 1&2 vs. 3&4	1	376.5753572	376.5753572	134.79	<.0001

14.4 SUMMARY

This chapter illustrates SAS procedures for comparing three or more means in both an independent group setting and for repeated measures. In both cases, the chapter includes examples illustrating how to perform posthoc multiple comparisons analysis.

EXERCISES

14.1 **Perform multiple comparisons.**
   a. Modify the PROC ANOVA program (AANOVA1.SAS) by replacing the Tukey comparison code to perform the Scheffe, LSD, and Dunnett's tests. Use the following statements:

   ```
   MEANS BRAND/SCHEFFE;
   MEANS BRAND/LSD;
   MEANS BRAND/DUNNETT ('1');
   ```

   Compare the results.
   b. Replace PROC ANOVA with PROC GLM in this program and rerun it. Note that the answers are the same. PROC GLM is a more advanced procedure that can do the same analysis as PROC ANOVA as well as more complex analyses.

14.2 **Run a one-way ANOVA.**
   A researcher is considering three training videos on how to operate a medical device. To determine which training video is the most effective, he randomly selects students from a class and organizes them into three groups. Each group is shown the video and then is tested on the device. Scores range from 0 to 100. The data are in a snippet of SAS code (EX_14.2.SAS):

```
DATA DEVICE;
INPUT SUBJ $ GROUP SCORE;
DATALINES;
AE 1 99
DF 2 99
ED 1 82
FR 3 79
EE 1 89
EG 2 87
IS 3 69
OE 2 77
KY 1 100
WD 3 82
AD 2 89
TR 1 99
SS 2 83
WE 3 81
;
TITLE 'Exercise 14.2';
```

    a. Which type of ANOVA is appropriate for this analysis?

    b. Complete the code to perform the correct analysis.

    c. Perform a multiple comparison test.

    d. Produce box plots.

    e. What are your conclusions?

14.3 **Run a repeated measures ANOVA.**

Complete the code in the program file EX_14.3.SAS to perform a repeated measures ANOVA where SID is the subject ID, TIME includes three (repeated) times of interest observed on each subject (Baseline, Time1, and Time2), and OBS is an observed result from each subject at each time. Use AGLM1.SAS as a model. How would you interpret the results of this analysis?

REFERENCE

Elliott, A.C. and Woodward, W.A. (1986). Analysis of an unbalanced two-way ANOVA on the microcomputer. *Communications in Statistics – Simulation and Computation* 15 (1): 215–225.

15

ANALYSIS OF VARIANCE, PART II

LEARNING OBJECTIVES

- To be able to use SAS® procedures to perform an analysis of covariance
- To be able to use SAS procedures to perform two-factor ANOVA using PROC GLM
- To be able to perform two-factor ANOVA using PROC MIXED
- To be able to use SAS procedures to perform repeated measures with a grouping factor

SAS provides procedures that can analyze data from a wide range of experimental designs. This chapter illustrates three designs not discussed in Chapter 14 and provides you with a brief introduction to PROC MIXED.

15.1 ANALYSIS OF COVARIANCE

An analysis of covariance (ANCOVA) is a combination of analysis of variance and regression. The covariate is a quantitative variable that is related to the dependent variable. However, the covariate is not controlled by the investigator but is some value intrinsic to the subject (or entity). In ANCOVA, the group means are adjusted by the covariate, and these adjusted means are compared with each other. Including this covariate variable in the model may explain some of the variability, resulting in a more powerful statistical test.

SAS® Essentials: Mastering SAS for Data Analytics, Third Edition. Alan C. Elliott and Wayne A. Woodward.
© 2023 John Wiley & Sons, Inc. Published 2023 by John Wiley & Sons, Inc.
Companion website: https://www.alanelliott.com/SASED3/

Consider an experiment designed to compare three medications for lowering systolic blood pressure (SBP). A potentially useful covariate is AGE because it is known that there is a relationship between SBP and age. If SBP is the dependent variable in a model, it might be helpful to adjust SBP by the covariate age. Thus, in ANCOVA, the means adjusted for AGE are compared rather than the raw means.

In Hands-On Example 15.1, we consider a dataset in which a fifth-grade math teacher randomly assigns the 18 students in her class to three different teaching methods for individualized instruction of a certain concept. The outcome of interest is the score on an exam over the concept after the instruction period. An exam of basic math skills (EXAM) is given to the students before beginning the instruction. The data are given in Table 15.1. The variable FINAL is compared across the three teaching methods. The covariate EXAM is assumed to be linearly related to FINAL. Note that if there were no EXAM used to adjust FINAL scores, the analysis would be a standard one-way analysis of variance comparing FINAL scores across the three teaching methods. However, with the addition of the EXAM information, the analysis is potentially more powerful. The ANCOVA tests the null hypothesis that the FINAL means for the three methods adjusted by EXAM are not different from each other.

The procedure requires a multistep approach:

1. Perform a test to determine if the FINAL-by-EXAM linear relationships by METHOD are parallel. That is, when you compare the regression lines for each of the three METHODs, the lines should theoretically be parallel. If the lines are sufficiently nonparallel, ANCOVA is not the appropriate analysis to perform. The statistical test used for parallelism is an F-test.

TABLE 15.1 Data from Math Course Analysis

Method	EXAM	FINAL
1	16	38
1	17	39
1	15	41
1	23	47
1	12	33
1	13	37
2	22	41
2	10	30
2	19	45
2	6	28
2	2	25
2	17	39
3	3	30
3	20	49
3	14	43
3	10	38
3	5	32
3	11	37

2. If the *F*-test in step 1 does not reject parallelism, then another *F*-test is used to compare the adjusted means across METHOD.
3. If there are differences in means, then appropriate multiple comparisons can be performed to determine which groups differ.

HANDS-ON EXAMPLE 15.1

This example uses an ANCOVA to compare three methods of teaching a fifth-grade math concept.

1. Open the program file AGLM ANCOVA.SAS.

```
DATA ANCOVA;
INPUT METHOD EXAM FINAL @@;
DATALINES;
1  36  38  1  37  39  1  35  41
1  43  47  1  32  33  1  33  37
2  42  41  2  30  30  2  39  45
2  26  28  2  22  25  2  37  39
3  23  30  3  40  49  3  34  43
3  30  38  3  25  32  3  31  37
;
RUN;
* CHECK THE ASSUMPTION THAT SLOPES ARE PARALLEL;
PROC GLM;
   CLASS METHOD;
     MODEL FINAL=EXAM METHOD EXAM*METHOD;
        TITLE 'Analysis of Covariance Example';
RUN;
* IF SLOPES ARE PARALLEL, DROP THE INTERACTION TERM;
PROC GLM;
   CLASS METHOD;
     MODEL FINAL=EXAM METHOD;
        LSMEANS METHOD/STDERR PDIFF ADJUST=SCHEFFE;
RUN;
QUIT;
```

Note that PROC GLM is run twice. The first instance is to test that the slopes are parallel. This is tested with the following MODEL statement:

MODEL FINAL=EXAM METHOD EXAM*METHOD;

The EXAM*METHOD component in the model is the interaction (or parallel slopes) term. When the test for this component is not significant, it suggests that the slopes among types of METHOD can be treated as parallel.

2. Run the program. As mentioned in Chapter 14, we recommend using Type III sums-of-squares. Observe these sums-of-squares in the output given in Table 15.2.

(*Continued*)

(*Continued*)

TABLE 15.2 Test of Slopes for Analysis of Covariance

Source	DF	Type III SS	Mean Square	F Value	Pr>F
EXAM	1	476.4504266	476.4504266	114.76	<.0001
METHOD	2	0.8199072	0.4099536	0.10	0.9067
EXAM*METHOD	2	3.2623239	1.6311619	0.39	0.6835

The row for the test of the interaction term (EXAM*METHOD) is nonsignificant ($p = 0.6835$). This provides evidence that the slopes can be treated as parallel and that the ANCOVA is an appropriate analysis.

3. Because the slopes are not significantly nonparallel, the second PROC GLM analysis is appropriate. Note that the interaction term has been removed from the MODEL statement

```
MODEL FINAL=EXAM METHOD;
```

forcing the model to assume parallel regression lines.

Table 15.3 shows Type III SS output table for the second PROC GLM. The factor of interest in this table is METHOD, and the corresponding F is a statistical test comparing the three methods adjusting for EXAM. Because $p = 0.0005$, we reject the null hypothesis that the adjusted means are equal and conclude that at least one pair of adjusted means are different from each other.

TABLE 15.3 Analysis of Covariance Test of Main Effects

Source	DF	Type III SS	Mean Square	F Value	Pr>F
EXAM	1	627.9168352	627.9168352	165.60	<.0001
METHOD	2	105.9808664	52.9904332	13.98	0.0005

4. Because we found a difference in adjusted means, pairwise comparisons can be used to identify which adjusted means are different from each other. The statement that provides that comparison is

```
LSMEANS METHOD/STDERR PDIFF ADJUST=SCHEFFE;
```

In this case, we have chosen to use the Scheffe multiple comparisons, and the results are given in Table 15.4. (Ignore the Pr>|*t*| column, it is testing that the mean is different from zero, which is of no interest in this analysis.)

The Scheffe results provide *p*-values for each pairwise comparison. In this case, the comparisons of METHODs 1 and 3 ($p = 0.0056$) and 2 and 3 ($p = 0.0007$)

ANALYSIS OF COVARIANCE

TABLE 15.4 Scheffe Multiple Comparisons for ANCOVA

METHOD	FINAL LSMEAN	Standard Error	Pr > |t|	LSMEAN Number
1	36.0946098	0.8300185	<.0001	1
2	35.0724100	0.7955731	<.0001	2
3	40.8329802	0.8215055	<.0001	3

Least Squares Means for effect METHOD
Pr > |t| for H0: LSMean(i)=LSMean(j)
Dependent Variable: FINAL

i/j	1	2	3
1		0.6837	0.0056
2	0.6837		0.0007
3	0.0056	0.0007	

indicate significant differences. METHODs 1 and 2 are not significantly different ($p = 0.68$).

Because a high FINAL score is the goal, there is evidence to support the contention that the adjusted mean of 40.82 for METHOD 3 is a significantly higher score than for METHODs 1 and 2, and thus that METHOD 3 is the preferred method.

5. Along with the tabled output, SAS includes several graphs. For the first PROC GLM (that included the interaction term) you get the plot shown in Figure 15.1. It shows the three regression lines by METHOD. Note that the lines are nearly parallel, which confirms the findings in the first PROC GLM procedure. (We show these plots as created using the ODS STYLE=JOURNAL option. See Chapter 8 for more on ODS. Your plots may appear in color.)

   For the model without the interaction term, we assume that the lines are parallel. Thus, the plot associated with the second PROC GLM uses a common slope estimate, and the lines in Figure 15.2 represent the adjusted means by METHOD.

   These plots support our overall finding that the adjusted mean for METHOD 3 is higher than the means for the other methods. In addition, SAS provides a graphical display of the Scheffe results, as shown in Figure 15.3. This plot shows confidence intervals for the adjusted means indicated by lines at the intersection of the three methods. This graph indicates that the comparisons of 3 versus 2 and 3 versus 1 are both significant because their lines do not cross the diagonal line. The comparison of 1 versus 2 is not significant.

6. Replace the Scheffe test in the LSMEANS line with a Bonferroni or Tukey test and compare the output and results.

(Continued)

368 CHAPTER 15: ANALYSIS OF VARIANCE, PART II

(*Continued*)

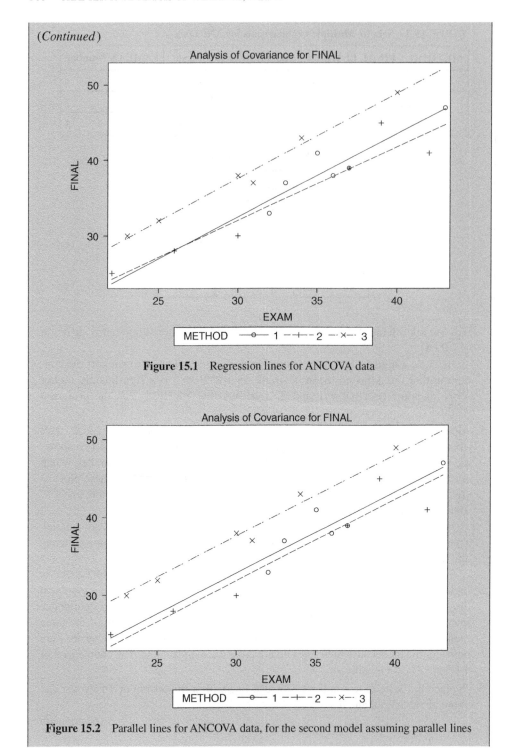

Figure 15.1 Regression lines for ANCOVA data

Figure 15.2 Parallel lines for ANCOVA data, for the second model assuming parallel lines

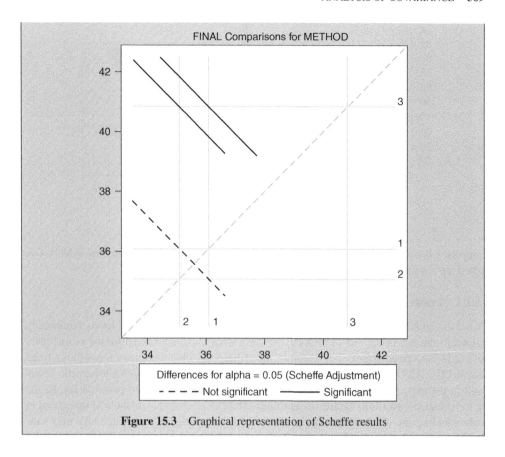

Figure 15.3 Graphical representation of Scheffe results

15.1.1 Two-Factor ANOVA Using PROC GLM

A two-way ANOVA is an analysis that allows you to simultaneously evaluate the effects of two experimental variables (factors). Each factor is a "grouping" variable such as type of treatment, gender, brand, and so on. The two-way ANOVA tests determine whether the factors are important (significant) either separately (called main effects) or in combination (via an interaction).

As an example of interaction effects, suppose you are observing annual salary where the two factors are gender and job category. Figure 15.4 illustrates the difference between an interaction effect and no interaction effect. Figure 15.4a indicates that female salaried workers have a different pattern of income from males across hourly and salary categories from males. Figure 15.4b indicates that the mean incomes for hourly and salaried personnel are parallel (noninteracting) across gender. When an interaction effect is present (Figure 15.4a), you cannot easily compare means between male and female workers because they have different patterns. When there is no interaction effect (Figure 15.4b), it makes sense to compare overall means between male and female workers. In Figure 15.4a, it appears that while there is no gender difference in salaries for hourly employees, male salaries are higher for salaried employees.

The dimensions of a factorial design depend on how many levels of each factor are used. For example, a design in which the first factor has two categories and the second has three

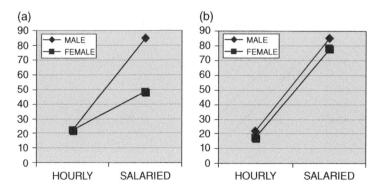

Figure 15.4 (a, b) Example of an interaction effect

categories is called a 2×3 (2 by 3) factorial design. Generally, the two-way ANOVA is called a $p \times q$ factorial design.

15.1.2 Understanding Fixed and Random Factors

Factors in an ANOVA are classified as fixed or random. When the factor levels completely define all the classifications of interest, we have what is called a **fixed factor** (sometimes referred to as a **fixed effect**). For example, GENDER (classified as male or female) is a fixed factor. (The ANOVA examples in Chapter 14 all involve fixed factors.) When the levels used in the experiment are randomly selected from the population of possible levels, the factor is called a **random factor** (or **random effect**). For example, a chemist interested in understanding the differences among laboratories that perform a certain assay may randomly select a sample of laboratories from among the large population of laboratories for his experiment. In this case, LAB would be a random effect. Two-way ANOVA models are classified with the following scheme:

- Both factors fixed: Model I ANOVA
- Both factors random: Model II ANOVA
- One random, one fixed: Model III ANOVA

In this section, we will assume that both factors (A and B) are fixed, and we illustrate an analysis for a Model I ANOVA. We will not provide an example of a Model II analysis, but a Model III ANOVA will be described in Section 15.2.

When gathering data for this fixed-factor ANOVA design, within the $p \times q$ possible combinations of factors (cells), subjects should be randomly assigned to treatment in such a way as to balance the number of subjects per cell. However, if the data are not balanced, an analysis can still be performed using PROC GLM (but not with PROC ANOVA).

There are some additional calculation issues involved with the unbalanced case that are not discussed here. Generally, in the unbalanced case, Type III sums-of-squares from the PROC GLM output are recommended. Moreover, the dependent variable should be a quantitative variable whose distribution within each of the $p \times q$ (p by q) combinations is approximately normal with equal variances among combinations. (Some of these issues are relaxed when you are using PROC MIXED.)

There are two types of hypotheses used in a two-factor Model I design: hypotheses about interaction effects and main effects. The interaction effect should be examined first as follows:

H_0: There is no interaction effect.
H_a: There is an interaction effect.

Interaction implies that the pattern of means across groups is inconsistent (as illustrated in Figure 15.4a). If there is an interaction effect, main effects cannot be examined directly because the interaction effect shows that differences across one main effect are not consistent across all levels of the other factor. Thus, if there is no interaction effect, it makes sense to test hypotheses about the main effects. The "main effects" hypotheses for factor A are as follows:

H_0: Means are equal across levels of A summed over B.
H_a: Means are not equal across levels of A summed over B.

Similarly, for factor B:

H_0: Means are equal across levels of B summed over A.
H_a: Means are not equal across levels of B summed over A.

The results of these three tests (interaction and two main effects tests) are given in an analysis of the variance table as F-tests. A low p-value (usually <0.05) for a test indicates evidence to reject the null hypothesis in favor of the alternative.

HANDS-ON EXAMPLE 15.2

This example (adapted from dataset 404 in Hand et al. 1994) concerns a study on weight loss for patients attending meetings about dieting. A variable called CONDITION indicates whether the patient received a weight loss manual (CONDITION=1) or not (CONDITION=2). The STATUS variable indicates whether the patient had already been trying to lose weight before the series of meetings (STATUS=1) or not (STATUS=2).

Both factors are assumed to be fixed effects for this example, and there are an unequal number of subjects in each category. For this case, PROC GLM can be used.

1. Open the program file AGLM 2FACTOR.SAS.

```
DATA SLIM;
INPUT CONDITION STATUS RESPONSE @@;
DATALINES;
1 1 -13.67 1 1 -12.85 1 1 -9.55 1 1 -17.03 1 1 13.61
1 2 .91 1 2 2.48 1 2 2.84 1 2 3.46 1 2 2.20 1 2 -.73
1 2 -3.05 1 2 -5.68 1 2 -3.44 1 2 -7.18 1 2 -3.40 1 2 -.74
2 1 -3.29 2 1 -4 2 1 -2.31 2 1 -3.4 2 1 -7.49 2 1 -13.62
```

(*Continued*)

(*Continued*)

```
2 1 -7.34 2 1 -7.39 2 1 -1.32 2 1 -12.01 2 1 -8.35
2 2 5.94 2 2 1.91 2 2 -4.0 2 2 -5.19 2 2 0 2 2 -2.8
;
RUN;
TITLE "Two-Way ANOVA Example";
PROC GLM;
CLASS CONDITION STATUS;
    MODEL RESPONSE=CONDITION STATUS CONDITION*STATUS;
        MEANS CONDITION STATUS CONDITION*STATUS;
RUN;
QUIT;
```

Recall that in the MODEL statement, the dependent variable (RESPONSE) appears to the left of the equal sign and the independent variables (CONDITION and STATUS) appear to the right. The CONDITION*STATUS component of the MODEL statement is the interaction term. The two factors, CONDITION and STATUS, are listed in the CLASS statement because they are categorical variables.

2. Run this program and observe the output, partially shown in Table 15.5. This is the output from the first PROC GLM statement in the code. Note that the overall model is significant ($p = 0.04$), indicating the presence of some differences in means. Note also that because of the unequal number of observations per cell, Type I and Type III sums-of-squares (SS) give different results. It is recommended that Type III SS be used, and only the SS table for Type IIII SS is given in Table 15.5.

3. For this analysis, first examine the interaction effect (the CONDITION*STATUS row) in Type III ANOVA table. Because $p = 0.7872$, there is no evidence of interaction. As mentioned previously:

- If there is no interaction, examine the main effects tests in the ANOVA table by comparing marginal means.
- If there are interaction effects, the factor effects should not be interpreted in isolation from each other. It is usually more appropriate to compare the effects of the first factor within levels of the second factor and vice versa. That is, compare cell

TABLE 15.5 ANOVA Results from a Two-Factor Analysis Using PROC GLM

Source	DF	Sum of Squares	Mean Square	F Value	Pr > F
Model	3	309.009404	103.003135	3.13	0.0402
Error	30	987.604196	32.920140		
Corrected Total	33	1296.613600			

Source	DF	Type III SS	Mean Square	F Value	Pr > F
CONDITION	1	6.1546978	6.1546978	0.19	0.6686
STATUS	1	293.1057835	293.1057835	8.90	0.0056
CONDITION*STATUS	1	2.4432188	2.4432188	0.07	0.7872

means rather than the marginal means using a posthoc analysis. Because there is no significant interaction in this example, it is appropriate to examine the main effects tests (STATUS and CONDITION):

- The test for the CONDITION main effect, $p = 0.6686$.
- The test for STATUS, $p = 0.0056$.

Thus, you can reject the null hypothesis that there is no STATUS effect and conclude that evidence supports the hypothesis that STATUS (i.e., whether the subject was already trying to lose weight) is a statistically significant factor in realized weight loss. Table 15.6 shows the marginal means for the two levels of STATUS. Because there are only two categories of STATUS, we do not need to perform multiple comparisons, and the clear conclusion is that STATUS=1 (i.e., those who had previously been trying to lose weight) had significantly more weight loss on average than those in STATUS=2.

Because $p = 0.67$ for the CONDITION factor, we conclude that there is no significant CONDITION effect (i.e., the manual seemed to have no effect).

It is often helpful to use graphs to help visualize the results of a two-factor analysis. Along with the tabled output, SAS produces the plot shown in Figure 15.5. This plot

TABLE 15.6 Marginal Means for Levels of STATUS

Level of STATUS	N	RESPONSE	
		Mean	Std Dev
1	16	-6.87562500	7.17206058
2	18	-0.91500000	3.63062829

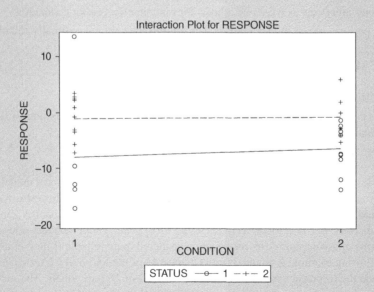

Figure 15.5 Interaction plot for the two-factor ANOVA example

(*Continued*)

> (*Continued*)
> gives dot plots of the data for the four combinations of STATUS and CONDITION. The solid line connects the mean RESPONSE for subjects in STATUS=1 across CONDITION. The dashed line provides the same information for those in STATUS=2.
>
> It can be seen that the lines are approximately parallel, supporting the finding of no significant interaction and indicating that RESPONSE scores are consistently higher for STATUS=2 across levels of CONDITION. Because RESPONSE is weight loss, this graphical representation supports the finding above that those who had previously been trying to lose weight (STATUS=1) had more success in the weight loss program.
>
> Also included in the output are boxplots by STATUS and CONDITION (not shown here) that provide a visual look at the distribution and location for those variables.

15.2 GOING DEEPER: TWO-FACTOR ANOVA USING PROC MIXED

PROC MIXED performs mixed-model analysis of variance and repeated measures analysis of variance with covariance structure modeling. PROC MIXED has features not available in PROC GLM. Briefly, here are some of the major differences between PROC GLM and PROC MIXED in relation to two-factor ANOVA analysis:

PROC GLM

- Is designed primarily for fixed effects models.
- Performs calculations based on ordinary least squares and method of moments.
- Defines all effects as fixed and adjusts for random effects after estimation.
- Can perform mixed model analysis, but some results are not optimally calculated.

PROC MIXED

- Is designed for mixed effects models.
- Uses generalized least squares for fixed effects and Restricted Maximum-Likelihood Estimation (RMLE) to estimate variance components.
- Allows selection of correlation models.

Both PROC GLM and PROC MIXED can analyze fixed effects and mixed models. When all factors are fixed, the two approaches produce the same results. However, for mixed models (and some repeated measures models), the two different approaches can lead to differing answers. Because PROC MIXED uses a more general approach that is less restrictive (in terms of assumptions), its results are considered more reliable (Littell et al. 1996).

The following example is adapted from Chapter 41 of the *SAS/STAT User's Guide*. The data include the response variable HEIGHT (in inches) of 18 individuals. The individuals are classified according to FAMILY and GENDER. Because FAMILY consists of some selection of families, it is reasonable to consider FAMILY to be a random effect. Moreover, the interaction FAMILY*GENDER is a random effect. With this information in mind, we illustrate the use of PROC MIXED to analyze this Model III ANOVA in Hands-On Example 15.3. In this example, the researcher wants to know if there is a difference in HEIGHT by GENDER and FAMILY.

> **HANDS-ON EXAMPLE 15.3**
>
> This example illustrates the use of PROC MIXED and assumes that FAMILY and FAMILY*GENDER are random factors.
>
> 1. Open the program file AMIXED1.SAS.
>
> ```
> DATA HEIGHTS;
> INPUT FAMILY GENDER$ HEIGHT @@;
> DATALINES;
> 1 F 67   1 F 66   1 F 64   1 M 71   1 M 72   2 F 63
> 2 F 63   2 F 67   2 M 69   2 M 68   2 M 70   3 F 63
> 3 M 64   4 F 67   4 F 66   4 M 67   4 M 67   4 M 69
> ;
> RUN;
> PROC MEANS MAXDEC=2 MEAN DATA=HEIGHTS;
>    CLASS FAMILY GENDER;
>    VAR HEIGHT;
> PROC MIXED;
>      CLASS FAMILY GENDER;
>      MODEL HEIGHT = GENDER;
>         RANDOM FAMILY FAMILY*GENDER;
> RUN;
> ```
>
> Run this program and observe the output given in Table 15.7. A table of means is created first using PROC MEANS. Note that the number of observations (N Obs) that go into calculating the means is not balanced. They range from 1 observation to 3.
>
> **TABLE 15.7 PROC MEANS Output**
>
> | Analysis Variable: HEIGHT | | | |
> |---|---|---|---|
> | FAMILY | GENDER | N Obs | Mean |
> | 1 | F | 3 | 65.67 |
> | | M | 2 | 71.50 |
> | 2 | F | 3 | 64.33 |
> | | M | 3 | 69.00 |
> | 3 | F | 1 | 63.00 |
> | | M | 1 | 64.00 |
> | 4 | F | 2 | 66.50 |
> | | M | 3 | 67.67 |
>
> (*Continued*)

(*Continued*)
2. Examine how the following PROC MIXED statement is constructed:

```
CLASS FAMILY GENDER;
MODEL HEIGHT = GENDER;
RANDOM FAMILY FAMILY*GENDER;
```

- FAMILY and GENDER appear in the CLASS statement because they are both grouping-type factors.
- The MODEL statement includes only the fixed factor GENDER.
- The RANDOM statement includes the random factors FAMILY and FAMILY*GENDER.

Observe the output table titled "Type 3 Tests of Fixed Effects" shown in Table 15.8. This shows the results of a hypothesis test that there is no difference in means by GENDER, which is marginally nonsignificant ($p = 0.07$). SAS reports no test involving FAMILY.

TABLE 15.8 PROC MIXED Output Testing the Fixed Factor

Type 3 Tests of Fixed Effects				
Effect	Num DF	Den DF	F Value	Pr > F
GENDER	1	3	7.95	0.0667

3. To illustrate why it is important to properly classify a random factor as random, change the MODEL to where there are only fixed effects (Remove the RANDOM statement.):

```
MODEL HEIGHT = GENDER FAMILY GENDER*FAMILY;
```

Run the changed program and observe the (abbreviated) output given in Table 15.9. Note that in this case, the test for a GENDER effect is significant ($p = 0.0018$). However, in the fixed effects model, the FAMILY effect applies only to the families in the study (fixed effect). When FAMILY is classified as a random factor, the test results are applicable to families in general, not just to the ones in the study.

TABLE 15.9 Model Run as If All Factors Were Fixed

Type 3 Tests of Fixed Effects				
Effect	Num DF	Den DF	F Value	Pr > F
GENDER	1	10	17.63	0.0018
FAMILY	3	10	5.90	0.0139
FAMILY*GENDER	3	10	2.89	0.0889

15.3 GOING DEEPER: REPEATED MEASURES WITH A GROUPING FACTOR

A common design in medical research and other settings is to observe data for the same subject under different circumstances or over time (longitudinal). The one-way repeated measures design is illustrated in Chapter 14. When the design also includes a grouping variable, the analysis becomes a little more complicated. The "repeated measures design with a grouping factor" can be analyzed in SAS using PROC MIXED. Although it is also possible to perform this analysis using PROC GLM, the approach used in PROC MIXED is preferred. In addition, a distinct advantage of using PROC MIXED is that it allows missing values in the design, whereas PROC GLM deletes an entire record from analysis if there is one missing observation in that record.

Suppose you observe a response to a drug over 4 hours for seven subjects, three male and four female. In this case, the repeated measure, HOUR, is longitudinal. The effect across hours, because it is a measurement on the same individual four times, is called a within-subject factor. GENDER is a between-subject factor because it is a measurement between independent groups. In this example, we'll assume that both GENDER and HOUR are fixed factors because they include all levels of interest. The observed data are given in Table 15.10. Note that there is one missing value and an unequal number of subjects by GENDER.

The dataset for this analysis is initially set up as in Table 15.10 (similar to most datasets in this book) with one subject's data per record. However, for repeated measures, the PROC MIXED procedure requires that the data be set up with one observation per record. Hands-On Example 15.4 illustrates how to rearrange the data and perform the repeated measures analysis using PROC MIXED.

TABLE 15.10 Original Dataset of Repeated Measures Data

SUB	GENDER	HOUR1	HOUR2	HOUR3	HOUR4
1	M	1.0	1.5	6.0	5.1
2	M	4.0	2.2	6.1	5.2
3	M	5.2	4.1	6.8	3.2
4	F	5.1	3.3	4.2	4.8
5	F	6.3	4.9	6.9	6.9
6	F	8.2	5.9	9.5	9.1
7	F	8.3	6.1	Missing	9.2

HANDS-ON EXAMPLE 15.4

This example uses PROC MIXED to analyze a repeated measures design that includes one grouping factor.

1. Open the program file AREPEAT1.SAS. The SAS code used to read in and rearrange the dataset for the analysis is shown here:

(*Continued*)

(*Continued*)
```
DATA REPMIXED(KEEP= SUBJECT GENDER TIME OUTCOME);
INPUT SUBJECT GENDER $ HOUR1-HOUR4;
OUTCOME = HOUR1;  TIME = 1; OUTPUT;
OUTCOME = HOUR2;  TIME = 2; OUTPUT;
OUTCOME = HOUR3;  TIME = 3; OUTPUT;
OUTCOME = HOUR4;  TIME = 4; OUTPUT;
DATALINES;
1  M  1     1.5  6    5.1
2  M  4     2.2  6.1  5.2
3  M  5.2   4.1  5.8  3.2
4  F  5.1   3.3  5.2  4.8
5  F  6.3   4.9  7.9  6.9
6  F  8.2   5.9  9.5  9.1
7  F  8.3   6.1  9.2
;
```

The dataset is initially set up with one subject per line. However, PROC MIXED requires there to be only one record per data line. That is, there should be one record per line with four records per subject. The SAS code above rearranges the data into the form required by PROC MIXED by

(a) assigning each of the four HOUR values to the variable OUTCOME
(b) creating a variable called TIME that contains the time marker (1–4)
(c) for each of these assignments, outputting the variables to the new dataset REPMIXED using the OUTPUT statement.

The KEEP statement in parentheses after REPMIXED in the DATA statement tells SAS which variables to include in the final dataset. The resulting dataset is in the form given in Table 15.11. Compare this dataset to the original data setup given in Table 15.10. Note that for each row in the original dataset, there are four rows in the REPMIXED dataset.

2. The following code is used to perform three repeated measures analyses using PROC MIXED:

TABLE 15.11 REPMIXED Data for Use in PROC MIXED

SUBJECT	GENDER	TIME	OUTCOME
1	M	1	1.0
1	M	2	1.5
1	M	3	6.0
1	M	4	5.1
2	M	1	4.0
2	M	2	2.2
2	M	3	6.1
Etc	etc	etc	etc

```
PROC MIXED DATA=REPMIXED;
    CLASS GENDER TIME SUBJECT;
    MODEL OUTCOME=GENDER TIME GENDER*TIME;
    REPEATED / TYPE=UN SUB=SUBJECT;
RUN;
PROC MIXED DATA=REPMIXED;
    CLASS GENDER TIME SUBJECT;
    MODEL OUTCOME=GENDER TIME GENDER*TIME;
    REPEATED / TYPE=CS SUB=SUBJECT;
RUN;
PROC MIXED DATA=REPMIXED;
    CLASS GENDER TIME SUBJECT;
    MODEL OUTCOME=GENDER TIME GENDER*TIME;
    REPEATED / TYPE=AR(1) SUB=SUBJECT;
RUN;
```

These three PROC MIXED analyses are identical except for the TYPE= statement. Each TYPE statement includes a different specification for the within-subject covariance matrix. A few commonly used structures include the following:

- VC (Variance components) is the default and simplest structure with all off-diagonal variances equal to 0.
- AR(1) (Autoregressive) assumes that nearby measurements are correlated and the correlations decline exponentially with time. That is, measurements at TIME1 and TIME2 are more highly correlated than are measurements at TIME1 and TIME3.
- CS (Compound symmetry) assumes homogeneous variances that are constant regardless of how far apart the measurements are.
- UN (Unstructured) allows all variances to be different.

See the SAS/STAT Manual (PROC MIXED chapter) for more options and details regarding covariance specifications.

The CLASS option indicates that the three factor variables are all classification (categorical) variables. Categorical variables can be discrete numeric or character. (We include SUBJECT here even though we are not generally interested in a SUBJECT effect, in order to fully define the model.) The MODEL statement indicates that OUTCOME is the dependent variable and GENDER, TIME, and the interaction term GENDER*TIME are independent variables used to predict the outcome.

The REPEATED statement indicates the TYPE of covariance structure to use in the analysis. As mentioned, the TYPE of covariance structures is the only difference among the three calls to PROC MIXED. SUB=SUBJECT indicates the name of the subject variable.

3. Run the program. A lot of output is created by SAS for this analysis, much of it dealing with the iterations required to fit the model. This output will not be discussed here. Initially, take note of the value of AIC (Akaike Information Criterion) in the Fit Statistics output table. This is a measure of the fit of the model, with smaller AIC values considered best. (There are other criteria you could also use, such as BIC and AICC, but we'll limit our discussion to AIC.)

(Continued)

(*Continued*)

You can fit any number of covariance structures to the data with the TYPE= statement and you then typically select the best-fitting model. In this example, based on the AIC criterion, the AR(1) specification best fits the data. For these three models, the AIC results from the Fit Statistics tables associated with each covariance structure considered are as follows:

- AIC(unstructured) = 71.0
- AIC(compound symmetry) = 77.6
- AIC(autoregressive) = 70.2

4. The main tests for this model are found in the table titled "Type 3 Test of Fixed Effects." Because the AIC for the AR(1) model is the lowest, we'll examine the results in that table (the last table ion the output), as given in Table 15.12.

TABLE 15.12   Results from PROC MIXED for AR(1) Structure

Type 3 Tests of Fixed Effects				
Effect	Num DF	Den DF	F Value	Pr > F
GENDER	1	5	5.22	0.0710
TIME	3	14	29.67	<.0001
GENDER*TIME	3	14	1.48	0.2637

In this case, the results indicate that there is no interaction (GENDER*TIME, $p=0.26$) effect. In the presence of no interaction, it is appropriate to examine the main effects tests for TIME and GENDER. For these results, there is a significant TIME ($p < 0.0001$) effect and a nonsignificant (but marginal) GENDER ($p = 0.071$) effect.

The other two models (unstructured and compound symmetry) report a significant GENDER effect (both $p < 0.05$). Thus, the structure selected can affect the outcome of the statistical tests.

Using the AR(1) results, we conclude that there is at most marginal evidence of a GENDER effect. There is (in all models) evidence of a significant TIME effect. To examine these results more closely, we'll look at plots of the means.

5. Open the program file AREPEAT2.SAS. This program is based on the same dataset analyzed previously in this example using the AREPEAT1.SAS code, and it produces two useful graphs. (SAS Graph procedures are discussed more thoroughly in Chapter 19.)

```
*-----------------PRODUCE GRAPHS OF MEANS;
PROC SORT DATA=REPMIXED;BY GENDER TIME;
PROC MEANS noprint; BY GENDER TIME;
   OUTPUT OUT=FORPLOT MEAN=;
RUN;
PROC GPLOT;
PLOT OUTCOME*GENDER=TIME;
SYMBOL1 V=CIRCLE I=JOIN L=1 C=BLACK;
SYMBOL2 V=DOT I=JOIN L=2 C=BLUE;
SYMBOL3 V=STAR I=JOIN L=2 C=RED;
```

GOING DEEPER: REPEATED MEASURES WITH A GROUPING FACTOR 381

```
SYMBOL4 V=SQUARE I=JOIN L=2 C=GREEN;
RUN;
PROC SORT DATA=REPMIXED;BY TIME GENDER;
PROC MEANS noprint; BY TIME GENDER;
   OUTPUT OUT=FORPLOT MEAN=;
RUN;
PROC GPLOT;
PLOT OUTCOME*TIME=GENDER;
SYMBOL1 V=CIRCLE I=JOIN L=1 C=BLACK;
SYMBOL2 V=DOT I=JOIN L=2 C=BLUE;
RUN;
```

This code calculates the means by GENDER and TIME (then TIME and GENDER) and plots the results. For more information about PROC GPLOT, see Chapter 19. Run the program and observe the plots, shown in Figure 15.6a, b.

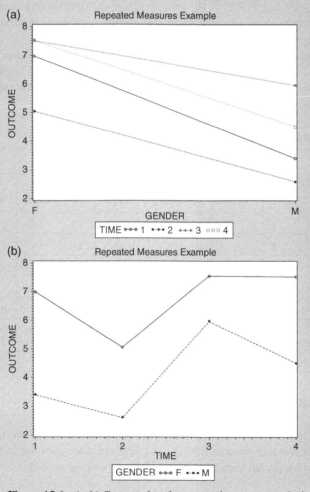

Figure 15.6 (a, b) Custom plots for repeated measures analysis

(*Continued*)

(*Continued*)

Although there (visually) appears to be a difference in means across GENDER, it is at most marginally significant (possibly because of the small sample size). From the graph, observe that the mean at TIME3 is the highest and the mean at TIME2 is the lowest.

6. To find any significant differences among TIMEs, include the following option in the section of PROC MIXED code related to the AR(1) covariance structure.

LSMEANS TIME/PDIFF;

This statement produces confidence limits on the differences in the (least square) means. Open and run the SAS program AREPEAT3.SAS. This program is similar to the original but includes only the AR(1) model (and the LSMEANS statement). Observe the results in Table 15.13. As we observed in Figure 15.6a, b, the difference between TIME2 and TIME3 is the greatest difference ($p < 0.0001$) as can be seen in the Pr>|t| column. Note also that there are significant differences between all other TIMEs except between TIME1 and TIME4 ($p = 0.17$).

This example illustrates the iterative nature of repeated measures analyses. It requires that you determine which variance structure best fits the model. It is also advantageous to plot the means and, when there are more than two means in a significant factor, to compare means using the LSMEANS statement.

TABLE 15.13 Comparing Pairwise Times

Differences of Least Squares Means							
Effect	TIME	_TIME	Estimate	Standard Error	DF	t Value	Pr>|t|
TIME	1	2	1.3625	0.3507	14	3.89	0.0016
TIME	1	3	−1.7890	0.4879	14	−3.67	0.0025
TIME	1	4	−0.8125	0.5651	14	−1.44	0.1725
TIME	2	3	−3.1515	0.3638	14	−8.66	<.0001
TIME	2	4	−2.1750	0.4781	14	−4.55	0.0005
TIME	3	4	0.9765	0.3638	14	2.68	0.0178

15.4 SUMMARY

This chapter discusses the use of PROC GLM to perform ANCOVA and two-factor ANOVA with fixed effects. We also introduce PROC MIXED and discuss its use in the two-factor mixed model and a repeated measures analysis with a grouping variable.

EXERCISES

15.1 **Use multiple comparisons in PROC MIXED.**

Another way to determine differences in means in a PROC MIXED model such as in AREPEAT3.SAS is to request multiple comparison tests using a statement such as

LSMEANS/ADJUST=TUKEY;

(You can optionally use BON, SCHEFFE, or SIDAK as the comparison method.) Replace the PDIFF command in AREPEAT3.SAS with one of these options and rerun the analysis.

15.2 **Use PROC MIXED with missing data.**

Note that the analysis using PROC MIXED was able to handle a missing value in the data as well as unbalanced data (more females than males). To perform this analysis in PROC GLM, use the program EX_15.2.SAS. Run this program and observe the interaction and main effects tests for GENDER and TIME in Type III GLM results tables ($P > F$ column). Although the results are similar to those of PROC MIXED, note that one entire subject was eliminated from the analysis because it contained a missing value. The loss of subjects with missing values plus the fact that PROC GLM uses a less sophisticated calculation for random effects makes it desirable to use PROC MIXED.

REFERENCES

Hand, D., Daly, F., McConway, K. et al. (1994). *A Handbook of Small Data Sets*. Boca Raton, FL: Chapman and Hall.

Littell, R.C., Milliken, G.A., Stroup, W.W., and Wolfinger, R.D. (1996). *SAS System for Mixed Models*. Cary, NC: SAS Institute, Inc.

16

NONPARAMETRIC ANALYSIS

LEARNING OBJECTIVES

- To be able to use SAS® procedures to compare two independent samples (Wilcoxon)
- To be able to use SAS procedures to compare k independent samples (Kruskal–Wallis)
- To be able to use SAS procedures to compare two dependent (paired) samples
- To be able to use SAS procedures to compare k-dependent samples (Friedman's test)
- To be able to use SAS procedures to perform multiple comparisons following a significant Kruskal–Wallis test (macro)

An assumption for many statistical tests, such as the t-tests and ANOVA, is that the data are normally distributed. When this normality assumption is suspect or cannot be met, there are alternative techniques for analyzing the data.

Statistical techniques based on an assumption that data are distributed according to some parameterized distribution (such as the normal distribution) are referred to as parametric analyses. Statistical techniques that do not rely on this assumption are called nonparametric procedures. This chapter illustrates several nonparametric techniques. One nonparametric procedure, Spearman's rank correlation, was mentioned in Chapter 13.

SAS® Essentials: Mastering SAS for Data Analytics, Third Edition. Alan C. Elliott and Wayne A. Woodward.
© 2023 John Wiley & Sons, Inc. Published 2023 by John Wiley & Sons, Inc.
Companion website: https://www.alanelliott.com/SASED3/

16.1 COMPARING TWO INDEPENDENT SAMPLES USING NPAR1WAY

A nonparametric test can be used to compare two independent groups when you cannot make the assumptions associated with the *t*-test. Using a nonparametric test is also a useful technique if you do not have exact data values for the observations but you do have order statistics – that is, you don't know the actual response values, but you know which is the largest, the next largest, and so forth, to the smallest. In this case, the smallest value is recoded as 1 and the next to the smallest is recoded as 2, and so forth. Many nonparametric procedures are based on these ranks.

The primary nonparametric test for comparing two groups discussed in this chapter is the Wilcoxon (sometimes called the Wilcoxon–Mann–Whitney) test. The hypothesis tested is as follows:

H_0: The two groups have the same distribution (they come from the same population).
H_a: The two groups do not have the same distributions (they come from different populations).

The SAS procedure for testing these hypotheses is PROC NPAR1WAY. The procedure outputs several statistical test results, but the focus will be on the Wilcoxon test. The syntax for NPAR1WAY is as follows:

PROC NPAR1WAY <options>;

Table 16.1 lists common options and statements used with PROC NPAR1WAY.

TABLE 16.1 Common Options for PROC NPAR1WAY

Option	Explanation
DATA = datasetname	Specifies which dataset to use
WILCOXON	Limits output to Wilcoxon-type tests
MEDIAN	Requests median test
VW	Requests Van der Waerden test
NOPRINT	Suppresses output
Common statements for PROC NPAR1WAY	
CLASS vars;	Specifies grouping variable(s)
VAR vars;	Specifies dependent variable(s)
EXACT	Requests exact tests
BY, FORMAT, LABEL, WHERE	These statements are common to most procedures and may be used here

HANDS-ON EXAMPLE 16.1

This example illustrates how to compare two independent groups using nonparametric methods. In it, observations from two brands of fertilizers are compared to determine if one provides better overall growth (measured by height) than the other. (This is the same data used in ATTEST2.SAS, which was discussed in Hands-On Example 12.2.

(Continued)

(*Continued*)
1. Open the program file `ANPAR1.SAS`.

   ```
   DATA FERTILIZER;
   INPUT BRAND $ HEIGHT;
   DATALINES;
   A    20.00
   A    23.00
   A    32.00
   A    24.00
   A    25.00
   A    28.00
   A    27.50
   B    25.00
   B    46.00
   B    56.00
   B    45.00
   B    46.00
   B    51.00
   B    34.00
   B    47.50
   ;
   PROC NPAR1WAY WILCOXON;
        CLASS BRAND;
        VAR HEIGHT;
          EXACT;
        Title 'Compare two groups using NPAR1WAY';
   RUN;
   ```

   The `WILCOXON` option is included in the `PROC NPAR1WAY` statement to limit the output to Wilcoxon-type analyses. The `EXACT` statement produces additional Exact tests.

2. Run the program and observe the output in Table 16.2.

   The "Wilcoxon Scores" table provides summary statistics for the data. The "Wilcoxon Two-Sample Test" table reports a variety of approximate *p*-values for the Wilcoxon test (both one-sided and two-sided). The Kruskal–Wallis (KW) test is an extension of the Wilcoxon test that is applicable to two or more independent samples. The Kruskal–Wallis *p*-value reported in this table is a result of a chi-square approximation.

   We recommend using the Exact test *p*-values obtained as a result of the information statement `EXACT`. The two-sided *p*-value (appropriate for testing the hypotheses as stated earlier) is 0.0025, suggesting a difference between the two distributions (i.e., that the effects of the fertilizers are different). This result is in the Wilcoxon Two-Sample Test table, under the title "Exact," SAS also produces side-by-side boxplots (not shown here) that illustrate the difference between the two distributions.

3. Also note that the boxplots produced by this analysis (not shown here) provide visual evidence that the distribution of observations in Group B tends to be higher than

TABLE 16.2 Output from NPAR1WAY for a Two-Group Analysis

Wilcoxon Scores (Rank Sums) for Variable HEIGHT Classified by Variable BRAND

BRAND	N	Sum of Scores	Expected Under H0	Std. Dev. Under H0	Mean Score
A	7	31.50	56.0	8.625543	4.50000
B	8	88.50	64.0	8.625543	11.06250

Average scores were used for ties.

Wilcoxon Two-Sample Test

Statistic (S)	Z	Pr < Z	Pr > |Z|	t Approximation Pr < Z	t Approximation Pr > |Z|	Exact Pr <= S	Exact Pr >= |S-Mean|
31.5000	−2.7824	0.0027	0.0054	0.0073	0.0147	0.0012	0.0025

Z includes a continuity correction of 0.5.

Kruskal-Wallis Test

Chi-Square	DF	Pr > ChiSq
8.0679	1	0.0045

that of group A. (Since this is a nonparametric test, the comparison is not of means, but rather a comparison of the distributions.)

Note for comparison that a parametric two-sample test was performed on the same data in Chapter 12. The results were similar, with a reported p-value of 0.0005 for that test.

A number of other tests are reported in the SAS output (when the WILCOXON option is not used), but they are not described here. They include a median test and a Van der Waerden two-sample test. Which test is appropriate? Different disciplines often favor one test over another, or you may find that the type of analysis you are performing is commonly reported using one or another of these tests. It is also useful to run all available tests and check to see if the results agree. Such agreement produces a measure of assurance concerning the results.

16.2 COMPARING *K* INDEPENDENT SAMPLES (KRUSKAL–WALLIS)

If more than two independent groups are being compared using nonparametric methods, SAS uses the Kruskal-Wallis (KW) test to compare groups. The hypotheses being tested are as follows:

H_0: There is no difference among the distributions of the groups.
H_a: There are differences among the distributions of the groups.

This test is an extension of the Wilcoxon test described earlier. It uses the `WILCOXON` procedure to calculate mean ranks for the data and uses that information to calculate the test statistic for the KW test.

HANDS-ON EXAMPLE 16.2

This example illustrates how to compare three or more independent groups using nonparametric procedures. The data are weight gains of 28 animals that are randomly assigned to four feed treatments with seven animals to a treatment type. Suppose you want to test whether there are differences among the effects of the different treatments. If you have reason to believe that the normality assumptions associated with the one-way ANOVA are not met, then the KW test would be appropriate. The data for this example are as follows:

Group 1	Group 2	Group 3	Group 4
50.8	68.7	82.6	76.9
57.0	67.7	74.1	72.2
44.6	66.3	80.5	73.7
51.7	69.8	80.3	74.2
48.2	66.9	81.5	70.6
51.3	65.2	78.6	75.3
49.0	62.0	76.1	69.8

1. Open the program file `ANPAR2.SAS`. (Note that the `DATA` step uses the compact data entry technique; see the "@@" in the `INPUT` line.)

```
DATA NPAR;
INPUT GROUP WEIGHT @@;
DATALINES;
1 50.8 1 57.0 1 44.6 1 51.7 1 48.2 1 51.3 1 49.0
2 68.7 2 67.7 2 66.3 2 69.8 2 66.9 2 65.2 2 62.0
3 82.6 3 74.1 3 80.5 3 80.3 3 81.5 3 78.6 3 76.1
4 76.9 4 72.2 4 73.7 4 74.2 4 70.6 4 75.3 4 69.8
;
PROC NPAR1WAY WILCOXON;
    CLASS GROUP;
    VAR WEIGHT;
    Title 'Four group analysis using NPAR1WAY';
RUN;
```

The SAS code for this program is very similar to that for the Wilcoxon test in Hands-On Example 16.1. Note that you request the KW test using the `WILCOXON` option. KW is a k-group version of the Wilcoxon test based on Wilcoxon rank sums. The only real difference is that this dataset includes four categories for the grouping variable. Also note that again, use of the `WILCOXON` option in the `NPAR1WAY` statement limits the amount of output.

2. Run the program and observe the output in Table 16.3. The first table provides summary information for the mean scores based on the ranks. The KW test reports $p < 0.0001$, indicating that there is a significant difference in the distribution of the groups. (Note that because in this case there are four groups, `NPAR1WAY` does not produce output for the two-group Wilcoxon Rank Sums test as in Table 16.2.) Moreover, you can request exact p-values using an `EXACT` statement, but the computing time involved will be extensive even in the current case of four groups of seven.

TABLE 16.3 Output from NPAR1WAY for a Four-Sample Group Comparison

Wilcoxon Scores (Rank Sums) for Variable WEIGHT Classified by Variable GROUP					
GROUP	N	Sum of Scores	Expected Under H0	Std. Dev. Under H0	Mean Score
1	7	28.00	101.50	18.845498	4.000000
2	7	77.50	101.50	18.845498	11.071429
3	7	171.00	101.50	18.845498	24.428571
4	7	129.50	101.50	18.845498	18.500000
Average scores were used for ties.					

Kruskal-Wallis Test		
Chi-Square	DF	Pr > ChiSq
24.4807	3	<.0001

Unlike in the `PROC ANOVA` procedure, there are no multiple comparison tests provided with `NPAR1WAY`. However, in Section 16.5, a method is provided to perform this follow-up procedure.

If you prefer not to use the method for multiple comparisons shown in Section 16.5, and the test for k-samples is significant, an acceptable procedure is to perform multiple pairwise comparisons across the groups and adjust the significance level from 0.05 using the Bonferroni technique of dividing the significance level by the total number of comparisons. For example, in the case with four groups, there are six pairwise comparisons, so you would perform six Wilcoxon comparisons. For each test, you would use the significance level adjusted to $(0.05/6 = 0.0083)$ as your criterion for rejecting the null hypothesis.

SAS displays comparative box plots in the output that provide a visual comparison of the groups (see Figure 16.1). Results of multiple comparison tests for these data will be given in Hands-On Example 16.5.

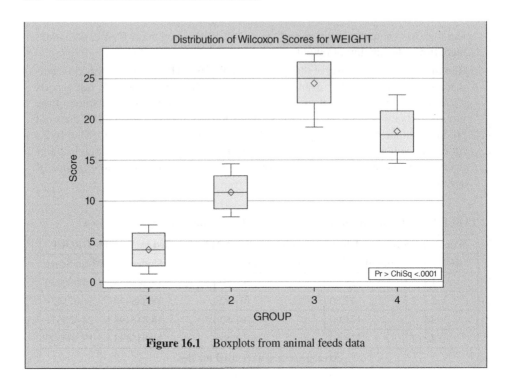

Figure 16.1 Boxplots from animal feeds data

16.3 COMPARING TWO DEPENDENT (PAIRED) SAMPLES

Two nonparametric tests for paired data are available when using the PROC UNIVARIATE procedure: the sign test and the signed-rank test. These tests are nonparametric counterparts of the paired *t*-test. The sign test is simple. It counts the number of positive and negative differences and calculates a *p*-value based on this information. The (Wilcoxon) signed-rank test bases its decision on the magnitude of the ranked differences. The hypotheses for these tests are as follows:

H_0: The probability of a positive difference is the same as that of a negative difference.
H_a: The probability of a positive difference is different from that of a negative difference.

Hands-On Example 16.3 illustrates how to perform these tests in SAS.

HANDS-ON EXAMPLE 16.3

This example illustrates how to compare paired observations using nonparametric techniques.

1. In the Editor window, open the program AUNIPAIRED.SAS.

```
DATA WEIGHT;
INPUT WBEFORE WAFTER;
* Calculate WLOSS in the DATA step *;
WLOSS=WBEFORE-WAFTER;
DATALINES;
200 185
175 154
188 176
198 193
197 198
310 275
245 224
202 188
;
PROC UNIVARIATE;
    VAR WLOSS;
    TITLE "Paired comparison using PROC UNIVARIATE";
RUN;
```

Note that the difference WLOSS is calculated in the DATA step. Because the test is based on the difference between WBEFORE and WAFTER, the variable WLOSS is the variable referenced in PROC UNIVARIATE.

2. Run the program and observe the (abbreviated) output in Table 16.4. The PROC UNIVARIATE procedure produces additional output, but this is the table we are interested in for the paired comparison.

TABLE 16.4 PROC UNIVARIATE to Test Paired Difference

Tests for Location: Mu0=0				
Test		Statistic		p Value
Student's t	t	3.943406	$Pr > \|t\|$	0.0056
Sign	M	3	$Pr >= \|M\|$	0.0703
Signed Rank	S	17	$Pr >= \|S\|$	0.0156

In the "Tests for Location" table, three test results are given. The p-value for the Student's t-test is the same that is given in the PROC MEANS example in Chapter 12 ($p = 0.0056$). The two nonparametric tests reported are the sign and signed-rank tests. The sign test reports a p-value of 0.0703 and the signed-rank test reports a p-value of 0.0156. All of the p-values are two-sided. The higher p-value associated with the sign test is typically based on the fact that it tends to be less powerful than the signed-rank test. Note that since the test is one-sided and the results support the alternative hypothesis (of positive weight gain), all the p-values in the table should be divided by 2, and consequently they all lead to the rejection of H_0.

16.4 COMPARING K-DEPENDENT SAMPLES (FRIEDMAN'S TEST)

When your data include more than two repeated measures and the normality assumption is questioned, you can perform a Friedman procedure to test the following hypotheses:

H_0: The distributions are the same across the repeated measures.
H_a: There are some differences in distributions across the repeated measures.

Although no procedure in SAS performs a Friedman's test directly, you can calculate the statistic needed to perform the test using PROC FREQ.

Recall the repeated measures example in Hands-On Example 14.2, in which four drugs were given to five patients in random order. In this repeated measures design, the observed value is a reaction to the drug, where a larger value is beneficial. The data are as follows:

Subj	Drug1	Drug2	Drug3	Drug4
1	31	29	17	35
2	15	17	11	23
3	25	21	19	31
4	35	35	21	45
5	27	27	15	31

Hands-On Example 16.4 shows how these data are coded for input so that Friedman's test may be performed, and how to perform Friedman's test using PROC FREQ.

HANDS-ON EXAMPLE 16.4

This example illustrates how to compare three or more repeated measures using nonparametric techniques.

1. Open the program file AFRIEDMAN.SAS.

```
DATA TIME;
INPUT SUBJ DRUG OBS;
DATALINES;
1   1   31
1   2   29
1   3   17
1   4   35
2   1   15
... ETC ...
5   2   27
5   3   15
5   4   31
;
Title "Friedman Analysis";
```

```
PROC FREQ;
  TABLES SUBJ*DRUG*OBS / CMH2 SCORES=RANK NOPRINT;
RUN;
```

Note that in this analysis, the dataset is structured differently from most datasets in this book, although it is set up the same here as for the repeated measures ANOVA analysis in Hands-On Example 14.2. In a repeated measures analysis, each line contains the observed reading for a single experimental condition, in this case for each subject and each drug administered. Therefore, because each subject received the four drugs (in random order and with a washout period between drugs), there are four lines for each subject.

The PROC FREQ statement produces a three-way table. Statement options include CMH2, which requests Cochran–Mantel–Haenszel Statistics, and the SCORES=RANK option indicates that the analysis is to be performed on ranks. The NOPRINT option is used to suppress the output of the actual summary table because it is quite large and provides nothing needed for this analysis.

2. Run the analysis and observe the output given in Table 16.5.

The test statistic for a Friedman's test is found in the "Row Mean Scores Differ" row, where the Friedman chi-square = 14.125 with three degrees of freedom and $p = 0.0027$. Because the p-value is <0.05, you would conclude that there is a difference in the distributions across DRUGs.

As with the KW test, SAS provides no built-in follow-up multiple comparison test for this procedure. If you reject the null hypothesis, you can perform comparisons for each drug pair using the Wilcoxon test on the differences as discussed in the earlier section on paired comparisons. As in the KW multiple comparisons discussed earlier, you should adjust your significance level using the Bonferroni technique. For example, in this case with four groups (measured in repeated readings), you would perform six pairwise comparisons, and for an overall 0.05 level test, you would use the $0.05/6 = 0.0083$ level of significance for each individual pairwise comparison.

TABLE 16.5 Results for Friedman's Test

Cochran-Mantel-Haenszel Statistics (Based on Rank Scores)				
Statistic	Alternative Hypothesis	DF	Value	Prob
1	Nonzero Correlation	1	1.2250	0.2684
2	Row Mean Scores Differ	3	14.1250	0.0027

16.5 GOING DEEPER: NONPARAMETRIC MULTIPLE COMPARISONS

Because SAS is a flexible language, SAS programmers often create their own procedures for performing analyses that are not already in the SAS language. For the KW nonparametric analysis of variance described earlier in this chapter, SAS provides no specific posthoc pairwise comparisons when the overall KW test is significant, indicating that there are some differences among groups.

Hands-On Example 16.5 illustrates how a procedure written in a SAS macro can be used to perform an analysis not included in the SAS program. This example utilizes a SAS macro implementation of a multiple comparison test (Nemenyi [Tukey-type] or Dunn's test) to be used along with NPAR1WAY (see Elliott and Hynan, 2007).

HANDS-ON EXAMPLE 16.5

This example shows how you can use a macro to calculate multiple comparisons following a KW test.

This example is based on the previous independent four-group comparison in Hands-On Example 16.2, with the addition of code to run a macro procedure performing multiple comparisons.

1. Open the program file ANPAR3.SAS.

```
%INCLUDE C:\SASDATA\KW_MC.SAS;
DATA NPAR;
INPUT GROUP WEIGHT @@;
DATALINES;
1 50.8 1 57.0 1 44.6 1 51.7 1 48.2 1 51.3 1 49.0
2 68.7 2 67.7 2 66.3 2 69.8 2 66.9 2 65.2 2 62.0
3 82.6 3 74.1 3 80.5 3 80.3 3 81.5 3 78.6 3 76.1
4 76.9 4 72.2 4 73.7 4 74.2 4 70.6 4 75.3 4 69.8
;
* BEGIN INFORMATION FOR USING THE SAS MACRO KW>MC; %LET
NUMGROUPS=4;
%LET DATANAME=NPAR;
%LET OBSVAR=WEIGHT;
%LET GROUP=GROUP;
%LET ALPHA=0.05;
Title "Kruskal-Wallis Multiple Comparisons";
***********************************************************
*invoke the KW_MC macro
***********************************************************;
    %KW_MC(SOURCE=&DATANAME, GROUPS=&NUMGROUPS,
    OBSNAME=&OBSVAR, GPNAME=&GROUP, SIG=&ALPHA);
```

Note the %INCLUDE statement at the top of the program. This statement tells SAS to get additional SAS code that is stored in a file in the indicated location. Thus, the file C:\SASDATA\KW_MC.SAS must be at the designated location, or you must change the file name to match the appropriate location.

The dataset used in this program is the same as the one used in the previous KW example on weight gains of animals assigned randomly to four treatments.

The series of %LET statements are used to tell the SAS macro the values of certain parameters needed by the macro. In this case, the number of groups (NUMGROUPS) is 4. The name of the dataset containing the data for analysis is NPAR. The variable containing the dependent variable is WEIGHT, and the grouping variable is named GROUP. The significance level used to perform the test is indicated by ALPHA=0.05.

The statement that begins with %KW_MC(source= is the statement that calls the SAS macro. You do not need to change anything about this statement to use the macro for any other dataset. If you use this procedure for another analysis, the only items you must change are the four values in the %LET statements.

2. Run the program and observe the results. The KW results are the same as those in Hands-On Example 16.2. The multiple comparison test results are given in Table 16.5.

   Table 16.6 displays the results of a series of multiple comparison tests. (This test is named Dunn's test and it is similar to the Tukey multiple comparison procedure illustrated in Chapter 14.) The test is performed at the overall 0.05 significance level. For example, the test indicates that for the comparison of groups 3 and 1, you reject the null hypothesis that the two groups are from the same distribution. The same conclusion is reached for the comparison of groups 3 and 2. Thus, your conclusion based on this test would be that weight gain for group 3 is different from that for groups 1 and 2, but not different from that for group 4. Weight gain for group 4 is significantly different from that for group 1 but not from that for group 2. Finally, weight gain for group 2 is not different from that for group 1. SAS also displays side-by-side boxplots that are the same as shown in Figure 16.1. This graph provides a visual confirmation that treatment 3 is preferred over treatments 1 and 2.

TABLE 16.6 Multiple Comparison Test for a Kruskal–Wallis Analysis

Compare	Diff	SE	q	q(0.05)	Conclude
3 vs 1	20.43	4.4	4.64	2.638	Reject
3 vs 2	13.36	4.4	3.04	2.638	Reject
3 vs 4	5.93	4.4	1.35	2.638	Do not reject
4 vs 1	14.50	4.4	3.30	2.638	Reject
4 vs 2	7.43	4.4	1.69	2.638	Do not reject
2 vs 1	7.07	4.4	1.61	2.638	Do not reject

Group sample sizes not equal, or some ranks tied. Performed Dunns test, alpha = 0.05; comparison group = GROUP. Reference: Biostatistical analysis, 4th Edition, J. Zar, 2010.

16.6 SUMMARY

This chapter introduces several nonparametric alternatives to standard parametric analyses. These procedures are useful when the normality assumption is questionable and particularly important when sample sizes are small.

EXERCISES

16.1 **Run a median test.**

   a. Using the SAS program in ANPAR2.SAS, change the selection from the Wilcoxon (KW) test to the median test, as shown here:

   PROC NPAR1WAY MEDIAN;

   b. Run the program and observe the results. Do the conclusions differ?

16.2 **Run a nonparametric analysis.**

A retail franchise is interested in stopping shrinkage (the loss of merchandise through theft) in its 11 stores in Denver, Colorado. The management believes that the presence of a uniformed police officer near the entrance will deter theft. Six of the 11 stores are randomly selected to have an officer present for a month. In the following month, the officers are assigned to the other five stores for a month. Shrinkage is determined at the end of each month, yielding the data given in Table 16.7. (Note: The two months selected have historically had about the same monthly sales volume.)

a. Write a SAS program to analyze these data using an appropriate nonparametric procedure. Start with the following code:

```
DATA STORES;
INPUT STORE $ WITHOUT WITH;
LOSSDIFF=WITH-WITHOUT;
DATALINES;
001 1.3 .9
002  .9 .46
... etc
;
PROC _____
VAR _____
RUN;
```

b. Observe the output and make a decision. Is there evidence that shrinkage is lower when an officer is present?

TABLE 16.7 Store Shrinkage, in Thousands

Store Number	Without Officer	With Officer
001	1.3	0.9
002	0.9	0.46
003	2.6	2.5
004	8.2	4.1
005	2.2	2.1
006	0.89	0.9
007	1.6	1.5
008	2.4	2.2
009	3.2	1.1
010	0.5	0.4
011	1.1	0.5

REFERENCES

Elliott, A.C. and Hynan, L.S. (2007). A SAS macro implementation of a multiple comparison post hoc test. *Proceedings of the Joint Statistical Meetings*.

Zar, J.H. (2010). *Biostatistical Analysis*, 5e. Upper Saddle River, NJ: Prentice-Hall/Pearson.

17

LOGISTIC REGRESSION

LEARNING OBJECTIVES

- To be able to perform a logistic analysis using PROC LOGISTIC
- To be able to create a SAS® program that will perform a simple logistic analysis
- To be able to create a SAS program that will perform multiple logistic analysis
- To be able to use SAS to assess a model's fit and predictive ability

Binary logistic regression models are based on a dependent variable that can take on only one of two values, such as presence or absence of a disease, deceased or not deceased, married or unmarried, and so on. In this setting, the independent (sometimes called explanatory or predictor) variables are used for predicting the probability of occurrence of an outcome (such as mortality).

The logistic regression model is sometimes called a logit model. Logistic analysis methods are available for cases in which the dependent variable takes on more than two values, but this topic is beyond the scope of this book, and we will discuss only the binary logistic model. Logistic analysis is used to create an equation that can be used to predict the probability of occurrence of the outcome of interest, assess the relative importance of independent variables, and calculate odds ratios (OR) that measure the importance of an independent variable relative to the response. The independent variables can be either continuous or categorical.

SAS® Essentials: Mastering SAS for Data Analytics, Third Edition. Alan C. Elliott and Wayne A. Woodward.
© 2023 John Wiley & Sons, Inc. Published 2023 by John Wiley & Sons, Inc.
Companion website: https://www.alanelliott.com/SASED3/

17.1 LOGISTIC ANALYSIS BASICS

Before describing the SAS implementation of logistic regression, we briefly discuss some of the basic ideas underlying logistic analysis. We show the actual logistic regression equations here because we believe they provide insights into how logistic regression works.

17.1.1 The Logistic Regression Model

The basic form of the logistic equation is

$$p = \frac{e^{\beta_0 + \beta_1 X_1 + \beta_2 X_2 + \cdots + \beta_k X_k}}{1 + e^{\beta_0 + \beta_1 X_1 + \beta_2 X_2 + \cdots + \beta_k X_k}}$$

where $X_1, \ldots, X_k$ are the k independent variables, p is the probability of occurrence of the outcome of interest (which lies between 0 and 1), β_i is the coefficient on the independent variable X_i, and β_0 is a constant term. As in linear regression, the parameters of this theoretical model are estimated from the data, resulting in the prediction equation

$$\hat{p} = \frac{e^{b_0 + b_1 X_1 + b_2 X_2 + \cdots + b_k X_k}}{1 + e^{b_0 + b_1 X_1 + b_2 X_2 + \cdots + b_k X_k}}$$

where the b_i's are maximum-likelihood estimates of the β_i's (calculated by SAS) and p is the estimated probability of occurrence of the outcome of interest. Of course, any variable with a zero coefficient in the theoretical model is not useful in predicting the probability of occurrence. SAS reports tests of the null hypothesis that all of the β_i's, $i = 1, \ldots, k$ are zero.

If this null hypothesis is not rejected, then there is no statistical evidence that the independent variables as a group are useful in the prediction. If the overall test is rejected, then we conclude that at least one of the variables is useful in the prediction. SAS reports individual tests of the importance of each of the independent variables. That is, for each $i = 1, \ldots, k$, SAS reports the results of the tests.

$H_0: \beta_i = 0$: The ith independent variable is not predictive of the probability of occurrence.
$H_a: \beta_i \neq 0$: The ith independent variable is predictive of the probability of occurrence.

17.1.2 Understanding Odds and Odds Ratios

Another use of the logistic model is the calculation of the OR for each independent variable. The odds of an event measure the expected number of times an event will occur relative to the number of times it will not occur. Thus, if the OR of an event is 5, this indicates that we expect five times as many occurrences as non-occurrences. An odds of 0.2 (=1/5) would indicate that we expect five times as many non-occurrences as occurrences.

Suppose you have a dichotomous dependent variable such as mortality that is predicted using an independent variable that measures whether or not a person had a penetrating wound in an automobile accident. If you calculate that the odds of dying given a penetrating wound is 0.088 and the odds of dying given a nonpenetrating wound is 0.024, then the OR of dying given a penetrating wound is OR = 0.088/0.0247 = 3.56. Thus, the odds of dying given a penetrating wound relative to a nonpenetrating wound is about four times greater than if one does not have a penetrating wound. ORs for quantitative-independent variables (`AGE`, `SBP`, `WEIGHT`, etc.) are interpreted differently, and this interpretation is discussed in Hands-On Example 17.3.

17.2 PERFORMING A LOGISTIC ANALYSIS USING PROC LOGISTIC

`PROC LOGISTIC` is the SAS procedure that allows you to analyze the data using a binary logistic model. An abbreviated syntax for this statement is as follows:

```
PROC LOGISTIC < options >;
CLASS variables;
MODEL dependentvar < (variable_options) > =
< independentvars> < / options >;
```

The `PROC LOGISTIC` and `MODEL` statements are required, and only one `MODEL` statement can be specified. The `CLASS` statement (if used) must precede the `MODEL` statement. Note that the dependent variable must be binary. For simplicity, all of the examples in this chapter use a response variable that takes on the values 0 or 1.

By default, SAS assumes that the outcome predicted (with p) in the logistic regression equation corresponds to the case in which the dependent variable is 0. If, for example, you have a variable such as `DISEASE` with `DISEASE=0` indicating the disease is absent and `DISEASE=1` indicating the disease is present, then SAS will predict the probability of "disease absent" by default.

Table 17.1 lists common options and statements used with PROC LOGISTIC.

The `MODEL` statement is the heart of the logistic procedure. It specifies the dependent (outcome) variable as well as the independent variables. For example,

```
MODEL DEPVAR = INDVAR1 INDVAR2 etc/options;
```

The binary is often coded 0 and 1 (typically 0 means No, and 1 means Yes), but the variable may take on any two values such as 1 and 2, A and B, Alive and Dead, and so on.

Because the logistic model is based on predicting an event, care must be taken concerning how the `DEPVAR` is set up. By default, SAS bases its prediction on the smallest value of the dependent variable. Therefore, if your dependent variable is `FAIL` (where 0 means not failed and 1 means failed), then by default SAS will model the prediction for `FAIL=0` (not failed). To reverse the default prediction, use the `DESCENDING` option. When that option is included in the `PROC LOGISTIC` statement, `FAIL=1` will be modeled instead of `FAIL=0`. Another way to choose the value modeled is to explicitly define it in the `MODEL` statement. For example,

```
MODEL FAIL(EVENT='1') = independentvars;
```

will cause SAS to use 1 as the value to model for the dependent variable `FAIL`. The `EVENT=` option is designated a "Response Variable option." When the response variable is binary, `EVENT= FIRST` or `LAST` specifies that either the first- or last-ordered value is to be used as the prediction event. Or you can indicate a value in quotes (as in the example code above). (We recommend that you choose to use either the `DESCENDING` option (see Table 17.1) or the `EVENT=` option to specify a value of the response variable to predict,

TABLE 17.1 Common Options for PROC LOGISTIC

Option	Explanation
DATA = datasetname	Specifies which dataset to use
DESCENDING	Reverses the sorting order for the levels of the response variable. By default, the procedure will predict the outcome corresponding to the lower value of the dichotomous-dependent variable. So, if the dependent variable takes on the values 0 and 1, then by default SAS predicts the probability that the dependent variable is 0 unless otherwise specified, for example by using the DESCENDING option. (See information about the (EVENT=) option hereunder.)
EVENT='option'	Specifies which value of a binary response variable is treated as the one modeled. For example, to the left of the equal sign in a MODEL statement DEATH(Event='1').
ALPHA= value	Specifies significance level for confidence limits.
NOPRINT	Suppresses output.
SIMPLE	Displays descriptive statistics.
PLOTS= option	In current versions of SAS, the OR plot is displayed by default. Use PLOTS=NONE; to suppress this plot. PLOTS=ALL produces a number of plots including receiver-operating characteristic (ROC) and influencing diagnostics.

Common statements for PROC LOGISTIC

MODEL depvar=indvar(s);	Specifies the dependent and independent variables for the analysis. More specifically, it takes the form MODEL dependentvariable=independentvariable(s); More explanation follows
CLASS variable list;	Specifies classification (either categorical character or discrete numeric) variables for the analysis. They can be numeric or character. You can specify a reference value for a class variable by including the option (REF='value'). For example, CLASS RACE (REF='HISPANIC') URBAN; specifies HISPANIC as the reference value to RACE. For URBAN the reference value will be the last ordered (alphabetic or numbered) value.
ODDSRATIO 'label' var;	Creates a separate table with OR estimates and Wald confidence intervals. For example, ODDSRATIO 'Injury Severity' ISS; ODDSRATIO 'Age' AGE;
OUTPUT out=NAME;	Creates an output dataset with all predictors and response probabilities. For example, OUTPUT OUT=MYFILE P=PRED; where MYFILE is the name of the resulting data and the predictions are named PRED.
BY, FORMAT, LABEL, WHERE	These statements are common to most procedures and may be used here.

TABLE 17.2 Common MODEL Statement options for PROC Logistic

Option	Explanation
EXPB	Displays the exponentiated values of parameters. These exponentiated values are the estimated ORs for the parameters corresponding to the explanatory variables
SELECTION=type	Specifies variable selection method (examples are STEPWISE, BACKWARD, and FORWARD). More explanation below
SLENTRY=value	Specifies significance level for entering variables. Default is 0.05
SLSTAY=value	Specifies significance level for removing variables. Default is 0.05
LACKFIT	Requests Hosmer–Lemeshow goodness-of-fit test
RISKLIMITS	Requests confidence limits for ORs
CTABLE PPROB=(list)	CTABLE requests a classification table that reports how data are predicted under the model. Usually used with PPROB to specify a list of cutoff points to display in the table. For example, CTABLE PPROB= (0.2, 0.4 to 0.8 by 0.1)
INCLUDE=n	Includes first n independent variables in the model
OUTROC=name	Outputs ROC values to a dataset. Also causes the ROC curve to be included in the output

if needed.) If the values of the response variable are categorical, then "order" is based on alphabetical order.

As in the model statements in previous chapters, the dependent variable appears on the left of the equal sign, and independent variables appear on the right side. MODEL statement options appear after a slash (/). Table 17.2 lists some of the common MODEL statement options.

The SELECTION options specify how variables will be considered for inclusion in the model. The BACKWARD method considers all predictor variables and eliminates the ones that do not meet the minimal SLSTAY criterion until only those meeting the criterion remain. The FORWARD method brings in the most significant variable that meets the SLENTRY criterion and continues entering variables until none of the remaining unused variables meets the criterion. STEPWISE is a mixture of the two. It begins like the FORWARD method and uses the SLENTRY criterion to enter variables but reevaluates variables at each step and may eliminate a variable if it no longer meets the SLSTAY criterion. A typical model statement utilizing an automated selection technique would be as follows:

```
MODEL Y = X1 X2 ... Xk
            / EXPB
              SELECTION=STEPWISE
              SLENTRY=0.05
              SLSTAY=0.1
RISKLIMITS;
```

where X1, X2, ..., Xk are the k candidate-independent variables and Y is the binary dependent variable. In this example, the STEPWISE procedure is used for variable selection. The p-value for variable entry is 0.05. Once variables are considered for the model, they must have a *p*-value of 0.1 or less to remain in the model. RISKLIMITS specifies that an OR table is to be included in the output for the final model.

If a model includes independent variables that are categorical, they must be indicated in a CLASS statement. (However, if a dichotomous independent (predictor) variable is coded as 0 and 1, it does not need to be included in the CLASS statement.) For example, the following model is the same as the one above, except that it includes two categorical variables CAT1 and CAT2. CAT1 and CAT2 can be numeric (i.e., 1, 2, and 3) or of character type (i.e., A, B, and C). Because CAT1 and CAT2 are categorical, they must appear in the CLASS statement if you use them in the MODEL statement.

```
CLASS CAT1 CAT2;
MODEL Y = X1 X2 ...  Xk CAT1 CAT2
                / EXPB
                  SELECTION=STEPWISE
                  SLENTRY=0.05
                  SLSTAY=0.1
                  RISKLIMITS;
```

When a variable is defined as a classification variable, SAS sets up a default parameterization of $N-1$ comparisons (where N is the number of categories). The default reference value to which the other categories are compared is based on the last ordered (alphabetic or numeric) value. For example, if RACE categories are AA (African-American), H (Hispanic), C (Caucasian), and O (Other), then ORs will be reported for AA, H, and C, where the OR is based on the reference to O because O is the last ordered (alphabetic) value. If RACE is defined using discrete numeric codes such as 1, 2, 3, 4, and 5, then the last ordered (numeric) value is 5. You can change the reference category by including the options (REF= "value") after the name in the CLASS statement. For example, the statement CLASS RACE (REF="AA") would make AA the reference value rather than O. For the discrete numeric coding version of RACE, the designation (REF="1") would make 1 the reference value. (You can also use REF=FIRST or REF=LAST to define a reference value.)

Another way to handle categorical variables with three or more categories is to recode them into a series of dichotomous variables (sometimes called indicator or dummy variables). One reason to do this is that it may make ORs easier to interpret. For example, with the character category coding of RACE, you could create three 0/1 variables. In the DATA step, you could define

```
IF RACE="AA" THEN RACEA=1; ELSE RACEA= 0;
IF RACE="H" THEN RACEH=1; ELSE RACEH= 0;
IF RACE="C" THEN RACEC=1; ELSE RACEC= 0;
```

You need one less than the number of categories. Therefore, if RACEA, RACEH, and RACEC are all 0, the race must be OTHER.

17.3 USING SIMPLE LOGISTIC ANALYSIS

A simple logistic model is one that has only one predictor (independent) variable. This predictor variable can be either a binary or a quantitative measure. Hands-On Example 17.1 illustrates the simple logistic model.

HANDS-ON EXAMPLE 17.1

This example uses a simulated trauma dataset named ACCIDENTS to find a model to predict death (DEAD=1) using the variable PENETRATE (where a value of 1 represents a penetrating wound and 0 represents a nonpenetrating wound) observed at the trauma incident.

1. Open the program file ALOG1.SAS.

```
PROC LOGISTIC DATA="C:\SASDATA\ACCIDENTS" DESCENDING;
  MODEL DEAD=PENETRATE / RISKLIMITS;
  TITLE 'Trauma Data Model Death by Penetration Wound';
RUN;
```

In this example, the independent variable PENETRATE, which is a 0,1 variable, is used to predict death (DEAD=1), so the DESCENDING option is used.

2. Run this program. In the output, note the statement that reads

```
Probability modeled is dead=1.
```

This statement tells you the value of the dependent variable for which the predicted probabilities are based. Observe the abbreviated output as given in Table 17.3.

TABLE 17.3 Output from Logistic Regression

Response Profile		
Ordered Value	dead	Total Frequency
1	1	103
2	0	3580
Probability modeled is dead = 1.		

Analysis of Maximum Likelihood Estimates					
Parameter	DF	Estimate	Standard Error	Wald Chi-Square	Pr > ChiSq
Intercept	1	−3.6988	0.1111	1108.0853	<.0001
penetrate	1	1.2697	0.2584	24.1519	<.0001

Odds Ratio Estimates and Wald Confidence Intervals				
Effect	Unit	Estimate	95% Confidence Limits	
penetrate	1.0000	3.560	2.145	5.906

The Response Profile table indicates that there are 103 subjects with DEAD=1 (which is indicated as the modeled outcome) and 3580 subjects who did not die (DEAD=0). The Maximum-Likelihood Estimates table indicates that the variable PENETRATE is significantly associated with the outcome variable ($p < 0.0001$) and the

(*Continued*)

(*Continued*)

Odds Ratio Estimates and Wald Confidence Intervals table reports that the OR for PENETRATE is 3.56. This indicates that the odds of a person dying who had a penetrating wound is 3.56 times greater than that for a person who did not suffer this type of wound.

3. Change the binary variable PENETRATE to the quantitative variable ISS (injury severity score) using the following statement:

```
MODEL DEAD=ISS/ RISKLIMITS;
```

Rerun the program and observe the output as given in Table 17.4. In Table 17.4, the variable ISS is shown to be predictive of mortality ($p < 0.0001$) with an OR = 1.111. However, this OR is interpreted differently for PENETRATE than for ISS because ISS is a quantitative measure and PENETRATE is a binary measure. Interpret OR = 1.111 in this way: The odds of dying are estimated to be 1.111 times greater for *each unit increase* in ISS.

4. Change the PROC LOGISTIC statement by removing the DESCENDING option. Rerun the code. Observe the results. What is the probability modeled for this version of the program? Note that the new OR = 0.9 which equals 1/1.111 where 1.111 was the OR obtained when using DESCENDING.

TABLE 17.4 Output for a Continuous Measure from Logistic Regression

Analysis of Maximum Likelihood Estimates					
Parameter	DF	Estimate	Standard Error	Wald Chi-Square	Pr > ChiSq
Intercept	1	−5.4444	0.2105	668.7126	<.0001
ISS	1	0.1056	0.00721	214.5334	<.0001

Odds Ratio Estimates and Wald Confidence Intervals				
Effect	Unit	Estimate	95% Confidence Limits	
ISS	1.0000	1.111	1.096	1.127

17.3.1 Graphing Simple Logistic Results

It is informative to examine a graph of the simple logistic regression analysis to understand how predictions are made using the logistic equation. (More details about SAS graphs will be discussed in Chapter 19.)

HANDS-ON EXAMPLE 17.2

This example illustrates a simple logistic regression using a quantitative measure as the independent variable.

1. Open the program file ALOG2.SAS.

```
PROC LOGISTIC DATA="C:\SASDATA\ACCIDENTS" DESCENDING;
   MODEL DEAD=ISS / RISKLIMITS;
   OUTPUT OUT=LOGOUT PREDICTED=PROB;
   TITLE "Simple binary logistic regession with plot.";
RUN;
*-------------------------------LOGISTIC PLOT;
PROC SORT DATA=LOGOUT;BY ISS;
TITLE 'LOGISTIC PLOT';
PROC GPLOT DATA=LOGOUT;
   PLOT PROB*ISS;
RUN;
QUIT;
```

The DESCENDING option is used because death (i.e., DEAD=1) is the outcome of interest in the predictive model. Otherwise, the OR would by default be based on predicting DEAD=0. The OUTPUT statement

```
OUTPUT OUT=LOGOUT PREDICTED=PROB;
```

creates a SAS dataset named LOGOUT that contains the predicted values calculated by the logistic regression equation. The logistic equation based on estimates given in the Maximum Likelihood Estimates tables (Table 17.4) is

$$\hat{p} = \frac{e^{-5.44+0.1056 \times ISS}}{1 + e^{-5.44+0.1056 \times ISS}}$$

The variable PROB contains predictions, $\hat{p}$, for each subject in the dataset ACCIDENTS.

2. Run the program and observe the output. In particular, examine the "S-shaped" plot shown in Figure 17.1. The graph provides a method for estimating the probability of death for each ISS. For example, locate ISS = 50 on the horizontal axis. Draw a vertical line to the plotted curve and from there draw a horizontal line to the vertical axis. This shows that for an ISS = 50, the probability of death is about 0.45 (or 45%). Note that by substituting ISS = 50 in the prediction equation, we obtain

$$\hat{p} = \frac{e^{-5.44+0.1056 \times 50}}{1 + e^{-5.44+0.1056 \times 50}} = 0.459$$

which is consistent with the graphical estimate. It is typical that when $\hat{p} < 0.5$, you would predict non-occurrence of the event of interest (i.e., survival) and if $\hat{p} > 0.5$, you would predict occurrence (i.e., death). In this case, we are pleased to predict that the subject will survive.

(*Continued*)

(*Continued*)

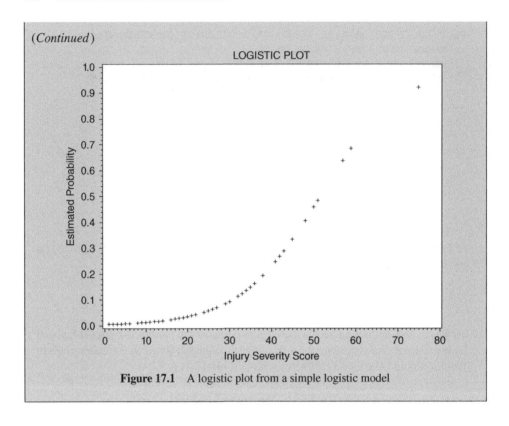

Figure 17.1 A logistic plot from a simple logistic model

17.4 MULTIPLE BINARY LOGISTIC ANALYSIS

A multiple binary logistic regression model has more than one independent variable. As such, it is analogous to a multiple regression model in the case in which the dependent variable is binary. In practice, it is common to have several potential predictor variables. One of the tasks of the investigator is to select the best set of predictors to create a parsimonious and effective prediction equation.

17.4.1 Selecting Variables for Multiple Logistic Analysis

The procedure used to select the best independent variables is similar to the one used in multiple linear regression. You can choose to select variables using manual or automated methods, or a combination of both. It is often desirable for the investigator to use his or her knowledge to perform a preliminary selection of the most logically (plausibly) important variables. Automated procedures can then be used to select other potential variables.

The importance of each variable as a predictor in the final model depends on the other variables in the model. Confounding and interaction effects may need to be addressed in certain models, but these topics are beyond the scope of this book.

HANDS-ON EXAMPLE 17.3

This example considers several potential predictors for mortality based on the simulated ACCIDENTS dataset. Subjects range in age from very young to 18 years old. Recall that in this dataset, the variable DEAD is coded as a 0/1 variable where 1 indicates that the subject died and 0 indicates that the subject survived. Moreover, ISS is an injury severity score where a larger score indicates a more severe injury, SBP is systolic blood pressure, and GCS is the Glasgow Coma Scale where a low value indicates a more severe injury.

1. Open the program file ALOG3.SAS.

    ```
    PROC LOGISTIC DATA="C:\SASDATA\ACCIDENTS" DESCENDING;
    CLASS GENDER;
    MODEL DEAD = PENETRATE ISS AGE GENDER SBP GCS
            / EXPB
            SELECTION=STEPWISE
            INCLUDE=1
            SLENTRY=0.05
            SLSTAY=0.05
            RISKLIMITS;
            TITLE 'LOGISTIC ON TRAUMA DATA WHERE AGE LE 18';
    RUN;
    QUIT;
    ```

    - Note that the DESCENDING option is used in order for the model to predict DEAD=1.
    - Seven possible independent variables are included in the MODEL statement: PENETRATE, ISS, AGE, RACE, GENDER, SBP, and GCS.
    - Because GENDER is a character variable, it is indicated in a CLASS statement before the MODEL statement. SAS uses this information to create a design variable so GENDER can be used in the model. (Another option would be to recode GENDER into a 0/1 variable and use the recoded variable in the model.)
    - Stepwise selection is requested with entry into the model set at 0.05 (SLENTRY) and removal from the model also set at 0.05 (SLSTAY).
    - The INCLUDE=1 option indicates that the first variable in the independent variable list (PENETRATE) should always be in the model.
    - RISKLIMITS requests ORs to be output.

2. Run this program. Note that Table 17.5 shows how the CLASS variable GENDER has been transformed to allow it to be used in the model.

TABLE 17.5 CLASS Variables

Class Level Information		
Class	Value	Design Variables
GENDER	Female	1
	Male	−1

(*Continued*)

(*Continued*)

3. In the output given in Table 17.6, we see that the variables GCS, ISS, and AGE are entered into the model and then no other variables meet the entry criterion. The final model and OR values are shown. Note that the OR for PENETRATE is different than it was in the simple logistic model because it is adjusted for the other variables in the model. The ORs for AGE and GCS indicate that, in general, older children and children with higher GCS scores are more likely to survive an injury.

Other model selection options include BACKWARD and FORWARD selections. These options do not always result in the same "final" model, and the decision concerning variables to be included in the final model should not be entirely based on the results of any automated procedure. The researcher's knowledge of the data should always be used to guide the model selection process even when automated procedures are used.

TABLE 17.6 Final Model

Analysis of Maximum Likelihood Estimates						
Parameter	DF	Estimate	Standard Error	Wald Chi-Square	Pr > ChiSq	Exp(Est)
Intercept	1	−0.4850	0.4661	1.0827	0.2981	0.616
penetrate	1	2.4072	0.4053	35.2776	<.0001	11.103
ISS	1	0.0673	0.00974	47.7172	<.0001	1.070
AGE	1	−0.1092	0.0245	19.9400	<.0001	0.897
GCS	1	−0.4193	0.0453	85.7115	<.0001	0.657

Odds Ratio Estimates and Wald Confidence Intervals			
Effect	Unit	Estimate	95% Confidence Limits
penetrate	1.0000	11.103	5.017 24.570
ISS	1.0000	1.070	1.049 1.090
AGE	1.0000	0.897	0.855 0.941
GCS	1.0000	0.657	0.602 0.719

17.5 GOING DEEPER: ASSESSING A MODEL'S FIT AND PREDICTIVE ABILITY

Once you decide on a final model, you should analyze that model to assess its predictive ability. One method of analyzing the predictive capabilities of the logistic equation is to use the Hosmer and Lemeshow test. This test is based on dividing subjects into deciles on the basis of predicted probabilities (see Hosmer and Lemeshow 2000). SAS reports chi-square statistics based on the observed and expected frequencies for subjects within these 10 categories.

The Hosmer and Lemeshow test is requested using the LACKFIT option in the MODEL statement. Another measure used to assess the fit is an ROC curve, which provides a measure of how well the model predicts outcomes.

HANDS-ON EXAMPLE 17.4

This example illustrates the Hosmer and Lemeshow test and an ROC curve.

1. Open the program file ALOG4.SAS.

   ```
   PROC LOGISTIC DATA="C:\SASDATA\ACCIDENTS" DESCENDING;
   CLASS GENDER RACE;
   MODEL DEAD = PENETRATE ISS AGE GCS
           / EXPB
             LACKFIT
             RISKLIMITS
             CTABLE
             OUTROC=ROC1;
   TITLE 'Assess models predictive ability';
   RUN;
   QUIT;
   ```

   - The LACKFIT option requests a Hosmer and Lemeshow analysis.
   - CTABLE requests predictions for various probability cutoff points.
   - The OUTROC=ROC1 statement produces an ROC curve.

2. Run the program. Observe the Hosmer and Lemeshow tables given in Table 17.7. SAS computes a chi-square from observed and expected frequencies in the table.

   Large chi-square values (and correspondingly small p-values) indicate a lack of fit for the model. In this case, we see that the Hosmer and Lemeshow chi-square test for the final model yields a p-value of 0.2532, thus suggesting a model with a satisfactory predictive value. Note that the Hosmer and Lemeshow chi-square test is not a test of the importance of specific model parameters. It is a separate posthoc test performed to evaluate an entire model.

3. Examine the Classification Table (partially) given in Table 17.8. As mentioned earlier, it is a common practice to use 0.5 as the cutoff for predicting occurrence. That is, to predict the non-occurrence of the event of interest whenever $\hat{p} < 0.5$ and to predict occurrence if $\hat{p} > 0.5$. The Classification Table indicates how many correct and incorrect predictions would be made for a wide range of probability cutoff points used for the model. In this case, 98% of the cases are correctly classified using the 0.50 cutoff point.

   Also note that at the 0.50 cutoff point, there is 38.8% sensitivity and 99.7% specificity. (Sensitivity is the probability of a correct prediction among subjects with the outcome of interest, and specificity is the probability of a correct prediction among subjects who do not exhibit the outcome.) If you desire to gain sensitivity, you can move the cutoff point to a lower probability level (at the sacrifice of specificity). Similarly, you can increase specificity by raising the cutoff probability. (For more information about selecting a cutoff point, see Cohen et al. (2002), p. 516 or Hosmer and Lemeshow (2000), p. 160ff.)

   (Continued)

(*Continued*)

TABLE 17.7 Hosmer and Lemeshow Results

Partition for the Hosmer and Lemeshow Test					
		dead = 1		dead = 0	
Group	Total	Observed	Expected	Observed	Expected
1	366	0	0.08	366	365.92
2	370	0	0.11	370	369.89
3	369	0	0.13	369	368.87
4	369	0	0.17	369	368.83
5	369	1	0.22	368	368.78
6	369	0	0.29	369	368.71
7	368	0	0.40	368	367.60
8	368	0	0.71	368	367.29
9	368	10	4.94	358	363.06
10	367	92	95.96	275	271.04

Hosmer and Lemeshow Goodness-of-Fit Test		
Chi-Square	DF	Pr > ChiSq
10.1716	8	0.2532

TABLE 17.8 Classification Table for Logistic Regression

Classification Table								
	Correct		Incorrect		Percentages			
Prob Level	Event	Non-Event	Event	Non-Event	Correct	Sensitivity	Specificity	Pos Pred	Neg Pred
0.000	103	0	3580	0	2.8	100.0	0.0	2.8	
0.020	97	3226	354	6	90.2	94.2	90.1	21.5	99.8
etc.									
0.460	43	3567	13	60	98.0	41.7	99.6	76.8	98.3
0.480	41	3569	11	62	98.0	39.8	99.7	78.8	98.3
0.500	40	3569	11	63	98.0	38.8	99.7	78.4	98.3
0.520	38	3570	10	65	98.0	36.9	99.7	79.2	98.2
etc.									
0.940	5	3580	0	98	97.3	4.9	100.0	100.0	97.3
0.960	1	3580	0	102	97.2	1.0	100.0	100.0	97.2
0.980	1	3580	0	102	97.2	1.0	100.0	100.0	97.2
1.000	0	3580	0	103	97.2	0.0	100.0		97.2

4. Examine the ROC curve, shown in Figure 17.2. This curve, which is a plot of 1 minus Specificity versus Sensitivity, measures the predictive ability of the model. Note that the "area under the curve" (AUC) is 0.9718. When the AUC is close to 1.0, a good fit

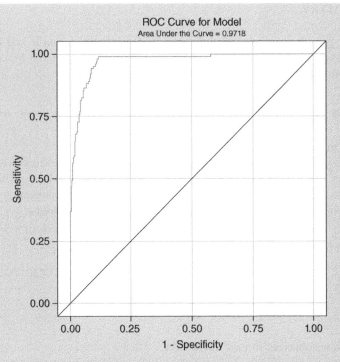

Figure 17.2 ROC curve

is indicated. The closer the curve extends to the upper left corner, the closer the AUC is to 1. AUC approaches 0.50 as the curve gets closer to the diagonal line and this would indicate a poor fit. The AUC statistic is often reported as an indicator of the predictive strength of the model. When you are considering competing "final" models, the Hosmer and Lemeshow test and the AUC (larger is better) are criteria often used.

Another way to assess the model is to use the information in the Model Fit Statistics table (not shown). For example, note that in the example, the value of -2 times the log likelihood for the model is given by $-2\log L = 417.470$ where 417.470 is given in the last row, second column of the "Model Fit Statistics" table (not shown) (for intercept and covariates). In general, when comparing models, the smaller the $-2\log L$ value, the better the fit. To determine whether the inclusion of an additional variable in a model gives a *significantly* better fit, you can use the difference in the $-2\log L$ values for the two models to compute a chi-square test statistic.

For example, for this model $-2\log L$ is 417.470 (on the SAS output). By removing the AGE variable from the equation and rerunning the analysis, we get $-2\log L = 438.094$ (try it). The difference has a chi-squared distribution with one degree of freedom, and if this value is larger than the α-level critical value for a chi-square with one degree of freedom, then this is evidence at the α-level of significance that the variable AGE should be included in the model. For this example, the difference is 20.61, and thus at the $\alpha = 0.05$ level (critical value is 3.84), we conclude that AGE should be included in the model. Other criteria that can be used to assess competing models are the AIC (Akaike information criterion) or SC (Schwarz criterion) values in the Model Fit Statistics table (smaller values indicate a better model in both cases).

17.6 SUMMARY

This chapter provides examples of the use of SAS for running simple and multivariate logistic regression analyses. Also techniques for selecting a model and for assessing its predictive ability are illustrated.

EXERCISES

17.1 **Run a multiple logistic regression.**

Data were collected on 200 males. All had no evidence of coronary disease. After 10 years, the subjects were examined for evidence of coronary disease. The researcher wants to know if the demographic information collected at the beginning of the study is helping in predicting if the subjects will develop coronary disease.

1. Using the following code as a beginning, use all three methods of model fitting to see what model looks the most promising.

```
PROC LOGISTIC DATA='C:\SASDATA\CORONARY' DESCENDING;
MODEL CORONARY= SBP DBP AGE BMI INSURANCE
    /_____;
RUN9;
```

2. What variables are in your final model?
3. Use SAS code to produce an ROC curve for the final model. What is the AUC for this model? How do you interpret this?
4. Use SAS code to produce a Hosmer and Lemeshow test on the final model. How do you interpret the results?
5. What are the ORs for the variables in the final model? How are they interpreted?
6. How well do you think this model predicts coronary disease in 10 years?
7. If a 60-year-old man has a body mass index (BMI) of 25, what is your best guess of his probability of having a coronary disease in 10 years? Hint: Use the following formula:

$$\hat{p} = \frac{e^{b_0+b_1X_1+b_2X_2}}{1+e^{b_0+b_1X_1+b_2X_2}}$$

17.2 **The appliance department in a big "box store" wants to know what helps them sell big ticket items. A portion of the data they've collected on past sales is given in Table 17.9.**

TABLE 17.9 Big Box Store Sales

Obs	Age	Price100	RACE	FREEBIE	GENDER	SOLD
1	45	75.0	AA	Yes	M	1
2	34	125.0	C	No	M	0
3	26	150.0	H	Yes	M	1
4	54	38.5	C	Yes	F	1
5	44	17.0	AA	No	F	0
6	19	28.0	C	Yes	M	0
7	82	44.5	H	No	M	1

1. Open the program file `EX_17.1.SAS`.
   a. Using the following preliminary code, perform a logistic analysis with BACKWARD selection to find a model for predicting whether an appliance is sold to a potential customer.

   ```
   PROC LOGISTIC DATA=mysaslib.boxstore;
   CLASS GENDER FREEBIE RACE;
   MODEL SOLD= GENDER FREEBIE RACE AGE PRICE100
           / EXPB
             SELECTION=BACKWARD;
   TITLE 'What helps sales?';
   RUN;
   QUIT;
   ```

   What probability is modeled? Sales=Yes or No?

   b. Change the probability modeled to Sales=Yes by adding a DESCENDING option to the `PROC LOGISTIC` statement. Examine the OR for FREEBIE. Because its reference is "No," the OR is interpreted as the OR for getting a sale given that no freebie is given. You want the reference to be "Yes." Add a (REF="Yes") after FREEBIE in the `CLASS` statement. Rerun and note the change in OR for FREEBIE. What is it now? How would you interpret this OR?

   c. Add these options after the / to perform diagnostics for the fit.

   ```
   LACKFIT
   RISKLIMITS
   OUTROC=ROC1
   ```

   Rerun the code.
   1. What does the Hosmer–Lemeshow test (`LACKFIT`) indicate?
   2. What additional information does the `RISKLIMITS` provide?
   3. What does the `OUTROC` option provide in the output? How does it help you assess how well this model fits the data?

   d. Run this model using FORWARD instead of BACKWARD selection. How does this change the outcome?

REFERENCES

Cohen, J., Cohen, P., West, S.G., and Aiken, L.S. (2002). *Applied Multiple Regression/Correlation Analysis for the Behavioral Sciences*, 3e. Mahwah, NJ: Lawrence Erlbaum.

Hosmer, D.W. and Lemeshow, S. (2000). *Applied Logistic Regression*, 2e. New York: Wiley.

18

FACTOR ANALYSIS

LEARNING OBJECTIVES

- To be able to perform an exploratory factor analysis using PROC FACTOR
- To be able to use PROC FACTOR to identify underlying factors or latent variables in a dataset
- To be able to use PROC FACTOR to rotate factors for improved interpretation
- To be able to use PROC FACTOR to compute factor scores

Factor analysis is a dimension reduction technique designed to express the actually observed variables (possibly many in number) using a smaller number of underlying latent (unobserved) variables. In this chapter, we will discuss the SAS® implementation of exploratory factor analysis, which involves identifying factors, determining which factors are needed to satisfactorily describe the original data, interpreting the meaning of these factors, and so on. Confirmatory factor analysis goes further and involves techniques for testing hypotheses to confirm theories. Although SAS contains procedures for running confirmatory factor analysis, we shall only discuss exploratory factor analysis in this chapter.

18.1 FACTOR ANALYSIS BASICS

The typical steps in performing an exploratory factor analysis are the following:

(a) Compute a correlation (or covariance) matrix for the observed variables.
(b) Extract the factors (this involves deciding how many factors to extract, the method to use, and the values to use for the prior communality estimates).
(c) Rotate the factors to improve interpretation.
(d) Compute factor scores (if needed).

Before proceeding, we mention that an exploratory factor analysis can be quite subjective and without unique solutions. Consequently, there is a certain amount of "art" involved in any factor analysis solution. This fact has caused the subject to be somewhat controversial. We believe that factor analysis methods can be useful in helping to understand the data, but that interpretations and conclusions should be made with caution.

18.1.1 Using PROC Factor

The SAS procedure used to perform exploratory factor analysis is PROC FACTOR. A simplified syntax for this procedure is as follows:

```
PROC FACTOR <options>;
    VAR variables;
    PRIORS communalities;
```

There are many options available for PROC FACTOR. A few of the common options along with some useful statements are listed in Table 18.1.

TABLE 18.1 Common Options for PROC FACTOR

Option	Explanation
DATA = datasetname	Specifies which dataset to use
METHOD=option	Identifies the estimation method. Options include ML and PRINCIPAL
MINEIGEN=n	Specifies the smallest eigenvalue for retaining a factor
NFACTORS=n	Specifies the maximum number of factors to retain
NOPRINT	Suppresses output
PRIORS= option	Specifies the method for obtaining prior communalities
ROTATE = name	Specifies the rotation method. The default is ROTATE=NONE. Common rotation methods are VARIMAX, QUARTIMAX, EQUAMAX, and PROMAX. All of the above are orthogonal rotations except for PROMAX
SCREE	Displays a Scree plot of the eigenvalues
SIMPLE	Displays means, standard deviations, and number of observations
CORR	Displays the correlation matrix

Common Statements for PROC FACTOR

VAR variable list;	Specifies the numeric variables to be analyzed. Default is to use all numeric variables
BY, FORMAT, LABEL, WHERE	These statements are common to most SAS procedures and may be used here

HANDS-ON EXAMPLE 18.1

Intelligence is a multifaceted quality. For example, some people may be good at reasoning with numbers, others may have outstanding memories, while others may have the ability to express themselves well, and may be very adept at dealing with other people. Researchers (see for example, Gardiner (2006)), have studied the issue of identifying types of intelligence. Two of the types of intelligence identified by Gardiner are Logical–Mathematical Intelligence (Math Intelligence) and Linguistic Intelligence. In addition to the existence of several types of intelligence, there are also a variety of aspects related to each type of intelligence. In this example, we examine a hypothetical dataset that contains six variables, each measured on a 0–10 scale and designed to measure an aspect of either Linguistic or Math Intelligence. The variables are as follows:

COMPUTATION – Test on mathematical computations
VOCABULARY – A vocabulary test
INFERENCE – A test of the use of inductive and deductive inference
REASONING – A test of sequential reasoning
WRITING – A score on a writing sample
GRAMMAR – A test measuring proper grammar usage.

Clearly, the variables are designed so that COMP, INFER, and REASON measure Math Intelligence while VOCAB, WRITING, and GRAMMAR seem to measure Linguistic Intelligence. We will use the techniques of factor analysis to determine the extent to which these two components of intelligence can be extracted from the data.

1. Open the program file AFACTOR1.SAS.

   ```
   PROC FACTOR DATA=MYSASLIB.INTEL SIMPLE CORR SCORE
   METHOD=PRINICPAL
   ROTATE=VARIMAX
   OUTSTAT=FS
   PRIORS=SMC
   PLOTS=SCREE;
   RUN;
   ```

2. Run the program and observe the several tables and plots of results. Because we did not specify variables in a VAR statement, the analysis will be based on all numeric variables in the dataset, that is, COMPUTATION through GRAMMAR. Table 18.2 shows the descriptive statistics and Table 18.3 shows the correlation matrix for the six variables. This output is obtained as a result of including SIMPLE and CORR as PROC FACTOR options. The high pairwise correlations among COMPUTATION, INFERENCE, and REASONING (to a lesser extent) seem to indicate some tendency to measure Math Intelligence while the variables VOCABULARY, WRITING, and GRAMMAR that seem to be measuring Linguistic Intelligence are also positively pairwise correlated. These observations and the discussion related to the construction of this example suggest the possibility of detecting two underlying factors.

TABLE 18.2 Descriptive Statistics for Intelligence Data

Means and Standard Deviations from 200 Observations		
Variable	Mean	Std Dev
COMPUTATION	5.0300000	1.7874774
VOCABULARY	5.2450000	1.6909380
INFERENCE	4.3400000	1.6758407
REASONING	4.6300000	1.9395643
WRITING	5.0000000	1.9949685
GRAMMAR	4.3800000	1.8446173

TABLE 18.3 Correlations for Intelligence Data

Correlations						
	COMPUTATION	VOCABULARY	INFERENCE	REASONING	WRITING	GRAMMAR
COMPUTATION	1.00000	0.13222	0.84374	0.53082	0.10569	0.10473
VOCABULARY	0.13222	1.00000	0.03962	0.09826	0.75227	0.67726
INFERENCE	0.84374	0.03962	1.00000	0.49961	0.05712	0.03765
REASONING	0.53082	0.09826	0.49961	1.00000	0.14805	0.15467
WRITING	0.10569	0.75227	0.05712	0.14805	1.00000	0.78519
GRAMMAR	0.10473	0.67726	0.03765	0.15467	0.78519	1.00000

3. Because we specified METHOD=PRINCIPAL and PRIORS=SMC, SAS uses the principal factors method where the prior communality estimate for each variable is the squared multiple correlation of it with all other variables. These prior communality estimates are given in Table 18.4.

   Table 18.5 displays eigenvalues associated with the factors based on the reduced correlation matrix. It is clear from the table that there are two dominant eigenvalues (2.319 and 1.725). Based on any reasonable criterion, it is clear that a two-factor solution should be used.

   The Scree plot in Figure 18.1 (obtained via the option PLOTS=SCREE) gives a visual illustration of the sizes of the eigenvalues. Again, it is clear that there are two dominant eigenvalues.

4. The communalities in Table 18.6 are the proportion of the variance in each of the original variables retained after extracting the factors. It seems that all six variables

TABLE 18.4 Prior Communalities Estimates Using PRIORS=SMC

Prior Communality Estimates: SMC					
COMPUTATION	VOCABULARY	INFERENCE	REASONING	WRITING	GRAMMAR
0.73727417	0.59614561	0.72279682	0.30703390	0.70916506	0.63690981

(*Continued*)

(Continued)

TABLE 18.5 Eigenvalues of the Reduced Correlation Matrix

	Eigenvalues of the Reduced Correlation Matrix: Total = 3.70932538 Average = 0.6182209			
	Eigenvalue	Difference	Proportion	Cumulative
1	2.31945916	0.59413444	0.6253	0.6253
2	1.72532473	1.72851659	0.4651	1.0904
3	−.00319186	0.07536140	−0.0009	1.0896
4	−.07855326	0.02936223	−0.0212	1.0684
5	−.10791548	0.03788243	−0.0291	1.0393
6	−.14579792		−0.0393	1.0000

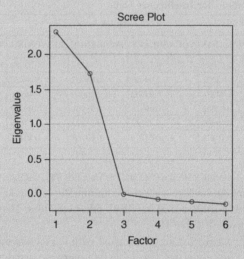

Figure 18.1 Scree plot

TABLE 18.6 Communalities Associated with the Two-Factor Solution

Final Communality Estimates: Total = 4.044784					
COMPUTATION	VOCABULARY	INFERENCE	REASONING	WRITING	GRAMMAR
0.80308734	0.64531015	0.77920274	0.33460577	0.78740819	0.69516970

are sufficiently well represented by the two factors, with variable REASONING having the smallest communality, 0.335.

5. Next we examine the two factors that are obtained and are described by the Factor Pattern Matrix given in Table 18.7.

FACTOR ANALYSIS BASICS

TABLE 18.7 Factor Pattern Matrix Associated with the Two-Factor Solution

Factor Pattern		
	Factor1	Factor2
COMPUTATION	0.55465	0.70389
VOCABULARY	0.69841	−0.39691
INFERENCE	0.48432	0.73800
REASONING	0.41302	0.40500
WRITING	0.77097	−0.43934
GRAMMAR	0.72423	−0.41310

In the table, it can be seen that for Factor 1, each variable has a positive coefficient ranging from .41 for `REASONING` to .77 for `WRITING`. A reasonable interpretation of this factor is that it is an overall measure of intelligence. The second factor has negative loadings on the variables measuring Linguistic Intelligence and positive coefficients on the others.

6. Based on the less than ideal interpretability of these factors, we use a rotation in hopes of producing more interpretable results. (Recall that by construction, there should be two factors: Math Intelligence and Linguistic Intelligence.) Using the option `ROTATE=VARIMAX`, we have instructed SAS to perform a Varimax rotation. SAS provides several rotation options, and Varimax is a popular "orthogonal rotation," which produces two orthogonal factors that are potentially easier to interpret. The resulting Factor Pattern Matrix after Varimax rotation is given in Table 18.8.

The rotated factor pattern matrix in Table 18.7 is much easier to interpret. For example, the coefficients for `COMPUTATION` are the correlations of the variable `COMPUTATION` with each of the two factors. In this case, there is a large positive

TABLE 18.8 Rotated Factor Pattern Matrix Using VARIMAX Rotation

Rotated Factor Pattern		
	Factor1	Factor2
COMPUTATION	0.06278	0.89395
VOCABULARY	0.80071	0.06457
INFERENCE	−0.01456	0.88260
REASONING	0.11376	0.56716
WRITING	0.88457	0.07029
GRAMMAR	0.83118	0.06570

(*Continued*)

(Continued)
correlation between COMPUTATION and Factor 2 and a very small correlation between COMPUTATION and Factor 1. Similar interpretations show that Factor 1 is highly correlated with the three variables measuring Linguistic Intelligence and Factor 2 tends to correspond to Math Intelligence. Consequently, the two rotated factors are consistent with the way in which the data were generated.

7. Suppose you want to calculate factor scores and save them in a temporary working file FSCORES. In order to accomplish this, add the following PROC FACTOR options before PLOTS=SCREE;

PROC SCORE DATA=MYSASLIB.INTEL SCORE=FS OUT=FSCORES;
RUN;

Then, after the RUN; statement add the code

PROC PRINT DATA=FSCORES;
    VAR FACTOR1 FACTOR2;
    RUN;

The first few lines of the output from PROC PRINT are given in Table 18.9. The 2-factor scores are given the default names FACTOR1 and FACTOR2 (the prefix "FACTOR" can be changed using the PREFIX= option). Recalling that Factor 1 is a measure of Linguistic Intelligence and Factor 2 measures Math Intelligence, from the factor scores it can be seen that Subject 1 has a higher Linguistic Intelligence score, Subject 2 seems to have High Math Intelligence, and Subject 3 unfortunately doesn't seem to have strength in either dimension.

TABLE 18.9 Output from PROC PRINT Showing the First 8 Factor Scores

Obs	Factor1	Factor2
1	0.51223	0.14558
2	−1.22246	1.92068
3	−1.31092	−0.80015
4	1.58737	0.78032
5	0.03969	2.64229
6	0.29565	−0.54992
7	−1.51494	1.09095
8	0.77701	−0.40789

HANDS-ON EXAMPLE 18.2

We consider a dataset containing scores of 193 athletes who completed all 10 decathlon events in the 1988 through 2012 Olympic Games. The 10 events in the decathlon are 100-m run, long jump, shot put, high jump, 400-m run, 100-m hurdles, discus, pole vault, javelin, and 1500-m run. These events measure a wide variety of athletic abilities, and in this example, we use this decathlon dataset to explore whether there are some underlying dimensions of athletic ability. It should be noted that the "times" in the running events are given negative signs, so that "larger" values are better than "smaller" values as is the case in the distance measurements. Moreover, the 1500-m results are given in (negative) seconds rather than the usual reporting of minutes and seconds. Additional variables in the dataset include YEAR (the year of the Olympic Games) and PTOTAL (the total number of points for that athlete based on the decathlon scoring system).

We use the procedure outlined in the previous example to run the factor analysis for the Olympic Athlete Data.

1. Open the program file AFACTOR2.SAS.

    ```
    PROC FACTOR SIMPLE CORR DATA MYSASLIB.OLYMPIC
    METHOD=PRINCIPAL MSA
       PRIORS=SMC
       ROTATE=VARIMAX
       OUTSTAT=FACT_ALL
       PLOTS=SCREE;
       VAR RUN100 LONGJUMP SHOTPUT HIGHJUMP RUN400
           HURDLES DISCUS POLEVAULT JAVELIN RUN1500S;
    RUN;
    ```

2. Run the program and again there are several pages of tables and plots. Because we want to run only the factor analysis on the 10-event results (and not YEAR or PTOTAL), we have included a VAR statement listing the variables related to the 10 decathlon events. Tables 18.10 and 18.11 show the descriptive statistics and correlation matrix obtained as a result of the options SIMPLE and CORR.

    Note the negative values in Table 18.9 for the running events: RUN100, RUN400, HURDLES, RUN1500S. In Table 18.10, note that as would be expected, there are positive correlations between speed events such as the 100-m run (RUN100) and 100-m hurdles (HURDLES) (0.692) and between strength events SHOTPUT and DISCUS (0.748). Also note that the 1500-m run (RUN1500S) is not highly correlated with any of the other events with the strongest correlation being with the medium distance 400-m run (0.368).

3. Because, as in Hands-on Example 18.1, we specified METHOD=PRINCIPAL and PRIORS=SMC, SAS uses the principal factors method where the prior communality estimate for each variable is the squared multiple correlation of it with all other variables. These prior communality estimates are given in Table 18.12.

(Continued)

(*Continued*)

TABLE 18.10 Descriptive Statistics for Olympic Athlete Data

Means and Standard Deviations from 193 Observations		
Variable	Mean	Std Dev
run100	−11.02399	0.280812
longjump	7.22306	0.350490
shotput	14.32907	1.236130
highjump	1.98575	0.085869
run400	−49.33124	1.283435
hurdles	−14.69803	0.555085
discus	43.58699	4.201578
polevault	4.73575	0.342200
javelin	59.03295	6.378388
run1500s	−278.74715	13.178349

Table 18.13 displays eigenvalues associated with the factors based on the reduced correlation matrix. `PROC FACTOR` selected three factors. It is clear from Table 18.13 and the Scree plot in Figure 18.2 that there are three dominant eigenvalues. Consequently, SAS selects three factors.

4. The communalities in Table 18.14 are the proportion of the variance in each of the original variables retained after extracting the factors. It seems that all 10 events are fairly well represented by the three factors, with all communalities above 0.33. However, `HIGHJUMP`, `POLEVALULT`, `JAVELIN`, and `RUN1500S` all have communalities below 0.4.

5. Next, we examine the three factors that are obtained and are described by the Factor Pattern Matrix given in Table 18.15.

    As was the case for the unrotated solution for the Intelligence Data, it can be seen that each variable has a positive coefficient in Factor 1, all of which are above 0.4 except for `RUN1500S`, which has a coefficient of 0.18. A reasonable interpretation is that Factor 1 measures overall athletic ability, primarily related to the first nine events. Factors 2 and 3 are more difficult to interpret.

6. Based on the confusing interpretations associated with the 3-Factor solution given in Table 18.15, we use a rotation to produce more interpretable results. Using the option `ROTATE=VARIMAX` results in the Rotated Factor Pattern Matrix given in Table 18.16.

TABLE 18.11 Correlations for Olympic Athlete Data

Correlations

		run100	longjump	shotput	highjump	run400	hurdles	discus	polevault	javelin	run1500s
run100	100 meters	1.00000	0.62456	0.46900	0.22067	0.62085	0.69179	0.36482	0.32626	0.15162	−0.00871
longjump	long jump	0.62456	1.00000	0.37162	0.46107	0.54245	0.54579	0.32381	0.43859	0.31312	0.23859
shotput	shot put	0.46900	0.37162	1.00000	0.36760	0.19800	0.46851	0.74831	0.38049	0.40673	−0.11159
highjump	high jump	0.22067	0.46107	0.36760	1.00000	0.21265	0.32034	0.33854	0.33966	0.30340	0.21658
run400	400 meter run	0.62085	0.54245	0.19800	0.21265	1.00000	0.51803	0.14483	0.31663	0.12819	0.36825
hurdles	110 meter hurdles	0.69179	0.54579	0.46851	0.32034	0.51803	1.00000	0.37030	0.43456	0.28326	0.04951
discus	Discus	0.36482	0.32381	0.74831	0.33854	0.14483	0.37030	1.00000	0.26674	0.35354	−0.10200
polevault	pole vault	0.32626	0.43859	0.38049	0.33966	0.31663	0.43456	0.26674	1.00000	0.35950	0.15728
javelin	Javelin	0.15162	0.31312	0.40673	0.30340	0.12819	0.28326	0.35354	0.35950	1.00000	0.16692
run1500s	1500 meter run	−0.00871	0.23859	−0.11159	0.21658	0.36825	0.04951	−0.10200	0.15728	0.16692	1.00000

(*Continued*)

(*Continued*)

TABLE 18.12 Prior Communalities Estimates Using PRIORS=SMC for Olympic Athlete Data

Prior Communality Estimates: SMC

run100	longjump	shotput	highjump	run400	hurdles	discus	polevault	javelin	run1500s
0.68603651	0.56671045	0.65394191	0.32430313	0.55219465	0.56191691	0.57223097	0.32024561	0.28455354	0.31622257

TABLE 18.13 Eigenvalues of the Reduced Correlation Matrix for Olympic Athlete Data

Eigenvalues of the Reduced Correlation Matrix: Total = 4.83835625 Average = 0.48383562				
	Eigenvalue	Difference	Proportion	Cumulative
1	3.67980012	2.63192235	0.7605	0.7605
2	1.04787777	0.40825292	0.2166	0.9771
3	0.63962484	0.52785571	0.1322	1.1093
4	0.11176913	0.04690376	0.0231	1.1324
5	0.06486537	0.12766893	0.0134	1.1458
6	−0.06280356	0.02819416	−0.0130	1.1329
7	−0.09099772	0.05539013	−0.0188	1.1140
8	−0.14638785	0.04594770	−0.0303	1.0838
9	−0.19233555	0.02072076	−0.0398	1.0440
10	−0.21305631		−0.0440	1.0000

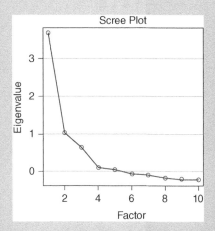

Figure 18.2 Scree plot for Olympic Athlete Data

The first rotated factor seems to focus on events 100-m long jump, 400-m run, and 110-m hurdles that involve the legs (speed and spring). Factor 2 seems to be primarily an arm strength factor with high coefficients for shot put and long jump and to a lesser extent javelin, pole vault, and (surprisingly) high jump. Finally, the only event with a large coefficient in Factor 3 is the 1500-m hurdles. This is consistent with our initial look at the correlation matrix that suggested the 1500-m run was "different" from the other events.

(*Continued*)

(*Continued*)

TABLE 18.14 Communalities Associated with the Three-Factor Solution for the Olympic Athlete Data

Final Communality Estimates: Total = 5.367303

run100	longjump	shotput	highjump	run400	hurdles	discus	polevault	javelin	run1500s
0.77536218	0.61125386	0.73515987	0.36009667	0.60920635	0.60227652	0.61756653	0.34739171	0.33404144	0.37494760

TABLE 18.15 Factor Pattern Matrix Associated with the Three-Factor Solution for the Olympic Athlete Data

	Factor Pattern		
		Factor1	Factor2	Factor3
run100	100 meters	0.75484	0.16862	−0.42089
longjump	long jump	0.74377	0.23169	0.06613
shotput	shot put	0.69665	−0.49932	−0.02274
highjump	high jump	0.50550	−0.03943	0.32095
run400	400-meter run	0.60300	0.48624	−0.09575
hurdles	110-meter hurdles	0.74637	0.09212	−0.19164
discus	Discus	0.59898	−0.50870	0.00398
polevault	pole vault	0.55685	0.02540	0.19149
javelin	javelin	0.44551	−0.16759	0.32783
run1500s	1500 meter run	0.16723	0.42715	0.40562

TABLE 18.16 Rotated Factor Pattern Matrix for the Olympic Athlete Data Using VARIMAX Rotation

	Rotated Factor Pattern		
		Factor1	Factor2	Factor3
run100	100 meters	0.84580	0.23258	−0.07671
longjump	long jump	0.60647	0.36590	0.33100
shotput	shot put	0.30160	0.77719	−0.20041
highjump	high jump	0.17997	0.47975	0.31233
run400	400-meter run	0.70075	0.04678	0.34054
hurdles	110-meter hurdles	0.68461	0.36227	0.04849
discus	discus	0.21230	0.72866	−0.20384
polevault	pole vault	0.31426	0.42324	0.26363
javelin	javelin	0.07697	0.52602	0.22676
run1500s	1500-meter run	0.09184	−0.01599	0.60519

If the `Methods=Principal` option is used, then the principal component analysis is performed when the `PRIORS=` option is not used or is set to `ONE` (the default). If you specify a `PRIORS=value` other than `PRIORS=ONE`, then a principal factor analysis is performed. A common usage is `PRIORS=SMC` in which case the prior communality for each variable is the squared multiple correlation of it with all other variables. After extracting the factors, the communalities represent the proportion of the variance in each of the original variables retained after extracting the factors.

Unless you are highly skilled at using factor analysis, we recommend that you check out references such as Stevens (2002), Johnson and Wichern (2007), and Johnson (1998) for more discussion of these concepts and the issues involved.

18.2 SUMMARY

In this chapter, we have discussed methods for using `PROC FACTOR` to perform exploratory factor analysis. In the Hands-On Examples, we have illustrated the use of rotation to obtain more understandable results.

EXERCISES

18.1 **Run a factor analysis on the intelligence data using principal components.**
   a. Modify the file `AFACTOR1.SAS` to run a principal component analysis by removing the `PRIORS=SMC` option.
   b. What are the initial communalities?
   c. Compare and contrast the results with those obtained using principal factors method used in Hands-On Example 18.1. Do you arrive at similar factors?
   d. Find the correlation between new variables `Factor1` and `Factor2`? Do the rotated factors appear to be uncorrelated (orthogonal)?

18.2 **Run a factor analysis on the Olympic Athlete Data using maximum likelihood.**
   a. Modify the file `AFACTOR2.SAS` to perform factor analysis on the Olympic Athlete Data using maximum likelihood.
   b. Compare and contrast the results with those obtained using the principal factors method used in Hands-On Example 18.2. Are the interpretations the same?
   c. Change the rotation method to `PROMAX` and rerun the analysis. Has this changed the basic interpretations of the factors?
   d. Modify the SAS program to print out factor scores for the three factors after `PROMAX` rotation. Name the factor scores `ML1`, `ML2`, and `ML3`. Find the correlation between new variables `FACTOR1` and `FACTOR2`? Do the rotated factors appear to be uncorrelated (orthogonal)?

REFERENCES

Gardiner, H. (2006). *Multiple Intelligences: New Horizons*. New York: Basic Books.
Johnson, D.E. (1998). *Applied Multivariate Methods for Data Analysis*. Pacific Grove, CA: Duxbury Press.
Johnson, R.A. and Wichern, D.W. (2007). *Applied Multivariate Statistical Analysis*, 6e. Englewood Cliffs, NJ: Pearson Prentice Hall.
Stevens, J.P. (2002). *Applied Multivariate Statistics for the Social Sciences*, 4e. Mahwah, NJ: Lawrence Erlbaum Associates, Publishers.

19

CREATING CUSTOM GRAPHS

LEARNING OBJECTIVES

- To be able to use SAS® to create scatterplots and line graphs using GPLOT
- To be able to use SAS to create charts using PROC GCHART
- To be able to use SAS to create boxplots (box-and-whiskers plots) using PROC BOXPLOT
- To be able to create basic plots using SGPLOT

SAS provides several procedures for creating graphs, which are an essential part of data analysis. In this chapter, graphs that are often used to complement the data analyses discussed in this book are illustrated by examples. Some of these graphical procedures were discussed in earlier chapters. This chapter provides more details and introduces graphs that have not been discussed earlier.

SAS provides a number of other sophisticated techniques for creating a variety of graphs that are not discussed in this chapter. However, you should be able to take what you learn here and adapt it to other SAS graphs.

19.1 CREATING SCATTERPLOTS AND LINE GRAPHS USING GPLOT

PROC GPLOT is the basic routine used to obtain several types of graphs in SAS. This section illustrates how to create scatterplots and line graphs using PROC GPLOT, with an

---

SAS® Essentials: Mastering SAS for Data Analytics, Third Edition. Alan C. Elliott and Wayne A. Woodward.
© 2023 John Wiley & Sons, Inc. Published 2023 by John Wiley & Sons, Inc.
Companion website: https://www.alanelliott.com/SASED3/

emphasis on creating graphs that are useful for data analysis. The abbreviated syntax for this procedure is as follows:

PROC GPLOT <options>; <PLOT plotrequest(s)/ option(s)>;

It is a good idea to include the following command to reset all SAS graphics options before you use PROC GPLOT or other SAS graph procedures:

GOPTIONS RESET=ALL;

Moreover, it is a good practice to include a QUIT statement after a plot to tell the SAS program that the graph is complete. Otherwise, you may note a "PROC GPLOT Running" warning in the Editor window, indicating that SAS is waiting for more information about the plot. Some of these options are illustrated in Hands-On Examples.

Table 19.1 lists some of the common options that can be used with PROC GPLOT and Table 19.2 lists some of the common statements used with PROC GPLOT.

The PLOT statement is the heart of PROC GPLOT. Using this statement, you request an X–Y plot from the PROC GPLOT procedure based on code such as the following:

```
PROC GPLOT DATA=sasdatafile;
    PLOT yvariable*xvariable/options;
RUN;
```

TABLE 19.1 Common Options for PROC GPLOT

Option	Explanation
DATA = datasetname	Specifies which dataset to use
ANNOTATE=	Allows you to change default graphical elements or add new elements to the graph. (This feature is not demonstrated in this chapter.)
UNIFORM	Specifies that the same axes be used in all of the graphs produced by the procedure

TABLE 19.2 Common Statements for PROC GPLOT

Statement	Explanation
PLOT (and PLOT2)	Specifies information about vertical and horizontal axes in a scatter/line plot, along with statement options that customize the look of the plot. These statements are discussed below. (PLOT2 is not discussed in this chapter.)
BUBBLE (and BUBBLE2)	Similar to the PLOT statement but specifies the information that creates a bubble plot. (An example BUBBLE is given in the SGGRAPH section in this chapter.)
AXIS, FOOTNOTE, PATTERN, TITLE	These are statements that can be used to enhance graphs. Examples in this chapter illustrate these options
BY, FORMAT, LABEL, WHERE	These statements are common to most procedures, and may be used with PROC GPLOT

Note: At least one PLOT or BUBBLE statement is required.

The `yvariable` is the variable that defines the (left) vertical axis and the `xvariable` defines the horizontal axis. These can be either numeric or character variables. If you specify

```
PLOT yvariable*xvariable=n;
```

the `n` indicates which symbol to use for the points on the plot. These symbols are described in Section A.6. You can also request multiple plots by including more than one variable within parentheses, such as in

```
PLOT (yvar1 yvar2)*xvariable;
```

This statement would produce two plots, one that is `yvar1*xvariable` and the other that is `yvar2*xvariable`. You could specify the same series of plots using

```
PLOT yvar1*xvariable yvar2*xvariable;
```

Another form of the `PLOT` statement allows you to use a third variable that is a grouping (classification) variable. For example,

```
PLOT yvariable*xvariable=group;
```

where `group` is some grouping variable. This will cause the points on the plot to be styled differently for each value of the grouping variable. (You can specify the symbol to use with the `SYMBOL` option.)

Options that appear after the `PLOT` specification are used to enhance the plot. For example,

```
PLOT yvariable*xvariable/CAXIS=color;
```

would cause the axes in the plot to be displayed in the specified color. Table 19.3 lists some of the most common statement options for the `PLOT` statement including those used in the examples in this chapter. (For more options, refer to SAS documentation.)

> To avoid possible confusion we note that PROC GPLOT is a SAS procedure while PLOT is a statement used by PROC GPLOT to specify information about graphs specified by PROC GPLOT.

19.1.1 Creating a Simple Scatterplot

The basic plot generated by `PROC GPLOT` is a scatterplot. The other plots are based on this framework, so Hands-On Example 19.1 shows how to create a simple *x–y* scatterplot.

HANDS-ON EXAMPLE 19.1

This example illustrates how to produce a simple scatterplot using `PROC GPLOT`.

1. Open the program file `AGRAPH1.SAS`.

   ```
   GOPTIONS RESET=ALL;
   Title "Simple Scatterplot";
   ```

(Continued)

432 CHAPTER 19: CREATING CUSTOM GRAPHS

(*Continued*)

```
PROC GPLOT DATA="C:\SASDATA\CARS";
    PLOT HWYMPG*ENGINESIZE;
RUN;
QUIT;
```

This simple program requests a scatterplot of the variables HWYMPG (highway miles per gallon) by ENGINESIZE in the CARS dataset.

2. Run the program and observe (in the Graph window) the results shown in Figure 19.1. This graph shows a decrease in HWYMPG as engine size increases. Note that the variable names are displayed on the *x*- and *y*-axes, and that the limits for each axis are selected by the program.
3. To make a simple change in the plot, for example, to change the color of the text to red, add the statement CTEXT=RED so the PLOT statement reads

```
PLOT HWYMPG*ENGINESIZE/CTEXT=RED;
```

Resubmit the code and observe the results. Note what text the CTEXT= option changes.

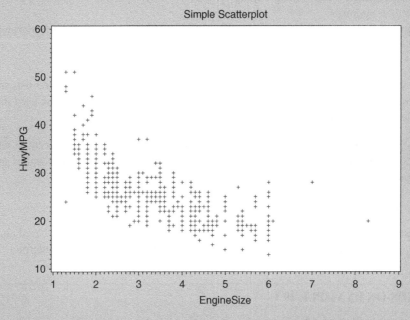

Figure 19.1 Default scatterplot using PROC GPLOT

TABLE 19.3 Common Options for PROC GPLOT's PLOT Statement

Option	Explanation
CAXIS=color	Specifies a color to use for the axis line and tick marks. Colors are described in Appendix A. For example, CAXIS=BLUE
CFRAME= color	Specifies color for the background frame. Colors are described in Appendix A
CTEXT= color	Specifies color for text. Colors are described in Appendix A
HAXIS= or VAXIS=	Specifies information about the horizontal or vertical axis including values and spacing of tick marks. For example,
	HAXIS=AXISk – Indicates which AXISk definition to apply to the horizontal axis. For example, HAXIS=AXIS1 indicates that the AXIS1 definition is to be applied
	VAXIS=AXISk – Indicates which AXISk definition to apply to the vertical axis. For example, VAXIS=AXIS2 indicates that the AXIS2 definition is to be applied
HREF=value list or VREF= value list	Draws a horizontal or vertical line at the indicated axis value or values
LEGEND=legend<n>	Specifies which legend definition to use for a plot. Hands-On Example 19.7 illustrates the use of a legend definition
NOLEGEND	Suppresses the standard plot legend
SYMBOL<n>	Allows you to specify which symbol to use for points, and attributes of that symbol such as color and size. See Appendix A for a list of symbols. See the Section 19.1.3. below
NOFRAME	Suppresses the (default) frame around the plot
GRID	Draws grid line at major tick marks

These options appear after a /.

19.1.2 Managing Your SAS Graph

As mentioned earlier, when you create a graph in SAS, the results are placed in the **Results Viewer** window. Here are some handy tips for working in the **Results Viewer** window.

Copy Graph: To copy the graph to the Clipboard, it can be pasted into Word, PowerPoint, or some other program, click on the graph and select **Edit → Copy** from the menu bar (or right-click on the graph and select **Copy**).

Export Graph: To export the graph to a standard graphics format file right-click on the graph, select **Save Picture as … ,** and select **Files of Type** according to the output type you need. Options are PNG and BIT (bitmapped).

Print: Use the **File → Print** menu option to print a hard copy of the graph. (or select right-click **Print Picture …**)

19.1.3 Enhancing a GPLOT Using the SYMBOL Statement

Often, the purpose of displaying a scatterplot is to illustrate an association between two variables. You may also want to include a regression line through the scatter of points to emphasize the relationship, or you may want to make improvements to the plot that are also available. The following example shows how to use the SYMBOL statement to enhance your plot. This statement may appear before or within the PROC GPLOT statement. The abbreviated syntax for the SYMBOL statement is as follows:

```
SYMBOL<1 ... 255>
<COLOR=colorsymbol>
<appearanceoption(s)>
<INTERPOLATION=type>;
```

An example statement is

```
SYMBOL1 V=STAR I=RL C=BLUE L=2;
```

The symbol number indicates the plot to which the options will be applied. Some commonly used appearance options for SYMBOL are given in Table 19.4.

Hands-On Example 19.2 illustrates how to use the SYMBOL statement and some of its options.

TABLE 19.4 Common Options for PROC GPLOT's SYMBOL Statement

Option	Explanation
COLOR=colorsymbol (or C=)	Indicates the color used for the dot in the plot. For example, C=BLUE causes the dots in the graph to be blue. Other common color options are RED, GREEN, and BLACK. See Appendix A for a more extensive list of available colors
INTERPOLATION=type (or I=)	Instructs SAS to include an interpolation curve in the plot. Several types of interpolation curves are available. A commonly used option is the RL or "regression, linear" fit. Common options are as follows: RL – Linear Regression RQ – Quadratic regression RC – Cubic regression STDk – Standard error bars of length k STDEV (add J [join at mean], B [include error bars], and T [include tops on error bars]). For example, STDkJ to join the bars STDM – Same as STDk but uses standard error of the mean JOIN – Joins dots BOXT – Indicates the pattern of the line drawn in the graph. For example, 1 is a solid line, 2 is small dots, 3 is larger dots, and so on. See Appendix A for other options
LINE=n (or L=)	Indicates the pattern of the line drawn in the graph. For example, LINE=1 produces a solid line. A value of 2 produces small dots, 3 is larger dots, and so on. See Appendix A for other options
VALUE=n (or V=)	Indicates the symbol to be used for the dots on the plot. For example, V=STAR displays a star instead of the default character (+). Some commonly used symbols are SQUARE, STAR, CIRCLE, PLUS, and DOT. See Appendix A for other options.

HANDS-ON EXAMPLE 19.2

This example illustrates how to change the style of a plot's points and regression line.

1. Open the program file AGRAPH2.SAS.

   ```
   GOPTIONS RESET=ALL;
   Title "Simple Scatterplot, Change symbol";
   PROC GPLOT DATA=MYSASLIB.CARS
       PLOT HWYMPG*CITYMPG;
     SYMBOL1 V=STAR I=RL C=BLUE L=2;
   RUN;
   QUIT;
   ```

2. Run this program and observe the output shown in Figure 19.2.
   Note on your screen that the following items are specified for this plot (see Appendix A for more information).
   - The displayed symbol is a star (V=STAR).
   - The displayed symbol is blue (C=BLUE).
   - A linear regression line is displayed (I=RL).
   - The displayed line is dotted (L=2).
3. Change the C, V, and L options by choosing other values from the definitions listed in Appendix A and rerun the plot to observe how changing these options modifies the plot.

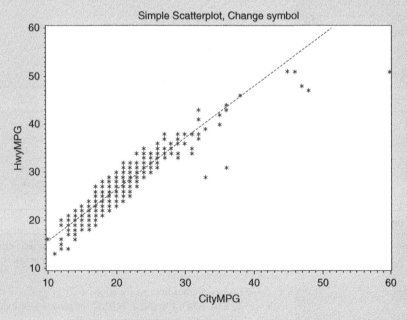

Figure 19.2 PROC GPLOT with dots and line specified

19.1.4 Customizing Axes in a Plot

If you are preparing a plot for use in a project presentation or publication, you may want to change the characteristics of the plot axes. Hands-On Example 19.3 illustrates some of the basic techniques. The syntax for the AXIS statement is

```
AXIS<n> <options>;
```

where *n* defines options for a specific axis such as AXIS1, AXIS2, and so on. Applying an axis definition to a plot requires the following two steps:

Step 1: Define the axis using an AXIS<n> statement.
Step 2: Associate the defined axis to a specific axis in the plot. GPLOT and GCHART (discussed in an upcoming section) have different names for the plot axes. Axis names for GPLOT and GCHART are listed in Table 19.5.

Axis options define the "look" of the axis: its scale, appearance, tick marks, and text. As an example, for the GPLOT procedure, the following code defines AXIS1. This is step 1 in using an axis.

```
AXIS1 C=RED ORDER=(0 TO 80 BY 5)
LABEL=('City Miles Per Gallon' FONT=SWISS)
OFFSET=(5) WIDTH=2 VALUE=(H=2);
```

To apply this axis to a plot (step 2 in using an axis), you would use code such as this example:

```
PROC GPLOT DATA=CARS;
    PLOT HWYMPG*CITYMPG/HAXIS=AXIS1;
```

TABLE 19.5 AXIS Names for GPLOT and GCHART

Axis Name for GPLOT	Refers to Which Axis?
HAXIS	HAXIS=AXIS<n> applies the specifications defined by AXIS<n> to the *horizontal* axis, also called the *X*-axis in a standard *X–Y* plot
VAXIS	VAXIS=AXIS<n> applies the specifications defined by axis<n> to the *vertical* axis, also called the *Y*-axis in a standard *X–Y* plot
Axis Name for GCHART	**Refers to**
MAXIS=	MAXIS=AXIS<n> applies the specifications defined by axis<n> to the *midpoint* axis, which is the axis associated with a group (categorical) specification
GAXIS=	GAXIS=AXIS<n> applies the specifications defined by AXIS<n> to the *group* axis, when you use a GROUP= statement (see example below)
RAXIS=	RAXIS=AXIS<n> applies the specifications defined by AXIS<n> to the *response* axis, which is the axis associated with the count, percent, or frequency

TABLE 19.6 Common Options for PROC GPLOT's AXIS Statement

Option	Explanation
COLOR=color (or C=)	Indicates the color used for the axis in the plot. For example, C=BLUE causes the axis to be blue. Other common color options are RED, GREEN, and BLACK. See Appendix A for a comprehensive list of available colors
LABEL=(specifications)	Specifies characteristics of an axis label. The text arguments define specific parts of the label's appearance. These include angle, color, font, justification, and rotation angle (A=), where A=90 means to display the label at a 90° angle (vertical). LABEL=NONE suppresses a label. An example of a label statement is LABEL=(H=4 J=C C=BLUE "My Axis"); where H=4 specifies Height of the text J=C means Justify Center C=BLUE specifies color of label text
ORDER=(list)	Specifies the order of the values on the axis. For example, ORDER=(1 to 100 by 5) specifies that the axis will contain the numbers 1–100 with axis label increments of 5
VALUE=(options)	Specifies the height of text (among other items) – for example, VALUE=(H=4)
WIDTH=thickness	Specifies the width of the axis (affects the width of the entire frame unless a NOFRAME option is used as defined below)
OFFSET=(x,y) <UNITS>	Moves the axis away from the edge by a certain amount. For example, OFFSET=(3,4) moves the *x*-axis 3% horizontally, and the *y*-axis 4% vertically. You can specify the offset in percent, centimeters, or inches using OFFSET=(3,4)PCT OFFSET=(3,4)cm OFFSET=(3,4)IN respectively, where percent is the default.
A=n	Specifies text angle. For example, use A=90 to place a horizontal axis at a 90° angle so it aligns with the axis
MAJOR=(options)	Controls major tick marks. For example, MAJOR=(COLOR=BLUE HEIGHT=3 WIDTH=2)
MINOR=(options)	Controls minor axis tick marks. MINOR=NONE causes no minor ticks to be displayed. For example, MINOR=(NUMBER=4 COLOR=RED HEIGHT=1)
JUSTIFICATION=type (J=)	Justification, L=Left, C=Center, R=Right

The AXIS1 statement creates a horizontal axis (HAXIS) that is red, whose tick marks are from 0 to 80 and labeled by multiples of 5. The label for the axis is "City Miles Per Gallon" and is displayed in a Swiss font. Table 19.6 lists common AXIS statement options:

> For Windows SAS version 9.3 and later, you may need to turn off automatic ODS HTML type graphics (which is the default in those versions) in order for some of the upcoming graphics examples to work (particularly the colors). To turn off ODS graphics, put this statement at the beginning of your code:
>
> ```
> ODS GRAPHICS OFF;
> ```

To turn ODS graphics back on, use the code

```
ODS GRAPHICS ON;
```

Another way to return ODS graphics back to their normal default state in SAS versions 9.3 and later is

```
ODS PREFERENCES;
```

HANDS-ON EXAMPLE 19.3

This example illustrates how to format axes using PROC GPLOT.

1. Open the program file AGRAPH2A.SAS.

```
GOPTIONS RESET=ALL;
AXIS1 C=RED ORDER=(0 TO 80 BY 5)
      LABEL=('City Miles Per Gallon' FONT=SWISS)
      OFFSET=(5) WIDTH=2 VALUE=(H=2);
AXIS2 C=GREEN ORDER=(0 TO 80 BY 5)
      LABEL=(A=90 'Highway Miles Per Gallon' FONT=SWISS)
      OFFSET=(5,5)PCT WIDTH=4 VALUE=(H=1);
TITLE "Enhanced Scatterplot";
PROC GPLOT DATA=MYSASLIB.CARS;
      PLOT HWYMPG*CITYMPG/HAXIS=AXIS1 VAXIS=AXIS2 NOFRAME;
      SYMBOL1 V=DOT I=RL C=PURPLE L=2;
RUN;
QUIT;
```

2. Run the program and observe the color version shown on your screen. A black and white version is illustrated in Figure 19.3. This program obviously creates a plot with strange color combinations, but it illustrates the possibilities. Note that there are two axis definitions (AXIS1 and AXIS2) and that these axis definitions are applied in the PLOT statement where VAXIS=AXIS2 and HAXIS=AXIS1.

   Because of the NOFRAME option, there is no frame around the entire plot (the frame is the default border that surrounds the plot as in Figure 19.2), but the colors of the axes are defined in the AXIS statements.

3. Clean up this plot by doing the following:
   - Remove the definition for axis color in the AXIS1 and AXIS2 statements.
   - Remove the NOFRAME option.
   - Change the COLOR for the dots to GRAY.
   - Change the LINE style (L=) to 5 (long dashes).
   - Change the OFFSET to 2.
   - Make the WIDTH 4 for each axis in both cases.
   - Specify HEIGHT (H=1) for AXIS1.
   - Rerun the program and observe the changes in the graph.

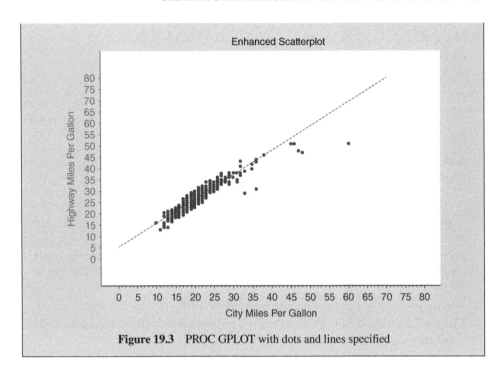

Figure 19.3 PROC GPLOT with dots and lines specified

19.1.5 Displaying Error Bars in GPLOT

You may want to display data by groups with error bars on the graph to indicate the variability of data within a group. Displaying error bars (and other types of bars) is accomplished with the `I=` option in the `SYMBOL` statement (see Table 19.4 for more options). For example,

```
SYMBOL<n> INTERPOLATION=specification (or I=)
```

To employ this option for displaying error bars, use one of the following techniques:

- `I=STD1JT` produces error bars that are 1 standard deviation (`STD1`) and that are joined (`J`) at the means with tops (`T`) on each bar. `STD` also supports values of 2 and 3.
- `I=BOXT` includes boxplots with tops (`T`) on each whisker.

> **HANDS-ON EXAMPLE 19.4**
>
> In this example, you will create a graph that plots means and error bars.
>
> 1. Open the program file `AGRAPH3.SAS`.
>
> ```
> GOPTIONS RESET=ALL;
> AXIS1 ORDER=(4 TO 12 BY 1)
>       LABEL=('Number of Cylinders' FONT=SWISS HEIGHT=4)
>       OFFSET=(5);
> ```
>
> *(Continued)*

(*Continued*)

```
AXIS2 ORDER=(0 TO 50 BY 5)
     LABEL=(A=90 'Highway Miles Per Gallon' FONT=SWISS
     HEIGHT=4)
     OFFSET=(5);
PROC GPLOT DATA=MYSASLIB.CARS
     PLOT HWYMPG*CYLINDERS/HAXIS=AXIS1 VAXIS=AXIS2;
     SYMBOL1 I=STD1JT H=1 LINE=1 VALUE=NONE;
RUN;
QUIT;
```

This plot is for highway miles per gallon (HWYMPG) by the number of CYLINDERs in the car's motor. Note that the I=STD1JT interpolation option in the SYMBOL1 statement requests 1 standard deviation error bars. The VALUE=NONE option suppresses the graph of individual points. The AXIS1 statement requests that the *x*-axis contain only the values from 4 to 12 because that is the number of CYLINDERs in the dataset.

2. Run the program and observe the output in Figure 19.4.
3. Change the interpolation option to request boxplots with tops on the ends of the whiskers:

I=BOXT

Run the revised program and observe the output.

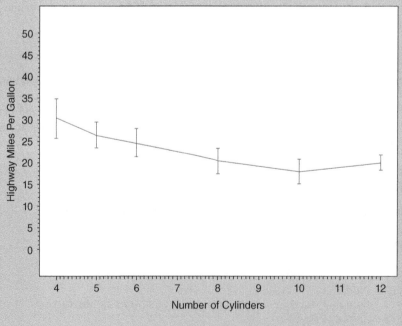

Figure 19.4 PROC GPLOT showing error bars

There are several other methods for customizing a SAS GPLOT. Refer to Appendix A and the SAS documentation for additional options.

19.2 CREATING BAR CHARTS AND PIE CHARTS

You can produce bar and pie charts for displaying counts or percentages for frequency data. These charts are created with the SAS GCHART procedure. The abbreviated syntax is as follows:

```
PROC GCHART <DATA=datasetname>;
HBAR|HBAR3D|VBAR|VBAR3D chartvar(s) </ chartoption(s)>;
PIE|PIE3D|DONUT|STAR chartvar(s) </ chartoption(s)>;
```

Chart-type specifications such as HBAR and VBAR refer to horizontal and vertical bar charts, pie charts, and three-dimensional (3D) versions of the charts. An example is

```
PROC GCHART DATA=MYDATA;
HBAR STATUS/DISCRETE COUTLINE=BLUE WOUTLINE=3;
```

Some common GCHART statements are listed in Table 19.7. Common bar chart options are listed in Table 19.8. Refer to SAS documentation for other statements and options.

A simple example of PROC GCHART is a horizontal bar chart (HBAR), shown in Hands-On Example 19.5.

TABLE 19.7 Common Options for PROC GCHART

Options	Explanation
DATA=datasetname	Specifies which dataset to use
HBAR, HBAR3D, VBAR, VBAR3D, PIE, PIE3D, DONUT, or STAR	Specify which type of plot to create. The option / HTML=variable specifies a variable that contains an HTML link address.
AXIS, FOOTNOTE, PATTERN, TITLE	These are statements that can be used to enhance graphs. Examples in this chapter illustrate these options
BY, FORMAT, LABEL, WHERE	These statements are common to most procedures, and may be used here

TABLE 19.8 Common GCHART Options for Bar Charts

Option	Explanation
CAXIS=color CFRAME=color CTEXT=color	Specifies a color to use for the axis, frame, or text. For example, CAXIS=BLUE
COUTLINE=color WOUTLINE=n	COUTLINE specifies color for bar outline (or SAME) and WOUTLINE specifies outline width. Default width is 1. For example, COUTLINE=BLUE WOUTLINE=3
DISCRETE	Treats numeric values as discrete (categorical)
GROUP variable	Splits bars into groups using specified classification variable

(Continued)

TABLE 19.8 (*Continued*)

Option	Explanation
SUBGROUP variable	Specifies qualities of bars according to subgroups within the group defined by the GROUP variable (i.e., displays a different color for each subgroup)
WIDTH=n SPACE=n	WIDTH specifies BAR WIDTH and SPACE= specifies the space *between bars*
INSIDE=statistic OUTSIDE=statistic	Determines where a statistic will be displayed in a chart. Statistics are MEAN, SUM, PERCENT (PCT), or FREQ
TYPE=statistic	Defines the statistic to be computed (to display on the chart). Options include SUM, MEAN, PERCENT (PCT), and FREQ(default)
SUMVAR=varname	Defines the response variable. When you use this option, TYPE= should be MEAN or SUM
NOLEGEND	Suppresses chart legend
CLM=p	Specifies the level for a confidence interval to be displayed on the bars. For example, CLM=95
ERRORBAR=type	Specifies TOP, BARS, or BOTH
HTML=	Defines an HTML link for a displayed category. When the graph image representing the group (represented by a bar, pie slice, etc) is clicked, the HTML address defined for that group is activated. For example HTML="http://www.cnn.com";

These options appear after the /.

HANDS-ON EXAMPLE 19.5

This example illustrates how to create a horizontal bar chart.

1. Open the program file AGRAPH4.SAS.

   ```
   GOPTIONS RESET=ALL;
   Title "Horizontal Bar Chart";
   PROC GCHART DATA="C:\SASDATA\SOMEDATA";
   HBAR STATUS/DISCRETE;
   RUN;
   QUIT;
   ```

   This code requests a horizontal bar chart for the variable STATUS. The /DISCRETE option tells SAS that the numeric values for the STATUS variable should be treated as discrete values. Otherwise, SAS may assume that the numeric values are a range and create a bar chart using midpoints instead of the actual values.

2. Run the program and observe the output in Figure 19.5.
3. Change HBAR to VBAR and rerun the program and observe the output.
4. Change the statement by adding WIDTH=10,

   ```
   VBAR STATUS/DISCRETE WIDTH=10;
   ```

   Rerun the program and observe the output. How did the WIDTH statement change the chart?

5. Remove the WIDTH=10 statement and change VBAR to PIE and rerun the program. Observe the output. Experiment with HBAR3D and PIE3D.

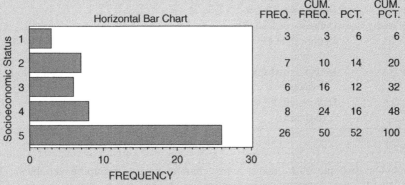

Figure 19.5 Horizontal bar chart using PROC GCHART

6. Change back to HBAR and include a PATTERN and COLOR statement to change the look of the graph. For example,

   HBAR STATUS/DISCRETE;
   PATTERN1 VALUE=R1 COLOR=GREEN;

   Run the revised program and observe the output.

7. Remove the /DISCRETE option and rerun the program. How does that change the output?

19.3 DEFINING GRAPH PATTERNS

As illustrated in the above example, the PATTERN statement allows you to change the color and pattern of a bar in a bar chart. A simplified syntax for the statement is

PATTERN<n> <COLOR=color> <VALUE= pattern >;

For example,

PATTERN1 VALUE=R1 COLOR=GREEN;

Patterns are discussed further in Appendix A. COLOR refers to standard SAS colors. The <n> in the statement determines which category the pattern will be defined for, where the categories are the order in which categories are displayed for the chart. These are listed in Table 19.9.

For example, the statement

PATTERN1 VALUE=SOLID C=LIGHTBLUE;

results in a solid light blue bar related to the first category in the chart.

TABLE 19.9 Pattern Definitions for the PATTERN Value Statement

Option	Explanation
COLOR=color	Specifies color for a pattern. See color options in Appendix A
EMPTY (E)	Defines EMPTY bar
SOLID (S)	Defines a SOLID bar
R1 to R5	Defines right 45° stripes. Thickness is specified by R1 (lightest) to R5 (heaviest)
X1 to X5	Defines crosshatch pattern. Thickness is specified by X1 (lightest) to X5 (heaviest)
L1 to L5	Defines left 45° stripes. Thickness is specified by R1 (lightest) to R5 (heaviest)

Hands-On Example 19.6 creates a bar chart using two grouping variables. This is accomplished with a /GROUP option as illustrated here:

```
VBAR INJTYPE/GROUP=GENDER;
```

HANDS-ON EXAMPLE 19.6

This example illustrates how to modify axes in a bar chart.

1. Open the program file AGRAPH5.SAS.

   ```
   GOPTIONS RESET=ALL;
   AXIS1 LABEL=NONE VALUE=(H=1.2);
   AXIS2 LABEL=(A=90 "Count");
   TITLE "Vertical Bar Chart";
   PROC GCHART DATA=MYSASLIB.SOMEDATA;
   VBAR GENDER/GROUP=GP MAXIS=AXIS1 RAXIS=AXIS2;
   PATTERN1 VALUE=R1 COLOR=GREEN;
   RUN;
   QUIT;
   ```

   Note that AXIS= controls the frequency (vertical) axis for GCHART and MAXIS= controls the appearance of the midpoint (horizontal) axis.

2. Run the program and observe the output in Figure 19.6. This type of plot can be used to display cross-tabulation analyses.
3. If you want the Female and Male bars to be in different colors, use a SUBGROUP=GENDER command and define the second bar color using a PATTERN2 statement.
   Open the program file AGRAPH6.SAS.

   ```
   GOPTIONS RESET=ALL;
   TITLE 'Bar Chart by Group with different bar colors';
   PROC GCHART DATA="C:\SASDATA\SOMEDATA";
   ```

```
VBAR GENDER/GROUP=GP SUBGROUP=GENDER;
PATTERN1 VALUE=R1 COLOR=RED;
PATTERN2 VALUE=R1 COLOR=BLUE;
RUN;
QUIT;
```

Note the added SUBGROUP= statement. This tells SAS to color the bars for each subgroup using a different color. The PATTERN2 statement specifies the color for males because the patterns are assigned in alphabetical order.

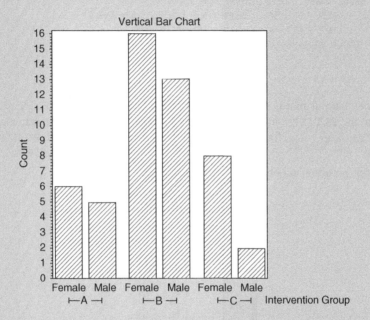

Figure 19.6 PROC GCHART showing groups bars

4. Run the program and observe the output. Note that the bars for each GENDER are now in different colors. The PATTERN remains the same because of the VALUE=R1 specification in both PATTERN statements.
5. Change the statement in PATTERN1 to VALUE=X4 and rerun the program to see how that change alters the output.
6. Change VBAR to VBAR3D. Note that the colors for the bars remain, but the PATTERN option no longer has any effect.

If you want the bars to appear in an order other than alphabetical, you can create a format for the variable and apply that format to the variable for the PROC GCHART statement.

19.4 CREATING STACKED BAR CHARTS

Another often-used type of bar chart is a stacked bar chart. To create stacked bar charts, you *do not* use the GROUP option. This is illustrated in Hands-On Example 19.7.

HANDS-ON EXAMPLE 19.7

This example illustrates how to create stacked bar charts using PROC GCHART.

1. Open the program file AGRAPH7.SAS.

```
GOPTIONS RESET=ALL;
PATTERN1 V=R1 C=RED;   * FOR BUS;
PATTERN2 V=R2 C=BLUE;  * FOR CAR;
PATTERN3 V=R3 C=BLACK; * FOR WALK;
TITLE C=RED 'Stacked Bars, Method of Arrival';
PROC GCHART DATA=MYSASLIB.SURVEY;
     VBAR GENDER / SUBGROUP=ARRIVE WIDTH=15;
RUN;
QUIT;
```

Note that this plot is set up a little differently from previous GCHART examples, in that the PATTERN statement occurs before the PROC GCHART statement. The PATTERNs are set up for the three arrival types that will be graphed: BUS, CAR, and WALK.

2. Run this program and observe the output in Figure 19.7.

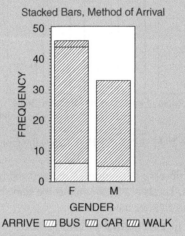

Figure 19.7 Stacked bar chart

3. Suppose that for a publication that doesn't print in more than one color, you want the bars to be in different shades of gray. Change the PATTERN and V statements to read as follows:

```
PATTERN1 V=S C=GRAYAA;  * FOR BUS (Dark Gray);
PATTERN2 V=S C=GRAYCC;  * FOR CAR (Medium Gray);
PATTERN3 V=S C=GRAYEE;  * FOR WALK (Light GRAY);
```

V=S indicates solid, and the various shades of gray are given as GRAYAA, GRAYCC, and so on to GRAYFF, which is white.

DEFINING GRAPH PATTERNS

4. Change the GCHART code to

```
PROC GCHART DATA=MYSASLIB.SURVEY;
    VBAR GENDER / GROUP=ARRIVE SUBGROUP=GENDER WIDTH=10;
RUN;
```

Rerun the program and observe the difference.

5. For further enhancement to the bar chart, open the program file AGRAPH7A.SAS and observe the additional code defining a plot key and an enhancement to the vertical axis.

```
GOPTIONS RESET=ALL;
PATTERN1 V=R1 C=RED; * FOR BUS;
PATTERN2 V=R2 C=BLUE; * FOR CAR;
PATTERN3 V=R3 C=BLACK; * FOR WALK;
LEGEND1
 ACROSS=1 LABEL=("Gender")
 POSITION=(top inside left)
 VALUE=(H=1 'Male' 'Female');
AXIS1 label=(a=90 Color=RED H=2 f="Swiss/Bold" "Count");
TITLE C=RED 'Example Bar Chart';
PROC GCHART DATA=MYSASLIB.SURVEY;
    VBAR GENDER / GROUP=ARRIVE SUBGROUP=GENDER WIDTH=10
                LEGEND=LEGEND1 RAXIS=AXIS1
                OUTSIDE=FREQ;
RUN;
QUIT;
```

- The LEGEND1 and AXIS1 statements: The LEGEND1 statement creates a key named "Gender" that appears at the top, inside left in the graph frame, with the key items 'Male' and 'Female' listed. The ACROSS option specifies that the key has items listed one across.
- The AXIS1 statement defines an axis to be used as the vertical axis where the name "Count" will be displayed at a 90° angle using the "Swiss/Bold" font, in red, with a height of 2 units (default is 1).
- In the PROC GCHART statements, note the LEGEND=LEGEND1 and RAXIS=AXIS1 statements. These statements apply the legend and axis to the chart.
- Note the statement OUTSIDE=FREQ. This tells the program to display the count (FREQ) for each bar, and to display it outside the bar.
  a. Run this program and observe the results shown in Figure 19.8.
  b. Take out the ACROSS=1 option and rerun. How does this change the legend?
  c. Change the legend position to the top inside right. Rerun the program to see if your change worked.

(*Continued*)

448 CHAPTER 19: CREATING CUSTOM GRAPHS

(*Continued*)

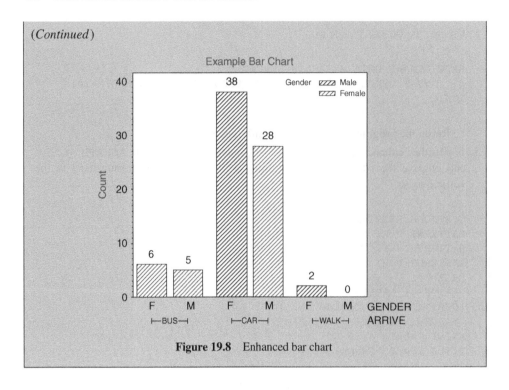

Figure 19.8 Enhanced bar chart

There are many options for customizing these plots, only a few of which have been discussed here. Several of the options (e.g., AXIS and LEGEND) are similar to those used in PROC GPLOT.

19.5 CREATING MEAN BARS USING GCHART

An often-used analysis graphic is one that displays means as bars with error bars around the mean indicating a level of variability. PROC GCHART can produce this graph with a little help. Hands-On Example 19.8 shows how this is accomplished.

HANDS-ON EXAMPLE 19.8

In this example, you will produce a bar chart with error bars representing means by group.

1. Open the program file AGRAPH8.SAS.

```
PATTERN1 V=L3 COLOR=LIGHTBLUE;
AXIS1 LABEL=(A=90 H=1.5 F="SWISS" "OBSERVATION" C=BLUE);
AXIS2 LABEL=(H=1.5 F="SWISS" "GROUP");
TITLE1 'Mean Displayed with Error Bars';
PROC GCHART DATA=MYSASLIB.SOMEDATA;
```

```
VBAR GP /
WIDTH=10
TYPE=MEAN
SUMVAR=TIME1
INSIDE=MEAN
CLM=99
ERRORBAR=TOP
RAXIS=AXIS1 MAXIS=AXIS2
COUTLINE=SAME WOUTLINE=2;
RUN;
QUIT;
```

Note the following statements and how they are used in this code:

- The PATTERN statement makes the bars light blue with an L3 (left stripe pattern).
- The AXIS1 statement is used to specify the RAXIS (which is the Response axis), and the AXIS2 statement is used to specify the MAXIS (Midpoint Axis). The RAXIS is displayed rotated by 90°. Both are displayed with a height of 1.5.
- The SUMVAR tells SAS which variable is used to calculate the mean (as specified by INSIDE=).

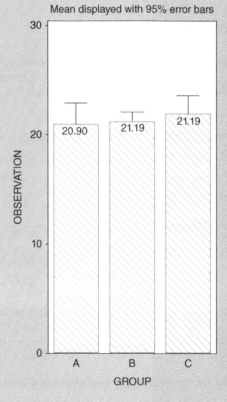

Figure 19.9 Means with error bars

(*Continued*)

> (*Continued*)
> - INSIDE=MEAN causes the mean values to be displayed within the bars.
> - The CLM=99 specifies a 99% confidence interval on the mean.
> - The TYPE= option specifies the type of statistic used in the ERRORBAR option.
> - The ERRORBAR=TOP causes the CI error bars to be displayed on the top of the bars.
> - COUTLINE and WOUTLINE specify a black outline with width 1.
>   Run this program and observe the output, as shown in Figure 19.9.
> 2. Change the AXIS1 statement to control attributes of the major and minor axes on the response scale.
>
>    ```
>    AXIS1 LABEL=(A=90 H=1.5 F="SWISS" "OBSERVATION" C=BLUE)
>    MAJOR=(COLOR=BLUE HEIGHT=3 WIDTH=2)
>    MINOR=(NUMBER=4 COLOR=RED HEIGHT=1);
>    ```
>
> 3. Change the ERRORBAR specification to BOTH, rerun and observe the changes.
> 4. Take off the ERRORBAR, TYPE, and CLM statements and change INSIDE=MEAN to OUTSIDE=FREQ (ERRORBAR, TYPE, and CLM do not work with FREQ), rerun and observe the changes.

19.6 CREATING BOXPLOTS

Examples in earlier chapters have shown how SAS output creates boxplots in some cases by default. This section provides information on methods for creating and enhancing boxplots. The abbreviated syntax for this procedure is as follows:

```
PROC BOXPLOT <DATA=datasetname>; <options>;
        PLOT depvar*groupvar(s)</options>;
<statements>;
```

The PLOT statement is required. The following is a simple PROC BOXPLOT command. (The data must first be sorted by GROUP):

```
PROC BOXPLOT DATA = MYDATA;
    PLOT HEIGHT*GROUP;
RUN;
```

There are numerous statements and options for PROC BOXPLOT. Some commonly used statements are listed in Table 19.10.

Options for the PLOT statement appear after a slash. For example,

```
PROC BOXPLOT DATA=CARDATA;
    PLOT HWYMPG*CYLINDERS/CBOXFILL=LIGHTGREEN;
```

Some common options for the PLOT statement associated with PROC BOXPLOT are listed in Table 19.11.

A number of other options are available, many similar to those listed for PROC GPLOT (see SAS documentation for more details).

CREATING BOXPLOTS

TABLE 19.10 Common Statements for PROC BOXPLOT

Statement	Explanation
`DATA=`*datasetname*	Specifies which dataset to use
`PLOT <`*depvar*`*>`*groupvar(s)*`;`	Specifies the variables to use in the plot. Typically, a dependent variable and grouping variable such as `PLOT AGE*GROUP;`
`INSET`	Specifies an inset key similar to those previously described for histograms. See Chapter 10
`ID` *variables;*	Specifies which variables will be used to identify observations
`BY, FORMAT, LABEL, WHERE`	These statements are common to most procedures and may be used here

TABLE 19.11 Common Options the BOXPLOT PLOT Statement (Follows a Slash (/) in PLOT Statement)

Option	Explanation
`CBOXES=`*color*	Specifies outline color for boxes. For example, `CBOXES=RED`
`CBOXFILL=`*color*	Specifies the fill pattern for boxes. For example, `CBOXFILL=LIGHTBLUE`
`NOCHES`	Species that boxes will be notches. For an explanation for notches, see McGill et al. (1978)
`BOXSTYLE=`	Specifies the style of the boxplot. Options include `SCHEMATIC`, `SCHEMATICID`, `SKELETAL` (default). `SCHEMATIC` displays outliers. `SCHEMATICID` displays labels for outliers according to the `ID` variable specified in the `ID` statement. `SKELETAL` does not display outliers. (See example that follows for more explanation.)

19.6.1 Creating a Simple Boxplot

The PLOT statement indicates the variables used for the analysis and grouping in the statement:

`PLOT` *analysisvariable*grouping variable;*

HANDS-ON EXAMPLE 19.9

This example illustrates a simple use of `PROC BOXPLOT`.

1. In the Editor window, open the file `AGRAPH9.SAS`.

```
GOPTIONS RESET=ALL;
Title "Simple Box Plot";
* You must sort the data on the grouping variable;
```

(*Continued*)

```
PROC SORT DATA=MYSASLIB.CARS
    OUT=CARDATA;BY CYLINDERS;
PROC BOXPLOT DATA=CARDATA;
    PLOT HWYMPG*CYLINDERS/CBOXFILL=LIGHTGREEN;
    WHERE CYLINDERS GT 1;
RUN;
```

Note that when creating side-by-side boxplots, you must sort the data on the grouping variable before running the PROC BOXPLOT procedure. Also note that the WHERE statement is necessary because one car (the Mazda RX-8) is listed as −1 cylinders because it uses a rotary engine. You could have also recoded the −1 as a missing value to obtain the same result.

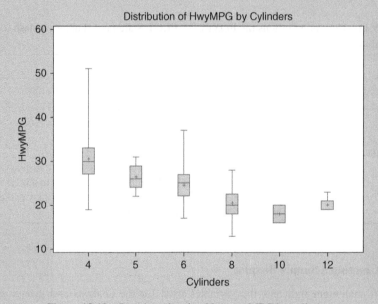

Figure 19.10 Boxplots showing highway MPG by cylinders

2. Run this program and observe the output in Figure 19.10.
3. Change the PLOT statement to include the following:

```
PLOT HWYMPG*CYLINDERS/
    CBOXFILL=VLIGB
    CBOXES=BLACK
    BOXSTYLE=SCHEMATIC
    NOTCHES;
```

The CBOXFILL and CBOXES options specify the fill color and outline color of the boxes, respectively. BOXSTYLE requests the SCHEMATIC style plot, which specifies the type of outliers displayed in the plot. The SCHEMATIC option defines outliers as those values greater than the absolute value of 1.5*(interquartile range) from the top or bottom of the box. Run the program and observe the output in Figure 19.11. Note that outliers are marked at the top of two of the boxplots.

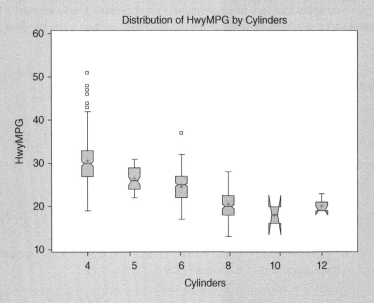

Figure 19.11 Boxplots with enhancements

4. To display the ID information for the outliers, change the style to SCHEMATICID and include the statement ID MODEL using the code shown here. Rerun and observe the results. (In this case, some of the ID text overlaps so that it is hard to read some of the results.)

```
PLOT HWYMPG*CYLINDERS/
        CBOXFILL=VLIGB
        CBOXES=BLACK
        BOXSTYLE=SCHEMATICID
        NOTCHES;
        ID MODEL;
```

5. To display summary statistics on the graph, include an INSETGROUP statement as in the code shown here. Rerun and observe the results.

```
    INSETGROUP MEAN (5.1) MIN MAX /
        header = 'Mean and Extremes';
```

The notches are a way to measure the difference between two medians. If there is no overlap in the corresponding notches for two plots, it indicates a significantly different in the group medians at approximately the 95% confidence level.

454 CHAPTER 19: CREATING CUSTOM GRAPHS

19.7 GOING DEEPER: CREATING AN INTERACTIVE BAR USING ODS

A useful feature of ODS introduced in Chapter 8 is its ability to create interactive links within graphs in HTML mode. To do this, you create a variable that creates a hypertext link, and then you apply that link to a portion of a graph. Hands-On Example 19.10 uses a graph created using the SAS procedure PROC GCHART, where each bar is a link to a report.

HANDS-ON EXAMPLE 19.10

This example illustrates how to use HTML links in a SAS graph.

1. In the Editor window, open the file AGRAPH10.SAS.

   ```
   TITLE "Bar Chart";
   PROC GCHART DATA=MYSASLIB.SOMEDATA;
        HBAR GP/ DISCRETE;
   RUN; QUIT;
   ```

2. Run this program and observe that it creates a standard horizontal bar chart that consists of three categories of the variable GP. See Figure 19.12.

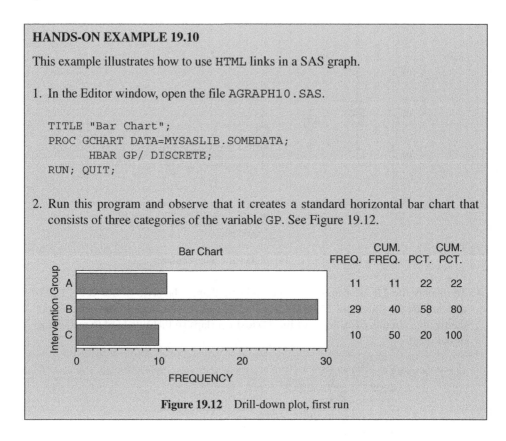

Figure 19.12 Drill-down plot, first run

Suppose that for each "group," represented by one of the bars on the plot, you want to make available a set of background statistics. You can associate a link for each bar on the graph to a separate Web page. Thus, each bar becomes a Web page link, and when you click on a bar on the graph, you open an HTML page. The code shown in Figure 19.13 creates an HTML file that contains the report (from PROC MEANS) you wish to associate with Group

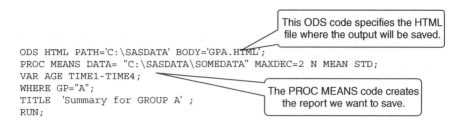

Figure 19.13 Code to create an HTML report file

TABLE 19.12 Report for Group A

	Summary for GROUP A The MEANS Procedure			
Variable	Label	N	Mean	Std Dev
AGE	Age on Jan 1, 2000	11	10.36	2.87
TIME1	Baseline	11	20.90	2.08
TIME2	6 Months	11	27.00	2.93
TIME3	12 Months	11	30.35	3.20
TIME4	24 Months	11	30.44	3.96

A. (We'll see this code again in Hands-On Example 19.11.) Note that the HTML report is saved in the file named GPA.HTM, and that it contains information about GP="A". (Remember that the format to specify the file in ODS HTML is BODY=.)

The report created by the code in Figure 19.13 is given in Table 19.12. We'll create this and other tables in Hands-On Example 19.11.

To define the link that will take the user to this report, use a standard IF statement in a DATA step such as the following:

```
IF GP="A" THEN HTMLLINK='HREF="GPA.HTM"';
```

This IF statement assigns the text 'HREF="GPA.HTM"' to a variable named HTMLLINK. The HREF statement is a standard HTML statement used in pages on the web to create links to other pages. (Note how the HREF is included in single quotation marks and the file name is in double quotation marks.) In this example, a link is defined when the group (GP) is "A". Once a link variable (HTMLLINK) is created, it can be applied to a portion of a graph as illustrated in the Hands-On Example 19.11.

HANDS-ON EXAMPLE 19.11

1. Open the program file AGRAPH11.SAS. In Part 1 of the code, note that there are three PROC MEANS calls that create three files containing reports for Groups A, B, and C. These three PROC MEANS calls are similar to the code shown in Figure 19.13 and each creates an HTML report file.

```
*......PART 1: CREATE REPORTS FOR GROUPS A, B, and C;
ODS HTML PATH='C:\SASDATA' BODY='GPA.HTML';
PROC MEANS DATA= "C:\SASDATA\SOMEDATA" MAXDEC=2 N MEAN STD;
VAR AGE TIME1-TIME4;
WHERE GP="A";
TITLE 'Summary for GROUP A';
RUN;
ods html path='c:\SASDATA' BODY='GPB.HTML';
PROC MEANS DATA= "C:\SASDATA\SOMEDATA" MAXDEC=2 N MEAN STD;
VAR AGE TIME1-TIME4;
WHERE GP="B";
TITLE 'Summary for GROUP B';
RUN;
ods html path='c:\SASDATA' BODY='GPC.HTML';
```

(*Continued*)

(*Continued*)

```
  PROC MEANS DATA= "C:\SASDATA\SOMEDATA" MAXDEC=2 N MEAN STD;
  VAR AGE TIME1-TIME4;
  WHERE GP="C";
  TITLE 'Summary for GROUP C';
  RUN;
  *.......PART 2: CREATE THE LINKS IN  A DATA STEP;
  DATA BARCHART;SET "C:\SASDATA\SOMEDATA";
  LENGTH HTMLLINK $40;
  IF GP="A" THEN HTMLLINK='HREF="GPA.HTML"';
  IF GP="B" THEN HTMLLINK='HREF="GPB.HTML"';
  IF GP="C" THEN HTMLLINK='HREF="GPC.HTML"';
  ODS HTML PATH='c:\SASDATA' BODY='GCHRT.HTML' GPATH=
  "C:\SASDATA\";
  *.......PART 3: CREATE THE GRAPH;
  PROC GCHART;
       HBAR GP/ HTML=HTMLLINK;
        TITLE 'Summary information for each GP.';
  RUN; QUIT;
  ODS HTML CLOSE;
  ODS PREFERENCES; *Reset HTML output;
```

Part 2 of the code creates a dataset called BARCHART that includes a new variable named HTMLLINK that is defined for each of the three groups. The third section of the code creates the bar chart using GCHART. In PROC GCHART, the statement option /HTML= in the line

```
HBAR GP/ HTML=HTMLLINK;
```

relates the HTML link (HTMLLINK) to each category of GP. Part 3 of the code creates the bar chart.

Run the program and observe the output in the **Results Viewer** (or view GCHART.HTM in a browser). The graph is virtually identical to the one illustrated in Figure 19.12. However, there is a difference.

2. Place your cursor on one of the bars. The icon turns into a pointing hand indicating a link. Click on the bar for group A and the information given earlier in Table 19.12 appears.
3. To return to the graph, locate the left arrow at the top of the window and click it once. Try the links for the other two bars. Using this technique, you can create graphs with drill-down capabilities that allow you to display background information for each category in a graph.
4. Change the last IF statement in the code to

```
IF GP="C" THEN HTMLLINK='HREF="http://www.google.com"';
```

or to some other Web page. (Be careful to include the quote marks correctly in the HTMLLINK statement. Resubmit the program and click on the third bar. If you specified a correct http page address, the contents of the indicated Web page should be displayed when you click on the bottom bar.)

19.8 GOING DEEPER: SGPLOTS

Beginning with SAS Version 9.2, a new way of creating high-quality graphs was introduced. These are called Statistical Graphics Procedures. These include `SGPLOT`, `SGSCATTER`, and `SGPANEL`. We'll illustrate the use of `SGPLOT` in this section. One advantage of these "SG" plots is that they can easily be created in standard graphical file types such as `JPG` (`JPEG`) or `TIF`. The basic syntax for `SGPLOT` is

```
PROC SGPLOT <DATA=datasetname>;
plotstatements/options;
```

At least one plot type statement is required. For example,

```
PROC SGPLOT;
  HISTOGRAM AGE;
  DENSITY AGE/TYPE=NORMAL;
```

A simplified list of these statements appears in Table 19.13:

TABLE 19.13 Common PLOT Statements for PROC SGPLOT

Plot Statement	Explanation
HBAR vname or VBAR vname	Specifies horizontal or vertical bar chart. For example, HBAR AGE; Options /GROUP= specifies group and GROUPDISPLAY=CLUSTER displays groups as clusters. (Default is stacked.)
HISTOGRAM vname	Specifies histogram. For example, HISTOGRAM AGE;
SCATTERPLOT	Specifies scatterplot. For example, SCATTER X=AGE Y=SBP;
DENSITY vname	Displays density plot. For example, DENSITY AGE </TYPE=KERNEL or NORMAL) NORMAL is default
KEYLEGEND/options;	LOCATION= specifies the location of legend (INSIDE or OUTSIDE) POSITION specifies the position of legend (topleft, topright, bottomleft, bottomright)
XAXIS options; and YAXIS options;	Common options: LABEL='Label name' MAX=, MIN=, VALUES=(Min to Max by interval) OFFSETMIN= OFFSETMAX= (from 0 to 1)
REG X=xvar Y=yvar/ options	Specify regression line for scatterplot, where options include ALPHA=, CLI and CLM and GROUP=
FILLATTRS=	Can be used to specify a fill color for elements in a plot. For example FILLATTRS=(COLORS=GREEN); See additional color choices in Section A.2.
TRANSPARENCY=pct	Specifies transparency for bubbles in a BUBBLE plot where pct ranges from 0 to 1. For example TRANSPARENCY=0.50.

CHAPTER 19: CREATING CUSTOM GRAPHS

Creating an SGPLOT HISTOGRAM: One of the main differences in using `SGPLOT` (as opposed to `GPLOT`) is that you can more easily overlay various components in the same plot. For example, to create a histogram, you could use the code

```
PROC SGPLOT;
   HISTOGRAM AGE;
RUN;
```

To add a density plot (default is a normal plot), you simply add another plot statement:

```
PROC SGPLOT;
   HISTOGRAM AGE;
   DENSITY AGE;
RUN;
```

HANDS-ON EXAMPLE 19.12

1. Open the program file `SGPLOT1.SAS`.

   ```
   PROC SGPLOT DATA=MYSASLIB.WOUND;
   HISTOGRAM SBP;
   TITLE "SGPLOT Histogram";
   RUN;
   ```

   Run this code and observe the output shown in Figure 19.14.

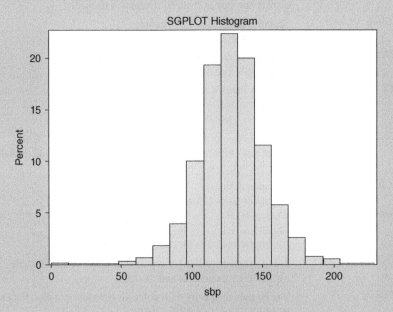

Figure 19.14 Histogram created using SGPLOTS

2. Add the following code after the `HISTOGRAM` line:

   ```
   DENSITY SBP/TYPE=KERNEL;
   ```

Rerun the code and observe that a normal curve is overlaid on the plot.
3. Add the following code after the DENSITY line:

```
DENSITY SBP/TYPE=NORMAL;
```

Rerun this code and observe that it contains all three plots: the histogram, the normal curve, and a kernel density curve as shown in Figure 19.15.

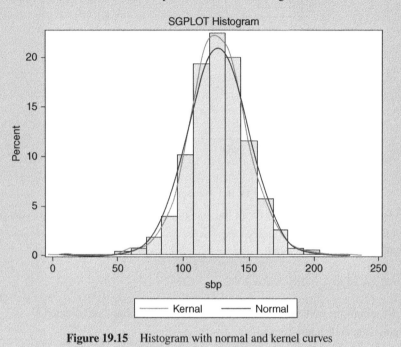

Figure 19.15 Histogram with normal and kernel curves

Creating an SGPLOT BAR CHART: You can also use PROC SGPLOT to create a Bar Chart. The code for a horizontal bar chart is

```
PROC SGPLOT;
   HBAR AGE;
RUN;
```

Hands-On Example 19.13 illustrates some of the options for bar charts.

HANDS-ON EXAMPLE 19.13

1. Open the program file SGPLOT2.SAS. Submit this code and observe the output shown in Figure 19.16.

```
PROC SGPLOT DATA=MYSASLIB.WOUND;
  HBAR RACE_CAT;
  TITLE "SGPLOT Bar Chart";
RUN;
```

(Continued)

(*Continued*)

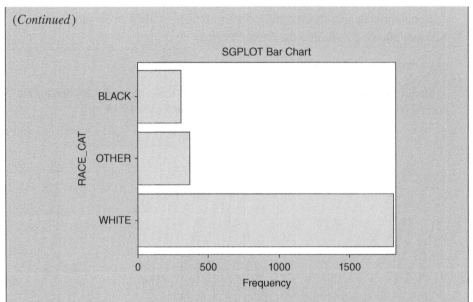

Figure 19.16 Horizontal bar chart

2. Change HBAR to VBAR and rerun the analysis to see a vertical bar chart version of this plot.
3. Change the bar code to read

```
VBAR RACE_CAT/ GROUP=GENDER;
```

Resubmit the code and observe the stacked vertical bar chart created.

4. Change the code again to read

```
VBAR RACE_CAT/ GROUP=GENDER GROUPDISPLAY=CLUSTER;
```

Submit this code and observe the vertical bar chart with bars clustered.

5. To illustrate manipulating the legend and axis labels add this code:

```
KEYLEGEND / LOCATION=INSIDE POSITION=TOPLEFT;
XAXIS LABEL="Race";
```

Resubmit this code and observe the results.

6. To produce a grayscale version of this plot, place the code.

```
ODS PDF STYLE=GRAYSCALEPRINTER;
```

before PROC SGPLOT, and

```
ODS PDF CLOSE;
```

after the RUN; statement, run the revised code. Observe the results as shown in Figure 19.17.

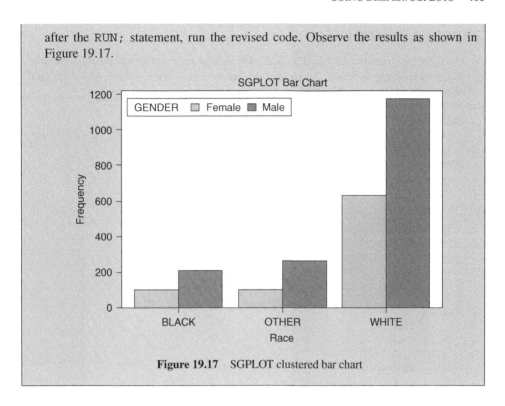

Figure 19.17 SGPLOT clustered bar chart

Creating a Scatterplot Using SGPLOT: You can also use SGPLOT to create a scatterplot. The code for a scatterplot is:

```
PROC SGPLOT DATA =sasfilename;
     SCATTER X=varname1 Y=varname2;
RUN;
```

An option that is helpful for the SCATTER plot is

```
MARKERATTRS=(specifications).
```

This option allows you to specify the type of symbol displayed at each point, along with size and color. For example,

```
MARKERATTRS=(SYMBOL=TRIANGLEFILLED SIZE=10PX COLOR=RED);
```

specifies a red-filled triangle of size 10 pixels. Other common ODS marker symbols (as these are called) include triangle (not filled), circle (filled or not), square (filled or not), and star (filled or not). Color options are listed in Appendix A. Refer to SAS documentation for other ODS marker symbols.

Hands-On Example 19.14 illustrates some of the options for scatterplots.

HANDS-ON EXAMPLE 19.14

1. Open the program file SGPLOT3.SAS. These data are pretest- and posttest for two different teaching methods. To look at the entire data at once, submit this code and observe the output shown in Figure 19.18.

```
PROC SGPLOT DATA =ANCOVA;
   SCATTER X=PRETEST Y=POSTTEST;
   TITLE "SGPLOT Scatter Plot";
RUN;
```

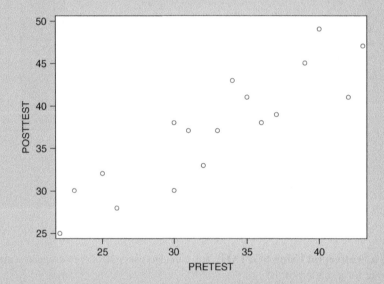

Figure 19.18 SGPLOT scatterplot

2. Change the SCATTER statement by adding a GROUP option:

```
SCATTER X=PRETEST Y=POSTTEST/GROUP=METHOD;
```

   Resubmit the code and observe that the groups are displayed in different colors.

3. Change the SCATTER statement by adding Marker Attributes (MARKERATTRS) using the following code:

```
SCATTER X=PRETEST Y=POSTTEST/GROUP=METHOD
        MARKERATTRS=(SYMBOL=TRIANGLEFILLED SIZE=10PX);
```

   (Other symbol types for SGPLOT are listed in Section A.6.)
   Note that no color option is used here because we want each group to be represented by a unique color. Resubmit this code and note how the "dots" are changed to triangles.

4. Add these lines immediately before the RUN; statement to control the legend and labels:

```
KEYLEGEND / LOCATION=INSIDE POSITION=TOPLEFT;
XAXIS LABEL="Pre Test Results";
YAXIS LABEL="Post Test Results";
```

Submit this code and observe the results.

5. Finally, add this code (just before the RUN; statement) to display a regression line for each group:

   REG X=PRETEST Y=POSTTEST/ GROUP=METHOD;

   Submit the code and observe the results shown in Figure 19.19. The three lines represent three linear regression lines, each fit through the data for one of the three groups.
6. Finally, change the line style in the REG statement by adding

   LINEATTRS=(PATTERN=LONGDASH)

   before the semicolon. See Appendix A for more options for line patterns.

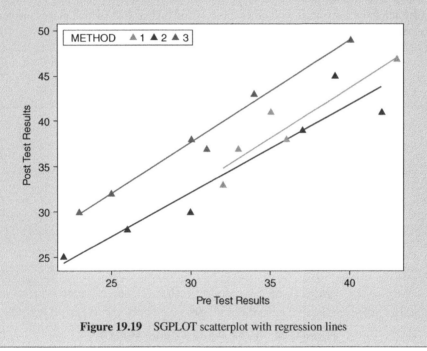

Figure 19.19 SGPLOT scatterplot with regression lines

Using SGPLOT to create a Bubble Plot: A bubble plot is similar to a scatterplot, but in which a third variable determines the size of the point (bubble). For example,

```
PROC SGPLOT DATA = MYDATA;
    BUBBLE X = AGE Y = SBP SIZE = WEIGHT;
RUN;
```

Hands-On Example 19.15 makes use of an option TRANSPARENCY=*value*, which is also available when creating bar graphs. Values range from 0 (opaque) to 1 (clear). The following example illustrates some of the options for SGPLOT bubble plots.

HANDS-ON EXAMPLE 19.15

1. Open the program file SGPLOT4.SAS. Observe the BUBBLE statement and the option following the slash. Run this program and observe the output shown in Figure 19.20.

```
PROC SGPLOT DATA =MYSASLIB.CORONARY;
    TITLE 'BUBBLE PLOT USING CORONARY DATA';
    TITLE2 'Bubble is BMI';
    BUBBLE X = AGE Y = SBP SIZE = BMI
        / TRANSPARENCY = 0.5;
    YAXIS GRID;
RUN;
```

Information you might learn from this graph other than that SBP tends to increase with AGE is that the bubbles get bigger toward the right and top meaning that higher BMI might be associated with older subjects and higher SBP (which makes medical sense).

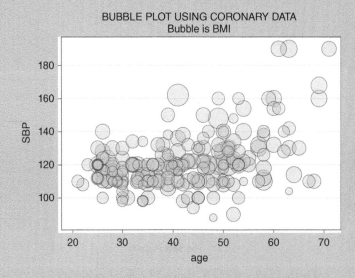

Figure 19.20 Bubble plot created using SGPLOT

2. Change the transparency to 0.9 and run the code. Change it to 0.1 and run the code. Note how transparency changes the appearance of the plot.
3. Change the transparency back to 0.5. Add the option GROUP=INSURANCE after the TRANSPARENCY statement (and before the semicolon). Rerun and note the different colors by group. Does this provide any additional information about how people with insurance might differ from those who don't have it?
4. Change the option in the BUBBLE statement to /TRANSPARENCY=0.5 FILLATTRS=(COLORS=GREEN) to change the color of the bubbles to green. Other possible colors are listed in Section A.2.

19.9 OTHER WAYS TO CUSTOMIZE PLOTS

There are at least two other techniques you can use to customize plots in SAS: Annotation Datasets and the Graph Template Language (GTL).

The Annotation technique can be used to customize plots that are created with GPLOT, GCHART, and other similar procedures. In this technique, you create an Annotation Dataset containing graphic commands that you then apply to an existing plot (or create an entirely custom plot).

The GTL is the SAS sublanguage used to specify the details of how an ODS graph is created. Each ODS graph has a default template, where these GTL commands are located. You can edit this template language to customize the graphs to meet specific needs.

Both of these techniques are beyond the scope of this book, but if you want to go deeper into the capabilities of SAS graphs, you can find more information about these techniques in SAS documentation.

19.10 SUMMARY

This chapter described a small portion of the many graphical options available in SAS. However, the discussion of graphs in this chapter (and elsewhere in the book) covered many of the common graphs used in data analysis.

EXERCISES

19.1 **Graph a comparison of means.**

Using the data in the file AANOVA1.SAS and the example illustrated by AGRAPH8.SAS, create a graph that displays means as bars and standard error markers.

19.2 **Display comparative box plots.**

Using AGRAPH9.SAS as a guide, create a plot of comparative box plots by CYLINDERS (where CYLINDERS > 1) for CITYMPG in the CARS dataset.

```
GOPTIONS RESET=ALL;
PROC SORT DATA="C:\SASDATA\CARS" OUT=MPGFILE;
    BY _____ ;
PROC _____ DATA=MPGFILE;
    PLOT _____*CYLINDERS;
    WHERE CYLINDERS GT _____ ;
RUN;
QUIT;
```

19.3 **Use an interactive ODS graph.**

a. Open AGRAPH11.SAS. Change the type of chart from HBAR to PIE and run the program. Test to verify that the HTML links work.

b. Change the type of chart to VBAR and run the program. Test to verify that the HTML links work.

c. Change the type of chart to a VBAR3D and run the program. Test to verify that the HTML links work.

d. Similarly, try the graph types BLOCK, HBAR3D, PIE3D, and STAR.

19.4 **Use SGPLOT to create a scatterplot.**

Using the skeleton SGPLOT code below, display a scatterplot of X=CITYMPG and Y=HWYMPG and fit a regression line through the scatter with an individual confidence limit (CLI).

```
PROC SGPLOT DATA ="C:\SASDATA\CARS";
SCATTER X=CITYMPG Y=_____;
XAXIS LABEL="City _____";
YAXIS _____;
REG X=_____ Y=HWYMPG / CLI;
TITLE "SGPLOT Scatter Plot for Miles Per Gallon";
RUN;
```

REFERENCE

McGill, R., Tukey, J.W., and Larsen, W.A. (1978). Variations of box plots. *The American Statistician* 32: 12–16.

20

CREATING CUSTOM REPORTS

LEARNING OBJECTIVES

- To be able to create custom reports using PROC TABULATE
- To be able to create custom reports using PROC REPORT
- To be able to create a custom report using the DATA step

Throughout the book, procedures such as `PROC PRINT` and `PROC FREQ` have been used to produce simple summaries of data. However, these procedures are limited. This chapter introduces two advanced reporting procedures that are designed to help you create professional reports.

20.1 USING PROC TABULATE

`PROC TABULATE` is used to make summary data tables. Unlike `PROC FREQ`, it can handle multiple levels in rows and columns. You can use `PROC TABULATE` to create professional reports and publication-ready tables. An abbreviated syntax for `PROC TABULATE` is as follows:

```
PROC TABULATE <options>;
CLASS variables < / options>;
```

SAS® Essentials: Mastering SAS for Data Analytics, Third Edition. Alan C. Elliott and Wayne A. Woodward.
© 2023 John Wiley & Sons, Inc. Published 2023 by John Wiley & Sons, Inc.
Companion website: https://www.alanelliott.com/SASED3/

```
VAR variables < / options>;
TABLE <page>,
 <row>,
 column
 < / options>;
... other statements ...;
RUN;
```

Table 20.1 lists commonly used options for PROC TABULATE. Moreover, Table 20.2 lists useful statements for PROC TABULATE.

The basic building blocks for PROC TABULATE are the TABLE, CLASS, and VAR statements:

> **The TABLE Statement**: This statement is the most important component in PROC TABULATE. It is used to define what information will be displayed in the table. For example, this simple PROC TABULATE statement defines a class variable (GP) and a table.
>
> ```
> PROC TABULATE DATA=sasdataset;
>   CLASS GP;
>   TABLE GP;
> RUN;
> ```
>
> **The CLASS Statement:** This statement identifies a categorical variable to be used in a table. You must identify the variables you are going to use before you use them in a TABLE statement.

TABLE 20.1 Common Options for PROC TABULATE

Option	Explanation
DATA=datasetname	Specifies which dataset to use
OUT=datasetname	Specifies an output dataset
STYLE=styletype	Specifies an ODS style format for the table
FORMAT=	Specifies a default overall format for each cell, Best 12.2 is the default
ORDER=option	Specifies the order that categories are displayed. Options are UNFORMATTED, DATA, FORMATTED, and FREQ
MISSING	Specifies that missing values are to be treated as a category

TABLE 20.2 Common Statements in PROC TABULATE

Statements	Explanation
VAR variables;	Specifies which variables you will use to calculate statistics. Must be numeric
CLASS variables;	Specifies categorical variables
TABLE specification;	Identifies the dimensions and other components of the table to create
BY, FORMAT, LABEL, WHERE	These statements are common to most procedures and may be used here

The TABLE statement above tells the procedure to create a table using the GP variable. As another example,

```
CLASS VNAME1 VNAME2;
TABLE VNAME1, VNAME2;
```

Commas separate the table dimensions. In the TABLE statement, commas separate rows from columns. Therefore, the code above creates a table, where the categories of VNAME1 define rows, and those of VNAME2 define columns. An asterisk (*) crosses two variables, where categories of the second variable are expanded within the first one. Thus,

```
CLASS VNAME1 VNAME2 VNAME3;
TABLE VNAME1, VNAME2*VNAME3;
```

creates a table where the categories of VAR1 define rows, VAR2 defines columns, and VAR3 defines columns within the categories of VAR2.

The VAR Statement: The VAR statement identifies quantitative (numeric) data. This variable may be used to indicate summary statistics in table cells, such as means, sums, or medians. For example,

```
CLASS VNAME1 VNAME2;
VAR OBS;
TABLE VNAME1, VNAME2*MEAN*OBS;
```

The MEAN*OBS addition to this code tells the procedure to place the mean of OBS, broken down by variables VNAME1 and VNAME2, in each cell. These building blocks for PROC TABULATE are illustrated in a series of upcoming Hands-On Examples.

Before going through the upcoming examples, it is helpful to know something about the dataset that will be used, MYSASLIB.REGION. To see the first few records of this SAS® dataset, run the following PROC PRINT code.

```
PROC PRINT DATA=MYSASLIB.REGION (FIRSTOBS=1 OBS=20); RUN;
```

The results are given in Table 20.3. Note that there are several categorical variables and one quantitative variable (SALES_K).

TABLE 20.3 The Regions Dataset

Obs	REGION	AREA	SITE	TYPE	MONTH	SALES_K
1	1	N	DOWNSTAIRS	KIOSK	NOV	33
2	1	N	DOWNSTAIRS	STORE	NOV	57
3	1	N	DOWNSTAIRS	STORE	NOV	41
4	1	S	DOWNSTAIRS	STORE	NOV	66
5	1	S	UPSTAIRS	STORE	NOV	72
6	1	S	DOWNSTAIRS	KIOSK	NOV	33
7	1	S	UPSTAIRS	KIOSK	NOV	34
8	1	E	DOWNSTAIRS	KIOSK	NOV	31
9	1	E	DOWNSTAIRS	STORE	NOV	49
10	1	W	UPSTAIRS	KIOSK	NOV	40

HANDS-ON EXAMPLE 20.1

1. Open the program file `ATAB1.SAS`. Observe that the variable `TYPE` is identified as a categorical variable (`CLASS`), and that a table is constructed using that variable. Because `TYPE` has two categories, we expect the table to consist of two cells, each with a count (`N`) that represents the number of items for that group. Submit the code and observe the output given in Table 20.4 where N represents the number of subjects

```
PROC TABULATE DATA=MYSASLIB.REGION;
   CLASS TYPE;
   TABLE TYPE;
RUN;
```

TABLE 20.4 Simple PROC TABULATE Table

TYPE	
KIOSK	STORE
N	N
27	45

2. Add another variable by modifying the `CLASS` and `TABLE` statements:

```
CLASS TYPE AREA;
TABLE TYPE*AREA;
```

where `AREA` values are the compass points N, S, E, and W. Submit this code and observe the results given in Table 20.5. Note that the first variable, `TYPE`, appears as the top category for the table followed by the next division specified by the `AREA` variable.

TABLE 20.5 PROC TABULATE Table Using Two Variables

TYPE							
KIOSK				STORE			
AREA				AREA			
E	N	S	W	E	N	S	W
N	N	N	N	N	N	N	N
9	5	8	5	9	17	8	11

3. Switch the variables in the `TABLE` statement so the statement reads

```
CLASS TYPE AREA;
TABLE AREA*TYPE;
```

Submit this code and observe the results. Note that this time AREA is the top (first) variable with TYPE below.

4. Add a third variable to the class statement (SITE) and to the TABLE statement, so they read

```
CLASS TYPE AREA SITE;
TABLE SITE*AREA*TYPE;
```

where site takes on the values "Upstairs" and "Downstairs."

Note that another layer is added to the table (and the table is very busy and hard to read).

5. Change the TABLE statement to the following. Note that the difference is in the comma between SITE and AREA (instead of an asterisk (*)).

```
CLASS TYPE AREA SITE;
TABLE SITE,AREA*TYPE;
```

Submit this code and observe the results. Note what the comma did. A comma separates row dimensions from column dimensions. In this case, SITE defines a row and AREA*TYPE defines columns. Submit this code and observe the results in Table 20.6.

TABLE 20.6 A PROC TABULATE Table with Rows and Columns

	AREA							
	E		N		S		W	
	TYPE		TYPE		TYPE		TYPE	
	KIOSK	STORE	KIOSK	STORE	KIOSK	STORE	KIOSK	STORE
	N	N	N	N	N	N	N	N
SITE								
DOWNSTAIRS	5	5	2	9	3	2	1	3
UPSTAIRS	4	4	3	8	5	6	4	8

Adding Labels and Formats to Variables: A label can be added to any variable by using an equal sign (=) followed by a quoted string. For example,

```
TABLE SITE="Store Site", AREA="Area"*TYPE="Type Store";
```

If you also want to format variables, you can create a format definition for N, S, E, and W using

```
PROC FORMAT;
VALUE $FMTCOMPASS "N"="1.North" "S"="2.South" "E"="3.East"
"W"="4.West";
RUN;
```

Note that you have indicated the order in which you want the compass points to appear. Using the format plus an ORDER= statement will make the category labels appear in the preferred order as illustrated in Hands-On Example 20.2:

HANDS-ON EXAMPLE 20.2

1. Open the program file `ATAB2.SAS`. Observe how the format `$FMTCOMPASS` along with the `ORDER=FORMATTED` statement cause the AREAs to appear in a preferred order. Labels are also assigned to each of the variables in the table. Submit this code and observe the results given in Table 20.7.

```
PROC FORMAT;
VALUE $FMTCOMPASS "N"="1.North" "S"="2.South" "E"="3.East"
"W"="4.West";
RUN;
PROC TABULATE DATA=MYSASLIB.REGION ORDER=FORMATTED;
  CLASS TYPE AREA SITE;
  TABLE SITE="Store Site", AREA="Area"*TYPE="Type Store";
  FORMAT AREA $FMTCOMPASS.;
RUN;
```

TABLE 20.7 A PROC TABULATE Table with Labels and Formats

	Area							
	1.North		2.South		3.East		4.West	
	Type Store		Type Store		Type Store		Type Store	
	KIOSK	STORE	KIOSK	STORE	KIOSK	STORE	KIOSK	STORE
	N	N	N	N	N	N	N	N
Store Site								
DOWNSTAIRS	2	9	3	2	5	5	1	3
UPSTAIRS	3	8	5	6	4	4	4	8

2. Take away the `ORDER=FORMATTED` statement and rerun the code. Note the order in which the AREA categories appear – in alphabetic order according to original values E N S W.

Reporting SUMS and MEANS: Currently, the table shows the number of stores for each breakdown of Area, Type Store, and Store Site. The dataset also includes sales figures (`SALES_K`) in thousands. To report the total sales, indicate the variable to sum using a VAR statement and cross that value with a category variable in the TABLE statement. For example,

```
VAR SALES_K;
TABLE SITE="Store Site", AREA="Area"*TYPE="Type
Store"*SALES_K;
```

SUM is the default statistic calculated for this method, but you can request other statistics. For example, to request the mean of SALES_K, change the statement to

```
VAR SALES_K;
TABLE SITE="Store Site", AREA="Area"*TYPE="Type Store"
*SALES_K*MEAN;
```

Statistics you can request are listed in Table 20.8.

TABLE 20.8 Statistics Available in PROC Tabulate

PCTSUM	NMISS	P5
COLPCTSUM	SUM	P10
MAX	PAGEPCTSUM	P99
ROWPCTN	PCTN	Q1 | P25
MEAN	VAR	QRANGE
ROWPCTSUM	MEDIAN | P50	P5
MIN	P1	PROBT
STDDEV / STD	Q3 | P75	T
N	P90	
STDERR	P95	

HANDS-ON EXAMPLE 20.3

1. Open the program file ATAB3.SAS. Note the new VAR SALES_K statement to define the numeric variable, and how it is crossed with the TYPE variable in the TABLE statement. Submit this code and observe the results given in Table 20.9. Note the value for Downstairs, North, Kiosk is 63, meaning that the sum of sales for that cell is $63,000.

```
PROC FORMAT;
VALUE $FMTCOMPASS "N"="1.North" "S"="2.South" "E"="3.East"
"W"="4.West";
RUN;
PROC TABULATE DATA=MYSASLIB.REGION ORDER=FORMATTED;
   CLASS TYPE AREA SITE;
   VAR SALES_K;
```

(Continued)

(Continued)

```
    TABLE SITE="Store Site",
        AREA="Area"*
        TYPE="Type Store"*
        SALES_K;
    FORMAT AREA $FMTCOMPASS.;
RUN;
```

TABLE 20.9 A PROC TABULATE Table with Sum

	Area							
	1.North		2.South		3.East		4.West	
	Type Store		Type Store		Type Store		Type Store	
	KIOSK	STORE	KIOSK	STORE	KIOSK	STORE	KIOSK	STORE
	SALES_K	SALES_K	SALES_K	SALES_K	SALES_K	SALES_K	SALES_K	SALES_K
	Sum	Sum	Sum	Sum	Sum	Sum	Sum	Sum
Store Site								
DOWNSTAIRS	63.00	495.00	110.00	127.00	139.00	256.00	32.00	211.00
UPSTAIRS	93.00	436.00	173.00	374.00	123.00	205.00	139.00	445.00

2. Provide a label for SALES by using SALES_K="Mean Sales $1000" in the TABLE Statement. Suppose you're interested in average sales per store. Change the TABLE statement to

```
TABLE SITE="Store Site",
    AREA="Area"*
    TYPE="Type Store"*
    MEAN="Mean Sales $1000"*SALES_K;
```

Resubmit the code and observe the results.

3. To add a format to a numeric value, use F=format. For example, MEAN="Sales $1000"*SALES_K *F=DOLLAR6.2. Change the TABLE statement to

```
TABLE SITE="Store Site",
    AREA="Area"*
    TYPE="Type Store"*
    MEAN="Mean Sales $1000"*SALES_K*F=DOLLAR6.2;
```

Resubmit the code. The results are shown in Table 20.10. The dollar values now include a dollar sign because of the format.

TABLE 20.10 A PROC TABULATE Table with Means

	Area							
	1. North		2. South		3. East		4. West	
	Type Store		Type Store		Type Store		Type Store	
	KIOSK	STORE	KIOSK	STORE	KIOSK	STORE	KIOSK	STORE
	Mean Sales $1000	Mean Sales $1000	Mean Sales $1000	Mean Sales $1000	Mean Sales $1000	Mean Sales $1000	Mean Sales $1000	Mean Sales $1000
	SALES_K	SALES_K	SALES_K	SALES_K	SALES_K	SALES_K	SALES_K	SALES_K
Store Site								
DOWNSTAIRS	$31.50	$55.00	$36.67	$63.50	$27.80	$51.20	$32.00	$70.33
UPSTAIRS	$31.00	$54.50	$34.60	$62.33	$30.75	$51.25	$34.75	$55.63

Dress up the Table: As seen in the previous example, you can dress up a value using a label, a statistic, and a format. For example,

```
*MEAN="Mean Sales $1000"*SALES_K*F=DOLLAR6.2;
```

Here is a breakdown of these components:

1. `*MEAN` – Report the Mean statistic
2. `="Mean Sales $1000"` – Apply a label
3. `*SALES_K` – Variable to use (must be in the VAR Statement)
4. `*F=DOLLAR6.2` – Apply a format

The table as created still looks busy and has unneeded information such as the actual names of the variables. For example, there are six lines of labels before the actual numbers in Table 20.10. To get rid of unwanted labels, change the labels in the TABLE statement to two quotation marks (" ") as in MEAN=" ". Add the same kind of label to SALES_K. The revised TABLE statement is

```
TABLE SITE="",
    AREA=""*
    TYPE=""*
    MEAN="Mean Sales $1000"*SALES_K=""*F=DOLLAR6.2;
```

476 CHAPTER 20: CREATING CUSTOM REPORTS

Moreover, we want to add new cells that reflect averages by SITE and by AREA. To add these cells, we use the ALL keyword. For example, we replace AREA with

    (AREA ALL)

Or if we don't want the label used for AREA, we'd use

    (AREA="" ALL)

Hands-On Example 20.4 will help clarify these comments.

HANDS-ON EXAMPLE 20.4

1. Open the program file ATAB4.SAS. The TABULATE portion of this code is as follows. Note that most variable labels have been replaced with " " and that a request for ALL means has been added to AREA. Submit this code and observe the results in Table 20.11.

```
PROC TABULATE DATA=MYSASLIB.REGION ORDER=FORMATTED;
CLASS TYPE AREA SITE;
VAR SALES_K;
TABLE SITE="",
   (AREA="" ALL)*TYPE=""
   *MEAN="Mean Sales $1000"*SALES_K=""*F=DOLLAR6.2;
  FORMAT AREA $FMTCOMPASS.;
RUN;
```

Note the "All" columns at the right of the table. These are the overall means by store type. Also the blank labels eliminated some of the labels and made the table less busy.

2. Flip the table by exchanging the SITE and AREA variables.

```
TABLE(AREA="" ALL),
   SITE=""*
   TYPE=""*
   MEAN="Sales $1000"*SALES_K*F=DOLLAR6.2;
```

Resubmit the code and note the changes. If you want to print the output in portrait orientation, this version would fit better.

3. Change SALES_K to a blank label with SALES_K="" and change the SITE variables to include means by group (SITE="" ALL).

```
TABLE(AREA="" ALL),
   (SITE="" ALL)*
   TYPE=""*
   MEAN="Sales $1000"*SALES_K=""*F=DOLLAR6.2;
```

Resubmit the code and note the differences.

TABLE 20.11 A PROC TABULATE with ALL

	1. North		2. South		3. East		4. West		All	
	KIOSK	STORE	KIOSK	STORE	KIOSK	STORE	KIOSK	STORE	KIOSK	STORE
	Mean Sales $1000	Mean Sales $1000	Mean Sales $1000	Mean Sales $1000	Mean Sales $1000	Mean Sales $1000	Mean Sales $1000	Mean Sales $1000	Mean Sales $1000	Mean Sales $1000
DOWNSTAIRS	$31.50	$55.00	$36.67	$63.50	$27.80	$51.20	$32.00	$70.33	$31.27	$57.32
UPSTAIRS	$31.00	$54.50	$34.60	$62.33	$30.75	$521.25	$34.75	$55.63	$33.00	$56.15

Enhancing Your Table: It is straightforward to add information to your table. For example, suppose you want to report more than one statistic, say N and MEAN. You can combine statistics by placing them in parentheses such as

```
(N MEAN)
```

Or to eliminate the label being put into the table, you could use

```
(N="" MEAN="")
```

Another enhancement is to put some information in the box that appears at the upper left of your table. You can do this by including a /BOX= "*label*" option. For example,

```
TABLE(AREA="" ALL="By Area"),(SITE="" ALL="By Site")*TYPE=""
       *(N="" MEAN="")*SALES_K=""*F=DOLLAR6.2
    / BOX="Fall Sales";
```

HANDS-ON EXAMPLE 20.5

1. Open the program file ATAB5.SAS. Submit this code and see that it contains all of the options discussed so far.

   ```
   PROC TABULATE DATA=MYSASLIB.REGION ORDER=FORMATTED;
      CLASS TYPE AREA SITE;
      VAR SALES_K;
      TABLE(AREA="" ALL),(SITE="" ALL)*TYPE=""
           *MEAN=""*SALES_K=""*F=DOLLAR6.2;
      FORMAT AREA $FMTCOMPASS.;
   RUN;
   ```

2. Add the N statistic to the table by changing the MEAN="" statement to (N="" MEAN=""). Note parentheses are important. Submit the code. Observe how both the count and mean are now reported.

3. Put a text in the upper-left box by adding the code

   ```
   / BOX="Fall Sales";
   ```

   Make sure the code is within the semicolon of the TABLE statement. With changes, it should be

   ```
   TABLE(AREA="" ALL="By Area"),(SITE="" ALL="By Site")*TYPE=""
          *(N="" MEAN="")*SALES_K=""*F=6.2
     / BOX="Fall Sales";
     FORMAT AREA $FMTCOMPASS.;
   ```

   Submit the code. Observe how the text appears in the box.

4. To save the table as a Word document using ODS, add the following lines. Before the TABULATE statement, add the line

```
ODS RTF STYLE=BANKER;
    Your PROC TABULATE;
RUN;
ODS RTF CLOSE;
```

Resubmit the code. The final table with these changes is given in Table 20.12 (with green and yellow background colors because of the BANKER style)

Note that if you submit the code that creates the RTF file again, you must first close the document in Word that you previously created.

TABLE 20.12   A PROC TABULATE with ALL

Fall Sales	DOWNSTAIRS				UPSTAIRS				By Site			
	KIOSK		STORE		KIOSK		STORE		KIOSK		STORE	
1. North	2.00	31.50	9.00	55.00	3.00	31.00	8.00	54.50	5.00	31.20	17.00	54.76
2. South	3.00	36.67	2.00	63.50	5.00	34.60	6.00	62.33	8.00	35.38	8.00	62.63
3. East	5.00	27.80	5.00	51.20	4.00	30.75	4.00	51.25	9.00	29.11	9.00	51.22
4. West	1.00	32.00	3.00	70.33	4.00	34.75	8.00	55.63	5.00	34.20	11.00	59.64
By Area	11.00	31.27	19.00	57.32	16.00	33.00	26.00	56.15	27.00	32.30	45.00	56.64

Saving Your Table in Excel: Once you create a table using PROC TABULATE, you may want to save it in an Excel format. As in the case of saving it to Word (RTF), you use an ODS procedure. The process is to open the Excel file using an ODS statement such as

```
ODS TAGSETS.EXCELXP BODY='C:SASDATA\TABLE.XLS';
    Your PROC TABULATE code;
    RUN;
ODS TAGSETS.EXCELXP CLOSE;
```

HANDS-ON EXAMPLE 20.6

(You must have Microsoft Excel installed on your computer to complete this example.)

1. Open the program file ATAB6.SAS. A portion of the code is shown here. Note the ODS statement before and after PROC TABULATE. Note the name of the file to be saved as the BODY= statement. If necessary, change the Windows patch to match an available folder on your computer. Submit the code.

```
ODS TAGSETS.EXCELXP
BODY='C:\SASDATA\TABLE.XLS';
```

(*Continued*)

(*Continued*)

```
PROC TABULATE DATA=MYSASLIB.REGION ORDER=FORMATTED;
  CLASS TYPE AREA SITE;
  VAR SALES_K;
TABLE(AREA="" ALL="By Area"),(SITE="" ALL="By Site")*TYPE=""
    *(N="" MEAN="")*SALES_K=""*F=6.2
  / BOX="Fall Sales";
  FORMAT AREA $FMTCOMPASS.;
RUN;
ODS TAGSETS.EXCELXP CLOSE;
```

This code produces a series of XML commands in the **Results Viewer** that begin with a line such as

```
<?xml version="1.0" encoding="windows-1252" ?>
```

This is the code (created by the SAS procedure) that copied the information to an Excel file. (You don't need to learn or understand this code.) To see the Excel file created by this program, begin Excel and open the file using the name you have given it. In this case, the resulting file (in Excel) is given in Table 20.13.

(Note: When you run this code, depending on your Excel installation, you may get a warning message with a Y/N option. If you choose Y (Yes) the table should appear correct.)

TABLE 20.13   A PROC TABULATE as Output to Excel

	A	B	C	D	E	F	G	H	I	J	K	L	M
1		DOWNSTAIRS				UPSTAIRS				By Site			
2	Fall Sales	KIOSK		STORE		KIOSK		STORE		KIOSK		STORE	
3	1. North	2	31.5	9	55	3	31	8	54.5	5	31.2	17	54.76
4	2. South	3	36.67	2	63.5	5	34.6	6	62.33	8	35.38	8	62.63
5	3. East	5	27.8	5	51.2	4	30.75	4	51.25	9	29.11	9	51.22
6	4. West	1	32	3	70.33	4	34.75	8	55.63	5	34.2	11	59.64
7	By Area	11	31.27	19	57.32	16	33	26	56.15	27	32.3	45	56.64

We have just scratched the surface regarding the features of PROC TABULATE. We recommend that you experiment with some of the other statements to see how they can be used to enhance your tabled report even more.

20.2   USING PROC REPORT

PROC REPORT is a procedure in SAS that allows you to create professional-quality data reports and summaries. It is widely used in companies and organizations where reports are produced on topics such as income, contributions, services rendered, sales, and so on. These reports are highly customizable using the many varied options within PROC REPORT. Although only a few of these options are illustrated here, our intent is that these examples will help you learn how to use the basics of PROC REPORT.

The syntax for PROC REPORT is a little different from most statistical procedures in SAS because it has a particular emphasis on defining the columns of the report. Here is a simplified syntax for PROC REPORT:

```
PROC REPORT DATA= sasdataset <options>;
   COLUMN variables and specifications;
   DEFINE column / attributes;
   COMPUTE column; compute statements; ENDCOMP;
RUN;
```

COLUMN: The COLUMN statement is used to identify variables of interest to be reported in columns and to specify headers and groups. Every column you want to display in the table must be indicated in the COLUMN statement. A simple report would be

```
PROC REPORT DATA=MYSASLIB.CARS NOFS;
COLUMN BRAND CITYMPG HWYMPG;
RUN;
```

which is about the same as a PROC PRINT. To enhance the report requires a DEFINE statement.

DEFINE: The DEFINE statement is used to specify the appearance of the columns. Its syntax is

```
DEFINE report-item / <option(s)>;
```

DEFINE statements allow you to specify attributes that determine how a variable in a column will be displayed. Options for the DEFINE statement follow a slash (/). These options control the appearance of the information in a column. They can be of several types including

- options that describe how to use a variable;
- options that control column headings;
- options that customize a report item;
- other options for a report item.

The options include DISPLAY, GROUP, ANALYSIS, COMPUTED, ORDER, and ACROSS. Table 20.14 includes explanations of each of these options.

TABLE 20.14   PROC REPORT DEFINE Statement Options

Name	Purpose
DISPLAY	Displays the indicated variable in the column. This is the default DEFINE SBP/DISPLAY;
ORDER	Sorts data and forms groups
GROUP	Group (consolidate) observations using the specified variable. For example, DEFINE BRAND/GROUP; Displays records grouped by BRAND
ANALYSIS	Defines a variable used to calculate a statistic within the cell of a report
COMPUTE and ENDCOMP	Begins and ends a code segment that calculates a data value
ACROSS	Creates groups across the page rather than down the page

TABLE 20.15 PROC REPORT DEFINE Statement Option Attributes

Attributes	Example
FORMAT = format	Example FORMAT=6.1
'LABEL'	Example 'Model Type' Or 'Model/Type' to split title
STATISTICNAME	Any statistic that is available in PROC MEANS. For example, N, MEAN, SUM, or STD
ORDER=ordertype	(DATA, FORMATTED, FREQ, or INTERNAL) same as in PROC FREQ. Use DESCENDING option to reverse the normal ascending order

Some common attributes you can include with the DEFINE options are listed in Table 20.15.

These appear after the slash and statement option named in a DEFINE statement. For example,

```
DEFINE var/DISPLAY statement options;
```

or a more specific example:

```
DEFINE AGE/DISPLAY 'A Label' FORMAT=6.1;
```

Just as the split character / in a label indicates to go to the next line, so the label 'Model/Type' is displayed on two lines. Note in the LABEL example that the default split character is a slash (/). However, it can be defined as a different character in the PROC REPORT statement. For example, SPLIT='*' makes the asterisk the split character. The following Hands-On Examples illustrate several uses of the DEFINE option.

DISPLAY Option: The DEFINE var/DISPLAY option in the DEFINE statement allows you to specify attributes for that variable including format and labels. This is illustrated in the following example.

HANDS-ON EXAMPLE 20.7

1. Open the program file AREPORT1.SAS.

```
PROC REPORT DATA=MYSASLIB.CARS NOFS;
   COLUMN BRAND MODEL CITYMPG HWYMPG;
   DEFINE CITYMPG/DISPLAY FORMAT=6.1 'CITY/MPG';
RUN;
```

In this instance, CITYMPG is displayed using a 6.1 format and with a column label CITY MGP on two lines. (The NOFS option turns off an older option that is rarely used nowadays. We do not discuss it here.) Submit this code and observe the output given (partially) in Table 20.16. Note how the CITYMPG column label is split, and the numbers are in a 6.1 format.

TABLE 20.16 PROC REPORT DISPLAY Option Example

Brand	Model	CITYMPG	HwyMPG
TOYOTA	Prius	60.0	51
HONDA	Civic Hybrid	48.0	47
HONDA	Civic Hybrid	47.0	48
HONDA	Civic Hybrid	46.0	51
HONDA	Civic Hybrid	45.0	51

2. Add a `DISPLAY` statement that creates a column for `HWYMPG` using the 6.1 format and with an appropriate label. Resubmit and observe changes.
3. Add another variable, `ENGINESIZE`, to the `COLUMN` statement and create a similar `DISPLAY` statement for it. Resubmit and observe the changes.

ORDER Option: The `DEFINE var/ORDER` option defines a column variable used to sort the data. You can control the `ORDER` by adding `DESCENDING` to reverse the order and use an additional `ORDER=` to specify `FORMATTED` (default), `FREQ`, `INTERNAL`, or `DATA` order. Hands-On Example 20.8 illustrates the use of the `ORDER` option.

HANDS-ON EXAMPLE 20.8

1. Open the program file `AREPORT2.SAS`.

   ```
   TITLE "ORDER OPTION EXAMPLE";
   PROC REPORT DATA=MYSASLIB.CARS NOFS;
     COLUMN BRAND CITYMPG HWYMPG;
     DEFINE BRAND/ORDER 'BRAND';
     DEFINE CITYMPG/DISPLAY FORMAT=6.1 'CITY MPG';
     DEFINE HWYMPG/DISPLAY FORMAT=6.1 'HIGHWAY MPG';
   RUN;
   ```

   In this example, the `BRAND/ORDER` option in the `DEFINE` statement causes the report to be ordered by the car brand. The other `DEFINE` statements are similar to the previous example. Submit this code and observe the output (partially) given in Table 20.17.

2. Change the order direction by putting `DESCENDING` after `BRAND/ORDER`. Rerun and verify that now `VOLVO` appears as the first brand in the report.

3. Replace the `DESCENDING` with `ORDER=DATA` so that it reads

   `DEFINE BRAND/ORDER ORDER=DATA 'Brand';`

   which orders the brands in the order they appear in the dataset. Rerun and verify that now `TOYOTA` (which is the first brand in the dataset) appears as the first brand in the report.

(*Continued*)

(*Continued*)

TABLE 20.17 PROC REPORT Using ORDER Option

Brand	City MPG	Highway MPG
ACURA	27.0	34.0
	25.0	34.0
	23.0	31.0
	22.0	31.0
	21.0	30.0
	20.0	29.0
	20.0	29.0
	18.0	26.0
	17.0	23.0
	17.0	24.0
	17.0	24.0
AUDI	24.0	32.0
	23.0	29.0
	23.0	34.0

4. Add the variable CYLINDERS after BRAND in the COLUMN statement so that it reads

   COLUMN BRAND CYLINDERS CITYMPG HWYMPG;

   After the BRAND/ORDER statement add a new statement

   DEFINE CYLINDERS /ORDER 'Cylinders';

   to specify that you want to sort by the number of cylinders within BRAND. Resubmit the code to verify that your changes produced the desired results. The output should show Toyota's four-cylinder cars first, then six-cylinder cars, then Honda, and so on. (The CYLINDERS column only lists the new number of cylinders the first time, and the column for the remaining cars with that cylinder count is blank until the next new cylinder count appears.)

5. Remove the =DATA option from the ORDER (for BRAND) statement, rerun, and now the order for brand starts with ACURA.

GROUP Option: The DEFINE *var*/GROUP option causes the table to be grouped by the specified variable. Here is an example using the GROUP option. This DEFINE GROUP option is illustrated in Hands-On Example 20.9.

HANDS-ON EXAMPLE 20.9

1. Open the program file AREPORT3.SAS.

   ```
   PROC REPORT DATA=MYSASLIB.CARS NOFS;
     COLUMN BRAND (CITYMPG HWYMPG), MEAN;
   ```

```
   DEFINE BRAND/GROUP;
   DEFINE CITYMPG/DISPLAY FORMAT=6.1 'CITY MPG';
RUN;
```

The GROUP option causes the values to be grouped by BRAND. Note that the MEAN statistic is applied to both variables listed in the parentheses. And the 6.1 format is applied to both CITYMPG and HWYMPG because they are grouped within the parentheses. Submit this code and observe the output given (partially) in Table 20.18.

TABLE 20.18   PROC REPORT by GROUP

	CITY MPG	HwyMPG
Brand	MEAN	MEAN
ACURA	20.6	28.6
AUDI	19.1	26.7
BMW	18.2	26.2
BUICK	18.0	25.1
CADILLAC	15.7	22.3
CHEVROLET	17.8	23.7
CHRYSLER	19.5	26.5

Observe that by using the GROUP option, all of the ACURA cars are combined and reported as the mean miles per gallon. If the MEAN option had not been used, the default statistic is SUM.

2. Redo the COLUMN statement to read

```
COLUMN BRAND CITYMPG,MEAN HWYMPG,MEAN;
```

Resubmit the code and note that the mean for HWYMPG is no longer using the 6.1 format.

3. Write a new DEFINE (with DISPLAY) statement that puts HWYMPG into a 6.1 format, and define an appropriate label for the column.

```
DEFINE HWYMPG/DISPLAY FORMAT=6.1 'Highway MPG';
```

Resubmit the code to verify that your changes produced the expected results.

4. Change BRAND to CYLINDERS in both the COLUMN and DEFINE statements. Rerun and verify that the report is now grouped by the number of CYLINDERS. (−1 means a rotary engine).

5. Add the variable N in the COLUMN statement after CYLINDERS and add the option ORDER=FREQ to the DEFINE statement after GROUP.

```
COLUMN CYLINDERS N CITYMPG,MEAN HWYMPG,MEAN;
DEFINE CYLINDERS/GROUP ORDER=FREQ;
```

Rerun the code. The order of the groups should now go from smallest N per group to largest.

ANALYSIS Option: The DEFINE var/ANALYSIS option allows you to display a statistic for a specified column. This is illustrated in Hands-On Example 20.10.

HANDS-ON EXAMPLE 20.10

1. Open the program file AREPORT4.SAS.

    ```
    PROC REPORT DATA=MYSASLIB.CARS NOFS;
      COLUMN BRAND CITYMPG HWYMPG;
      DEFINE BRAND/GROUP;
      DEFINE CITYMPG/ANALYSIS MEAN FORMAT=6.1 'CITY/MPG';
    RUN;
    ```

    Instead of defining MEAN in the COLUMN statement, the ANALYSIS option allows you to indicate MEAN for the CITYMPG variable in a DEFINE statement. (The default is SUM.) (You can leave off the ANALYSIS option in the DEFINE statement and it will still work.) Submit this code and observe the output given (partially) in Table 20.19. Note that the mean is reported for CITYMPG, but the SUM (3rd column) is reported for HWYMPG SUM.

    TABLE 20.19 PROC REPORT Using ANALYSIS Option

Brand	CITY MPG	HwyMPG
ACURA	20.6	315
AUDI	19.1	1094
BMW	18.2	1517
BUICK	18.0	401
CADILLAC	15.7	514
CHEVROLET	17.8	2252
CHRYSLER	19.5	981

2. Create a DEFINE statement (with ANALYSIS) that causes the mean HWYMPG to be displayed with a 6.1 format and with an appropriate label. Resubmit the code to verify that your changes produced the expected results.
3. Add the ENGINESIZE variable as the last variable in the COLUMN statement and write a DEFINE statement (with ANALYSIS) that displays the mean engine size with a 6.1 format and with an appropriate label.

    ```
    DEFINE ENGINESIZE/ANALYSIS FORMAT=6.1 'Engine Size';
    ```

    Resubmit the code to verify that your changes produced the expected results.

COMPUTE Option: The COMPUTE and ENDCOMP statements begin and end a segment of SAS code that allows you to calculate values of new variables. Hands-On Example 20.11 illustrates the COMPUTE option.

HANDS-ON EXAMPLE 20.11

1. Open the program file `AREPORT5.SAS`.

```
TITLE "ORDER OPTION EXAMPLE";
PROC REPORT DATA=MYSASLIB.CARS NOFS SPLIT="~";
  COLUMN BRAND CITYMPG, MEAN HWYMPG, MEAN RATIO_MPG;
  DEFINE BRAND/ORDER;
  DEFINE BRAND/GROUP;
  DEFINE CITYMPG/DISPLAY FORMAT=6.1 'CITY MPG';
  DEFINE HWYMPG/DISPLAY FORMAT=6.1 'HIGHWAY MPG';
  DEFINE RATIO_MPG/COMPUTED FORMAT=6.2 'Ratio~CITY/HWY';
  COMPUTE RATIO_MPG;
    RATIO_MPG=_C2_/_C3_;
  ENDCOMP;
RUN;
```

Note the `COMPUTE` statement, the calculation that follows, and the ending `ENDCOMP` statement. In order to use the `COMPUTE` statement to create a new variable, you also need to

- specify the new variable name in the `COLUMN` statement,
- include a `DEFINE` statement with the `/COMPUTED` option to define the new variable, and
- put the actual computation between `COMPUTE` and `ENDCOMP` statements.

The way you calculate variables between `COMPUTE` and `ENDCOMP` is the same as you would use in a `DATA` step. To name variables used for computing, you must refer to the variables by which column they represent as `_C2_` (`CITYMPG`) for the column 2 variable and `_C3_` (`HWYMPG`) for the column 3 variable. The output for this example is given in Table 20.20.

Also note in this example that a `SPLIT` character is defined in the `PROC REPORT` statement (`SPLIT="~"`), so the label for the ratio can be defined as `'Ratio~CITY/HWY'`, which is split with `RATIO` on one line and `CITY/HWY` on the other line of the label in the output. Because the default split character is a slash, a new one had to be defined in order for this label to work correctly.

TABLE 20.20 PROC REPORT Using COMPUTE Option

	City MPG	Highway MPG	
Brand	MEAN	MEAN	Ratio CITY/HWY
ACURA	20.6	28.6	0.72
AUDI	19.1	26.7	0.71
BMW	18.2	26.2	0.70
BUICK	18.0	25.1	0.72
CADILLAC	15.7	22.3	0.70
CHEVROLET	17.8	23.7	0.75
CHRYSLER	19.5	26.5	0.74

(Continued)

(*Continued*)

2. Create a new column named CITY ENGINE RATIO based on the calculation CITYMPG/ENGINESIZE. Resubmit your code to verify that your new column appears correctly. Your revised code should be similar to the following. Additions are in bold italic. Why is _C4_ used to represent ENGINESIZE?

```
TITLE "PROC REPORT COMPUTE Example";
PROC REPORT DATA=MYSASLIB.CARS NOFS SPLIT="~";
  COLUMN BRAND CITYMPG,MEAN HWYMPG, MEAN ENGINESIZE, MEAN
RATIO_MPG CER;
  DEFINE BRAND/ORDER;
  DEFINE BRAND/GROUP;
  DEFINE CITYMPG/DISPLAY FORMAT=6.1 'City MPG';
  DEFINE HWYMPG/DISPLAY FORMAT=6.1 'Highway MPG';
  DEFINE ENGINESIZE/DISPLAY FORMAT=6.1 'Engine Size';
  DEFINE RATIO_MPG/COMPUTED FORMAT=6.2 'Ratio~CITY/HWY';
  *CALCULATE CITYMPG/HWYMPG;
  COMPUTE RATIO_MPG;
     RATIO_MPG=_C2_/_C3_;
  ENDCOMP;
  DEFINE CER/COMPUTED FORMAT=6.2 'City Engine Ratio';
  * CALCULATE CITYMPG/ENGINESIZE;
  COMPUTE CER;
     CER=_C2_/_C4_;
  ENDCOMP;
RUN;
```

(Note that you cannot define a statistic on a computed variable in the COLUMN statement, so RATIO_MPG,MEAN or CER,MEAN will not work.)

ACROSS Option: The DEFINE *var*/ACROSS option allows you to create a column for each unique item for a categorical variable. Hands-On Example 20.12 illustrates the use of the ACROSS option.

HANDS-ON EXAMPLE 20.12

1. Open the program file AREPORT6.SAS.

```
PROC FORMAT;
VALUE FMTSUV 0="NOT SUV" 1="SUV";
PROC REPORT DATA=MYSASLIB.CARS NOFS;
  COLUMN BRAND CITYMPG,SUV HWYMPG,SUV;
  DEFINE BRAND/ORDER;
  DEFINE BRAND/GROUP;
  DEFINE CITYMPG/ANALYSIS MEAN FORMAT=6.1;
  DEFINE HWYMPG/ANALYSIS MEAN FORMAT=6.1;
  DEFINE SUV/ ACROSS 'BY SUV';
  FORMAT SUV FMTSUV.;
RUN;
```

In this example, the SUV variable has values 0 and 1. To display the MPG variables by SUV or not, the SUV variable is listed with those two variables separated by a comma as in CITYMPG,SUV. The SUV variable is then defined as an ACROSS variable and given a label. For this example, a format is created for the SUV variable and applied to that variable within PROC REPORT. The results are given in Table 20.21.

Note how the mean MPG is listed across the table as defined by SUV or NOT SUV for each of the two MPG variables. (Some car brands do not have SUVs or have all SUVs, so there are some missing values in the table.)

TABLE 20.21 PROC REPORT Using the ACROSS Option

	CityMPG		HwyMPG	
	BY SUV		BY SUV	
Brand	NOT SUV	SUV	NOT SUV	SUV
ACURA	21.0	17.0	29.2	23.0
AUDI	19.3	16.0	27.1	22.0
BMW	18.5	16.3	26.6	22.7
BUICK	19.1	16.9	27.4	22.8
CADILLAC	16.4	14.4	24.2	19.4
CHEVROLET	18.8	15.2	25.2	19.7

2. Create another column for ENGINESIZE where the sizes are listed across the table by SUV.
3. Change all of the SUV references to AUTOMATIC to compare MPG and ENGINESIZE for cars with and without automatic transmissions.

ADDITIONAL TECHNIQUES: In addition to the original six DEFINE statement options (options that appear after DEFINE var/), there are several other ways you can enhance the output of your report. The next two examples illustrate two of these additional techniques.

Reporting Multiple Statistics: In the COLUMN statement, you can request multiple statistics for a variable by placing a request in parenthesis. For example,

```
COLUMN BRAND CITYMPG,(N MEAN STD);
```

Hands-On Example 20.13 illustrates this technique.

HANDS-ON EXAMPLE 20.13

1. Open the program file AREPORT7.SAS.

```
TITLE "PROC REPORT Multiple Statistics Example";
PROC REPORT DATA=MYSASLIB.CARS NOFS;
COLUMN BRAND CITYMPG,(N MEAN STD) HWYMPG ENGINESIZE;
   DEFINE BRAND/GROUP;
   DEFINE CITYMPG/ FORMAT=6.1 'City MPG';
```

(*Continued*)

(*Continued*)
```
    DEFINE N / 'N' format=4.;
    DEFINE HWYMPG/ ANALYSIS MEAN FORMAT=6.1;
    DEFINE ENGINESIZE/ ANALYSIS MEAN FORMAT=6.1;
RUN;
```

Note the COLUMN statement where some statistics are listed with parentheses (N MEAN STD). Submit the code and observe the output given (partial) in Table 20.22. Note the N, MEAN, and STD in the CITYMPG column.

Note the two DEFINE statements for CITYMPG and N. The first statement defines the label and format for CITYMPG. The FORMAT=6.1 applies to all of the statistics in the parentheses (N MEAN STD). The second DEFINE specifies a separate format for N, so it appears as an integer.

```
DEFINE CITYMPG/ FORMAT=6.1 'City MPG';
DEFINE N / 'N' format=4.;
```

TABLE 20.22 PROC REPORT with Means

Brand	City MPG			HwyMPG	EngineSize
	N	MEAN	STD		
ACURA	11	20.6	3.4	28.6	2.8
AUDI	41	19.1	2.7	26.7	2.9
BMW	58	18.2	1.8	26.2	3.2
BUICK	16	18.0	1.8	25.1	3.9
CADILLAC	23	15.7	1.6	22.3	4.5
CHEVROLET	95	17.8	3.6	23.7	4.0
CHRYSLER	37	19.5	2.2	26.5	3.3
DODGE	42	17.1	3.7	23.0	4.0

2. Create similar columns for HWYMPG and ENGINESIZE.

Reporting Totals and Subtotals: As in PROC PRINT, you may want to display totals at the bottom of a column of numbers. To do this in PROC REPORT, use the BREAK and RBREAK statements. The BREAK statement produces a summary at a specified break, and RBREAK specifies a summary at the beginning or end of a report. A simplified BREAK statement syntax is

```
BREAK <BEFORE|AFTER> var/SUMMARIZE <SUPPRESS>;
```

For the RBREAK statement, the syntax is

```
RBREAK < BEFORE|AFTER >/SUMMARIZE;
```

HANDS-ON EXAMPLE 20.14

1. Open the program file AREPORT8.SAS.

   ```
   PROC REPORT DATA=MYSASLIB.CARS NOFS;
   COLUMN BRAND CYLINDERS AUTOMATIC SUV;
     DEFINE BRAND/GROUP;
     DEFINE CYLINDERS/GROUP;
     BREAK AFTER BRAND/SUMMARIZE SUPPRESS;
     RBREAK AFTER/ SUMMARIZE;
     WHERE CYLINDERS NE -1;
   RUN;
   ```

   This program lists cars by BRAND. For each BRAND, the number of cars is listed by CYLINDERS. Also, BREAK is used to count the number of cars that are AUTOMATIC or are SUVs and RBREAK creates a grand total. Submit the code and observe the output given (partially) in Table 20.23.

   TABLE 20.23 A Portion of the PROC REPORT Table with Means

Brand	Cylinders	Automatic	SUV
ACURA	4	2	0
	6	4	1
		6	1
AUDI	4	9	0
	6	11	2
	8	6	1
	12	1	0
		27	3
BMW	6	25	6
	8	9	1
etc.		*etc.*	
VOLVO	5	21	3
	6	2	1
	8	1	1
		24	5
		753	279

   Note the subtotals after the ACURA 6 cylinder line and the grand total after the VOLVO subtotal.

2. To see what the SUPPRESS option does in the BREAK statement, delete SUPRESS and rerun the code. Observe that without SUPPRESS, the name of the BRAND is repeated on the subtotal line.

20.2.1 Writing Reports in a Data Statement

Another way to create reports from a SAS dataset is to use the PUT statement within a DATA step. Typically, this is done using the following procedure:

1. Specify the name _NULL_ as the dataset name in a DATA step in order to be able to use the features of the DATA step without creating a SAS dataset as output (because you are interested in creating a report rather than creating a SAS dataset).
2. Define a destination (output file) for the report using a FILE statement. The statement FILE PRINT sends the report output to the SAS **Results Viewer**. When you use FILE "*file-specification*", the report output is sent to the specified text file location.
3. Use PUT statements within the DATA step to create lines of output for the report.

HANDS-ON EXAMPLE 20.15

The heart of a DATA statement report is the PUT statement. By default, a PUT statement displays information for each record in a dataset in the SAS **Log Window**. For example, open the program file APUT_REPORT1.SAS.

```
DATA _NULL_; SET MYSASLIB.SOMEDATA;
     PUT AGE;
RUN;
```

This simple program lists the values of AGE for each record in the specified SAS dataset. In this case, when no FILE option is used, the report output is sent to the **Log Window**.

1. Run this program and observe the output in the **Log Window**. Note that there are 50 lines of ages listed there.
2. You would probably want to display the results of your report in the **Results Viewer** or capture the results in a File instead of simply having them listed in the **Log Widow**. To display the output in the **Results Viewer**, include the statement FILE PRINT; in the code:

```
DATA _NULL_; SET MYSASLIB.SOMEDATA;
     FILE PRINT;
     PUT AGE;
RUN;
```

Rerun the code and observe the output. This time the values of AGE are displayed in the **Results Viewer**.

3. To make the report more interesting, change the PUT statement to read

```
PUT "The Age for Subject "ID "IS "AGE;
```

> Run the revised code. This time, the following lines are displayed in the **Results Viewer**:
>
> ```
> The Age for Subject 101 is 12
> The Age for Subject 102 is 11
> The Age for Subject 110 is 12
> The Age for Subject 187 is 11
> ... and so on...
> ```
>
> Note that text in quotes is output verbatim. However, a variable name in a PUT statement (that is not in quotes) causes the value for that variable for a particular record in the dataset to be listed.
>
> 4. Alternatively, you can specify an output file as the destination for the report. Change the FILE PRINT statement to
>
> ```
> FILE "C:\SASDATA\OUTPUT.TXT";
> ```
>
> Run the revised program. Go to Windows and look in the C:\SASDATA folder. Open the file named OUTPUT.TXT to verify that the report was written to that file.

There are a number of other ways to customize each line of output using the PUT statement. Table 20.24 lists common options for the PUT statement.

Below is listed a small dataset containing grades for a class. The goal is to create a summary report that includes a listing of grades, each student's grade average, and an average for the class. The following data are the results of three tests.

```
Alice, Adams   88.4   91    79
Jones, Steve   99     100   88.4
Zabar, Fred    78.6   87    88.4
```

To make student's grades available for the report, AVE is calculated when the dataset GRADES is created using the following code:

```
DATA GRADES;
INPUT NAME $15. G1 G2 G3;
AVE=MEAN(of G1-G3);
DATALINES;
Alice, Adams   88.4 91 79
Jones, Steve   99 100 88.4
Zabar, Fred    78.6 87 88.4
;
RUN;
```

Because we also want to include the class average in the report, PROC MEANS is used to calculate that value, using the following code:

```
PROC MEANS DATA=GRADES NOPRINT;OUTPUT OUT=GMEAN;RUN;
```

TABLE 20.24 Options for the PUT Statement

Option	Meaning
`'text'`	Information in quotes is output verbatim (unless macro variables are within the quotes). For example, `PUT 'Some Text';`
`variablename(s)`	The value of a variable for each record in the dataset is output where a variable name is included, not in quotes. For example, `PUT TIME1 TIME2 TIME3;`
`@n, @var, or @expression`	Moves the pointer to a specified column location. When using `@var` or `@expression`, they must be numeric. For example, `PUT @5 'Age =' AGE;`
`+n, +var, or +expression`	Moves the pointer to a relative location *n* columns to the right from its current location. When using `@var` or `@expression`, they must be numeric
`@`	Holds the output on the current output line. The default is to move to the next line of output
`var format.`	A format specification following a variable causes the variable to be displayed in the specified format. For example, `PUT AGE 5.1;` displays the value of `AGE` using a 5.1 format
`n*'char';`	Repeats a character n times. For example, `PUT 50*'-';` creates a dashed line 50 characters wide

TABLE 20.25 Results of PROC MEANS on the GRADE Data

	TYPE	_FREQ_	_STAT_	G1	G2	G3	AVE
1	0	3	N	3	3	3	3
2	0	3	MIN	78.6	87	79	84.666666667
3	0	3	MAX	99	100	88.4	95.8
4	0	3	MEAN	88.666666667	92.666666667	85.266666667	88.866666667
5	0	3	STD	10.202614044	6.6583281185	5.4270925304	6.0490586963

The overall class mean is captured in the output SAS dataset named `GMEAN`. The contents of this dataset are given in Table 20.25. To display this table, select the **Work** library in **Explorer**, and double-click on the table `GMEAN` to view this information in the **Viewtable**.

Note that the class mean is 88.8667. It appears in the `GMEAN` dataset on the line where the variable `_STAT_` is `"MEAN"`.

A final step before writing the report is to capture the overall average into a macro variable. This is accomplished with a `CALL SYMPUT` statement (see Appendix B) as shown here:

```
DATA _NULL_;SET GMEAN;
IF _STAT_="MEAN" THEN CALL SYMPUT('CLASSMEAN',ROUND(AVE,.01));
RUN;
```

In this case, DATA _NULL_ is used to create the macro variable named CLASSMEAN, which is assigned the value AVE from the dataset GMEAN. The name _NULL_ was used because we did not need to create any SAS dataset from the DATA step.

The report created using this information is illustrated in Hands-On Example 20.16:

HANDS-ON EXAMPLE 20.16

1. Open the program file APUT_REPORT2.SAS. Examine the code. The preliminary calculations appear at the beginning of the code. Note the section of code that produces the report:

```
TITLE "Grade Report for Course 1234";
DATA _NULL_; SET GRADES END=EOF;
TABLEHEAD="SUMMARY OF GRADES";
FILE PRINT;
IF _N_ = 1 then DO; * PUT HEADER;
      L=5+(60-LENGTH(TABLEHEAD))/2; * CENTER HEAD;
      PUT @L TABLEHEAD;
      PUT @5 60*'=';
      PUT @5 "NAME" @20 "| TEST 1" @30 "| TEST 2 "
          @40 "| TEST 3" @50 "| AVERAGE";
      PUT @5 60*'-';
END;
PUT @5 NAME @20 "|" G1 5.1 @30 "|" G2 5.1
        @40 "|" G3 5.1 @50 "|" AVE 5.1;
IF EOF =1 then do; * PUT BOTTOM OF TABLE;
      PUT @5 60*'-';
      PUT @5 "CLASS AVERAGE "@; * HOLDS LINE FOR NEXT PUT;
      PUT @50 "| %TRIM(&CLASSMEAN) ";
      PUT @5 60*'=';
END;
RUN;
```

A breakdown of this code is as follows:
- The report is created within a DATA_NULL_ step using the information from the GRADES SAS dataset. (The EOF variable is created to mark the last records in the GRADES dataset.)
- TABLEHEAD is a title to be displayed in the report.
- FILE PRINT sends the report to the SAS **Results Viewer**.
- Note the three parts of the program within the DATA step:
    1. The section beginning with IF_N_=1 is a header that is only displayed when the first record of the dataset is read. It centers and PUTs the title (TABLEHEAD) and PUTs several lines that make up the header for the report.
    2. The center of the report is a PUT statement that outputs each grade (G1–G3) in specified columns and with specified formats, and also displays the student average AVE.

(Continued)

(Continued)

3. The end of the report is output only when the last record is reached (`IF EOF=1`). PUT statements are used to output footer information that includes the class mean (using the macro variable &CLASSMEAN). The trim statement trims unwanted blanks.

Submit this code and note the output as given in Table 20.26.

TABLE 20.26 Grade Report Created Using PUT Statements

```
                   Grade Report for Course 1234
                        SUMMARY OF GRADES
===============================================================
NAME               | TEST 1  | TEST 2  | TEST 3  | AVERAGE
---------------------------------------------------------------
Alice, Adams       | 88.4    | 91.0    | 79.0    | 86.1
Jones, Steve       | 99.0    |100.0    | 88.4    | 95.8
Zabar, Fred        | 78.6    | 87.0    | 88.4    | 84.7
---------------------------------------------------------------
CLASS AVERAGE                                    | 88.87
===============================================================
```

4. Add a new student to the data, and rerun to see how the table adapts to new information. Rerun the code to verify that your change works correctly. Are the new averages calculated correctly?
5. Change the format for the student averages to 6.2. Rerun the code to verify that your change works correctly.

EXERCISES

20.1 Tabulate procedure

Open the program file `EX_20.1.SAS`. This file contains a program that creates a table showing average CITYMPG by BRAND and SUV.

```
PROC TABULATE DATA=MYSASLIB.CARS;
   CLASS BRAND SUV;
   VAR CITYMPG;
   TABLE BRAND, SUV*MEAN*CITYMPG;
RUN;
```

a. Add code to clean up this table by removing unnecessary labels.
b. Create a FORMAT for SUV that displays SUV and NON-SUV for labels.
c. Add code to save the table to a Word file (using RTF) and apply a style.
d. Add code to save the table to an Excel file.

20.2 **Report procedure**

Open the program file EX_20.2.SAS.

```
PROC REPORT DATA=MYSASLIB.SOMEDATA NOFS;
  COLUMN AGE STATUS TIME1-TIME4 MEANTIME;
  DEFINE AGE/GROUP;
  DEFINE TIME1/ANALYSIS MEAN;
  DEFINE MEANTIME/COMPUTED 'AVE TIME 1 TO 2';
  COMPUTE MEANTIME;
    MEANTIME=(_C3_+_C4_)/2; *MEAN OF TIME1 AND TIME2;
  ENDCOMP;
RUN;
```

a. Using the code above as a start, create a report using the SOMEDATA dataset that groups subjects by AGE, displays columns of means for TIME1 through TIME4, and summarizes the data at the end of the table.

b. Why does MEANTIME=(_C3_+_C4_)/2; calculate the average of TIME1 and TIME2?

c. Add format statements so that each TIME variable and MEANTIME are reported to one decimal place.

20.3 **Creating a report using PUT statements**

Using example APUT_REPORT2.SAS, make the following changes:

a. Add a standard deviation at the bottom of the table. This value is already in the GMEAN SAS dataset. Use the following code:

```
IF _STAT_="STD" THEN CALL SYMPUT('CLASSSTD',ROUND(STD,.01));
```

To create a macro variable named CLASSSTD add the following statement to the report: (Put the statement below the current CALL SYMPUT statement.)

```
PUT @5 "CLASS Standard Deviation" @;
  PUT @50 "| %TRIM(&CLASSSTD) ";
```

Insert this statement immediately following the PUT statement for CLASS AVERAGE. Rerun the revised code to verify that your new information is included in the tabled output.

b. Change the columns so that there is more room for the name by placing the test scores at columns 25, 35, and 45 and the average at column 55. Rerun to verify that your changes worked.

APPENDIX A

OPTIONS REFERENCE

This appendix is a brief reference to a number of SAS® elements used in graphs and other procedures. These elements allow you visual control over many of the graphs produced in SAS procedures. Not all of the options for these elements are listed in this appendix. Refer to the SAS documentation for further information. The elements included in this appendix are as follows:

- Fonts
- Color options
- Graph patterns
- Bar and block patterns
- Line styles
- Plot symbols
- Informats and formats

A.1 USING SAS FONTS

Several standard SAS fonts are illustrated in Figure A.1. These are typical fonts for a Windows computer. The fonts available to you may differ according to your Windows and SAS versions. They were created using the SAS program `APXA_SHOWFONTS.SAS`.

Use these fonts in any statement that allows the specification of a font. For example,

```
TITLE1 F="SWISS" "This is an example use of the SWISS font.";
```

SAS® Essentials: Mastering SAS for Data Analytics, Third Edition. Alan C. Elliott and Wayne A. Woodward.
© 2023 John Wiley & Sons, Inc. Published 2023 by John Wiley & Sons, Inc.
Companion website: https://www.alanelliott.com/SASED3/

What's the default font?

> **ABCabc Example SCRIPT**
> **ABCabc Example CENTURY**
> ABCabc Example COURIER
> ABCabc Example HERSEY
> **ABCabc Example SWISS**
> **ABCabc Example TIMES**
> ABCabc Example ZAPF

Figure A.1 Example SAS fonts

In Windows, you can also specify any installed font on your machine. For example, the following statement uses the Times New Roman font.

```
TITLE F="TIMES" "This is an example use of the TIMES font.";
```

You can also specify height and color in such statements. (The default height is usually H=1.) For example,

```
TITLE1 F="COURIER" H=2 C=RED "Specify the height of the font.";
```

There are also bold, italic, and wide versions of some fonts. Refer to the SAS documentation for more information.

A.2 SPECIFYING SAS COLOR CHOICES

Color choices that can be used in SAS procedures include the following:

Black	Lilac	Rose
Blue	Lime	Salmon
Brown	Magenta	Tan
Cream	Maroon	Violet
Cyan	Orange	White
Gold	Pink	Yellow
Gray		Purple
Green		Red

If you do not specifically indicate a color, the SAS default color is BLACK. You can modify these colors with prefixes such as LIGHT to create, for example, LIGHTBLUE. Other prefixes include the following:

Brilliant	Moderate
Dark	Pale
Deep	Strong
Light	Vivid
Medium	

Not all combinations work, so you may have to experiment to get the color you want.

Thousands of other SAS colors are available. If you know a standard RGB specification for a color (often institutions will have a specified RGB color for official documents), you can use that specification in a color option. An example would be CX3230B2 as the RGB code for "Brilliant Blue." Here are two examples (where H is the height option):

```
TITLE H=2 C=RED "This produces a Red title.";
TITLE2 H=1.5 C="CX3230B2" "This title is Brilliant Blue";
```

A.3 SPECIFYING PATTERNS FOR PROCS GPLOT AND PROC UNIVARIATE

To specify a pattern in SAS, you use a pattern code as in a histogram, typically in a PFILL statement. The form is

```
PFILL=Mabccc
```

where the definitions of these a, b, and ccc codes are listed in Table A.1. These patterns also apply to the VALUE= option in the PATTERN statement used to fill bars or under a curve for various SAS/GRAPH types.

TABLE A.1 Custom Pattern Codes

Code	Meaning
a	Ranges from 1 to 5 (thickness/density of line, 1=thinnest)
b	X indicates crosshatch and N indicates crosshatch
ccc	ccc is a number ranging from 0 to 360 where 0 indicates no angle, 45 is 45° angle, and so on.
EMPTY or E	Use PFILL=EMPTY to specify no fill
SOLID or S	Use PFILL=SOLID to specify solid fill

For example, open the program file APXA_PATTERN1.SAS. For some versions of SAS, you may need to include the statement

```
ODS GRAPHICS OFF;
```

for this example to work because it applies to standard SAS graphs, and not ODS graphs. The example code

```
PROC UNIVARIATE DATA="C:\SASDATA\SOMEDATA" NOPRINT;
  VAR AGE;
  HISTOGRAM /NORMAL PFILL=M3X45;
RUN;
```

produces (along with a number of tables) a histogram using a crosshatch pattern at a 45° angle as shown in Figure A.2. To control the color of the pattern, include a statement such as the following before the PROC statement:

```
PATTERN COLOR=RED;
```

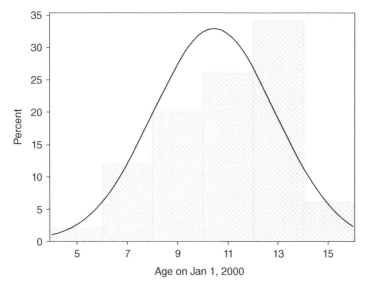

Figure A.2 Example crosshatch pattern

A.4 BAR AND BLOCK PATTERNS FOR BAR CHARTS, PIE CHARTS, AND OTHER GRAPHICS

Pattern choices can be used to specify colors and patterns for graphs in the PROC GCHART, PROC GPLOT, and PROC UNIVARIATE procedures with the PATTERN statement:

PATTERNn options;

where n (ranging from 1 to 30) specifies the graph element in which the pattern is to be used. The following options in the PATTERN statement allow you to specify patterns and colors.

```
V=bar/block pattern
    E              Empty pattern (empty box)
    R1 to R5       Right 45° stripes (light to heavy lines)
    X1 to X5       Crosshatched patterns (light to heavy lines)
    L1 to L5       Left 45° stripes (light to heavy lines)
C=color            See Section A.2
```

For example, open the program file GPATTERN2.SAS. For some versions of SAS, you may need to include the statement

ODS GRAPHICS OFF;

for this example to work because it applies to standard SAS graphs, and not ODS graphs. The example code (APXA_PATTERN2.SAS) is shown here.

502 APPENDIX A: OPTIONS REFERENCE

```
GOPTIONS RESET = ALL;
DATA BARS;
INPUT A B;
DATALINES;
1 1
2 2
3 3
4 4
5 5
;
PATTERN1 V=E C=BLUE;
PATTERN2 V=R1 C=BLACK;
PATTERN3 V=X2 C=BLACK;
PATTERN4 V=L3 C=BLACK;
PATTERN5 V=S C=BLACK;
PROC GCHART ;VBAR A
/DISCRETE WIDTH=10
SUBGROUP=B;
RUN;
QUIT;
```

This code produces the graph in Figure A.3, which shows examples of the various patterns. Each PATTERN statement is applied to the bars, Pattern1 to the first bar, Pattern2 to the second, and so on.

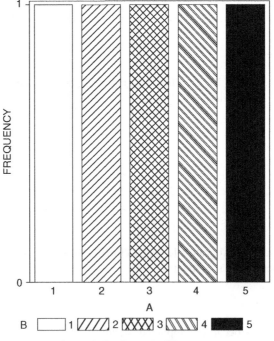

Figure A.3 Example fill patterns

A.5 SAS LINE STYLES

When drawing lines in a procedure such as PROC GPLOT, you can choose from one of the popular line styles shown in Table A.2. These are specified within a SYMBOLn statement where n refers to lines in the order they are specified in the plot by L=*linestyle number* (L=2 meaning short dash in the example code here).

For example, open the program file. APXA_LINE.SAS. For some versions of SAS, you may need to include the statement

```
ODS GRAPHICS OFF;
```

for this example to work because it applies to standard SAS graphs, and not ODS graphs. The example code

```
GOPTIONS RESET = ALL;
SYMBOL1 V=CIRCLE C=BLUE I=R L=2 W=5;
TITLE 'REGRESSION LINE TYPE SELECTED';
PROC GPLOT DATA="C:\SASDATA\CARS";
    PLOT HWYMPG*ENGINESIZE;
RUN;
QUIT;
```

produces a scatterplot and regression line with line style L=2 and width W=5 for emphasis to illustrate the selected line style, as shown in Figure A.4.

In some instances, such as in SGPLOT, line attributes may be specified using the *linestyle name*. For example, the following code specifies the PATTERN using the pattern name (LONGDASH) instead of the number (5).

```
LINEATTRS=(PATTERN=LONGDASH)
```

In SGPLOT, you can use either 5, in this case, or the phrase LONGDASH in the LINEATTRS statement.

TABLE A.2 Line styles

STYLE	LINESTYLE
1	Solid
2	ShortDash
4	MediumDash
5	LongDash
8	MediumDashShortDash
14	DashDashDot
15	DashDotDot
20	Dash
26	LongDashShortDash
34	Dot
35	ThinDot
41	ShortDashDot
42	MediumDashDotDot

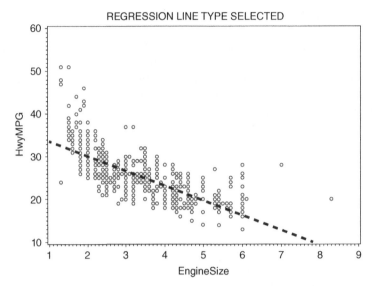

Figure A.4 Graph showing selected line style

A.6 USING SAS PLOTTING SYMBOLS

The symbol used for displaying a point on a graph, such as a scatterplot in `PROC GPLOT`, can be specified using a `SYMBOL` statement such as the following:

```
SYMBOL1 V=CIRCLE;
```

The standard SAS symbols from which you can select include:

```
ARROWDOWN      PLUS        Y
ASTERISK       SQUARE      Z
CIRCLE         STAR        CIRCLEFILLED
DIAMOND        TACK        DIAMONDFILLED
GREATERTHAN    TILDE       HOMEDOWNFILLED
HASH           TRIANGLE    SQUAREFILLED
HOMEDOWN       UNION       STARFILLED
IBEAM          X
```

These symbols may also be used in `PROC SGPLOT`.

A.7 USING ODS STYLE ATTRIBUTES

`ODS` style attributes allow you to specify how `ODS` Traffic Light values are displayed (see `ODS9.SAS` for an example). Values for these items are given in other sections of this appendix. For example, for all colors, refer to the color descriptions in Section A.2. For other attributes, refer to SAS documentation. Table A.3 lists some common options for `ODS` Styles.

A.8 COMMON (INPUT) INFORMATS

Informats are used to read data. The use of informats is introduced in Section 2.5. Table A.4 lists some common informats.

TABLE A.3 Common ODS Style Options

Style Item	What It Sets
BACKGROUND=	Sets the color of the cell background. For example, BACKGROUND=YELLOW
FLYOVER=	Displays text when the mouse is moved on top of the item. For example, FLYOVER="Summed Value"
FONT=	Defines a font definition for the item. For example, FONT=("SWISS",10pt)
FONT_FACE=	Specifies the font typeface. For example, FONT_FACE=SCRIPT
FONT_SIZE=	Specifies the font size. For example, FONT_SIZE=12
FONT_WEIGHT=	Specifies the weight of the font. For example, FONT_WEIGHT=BOLD
FOREGROUND=	Specifies the color for text and column labels not otherwise specified. For example, FOREGROUND=LIGHTBLUE
JUST=	Specifies cell justification (horizontal, left, right, or center). For example, JUST=LEFT

TABLE A.4 Common SAS Informats (Input Formats)

Informat	Meaning
5.	Five columns of data as numeric data
$5.	Character variable with width 5, removing leading blanks
$CHAR5.	Character variable with width 5, preserving leading blanks
COMMA6.	Six columns of numeric data and strips out any commas or dollar signs (i.e., $40,000 is read as 40,000)
COMMA10.2	Reads 10 columns of numeric data with two decimal places (strips commas and dollar signs). $19,020.22 is read as 19,020.22
MMDDYY8.	Date as 01/12/16. (Watch out for Y2K issue.)
MMDDYY10.	Date as 04/07/2016
DATE7.	Date as 20JUL016
DATE9.	Date as 12JAN2016. (No Y2K issue.)

A.9 COMMON (OUTPUT) FORMATS

Formats are similar to informats (and there are some overlaps). They are used to write out data values. The use of formats is introduced in Section 2.5. Table A.5 lists some common formats.

TABLE A.5 Common SAS (Output) FORMATs

Format	Meaning
`5.`	Writes data using five columns with no decimal points. For example, 1234
`5.1`	Writes data using five columns with 1 decimal point. For example, 123.4
`DOLLAR10.2`	Writes data with a dollar sign, commas at thousands, and with two decimal places. For example, $1,234.56
`COMMA10.1`	Writes data with commas at thousands, and one decimal place. For example, 1234.5
`MMDDYY10.`	Displays dates in common American usage as in 01/09/2016
`DATE9.`	Display time using a military-style format. For example, 09JAN2016
`WORDDATE12.`	Displays dates with abbreviated month names as in January 9, 2016
`WORDDATE18.`	Displays dates with full month names such as in January 9, 2016
`WEEKDATE29.`	Display dates with day of week as in Saturday, January 9, 2016
`DOWNAME.`	Outputs day of week as a name
`MONNAME.`	Outputs month as a name

APPENDIX B

SAS® FUNCTION REFERENCE

SAS functions can be used in a DATA statement to calculate new variables or as a part of a conditional statement. This appendix is more of a reference than a tutorial, but it does illustrate the use of a number of common functions in SAS. It is not an exhaustive list of functions, but it does include commonly used functions that are useful in manipulating data in the preparation for analysis. Some examples are provided for functions, whose use may not be apparent in a simple description. For a more extensive description of this material, please refer to the SAS documentation.

B.1 USING SAS FUNCTIONS

Here are some examples of how SAS functions are used:

```
DATA NEWDATA;SET OLDDATA;
TOTAL=SUM(A,B,C);
GENDER=UPCASE(SEX);
SQRAGE=SQRT(AGE);
RUN;
```

Note that in these cases, the function on the right side of the = sign does something to the argument(s) in the parentheses and returns a value, which is then assigned to the name on the left-hand side of the = sign.

The SUM function returns the total of the variables A, B, and C. The SQRT function returns the square root of the AGE variable. The UPCASE function makes all character values in the SEX variable uppercase. For example, if the SEX variable contains the value f, then the GENDER variable will contain the value F. The syntax for a function is

---

SAS® Essentials: Mastering SAS for Data Analytics, Third Edition. Alan C. Elliott and Wayne A. Woodward.
© 2023 John Wiley & Sons, Inc. Published 2023 by John Wiley & Sons, Inc.
Companion website: https://www.alanelliott.com/SASED3/

```
NEWVARIABLE = FUNCTIONNAME(argument1, argument2...argumentk);
```

where arguments can be constants, variables, expressions, or other functions. For functions that are not described in this appendix, refer to the SAS documentation.

B.2 ARITHMETIC/MATHEMATICAL FUNCTIONS

Using arithmetic functions provides a way to obtain common mathematical results using simple code. For example,

```
AVE = MEAN(OF OBS1-OBS5);
```

calculates the average of the variables `OBS1` through `OBS5`. Typically, using the function to calculate certain values is easier than coding the actual arithmetic calculation such as

```
AVE = (OBS1+OBS2+OBS3+OBS4+OBS5)/5;
```

Moreover, in this case, the `MEAN` function finds the mean of the non-missing values if there are missing values in the list, whereas the individually added variables will result in the `MEAN` being set to missing if any one of the `OBS` values is missing.

Also take note of how the various lists of arguments are used in the function. For example, both of these calls to the `SUM` function calculate the same results:

```
SUMVAL = SUM(PRE,MONTH6,MONTH12,MONTH24);
SUMVAL = SUM(OF PRE--MONTH24);
```

Note that when you use a range such as `PRE--MONTH24` or `OBS1-OBS50` as arguments in a SAS function, you must precede the list with an `OF` argument. Otherwise, you may get unexpected results (Table B.1).

TABLE B.1 Arithmetic and Mathematical Functions

Function	Description
ABS(x)	Absolute value. Example: Y = ABS(-9) returns the value 9
EXP(x)	Exponentiation. Example: Y = EXP(1) returns the value 2.71828
LOG(x)	Natural log. Example: Y = LOG(10) returns the value 2.30259
LOG10(x)	Log base 10. Example: Y = LOG10 returns the value 1.0
FACT(x)	"Factorial. Example: FACT(7) returns the value 5040, which is $(7 \times 6 \times 5 \times 4 \times 3 \times 2 \times 1)$."
MAX(x1,x2,x3...)	Returns maximum value in the list
MEAN(x1,x2,x3...)	Returns the arithmetic mean of non-missing values in list. Example: MEAN(2,4,.,6) returns the value 4. (Note that the value dot . in the list represents a missing value.)

ARITHMETIC/MATHEMATICAL FUNCTIONS

TABLE B.1 (*Continued*)

Function	Description
MEDIAN (x1,x2,x3...)	Returns the median of non-missing values in the list
MIN(x1,x2,x3...)	Returns the minimum in the list
MOD(x,k)	Remainder of x/k. Examples: MOD(5,3)=2, MOD(6,2)=0
N(x1,x2,x3...)	Returns the number of non-missing values. Example: N(1,2,.,4) returns the value 3
NMISS(x1,x2,x3...)	Returns the number missing. Example: NMISS(1,2,.,4) returns the value 1
ORDINAL(k,x1,x2,x3...)	Finds the kth number in the sorted list. ORDINAL(2,11,6,9,3) = 6 because it is second in the ordered list
RANUNI(0)	A different random number (uniform distribution) each time
RANUNI(x)	Returns the same random sequence (uniform) each time for that particular seed, where x is a seed value you enter
RANNOR(seed)	A random number from a normal distribution with mean 0 and standard deviation 1. (See seed info for RANUNI.)
SIGN(x)	Returns the sign of x as a 1 or −1 value
SQRT(x)	Square root. Example: Y = SQRT(9) returns the value 3
SUM(x1,x2,x3...)	Returns the sum of non-missing values in the list

HANDS-ON EXAMPLE B.1

Open the program file FUNC_ARITH.SAS. This code uses several arithmetic functions to summarize data for the time variables PRE to MONTH24. Run the code and observe the results given in Table B.2.

```
DATA ARITH; SET MYSASLIB.HASMISSING;
MAXTIME = MAX(PRE,MONTH6,MONTH12,MONTH24); MINTIME = MIN(OF
PRE--MONTH24);
MISSTIME = NMISS(OF PRE--MONTH24);
MEANTIME = MEAN(OF PRE--MONTH24);
;
TITLE "Using Arithmetic Functions";
PROC PRINT DATA = ARITH; RUN;
```

1. Using the SUM function with the argument list PRE, MONTH6, MONTH12, MONTH24, add code to calculate SUMTIME. Using the MEDIAN function with the argument list OF PRE—MONTH24, add code to calculate MEDIANTIME. Run the code and observe the results.

(*Continued*)

TABLE B.2 Using of Arithmetic Functions

Use Arithmetic Functions

Obs	ID	PRE	MONTH6	MONTH12	MONTH24	TEMP1	TEMP2	TEMP3	TEMP4	MAXTIME	MINTIME	MISSTIME	MEANTIME
1	00101	22.3	25.3	28.2	30.6	101	95	94	97	30.6	22.3	0	26.6000
2	00102	22.8	27.5	33.3		95	103	94	101	33.3	22.8	1	27.8667
3	00110	18.5	26.0	29.0	27.9	98	103	94	101	29.0	18.5	0	25.3500
4	00187	22.5	29.3	32.6	33.7	102	101	96	99	33.7	22.5	0	29.5250
5	00312	17.0	21.3	22.7	21.2	104	95	103	100	22.7	17.0	0	20.5500
6	00390	21.3	30.5	30.7	30.1	95	103	94	97	30.7	21.3	0	28.1500
7	00445	22.1	31.3	31.6	31.6	100	101	100	101	31.6	22.1	0	29.1500
8	00543	20.8	28.6	34.2	30.0	100	101	98		34.2	20.8	0	28.4000
9	00544	18.1	23.8	29.0	27.8	98	98	100	95	29.0	18.1	0	24.6750
10	00550	23.0		29.9	31.8	104	101	94	97	31.8	23.0	1	28.2333
11	00561	21.5	27.0	32.6	34.3	97		95	98	34.3	21.5	0	28.8500
12	00104	22.8	30.0	32.8	31.0	94	99	97	101	32.8	22.8	0	29.1500
13	00122	19.5	25.0	25.3	26.6	95	99	103	94	26.6	19.5	0	24.1000
14	00123	23.5	28.8	34.2	35.6	100	94	97	101	35.6	23.6	0	30.5250

(*Continued*)

TABLE B.3 Arithmetic and Trigonometric Functions

Function	Description
ARCOS(x)	Returns the value of arccosine
ARSIN(x)	Returns the value of arcsine
ATAN(x)	Returns the value of arctangent
COS(x)	Returns the value of cosine
COSH(x)	Returns the value of hyperbolic cosine
SIN(x)	Returns the value of sine
SINH(x)	Returns the value of hyperbolic sine
TAN(x)	Returns the value of tangent
TANH(x)	Returns the value of hyperbolic tangent

B.3 TRIGONOMETRIC FUNCTIONS

Each of these functions returns the specified trigonometric value for the value of the argument (in radians) entered into the function. For example,

```
Y = COS(3.14);
```

returns the value −1.000 (Table B.3).

B.4 DATE AND TIME FUNCTIONS

SAS date and time values are stored as integers and indicate the number of days since 1/1/1960. A positive number indicates a date after 1/1/1960, and a negative number indicates a date before 1/1/1960. Date values that contain both date and time are stored in SAS as the number of seconds since midnight on 1/1/1960. Dates are typically displayed using a SAS date format such as MMDDYY10., which displays a date as 03/09/2024.

For these functions, a SAS date is a variable designated as a date value. Here are two examples of how variables can be designated as dates:

- Assign a date value to a fixed date. A d at the end of the value tells SAS to interpret this quoted string as a date. For example (note that case does not matter either for the month abbreviation or for the "d"),

```
BDATE = '12DEC20244'd;
BEGINDATE = "1jan2011"D;
EXAMDATE = '13-APR-2024'd;
```

- Read a value as a date in an INPUT statement. For example,

```
INPUT BDATE MMDDYY10.;
```

In the following examples, a "date" is a SAS value that contains only a date, and a "*datetime*" value contains both date and time values.

When using date functions, beware of the "Y2K" problem. Any two-digit year lower than 40 is interpreted as being in the twenty-first century (e.g., the two-digit year 15 is interpreted as 2015) and two-digit year 40 or greater are interpreted as being in the twentieth century (e.g., the two-digit year 41 is interpreted as 1941). You can change this default using the statement YEARCUTOFF = 19xx in your program. For example, if you specify YEARCUTOFF = 1925, any two-digit year that is between 00 and 24 will be interpreted as being in the twenty-first century (e.g., 21 will be interpreted as 2021), and years from 25 to 99 will be interpreted as in the twentieth century (e.g., 44 will be interpreted as 1944):

DATDIF(sdate,edate,basis)

returns the number of days between two dates. BASIS is (default) '30/360', 'ACT/ACT', 'ACT/360', or 'ACT/365'. The BASIS option tells SAS what counting method to use to count days. (These are some day-count conventions used in investing, accounting, and other disciplines.)

- '30/360' option means to use a 30-day month and a 360-day year.
- 'ACT/ACT' means to use the actual number of days.
- 'ACT/360' means to use the actual number of days in each month and 360 days in a year.
- 'ACT/365' means to use the actual number of days in each month and 365 days in a year.

For example,

DAYS = DATDIF('07JUL1976'd,'01JAN2013'd,'30/360');

returns the value (number of days) as 13,134, whereas

DAYS = DATDIF('07JUL1976'd,'01JAN2013'd,'ACT/ACT');

returns the value (number of days) as 13,327. These integer values can be displayed in a date-readable manner by applying a date format to the variable such as

FORMAT DAYS MMDDYY10.;

Table B.4 lists some of the common functions used to manage dates in a SAS program. Vertical bars in the function definitions indicate choose one listed option "or" the other.

TABLE B.4 Date and Time Functions

Function	Description
DATDIF(sdate,edate,basis)	Calculates the difference between two dates using the indicated basis method (described earlier)
DATE()	Returns the current date (SAS date value)
DATEPART(datetime)	Extracts and returns only the date from a SAS datetime variable
DATETIME()	Returns the current date and time of day from the computer's date and time values

TABLE B.4 (*Continued*)

Function	Description
DAY(*date*)	Returns the day of month (a value from 1 to 31) from a SAS date
DHMS(*date,hour,minute,second*)	Returns a SAS *datetime* value from the argument date, hour, minute, and second in the function
HMS(*hour,minute,second*)	Creates a SAS time value from hour, minute, and second
HOUR(<*time* | *datetime*>)	Returns the hour value from a SAS time or *datetime* variable
INTCK('*interval*',*from,to*)	Returns the number of time intervals in a given time span, where the interval can be as follows: **Datetime intervals:** DTDAY, DTWEEKDAY, DTWEEK, DTTENDAY, DTSEMIMONTH, DTMONTH, DTQTR, DTSEMIYEAR, DTYEAR **Date intervals:** DAY, WEEKDAY, WEEK, TENDAY, SEMIMONTH, MONTH, QTR, SEMIYEAR, YEAR **Time intervals:** HOUR, MINUTE, SECOND For example, WEEKS = INTCK('WEEK','07JUL1976'd, '01JAN2013'd); Returns a value of 1904, indicating the number of weeks between the two indicated dates
INTNX('*interval*', *start-from*, *increment*<,'*alignment*'>)	Advances a date, time, or *datetime* value by a given interval, and returns a date, time, or *datetime* value – see intervals earlier
MDY(*month,day,year*)	Creates and returns a SAS date from month, day, and year
MINUTE(<*time* | *datetime*>)	Extracts and returns the minute value (1–60) from a SAS time or *datetime* variable
MONTH(*date*)	Extracts and returns the month value (1–12) from a SAS date value
QTR(*date*)	Extracts and returns the quarter value (1–4) of the year from a SAS date value
SECOND(<*time* | *datetime*>)	Extracts and returns the value for seconds (1–60) from a SAS time or *datetime* value
TIME()	Returns the current time of day as a *datetime* value (from the computer's system clock)
TIMEPART(*datetime*)	Extracts and returns the time value from a SAS *datetime* variable
TODAY()	Returns the current date as a SAS date (from the computer's system clock)
WEEKDAY(*date*)	Extracts and returns the day of the week (1–7) from a SAS date value
YEAR(*date*)	Extracts and returns the year value (such as 2019) from a SAS date value
YRDIF(*startdate,enddate, basis*)	Returns the difference in years between two dates. Basis is '30/360', 'ACT/CT', 'ACT/360', or 'ACT/365'. (See descriptions of these basic values in the DATDIF function earlier.)
YYQ(*year,quarter*)	Extracts and returns a SAS date from year and quarter

HANDS-ON EXAMPLE B.2

1. Open the program file `FUNC_DATETIME.SAS`. This program reads in a date and time arrived as separate variables and `DATETIMELEAVE` as a single variable.

```
DATA TIME;
FORMAT DTARRIVE MMDDYY10. TIMEARRIVE TIME10.
    ARRIVETIME DATETIMELEAVE DATETIME19.;
INPUT @1 DTARRIVE DATE10.
    @12 TIMEARRIVE TIME10.
        @23 DATETIMELEAVE DATETIME19.;
*MERGE DTARRIVE AND TIMEARRIVE;
    ARRIVETIME=DHMS(DTARRIVE,0,0,TIMEARRIVE);
    SECONDS1=DATETIMELEAVE-ARRIVETIME;
    SECONDS2=INTCK('SECONDS',ARRIVETIME,DATETIMELEAVE);
DATALINES;
10NOV2016   1:13 PM     10NOV2016:18:49:19
12DEC2016   9:20 AM     12DEC2016:13:23:22
;
PROC PRINT DATA=TIME;RUN;
```

To calculate the difference between the arrive and leave time, we want to first combine the date arrived with the time arrived. Note how this is done using the DHMS function:

```
ARRIVETIME=DHMS(DTARRIVE,0,0,TIMEARRIVE);
```

The first parameter is the date, and because we don't have *hours* and *minutes*, we put `TIMEARRIVE` in the *seconds* argument. Once we have `ARRIVETIME`, we can calculate the number of seconds between the two *datetimes*. This is done both by subtraction and with the INTCK function. Run the program and observe the results given in Table B.5.

2. Calculate minutes and hours using two methods as follows:

```
MINUTES1=SECONDS1/60;
HOURS1=MINUTES1/60;
MINUTES2=INTCK('MINUTE',ARRIVETIME,DATETIMELEAVE);
HOURS2=INTCK('HOUR',ARRIVETIME,DATETIMELEAVE);
```

Resubmit the code and observe the output. Note the difference in the results. INTCK gives the integer portion of the results, whereas a straight calculation preserves the fraction.

3. Extract the month and day of the month from `DTARRIVE` using the MONTH and DAY functions:

```
MONTH=MONTH(DTARRIVE);
DOM=DAY(DTARRIVE);
```

Run the code and observe the results. The month and day of the month are reported as numbers. See Hands-On Example B.7 for an example of how to display the month and day of the month in words.

TABLE B.5 Using Date and Time Functions

Obs	DTARRIVE	TIMEARRIVE	ARRIVETIME	DATETIMELEAVE	SECONDS1	SECONDS2
1	11/10/2016	13:13:00	10NOV2016:13:13:00	10NOV2016:18:49:19	20179	20179
2	12/12/2016	9:20:00	12DEC2016:09:20:00	12DEC2016:13:23:22	14602	14602

B.5 CHARACTER FUNCTIONS

Character functions allow you to manipulate text values. These functions are essential when you must extract information from text values, when you need to create a new text value from several pieces of information, or when you need to "clean up" text values to make them into better data values for analysis.

Table B.6 lists some of the common functions used to manipulate character/text variables.

TABLE B.6 Character Functions

Function	Description
CAT(text1, text2,...) CATS(text1, text2,...) CATT(text1, text2,...)	Concatenates text without removing leading or trailing blanks. CATS removes leading and trailing blanks and CATT removes trailing blanks. For example, RESULT=CAT("GEORGE"," WASHINGTON"); Returns the value GEORGE WASHINGTON, and For example, RESULT=CATS("GEORGE"," WASHINGTON"); Returns the value GEORGEWASHINGTON
CATX (separator, text1, text2,...)	Removes leading and trailing blanks and inserts a separator between concatenated components. For example, RESULT = CATX(":","GEORGE"," WASHINGTON"); Returns the value GEORGE:WASHINGTON
COMPBL(text)	Removes multiple blanks between words in a character string
COMPRESS(text<,characters-to-remove>)	Removes specified characters from a character string
DEQUOTE(text)	Removes quotation marks from text
FIND(text, texttofind <,'modifiers'> <,start>)	Finds an instance of the texttofind within the text. Start is an optional value indicating where in the text to start looking and modifiers are I for ignore case and T for ignore blanks. For example, RESULT=FIND("aBcDef", "DEF","I"); Returns a value of 4 because DEF begins at the fourth character of ABCDEF (and the case is ignored). If no match is found, a value of 0 is returned

(Continued)

TABLE B.6 (*Continued*)

Function	Description
INDEX(*text*,*texttofind*)	Searches text for specified *texttofind* string and returns a number where the text begins, or 0
LEFT(*text*)	Left-aligns a character string
LENGTH(*text*)	Returns the length of a text string
LOWCASE(*text*)	Converts all letters to lowercase
QUOTE(*text*)	Adds double quotation marks to text
REPEAT(*text*,*n*)	Repeats text *n* times. For example, R = repeat("ABC",3); Results in R = "ABCABCABC"
REVERSE(*argument*)	Reverses a text expression
RIGHT(*argument*)	Right-aligns text
SCAN(*text*,*n*,<*delimiters*>)	Returns the *n*th word from a text expression. For example, FIRST = SCAN("JOHN SMITH",1,' '); returns JOHN and LAST = SCAN("JOHN SMITH",2,' '); returns SMITH
SOUNDEX(*text*)	Turns text into a sounds-like argument. For example, SOUNDEX("ALLEN") = SOUNDEX("ALAN"); matches and SOUNDEX("ALLEN") = SOUNDEX("MELON"); does not
SUBSTR(*text*,*start*<,*n*>)	Searches for and replaces text. For example, VAR = SUBSTR("LINCOLN",3,2); returns VAR = "NC" and SUBSTR("LINCOLN",3,2) = "XX"; returns LIXX
TRANSLATE(*text*,*to1*,*from1* <,...*ton*,*fromn*>)	Replaces characters in text with specified new characters. For example, TRANSLATE('LINCOLN',"X","C"); returns LINXOLN
TRANWRD(*source*,*target*, *replacement*)	Replaces or removes all occurrences of a word in a text string. For example, TRANWRD('SPECIAL WED MEAL','WED','MON'); returns 'SPECIAL MON MEAL'
TRIM(*text*)	Removes trailing blanks from the text. (Returns one blank if the expression is missing.)
TRIMN(*text*)	Removes trailing blanks from the text. (Returns a null string if the expression is missing.)
UPCASE(*text*)	Converts text to uppercase
VERIFY(*source*,*excerpt1*<,... *excerpt-n*)	Searches text and returns the position of the first nonmatched character in the first argument. VERIFY("12345","54301") returns 2 because the second character doesn't match any character in the second argument

HANDS-ON EXAMPLE B.3

1. Open the program file `FUNC_CHAR.SAS`. This program illustrates how you could separate two numeric values from a text value. In this case, systolic and diastolic blood pressures were recorded as a single text value such as 120/80. Run this code and observe how the two values are extracted and made into two numeric values SBP and DBP.

```
DATA BP;
INPUT BP $10.;
I=FIND(BP,"/");
SBP=SUBSTR(BP,1,I-1); * get number before the /;
DBP=SUBSTR(BP,I+1); *get number after the /;
DROP I;
DATALINES;
120/80
140/90
110/70
;
PROC PRINT DATA=BP;RUN;
```

2. Note the code `I=FIND(BP,"/")`. What value is the variable I set to on the first record?
3. What does the code `SUBSTR(BP,1,I-1)` do?
4. What does the code `SUBSTR(BP,I+1)` do?

HANDS-ON EXAMPLE B.4

1. Open the program file `FUNC_SCAN.SAS`. The goal of this program is to separate a series of names into first and last names. In this case, we don't care about middle initials. Run the program and observe the results in Table B.7.

```
DATA NAMES;
FORMAT FIRST LAST MAYBE $20.;
INPUT @1 NAME $30.;
FIRST=SCAN(NAME,1," ");
LAST=SCAN(NAME, -1);
MAYBE=SCAN(NAME,-2);
IF MAYBE=FIRST THEN MAYBE="";
IF LENGTH(MAYBE)=1 THEN MAYBE="";
LASTNAME=CAT(MAYBE, LAST);
FULLNAME=CAT(FIRST,TRIM(LASTNAME));
```

(Continued)

(Continued)

```
DATALINES;
Alfred J. Prufrock
Benjamin Harrison
George H. W. Bush
Vincent Van Gogh
A. C. Elliott
;
RUN;
PROC PRINT DATA=NAMES;RUN;
```

2. Put your name in the data segment as first name, initial, and last name. Run the program to see if it separates your name correctly.
3. Look at the three uses of the SCAN function

```
FIRST=SCAN(NAME,1," ");
LAST=SCAN(NAME, -1);
MAYBE=SCAN(NAME,-2);
```

What does each do? What do the negative numbers as arguments mean?
4. What does the LENGTH function do? How is it used to tell if there is a middle name?
5. What does the CAT function do? Replace it with a CATS function when creating FULLNAME, rerun the code. What changes? Finally, note the TRIM function used in the FULLNAME statement. It is needed to trim off any extra blanks in LASTNAME.

TABLE B.7 SCAN, LENGTH, and CAT Functions

Obs	FIRST	LAST	MAYBE	NAME	LASTNAME	FULLNAME
1	Alfred	Prufrock		Alfred J. Prufrock	Prufrock	Alfred Prufrock
2	Benjamin	Harrison		Benjamin Harrison	Harrison	Benjamin Harrison
3	George	Bush		George H. W. Bush	Bush	George Bush
4	Vincent	Gough	Van	Vincent Van Gough	Van Gough	Vincent Gough
5	A.	Elliott		A. C. Elliott	Elliott	A. Elliott

B.6 TRUNCATION FUNCTIONS

Truncation Functions are used to extract certain portions of a number such as the smallest integer, largest integer, nearest integer, and so on. These are described in Table B.8.

B.7 FINANCIAL FUNCTIONS

Financial functions allow you to perform common financial calculations on values. There are over 50 such functions available in SAS, too many to describe here. A few of the most commonly used financial functions are listed in Table B.9.

TABLE B.8 Truncation Functions

Function	Description
CEIL(x)	Returns the smallest integer greater than or equal to the argument
FLOOR(x)	Returns the largest integer less than or equal to the argument
FUZZ(x)	Returns the nearest integer if the value is within 1E-12
INT(x)	Returns the decimal portion of the value of the argument
ROUND(x, roundoffunit)	Rounds a value to the nearest round-off unit. For example, ROUND(1.2345,.01) returns the value 1.23

TABLE B.9 Special Use Functions

Function	Description
IRR(period,cash0,cash1,...)	Returns the internal rate of return as a percentage
MORT(amount,payment,rate, number)	When you set the PAYMENT variable to missing, this returns the payment for a loan calculation. For example, when you run DPAYMENT.SAS DATA LOAN; PAYMENT= MORT(30000,.,.06/12,48); RUN; PROC PRINT;RUN; the value of (the monthly) PAYMENT is 704.55
NPV(rate,period,cash0, cash1,...)	Returns the net present value expressed as a percentage

HANDS-ON EXAMPLE B.5

1. Open the program file FUNC_MORT.SAS. This program uses the MORT function to calculate the payments for a loan. This example displays payments for different loans, interest, and the number of payments. Run the program and observe the results in Table B.10.

```
DATA LOAN;
FORMAT VALUE DOLLAR12.
       INTRATE1-INTRATE3 4.2
         NUMBER 4.;
```

(Continued)

(Continued)

```
  INPUT VALUE INTRATE1-INTRATE3 NUMBER;
  ARRAY INT(3) INTRATE1-INTRATE3;
  ARRAY PAY(3);
  DO I=1 to 3;
      PAY(I) = MORT(VALUE,.,INT(I)/12,NUMBER);
  END;
  LABEL MORT="Mortgage Payment";
  LABEL INTRATE1="Int. Rate 1 ";
  LABEL INTRATE2="Int. Rate 2 ";
  LABEL INTRATE3="Int. Rate 3 ";
  LABEL PAYMENT="Amount of Payment";
  LABEL VALUE="Loan Amount";
  LABEL NUMBER="Num. Payments";
  DROP I;
  DATALINES;
  200000 .05 .04 .03 360
  200000 .05 .04 .03 180
  350000 .05 .04 .03 360
  350000 .05 .04 .03 180
  ;
  RUN;
  PROC PRINT LABEL DATA=LOAN ;
  RUN;
```

2. Add lines to the data for $200,000 loan for a 30- and a 15-year mortgage with 0.05, 0.04, and 0.03 interest rates. Run the program and observe the results.

TABLE B.10 Example of MORT Function

Obs	Loan Amount	Int. Rate 1	Int. Rate 2	Int. Rate 3	Num. Payments	PAY1	PAY2	PAY3
1	$200,000	0.05	0.04	0.03	360	1073.64	954.83	843.21
2	$200,000	0.05	0.04	0.03	180	1581.59	1479.38	1381.16
3	$350,000	0.05	0.04	0.03	360	1878.88	1670.95	1475.61
4	$350,000	0.05	0.04	0.03	180	2767.78	2588.91	2417.04

B.8 SPECIAL USE FUNCTIONS

The term "special use function" is a catch-all phrase referring to other functions that are useful in SAS. They include functions that provide ways to convert variables from one type to another, create macro variables, create lag values, and determine differences in distance using zip codes. These functions are listed in Table B.11.

TABLE B.11 Special Use Functions

Function	Description
`INPUT(string,informat)`	The `INPUT` function converts a character variable to character or numeric

Character → INPUT → Character or numeric

For example,
`s = "1,212";`
`x = input(s,comma5.);`
returns x = 1212 (a numeric value).
Note: INPUTC(*string1,char-informat*) and INPUTN(*string1, num-informat*) are similar to `INPUT()`, but you can assign the format within the code

`PUT(x1,format)`	The `PUT` function converts variables from character or numeric to character

Character or numeric → PUT → Character

In the function, `x1` is a character or numeric. The PUT function returns a character value. For example,
`x = PUT(1234,comma6.);`
returns the string value 1,234
Note: PUTC(x1,char-informat) and PUTN(x1,num-informat) are similar to `PUT()`, but you can assign the format within the code

`CALL SYMPUT` and `CALL SYMPUTN`	Not strictly functions, but these routines assign a value produced in a `DATA` step to a macro variable. The formats for these routines are as follows:
	`CALL SYMPUT(macro-var,cval);`
	`CALL SYMPUTN(macro-var,nval);`
	where `cval` is some character value or variable and `nval` is some numeric value or variable name. For example, to store the value of the character variable MYVAR into the macro variable named LEVEL, use the command
	`CALL SYMPUT('LEVEL',MYVAR);`
	You can then reference the variable as &LEVEL elsewhere in your SAS code

`LAGn(x)`	Returns the value of the *n*th previous observation. Example: If your data are x = 10, 30, 15, then `LAG2(x)` for the third item would be 10. `LAG()` is the same as `LAG1()`
`DIFn(x)`	Returns the difference in the current value and the *n*th previous value. Example: If your data are x = 10, 30, 15, then `DIF2(x)` of the third item would be 5 (15 − 10 = 5). `DIF()` is the same as `DIF1()`

(*Continued*)

TABLE B.11 *(Continued)*

Function	Description
ZIPCITY(zipcode)	Converts a ZIP code to a city and state. For example, the expression CITY = ZIPCITY(75137); where the zipcode (in this case, 75137) is a character or numeric variable that returns a value of "Duncanville, TX" for the CITY variable
ZIPCITYDISTANCE(zipcode1,zipcode2)	Returns the distance (in miles) between the two ZIP codes
ZIPSTATE(zipcode)	Converts a ZIP code to a two-letter state code. For example, the expression STATE = ZIPSTATE("75137") ; where the zipcode (in this case, "75137") is a character or numeric variable that returns a value of "TX" for the STATE variable

HANDS-ON EXAMPLE (PUT AND INPUT) B.6

1. Open the program file FUNC_CONVERT.SAS. Run the code and observe the output in Table B.12. This program demonstrates the INPUT and PUT functions and illustrates how you can convert character variables to numeric and numeric variables to the character.

```
DATA TEMP;
LENGTH CHAR4 $ 4;
INPUT NUMERIC CHAR4;
* CONVERT CHARACTER VARIABLE TO NUMERIC; NEW_NUM = INPUT
(CHAR4,BEST4.);
* CONVERT NUMERIC VARIABLE TO CHARACTER; NEW_CHAR = PUT
(NUMERIC,4.0);
DATALINES;
789.1 1234
009.2 0009
1.5 9999
;
PROC PRINT;
FORMAT NEW_NUM 6.1 NEW_CHAR $8.;
RUN;
```

Note that the converted numbers (NEW_CHAR) have been rounded to integers because they were converted using a 4.0 format in the PUT statement.

TABLE B.12 Output from PUT and INPUT Example

Obs	CHAR4	NUMERIC	NEW_NUM	NEW_CHAR
1	1234	789.1	1234.0	789
2	0009	9.2	9.0	9
3	9999	1.5	9999.0	2

HANDS-ON EXAMPLE (PUT WITH MONNAME AND DONAME) B.7

1. Open the program file FUNC_SPECIAL.SAS. This example shows how to use the PUT function along with the MONNAME and DOWNAME formats to create variables that contain the name of the week and the name of the month. Run this program and observe the results given in Table B.13.

```
DATA DATES;
FORMAT BDATE WEEKDATE17.;
INPUT @1 BDATE MMDDYY8.;
BDATEDOW = PUT(BDATE,DOWNAME.);
BDATEMONTH = PUT(BDATE,MONNAME.);
DATALINES;
08212016
07041776
01011900
;
PROC PRINT DATA=DATES; RUN;
```

2. Add these lines after the BDATEMONTH line. Rerun the program and observe the output.

```
BDAY = DAY(BDATE);
BYEAR = YEAR(BDATE);
```

This illustrates how to extract various components from a date.

TABLE B.13 Convert Dates to Components

Obs	BDATE	BDATEDOW	BDATEMONTH
1	Sun, Aug 21, 2016	Sunday	August
2	Thu, Jul 4, 1776	Thursday	July
3	Mon, Jan 1, 1900	Monday	January

HANDS-ON EXAMPLE (LAG) B.8

1. Open the program file FUNC_LAG.SAS. This program uses the LAG function to calculate the difference in price of gold from the previous year. Run this program and observe the results in Table B.14.

```
DATA PRICE;
INPUT YEAR PRICE;
 LAG_PRICE = LAG(PRICE);
 DIFF_PRICE = PRICE - LAG_PRICE;
 PCT_INCREASE = (DIFF_PRICE/LAG_PRICE)*100;
```

(*Continued*)

(*Continued*)

```
DATALINES;
2010   1113.00
2011   1405.50
2012   1590.00
2013   1681.50
2014   1219.75
2015   1184.25
2016   1075.20
2017   1162.00
2018   1312.80
2019   1287.20
2020   1520.55
2021   1946.60
2022   1800.10;
RUN;
PROC PRINT;RUN;
```

2. Change the LAG function to LAG2 to find the 2-year changes. Rerun the program and observe the output.

TABLE B.14   January Open Prices, Yearly Change

Obs	YEAR	PRICE	LAG_PRICE	DIFF_PRICE	PCT_INCREASE
1	2010	1113.00			
2	2011	1405.50	1113.00	292.50	26.2803
3	2012	1590.00	1405.50	184.50	13.1270
4	2013	1681.50	1590.00	91.50	5.7547
5	2014	1219.75	1681.50	−461.75	−27.4606
6	2015	1184.25	1219.75	−35.50	−2.9104
7	2016	1075.20	1184.25	−109.05	−9.2084
8	2017	1162.00	1075.20	86.80	8.0729
9	2018	1312.80	1162.00	150.80	12.9776
10	2019	1287.20	1312.80	−25.60	−1.9500
11	2020	1520.55	1287.20	233.35	18.1285
12	2021	1946.60	1520.55	426.05	28.0195
13	2022	1800.10	1946.60	−146.50	−7.5259

HANDS-ON EXAMPLE (ZIP FUNCTIONS) B.9

A hospital is treating people with rare diseases from all over the world. They want to study the distance their patients have to travel for treatment based on the hospital and subject's zip code. The following program calculates that distance using U.S. ZIP code functions.

1. Open the program file FUNC_ZIP.SAS.

```
DATA ZIP;
INPUT BEGIN $ END $;
HOSPITAL = ZIPCITY(BEGIN);
SUBJECT = ZIPCITY(END);
DISTANCE = ZIPCITYDISTANCE(BEGIN, END);
DATALINES;
75275 75201
75275 10001
75275 96801
75275 96951
;
PROC PRINT;
RUN;
```

Run this program and the results are given in Table B.15. There you can see, for instance, that the distance from Honolulu to the hospital in Dallas is 3795 miles.

2. You are curious how far it is from the smallest zip code (Holtsville, NY, 00501) to the largest (Ketchikan, AK, 99950) (at the time of this publication.) Put those values into the program, run it, and observe the distance. (By the way, the 96951 U.S. zip code is for Sonsong (Rota) Mariana Islands (MP) which is northeast of Guam. It is an organized, unincorporated territory of the United States.)

TABLE B.15 Find Distances Using the ZIP Functions

Obs	BEGIN	END	HOSPITAL	SUBJECT	DISTANCE
1	75275	75201	Dallas, TX	Dallas, TX	4.0
2	75275	10001	Dallas, TX	New York, NY	1371.5
3	75275	96801	Dallas, TX	Honolulu, HI	3794.7
4	75275	96951	Dallas, TX	Rota, MP	7228.3

B.9 QUICK GUIDE TO SAS FUNCTIONS

The following Figures B.1 and B.2 provide a summary of the most commonly used function in SAS These sheets are also available to print out from www.alanelliott.com. Click on the Quick Tips link.

SAS Functions Reference

SAS ESSENTIALS Ed. 3
by Alan Elliott and Wayne Woodward

How to use SAS Functions

SAS functions can be used in a DATA statement to calculate new variables or as a part of a conditional statement. For example, here is an example of how some simple functions might be used:

DATA NEWDATA;SET OLDDATA;
    TOTAL=SUM(A,B,C);
    SQAGE=SQRT(AGE);
    GENDER=UPCASE(SEX);
RUN;

Notice that in these cases the function on the right side of the "=" sign does something to the parameter(s) in the parenthesis and returns some value, which is then assigned to the name on the left side of the "=" sign.

The SUM function returns the total of the number A, B and C. The SQRT function returns the square root of the AGE variable, UPCASE function makes all character values in the SEX variable upper case. For example, if the SEX variable contains the value "f" then the GENDER variable will contain the value "F."

The syntax for a function is

FUNCTIONNAME(argument-1, argument-2...argument-k)

Where arguments may be constants, variables, expressions or other functions.

NOTE: Not all SAS Functions are described here — check SAS manuals for complete list.

Arithmetic/Mathematical Functions

(Note: Arguments for functions can be variables, expressions or other functions. Most of the examples below use constants.)

ABS() - Absolute Value, Ex: X=ABS(-9) returns the value 9
EXP() - Exponentiation, Ex: X=EXP(100) returns the value 2.688
LOG() - Natural Log, Ex: X=LOG(10) returns the value 2.3026
LOG10() - Log Base 10, Ex: X=LOG10 returns the value 1.0
FACT() - Factorial, Ex: FACT(7) returns the value 5040, which is (7*6*5*4*3*2*1)
MAX(x1,x2,x3...) maximum in list
MEAN() Mean of non-missing values in list. Ex: MEAN(2,4,.,6) returns the value 4
MEDIAN() Median of non-missing values in list
MIN(x1,x2,x3 ...) minimum in list
MOD(x1, modulo) remainder of x1/modulo Ex: MOD(5,3)=2, MOD(6,2)=0
N() - Number of non-missing values, Ex: N(1,2,.,4) returns the value 3.
NMISS() - Number missing, Ex: NMISS(1,2,.,4) returns the value 1.
ORDINAL(k,x1,x2,x3...) finds the kth number in the sorted list. ORDINAL(2,11,6,9,3) = 6 since it is 2nd in the ordered list.
RANUNI(0) - A different random number (uniform distribution) each time
RANUNI(X) - Where X is a "seed" value you enter returns the same random sequence (uniform) each time for that particular seed.
RANNOR() - A random number from a normal distribution with mean 0 and standard deviation of 1 (See seed info for RANUNI)
SIGN(x) - The sign of x
SUM() - Sum of non-missing values in list
SQRT() - Square Root, Ex: X=SQRT(9) returns the value 3

Special Use Functions

Variable conversion/re-read functions
INPUT(string,informat) EX: s="1,212";x=input(s,comma5.); Returns x=1212 (a numeric value.) (Note: INPUTC(string1,char-informat) and INPUTN(string1,num-informat) are similar to INPUT(), but you can assign the format within the code.)

PUT (X1,format) where X1 is string or numeric, returns a character value. EX: x=PUT(1234,comma6.); returns the string value "1,234."(Note: PUTC(x1,char-informat) & PUTN(x1,num-informat) are similar to PUT(), but you can assign format within the code.)

Working with previous observations
LAGN() returns the value of the nth previous observation. Ex: Your data has values x=1,2,3. Then LAG2(x) for the third item would be 1. LAG() is the same as LAG1()

DIFN() returns the difference in the current value and the nth previous value. Ex: Your data has values x=1,2,3. DIF2(x) of third item would be 2 (3-1=2). DIF() is the same as DIF1().

IN Operator—Not a function but VAR IN(1 2 3) or VAR IN("A" "B" "C") returns TRUE if VAR is in the specified list.

Trignometric Functions

ARCOS(argument) -arccosine
ARSIN(argument) - arcsine
ATAN(argument) - arctangent
COS(argument) - cosine
COSH(argument) - hyperbolic cosine
SIN(argument)- sine
SINH(argument) - hyperbolic sine
TAN(argument) - tangent
TANH(argument) - hyperbolic tangent

Figure B.1 Quick Guide to SAS Functions, page 1

SAS Functions, Page 2

Date and Time Functions

DATDIF(sdate,edate,basis) - num days between two dates. Basis is (default) '30/360','ACT/ACT', 'ACT/360', or 'ACT/365'
DATE() - current date (SAS date value)
DATEJUL(julian-date) - converts Julian date to SAS date
DATEPART(datetime) - extracts date from SAS datetime
DATETIME() - current date and time of day
DAY(date) - day of month from a SAS date
DHMS(date,hour,minute,second) - SAS datetime value from date, hour, minute, and second
HMS(hour,minute,second) - SAS time value from hour, minute, and second
HOUR(<time | datetime>) - hour from a SAS time or datetime value
INTCK('interval',from,to) - number of time intervals in a given time span. 'Interval' can be:
  Date Intervals: DAY, WEEKDAY, WEEK, TENDAY, SEMIMONTH, MONTH, QTR, SEMIYEAR, YEAR
  Datetime Intervals: DTDAY, DTWEEKDAY, DTWEEK, DTTENDAY, DTSEMIMONTH, DTMONTH, DTQTR, DTSEMIYEAR, DTYEAR
  Time Intervals: HOUR, MINUTE, SECOND
INTNX('interval',start-from,increment<,'alignment'>) - advances a date, time, or datetime value by a given interval, and returns a date, time, or datetime value — see intervals above
JULDATE(date) - Julian date from a SAS date
MDY(month,day,year) - SAS date from month, day, year
MINUTE(time | datetime) - minute from a SAS time or datetime value
MONTH(date) - month from a SAS date value
QTR(date) - quarter of the year from a SAS date value
SECOND(time | datetime) - second from a SAS time or datetime value
TIME() - current time of day
TIMEPART(datetime) - time value from a SAS datetime
TODAY() - current date as a SAS date
WEEKDAY(date) - day of the week from a SAS date
YEAR(date) - year from a SAS date value
YRDIF(startdate,enddate,basis) - diff in years between 2 dates. Basis is '30/360','ACT/CT', 'ACT/360', or 'ACT/365'
YYQ(year,quarter) - SAS date from year and quarter

Character Functions

CAT(text1, text2, ...) concatenate text without removing leading or trailing blanks. CATX removes leading & trailing blanks, inserts separators. CATS removes leading & trailing blanks and CATT removes trailing blanks.
COMPBL(text) - removes multiple blanks between words in a character string
COMPRESS(text<,characters-to-remove>) - removes specified characters from a character string
DEQUOTE(argument) - removes quotation marks from text
FIND(text, texttofind <,'modifiers'> <,start>) - find the texttofine within the text. Example FIND(LINCOLN,"CO") returns a 4. Start is optional indicating where in text to start looking. Modifiers are I (ignore case) and T (ignore blanks). Example:
  `RESULT=FIND("aBcDef", "DEF","I")`
INDEX(text,excerpt) searches text for specified string — see also INDEXC and INDEXW
LEFT(text) - left-aligns a character string
LENGTH(text) - length of text
LOWCASE(text) - converts all letters to lowercase
QUOTE(text) - adds double quotation marks to text
REPEAT(text,n) repeats text n times
REVERSE(argument) - reverses a text expression
RIGHT(argument) - right-aligns a text
SCAN(argument,n<,delimiters>) -returns the nth word from a text expression. Example:
  First=scan("JOHN SMITH",1,' '); returns "JOHN"
  Last=scan("JOHN SMITH",2,' '); returns "SMITH"
SOUNDEX(text) - turns text into "sounds like" argument soundex("ALLEN")=soundex("ALAN"); matches soundex("ALLEN")=soundex("MELON"); does not
STRIP(text) - strips leading and trailing blanks
SUBSTR(text,start<,n>) Search and replace. Example:
  var=SUBSTR("LINCOLN",3,2); returns var ="NC"
  SUBSTR("LINCOLN",3,2)="XX"; returns LIXX
TRANSLATE(text,to1,from1<,...ton,fromn>) - replaces characters in text with specified new characters
  TRANSLATE('LINCOLN',"X","C"); returns LINXOLN
TRANWRD(source,target,replacement) replaces or removes all occurrences of a word in a text string
  TRANWRD('SPECIAL WED MEAL','WED','MON');
  Returns 'SPECIAL MON MEAL'
TRIM(argument) - removes trailing blanks from text. (Returns one blank if the expression is missing.) (Also see TRIMN)
UPCASE(argument) - converts text to uppercase
VERIFY(source,excerpt-1<,...excerpt-n) - searches text, returns position of 1st non-matched character in 1st argument. VERIFY ("12345","54301") returns 2 since 2nd character doesn't match any character in 2nd argument.

Selected Misc Functions

IRR(period,cash0,cash1,...) internal rate of return as a percentage
MORT(amount,payment,rate,number) loan calculation. Example carloan MORT(30000,.,.06/12,48); (Payment set at .) returns $704.55 monthly payment.
NPV(rate,period,cash0,cash1,..) net present value expressed as
ZIPSTATE(zipcode) convert ZIP codes to two-letter state codes
ZIPNAMEL(zipcode) convert ZIP codes to state name (upper and lowercase)

Truncation Functions

CEIL(value) - smallest integer GE argument
FLOOR(value) - largest integer LE argument
FUZZ(value) - nearest integer if value is within 1E-12
INT(value) - decimal portion of the value of the argument
ROUND(value,unit) - value to the nearest roundoff unit

© Alan Elliott and Wayne Woodward, 2022

SAS ESSENTIALS Ed. 3
by Alan Elliott and Wayne Woodward
www.alanellitt.com/sas

Figure B.2 Quick Guide to SAS Functions, page 2

APPENDIX C

CHOOSING A SAS® PROCEDURE

The guidelines in this appendix help you determine the proper analysis to perform based on your data. In these tables, the term *Normal* indicates that the procedure is based on a normality assumption. The term *at least ordinal* indicates that the data have an order.

C.1 DESCRIPTIVE STATISTICS

Make a decision by reading from the left to the right (Table C.1).

TABLE C.1 Decision Table for Descriptive Statistics

	What Is the Data Type?	Statistical Procedures/SAS PROC
You want to describe a single variable.	Normal	Descriptive statistics: mean, standard deviation, etc. `PROC MEANS, UNIVARIATE` (Chapter 10)
	Quantitative	Descriptive statistics, median, histogram, boxplots `PROC UNIVARIATE` (Chapters 10 and 19)
	Categorical	Frequencies `PROC FREQ, GCHART` (Chapters 11 and 19)
You want to describe two related or paired variables.	Both are normal	Pearson's correlation and graphs `PROC CORR, REG, GPLOT` (Chapters 13 and 19)
	Both are at least ordinal	Spearman's correlation and graphs `PROC CORR,GPLOT` (Chapters 13 and 19)
	Both are categorical	Cross-tabulation (Chapters 11 and 19)

SAS® Essentials: Mastering SAS for Data Analytics, Third Edition. Alan C. Elliott and Wayne A. Woodward.
© 2023 John Wiley & Sons, Inc. Published 2023 by John Wiley & Sons, Inc.
Companion website: https://www.alanelliott.com/SASED3/

C.2. COMPARISON TESTS

Make decision by reading from the left to the right (Table C.2).

TABLE C.2 Decision Table for Comparison Tests

	What Is the Data Type?	Procedure to Use/SAS PROC
You are comparing a single sample to a norm (gold standard).	Normal	Single-sample t-test `PROC MEANS, UNIVARIATE` (Chapter 12)
	At least ordinal	Sign test `PROC UNIVARIATE` (Chapters 10 and 16)
	Categorical	Goodness-of-fit `PROC FREQ` (Chapter 11)
You are comparing data from two independent groups.	Normal	Two-sample t-test `PROC TTEST` (Chapter 12)
	At least ordinal	Mann–Whitney `PROC NPAR1WAY` (Chapter 16)
	Categorical	Test for homogeneity chi-square `PROC FREQ` (Chapter 11)
You are comparing paired, repeated, or matched data	Normal	Paired t-test `PROC TTEST, UNIVARIATE` (on difference) (Chapters 10 and 12)
	At least ordinal	Sign test or Wilcoxon signed-rank test `PROC NPAR1WAY` (Chapter 16)
	Binary (dichotomous)	McNemar `PROC FREQ/AGREE` (Chapter 11)
More than two groups – INDEPENDENT	Normal	One-way ANOVA `PROC ANOVA, GLM` (Chapter 14)
	At least ordinal	Kruskal–Wallis `PROC NPAR1WAY` (Chapter 16)
	Categorical	$r \times c$ test for homogeneity/chi-square, or kappa for interrater reliability `PROC FREQ, FREQ/KAPPA` (Chapter 11)
More than two groups – REPEATED MEASURES	Normal	Repeated measures `PROC GLM` (only for one way) or `PROC MIXED` (Chapter 15)
	At least ordinal	Friedman ANOVA `PROC FREQ` (Chapter 16)
	Categorical	Cochran Q (not discussed)
Comparing means, model includes covariate adjustment	Normal	Analysis of covariance `PROC GLM` (Chapter 15)

C.3. RELATIONAL ANALYSES (CORRELATION AND REGRESSION)

Make a decision by reading from the left to the right (Table C.3).

TABLE C.3 Decision Table for Relational Analysis

	What Is the Data Type for the Dependent (Response) Variable?	Procedure to Use/SAS PROC
You want to analyze the relationship between two variables (if regression, one variable is classified as a response variable and the other a predictor variable).	Normal	Pearson correlation, simple linear regression PROC CORR, REG (Chapter 13)
	At least ordinal	Spearman correlation PROC CORR (Chapter 13)
	Categorical	$r \times c$ contingency table analysis PROC FREQ (Chapter 11)
	Binary	Logistic regression PROC LOGISTIC (Chapter 17)
You want to analyze the relationship between a response variable and two or more predictor variables.	Normal	Multiple linear regression PROC REG (Chapter 13)
	Binary	Logistic regression PROC LOGISTIC (Chapter 17)

In this table the term "Data type" applies to the dependent variable for regression procedures. For assessment of association (such as correlation or crosstabulation), the data type applies to both variables.

APPENDIX D

SAS CODE EXAMPLES

This appendix provides a series of brief code examples of basic SAS® tasks. Most of the tasks are listed here without explanation, except for a reference to a chapter.

Ways to name a dataset – can be used whenever a dataset is named:

```
DATA SOMEDATA;    * create working/temporary dataset;
DATA "C:\SASDATA\SOMEDATA"; *create permanent dataset
   in specified location;
LIBNAME MYSASLIB "C:\SASDATA"; * create a temporary SAS library;
DATA MYSASLIB.SOMEDATA;   * create permanent dataset in MYSASLIB
library;
```

Versions of the INPUT statement:

```
* freeform input;
INPUT ID $ GENDER $ SBP DBP WEIGHT;
* column numbers specify location of data;
INPUT ID 1-4 SEX $ 5-5 AGE 6-9;
* specify beginning column, name, informat.
INPUT @1 ID 4. @5 SEX $1. @10 BDATE DATE9.;
```

Read date using format specification (see format table, Appendix A):

```
DATA MYDATA;
INPUT @1 FNAME $11. @12 LNAME $12. @24 BDATE DATE9.;
DATALINES;
```

SAS® Essentials: Mastering SAS for Data Analytics, Third Edition. Alan C. Elliott and Wayne A. Woodward.
© 2023 John Wiley & Sons, Inc. Published 2023 by John Wiley & Sons, Inc.
Companion website: https://www.alanelliott.com/SASED3/

```
Bill    Smith    08JAN1952
;
RUN;
```

Ways to specify the source of data or enter the dataset:

```
DATALINES;          * data are listed in the code following
   this statement;
INFILE filename;    * data are in an ASCII file on disk;
SET dataset;        * data are in a SAS data file;
PROC IMPORT;        * imports data from a non-SAS filetype;
```

Common ways to manipulate data within the DATA step (Chapters 4–7):

```
* Conditionally specify missing value;
IF AGE LE 0 then AGE = .;
IF SBP GE 140 THEN HIGHBP=1; ELSE HIGHBP=0; * If-Then-Else;
IF GENDER="MALE";              * Subsetting if statement;
* Define label for a variable name;
LABEL ID="Identification";
AREA=LENGTH * WIDTH;           * Calculate a new variable;
```

Titles and footnotes (Chapter 5)

```
TITLE 'First title line';
TITLE2 'Up to 9 title lines';
FOOTNOTE 'First footnote line';
FOOTNOTE9 'Up to 9 footnote lines';
```

ODS – Output Delivery System (Chapter 8)

```
ODS TRACE ON;
ODS GRAPHICS ON;
ODS LISTING; *standard output;
ODS HTML [BODY='FILENAME.HTML'];
ODS PDF [FILE='FILENAME.PDF'];
ODS RTF [FILE='FILENAME.RTF'];
ODS PS [FILE='FILENAME.PS'];
  PROC MEANS * PUT PROCS HERE;
ODS TYPE CLOSE;
ODS TRACE OFF;
ODS GRAPHICS OFF;
```

SQL Examples (Chapter 9)

```
      * Sample SQL program;
      PROC SQL;
      SELECT ID, GP, AGE
              FROM MYSASLIB.SOMEDATA
      WHERE GENDER CONTAINS "Female"
```

```
        ORDER BY ID;
        QUIT;
        * Sample SQL code to create a SAS dataset;
        PROC SQL;
            CREATE TABLE MYQUERY AS
            SELECT * FROM MYSASLIB.SOMEDATA;
        QUIT;

*Sample SQL code show order of clauses;
        PROC SQL;
           SELECT column(s)
           FROM table-name | view-name
           WHERE expression
           GROUP BY column(s)
           HAVING expression
           ORDER BY column(s);
        QUIT;
```

PROC FORMAT – used to define custom formats for values (Chapter 4)

```
PROC FORMAT;
VALUE $FMTSEX  'F' = 'Female'
               'M' = 'Male';
VALUE   FMTYN  1='Yes'
               2='No';
PROC PRINT;      * apply a format to variables;
FORMAT SEX $FMTSEX.
       QUEST FMTYN.;
RUN;
```

PROC SORT – sort a dataset (Chapter 4)

```
PROC SORT DATA=dataset;
   BY AGE SBP DESCENDING;
RUN;
```

PROC PRINT – create a listing of the data (Chapter 5)

```
PROC PRINT DATA=dataset;
   VAR varlist;
   SUM sumvar;
RUN;
```

PROC MEANS – used to calculate statistics for quantitative data (Chapter 9)

```
PROC MEANS N MEAN MEDIAN MAXDEC=2 DATA=dataset;
  VAR varlist;
RUN;
PROC MEANS DATA=dataset; * Means by group;
  CLASS groupvar;
  VAR varlist);
RUN;
```

534 APPENDIX D: SAS CODE EXAMPLES

PROC UNIVARIATE – calculate detailed statistics on a variable (Chapter 10)

```
PROC UNIVARIATE DATA=dataset NORMAL PLOT;
   VAR SBP;
   HISTOGRAM SBP/NORMAL;
RUN;
```

PROC FREQ – frequencies and crosstabulations (Chapter 11)

```
PROC FREQ DATA=dataset; * basic frequency table;
   TABLES varlist;
RUN;
PROC FREQ DATA=dataset;
   TABLES var1*var2/CHISQ; * crosstabulation with chi-square;
RUN;
```

PROC TTEST – comparing means (Chapter 12)

```
PROC TTEST DATA=dataset H0=30; * Single sample t-test;
   VAR varname;
RUN;
PROC TTEST DATA=dataset; * independent group t-test;
   CLASS groupvar;
   VAR varname;
RUN;
PROC TTEST DATA=dataset; * paired t-test;
   PAIRED WBEFORE*WAFTER;
RUN;
```

PROC ANOVA, GLM, and MIXED – comparing more than three means (Chapters 14 and 15)

```
PROC ANOVA DATA=dataset; * one-way independent group;
   CLASS groupvar;
   MODEL depvar= groupvar;
   MEANS groupvar /TUKEY; RUN;
PROC GLM;             * two fixed-factors;
  CLASS factor1var factor2var;
  MODEL RESPONSE= factor1var factor2var factor1var*factor2var;
RUN;
PROC MIXED;          * two-factor, one random;
  CLASS randomfactor fixedfactor;
  MODEL depvar = fixedfactor;
  RANDOM randomfactor randomfactor*fixedfactor;
RUN;
```

PROC CORR – correlations between pairs of variables (Chapter 13)

```
PROC CORR DATA=SOMEDATA SPEARMAN PEARSON;
   VAR AGE TIME1-TIME4;
RUN;
```

PROC REG – linear regression analysis (Chapter 13)

```
PROC REG DATA=dataset; * simple linear regression;
    MODEL dependentvar = independentvar/R; * examine RESIDUALS;
RUN;
PROC REG DATA=dataset;    * multiple linear regresson;
    MODEL dependentvar = ind1 ind2 … etc
    /SELECTION=STEPWISE
    SLENTRY=0.05
    SLSTAY=0.05;
RUN;
```

PROC NPAR1WAY – nonparametric comparisons (Chapter 16)

```
PROC NPAR1WAY DATA=dataset;
    CLASS GROUP;       * compare independent groups;
    VAR SALARY;
RUN;
```

PROC LOGISTIC – logistic regression (Chapter 17)

```
PROC LOGISTIC DATA=dataset;
CLASS variables;
MODEL dependentvar = independentvar(s)
/ EXPB
  SELECTION=SETPWISE SLENTRY=0.05
  SLSTAY=0.1
  RISKLIMITS
  CTABLE
  OUTROC=ROC1;
RUN;
```

PROC FACTOR – factor analysis (Chapter 18)

```
PROC FACTOR DATA=dataset;
    METHOD=P PRIORS=SMC
    ROTATE=PROMAX SCREE CORR RES;
RUN;
```

PROC GPLOT – scatterplot with regression line (Chapter 19)

```
GOPTIONS RESET=ALL;
SYMBOL1 V=STAR I=RL;
PROC GPLOT DATA=dataset;
  PLOT yvar*xvar;
RUN;
QUIT;
```

PROC GCHART – barchart (Chapter 19)

```
PROC GCHART DATA=dataset;
  HBAR varname;
RUN;
```

PROC BOXPLOT – boxplot (Chapter 19)

```
PROC SORT DATA=dataset; BY groupvar;
PROC BOXPLOT DATA=dataset;
     PLOT varname*groupvar;
RUN;
```

PROC TABULATE – create tables (Chapter 20)

```
PROC TABULATE DATA=dataset ORDER=FORMATTED;
  CLASS TYPE AREA SITE;
  VAR SALES_K;
  TABLE(AREA="" ALL),(SITE="" ALL)*TYPE=""
        *MEAN=""*SALES_K=""*F=DOLLAR6.2;
  FORMAT AREA $FMTCOMPASS.;
RUN;
```

PROC REPORT – create reports (Chapter 20)

```
PROC REPORT DATA=MYSASLIB.CARS NOFS;
  COLUMN BRAND CITYMPG,SUV HWYMPG,SUV;
  DEFINE BRAND/ORDER;
  DEFINE BRAND/GROUP;
  DEFINE CITYMPG/ANALYSIS MEAN FORMAT=6.1;
  DEFINE HWYMPG/ANALYSIS MEAN FORMAT=6.1;
  DEFINE SUV/ ACROSS 'BY SUV';
  FORMAT SUV FMTSUV.;
RUN;
```

PUT STATEMENTS – create reports (Chapter 20)

```
DATA _NULL_; SET GRADES END=EOF;
TABLEHEAD="SUMMARY OF GRADES";
FILE PRINT;
IF _:=1 then DO; * PUT HEADER;
     L=5+(60-LENGTH(TABLEHEAD))/2; * CENTER HEAD;
     PUT @L TABLEHEAD;
     PUT @5 60*'=';
     PUT @5 "NAME" @20 "| TEST 1" @30 "| TEST 2 "
         @40 "| TEST 3" @50 "| AVERAGE";
     PUT @5 60*'-';
END;
PUT @5 NAME @20 "|" G1 5.1 @30 "|" G2 5.1
          @40 "|" G3 5.1 @50 "|" AVE 5.1;
IF EOF =1 then do; * PUT BOTTOM OF TABLE;
     PUT @5 60*'-';
     PUT @5 "CLASS AVERAGE " @; * HOLDS LINE FOR NEXT PUT;
     PUT @50 "| %TRIM(&CLASSMEAN) " ;
```

```
        PUT @5 60*'=';
END;
RUN;
```

Sample Complete SAS Program – This sample SAS program illustrates many of the components of the SAS DATA Step covered in this book. (See `SampleProgram.SAS`)

```
***** PROC FORMAT defines formats for categorical variables;
PROC FORMAT;
VALUE $fmtgender 'F' = 'Female'  'M' = 'Male';
VALUE  fmtversion  1='Test Version 1' 2='Test Version 2';
VALUE  fmtyn       1='Yes' 0='No';
RUN;
* Define up to 9 titles and footnotes;
TITLE 'Example SAS Program';
TITLE2 'From SAS Essentials by Elliott and Woodward';
FOOTNOTE 'Uses the SURVEY dataset';
FOOTNOTE3 'Illustrates typical SAS DATA statements';
* Begin a dataset by reading a current SAS dataset;
DATA NEW; SET MYSASLIB.SAMPLE;
****** apply formats to several variables;
FORMAT GENDER $fmtgender. VERSION fmtversion. SATISFIED
MARRIED FMTYN.;
FORMAT SCHOOLING $14.;
****** Create a new variable by calculation;
IF SATISFACTION GT 80 THEN SATISFIED=1; ELSE SATISFIED=0;
****** Create a new variable using IN statement;
IF ARRIVE IN("BUS" "CAR") THEN HOW="Arrive by Vehicle"; else
HOW='Arrived by Walking';
****** define labels;
LABEL  GENDER="Gender"
       MARRIED="Married"
       SATISFIED="Satisfied with Service"
       HOW="How Arrived at Clinic?"
       SCHOOLING="How Much School"
       AGE="Age in 2016";
****** set missing values;
IF TEMP GT 110 THEN TEMP=.;
IF GENDER="X" THEN GENDER="";
****** Use a subsetting IF to limit records in the dataset;
IF STAYMINUTES GE 20;
****** Recode a value using SELECT;
SELECT;
   WHEN (EDU LT 12 AND EDU GT 5) SCHOOLING="Less than 12";
   WHEN (EDU EQ 12) SCHOOLING="High School";
   OTHERWISE SCHOOLING="Some College";
END;
****** DROP (or KEEP) certain variables;
```

```
DROP SATISFACTION EDU;
RUN;
PROC PRINT LABEL DATA=NEW;
VAR GENDER AGE MARRIED HOW SATISFIED SCHOOLING;
RUN;
PROC MEANS MAXDEC=2 DATA=NEW;
CLASS SCHOOLING;
VAR TEMP STAYMINUTES;
RUN;
PROC FREQ DATA=NEW;
TABLES SCHOOLING*HOW/CHISQ NOPERCENT NOCOL;
RUN;
```

APPENDIX E

USING SAS® ONDEMAND FOR ACADEMICS WITH SAS ESSENTIALS

SAS **OnDemand for Academics** (ODA) is a free online (cloud) version of SAS software for data analytics, data mining, and forecasting. Although the interface, called **SAS Studio**, is different from a locally installed version of SAS, the program provides analytics software similar to that provided by the standard SAS version. In addition, **OnDemand** will run on a PC (Windows), Mac, or Linux workstation. The key difference is that in order to run the **OnDemand** version, you must have an active Internet connection. When you run(submit) a SAS program using **OnDemand**, the code is sent to a server (via the Internet), the code is processed there, and returned to your workstation.

This appendix provides a few examples of how you can use ODA to run the Hands-On Examples in *SAS Essentials*. The main differences between running a program in ODA compared to "regular SAS" is the user interface and how you access files. Not all examples from the book will run in **OnDemand**, so we recommend using the locally installed version of SAS when possible.

E.1 INSTALLING SAS ONDEMAND FOR ANALYTICS

Installation instructions for ODA may change (they have in the past), so we'll give a simple explanation, but recommend that you check the SAS website for specific installation information. Here is a summary of the instructions (for installation on Windows).

SAS® Essentials: Mastering SAS for Data Analytics, Third Edition. Alan C. Elliott and Wayne A. Woodward.
© 2023 John Wiley & Sons, Inc. Published 2023 by John Wiley & Sons, Inc.
Companion website: https://www.alanelliott.com/SASED3/

STEP 1: *Create a SAS Profile:* You must be a registered SAS user (i.e. create a SAS Profile) before you can use ODA. If you already have a SAS Profile, skip to the next step. Otherwise, go to the SAS ODA welcome page at
https://welcome.oda.sas.com/login
and select "Don't have a SAS Profile." Follow the instructions to create a profile.

STEP 2: *Register:* On the SAS **OnDemand for Academics** page, select your home region and click Submit. You'll get a confirmation screen, and will receive an email from SAS with a User ID. You'll use this User ID or your email address to sign in on the SAS ODA sign in page.

STEP 3: *Login:* From the ODS dashboard click on **SAS Studio**.

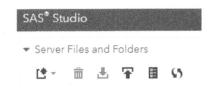

Figure E.1 SAS Studio server files and folders. If SAS Studio doesn't come up, you may need to choose a different browser. For example, use Microsoft Edge.

STEP 4: *Upload files:* At the left of the **SAS Studio** Screen, click on **Files (Home)**. Locate and click on the upload (up-arrow icon) as shown in Figure E.1 (third icon from the right). A dialog box shown as Figure E.2 appears. Click Choose File and select SECOND.SAS from the C:\SASDATA folder (or where-ever you have it stored.)

Figure E.2 Upload Files dialog box for SAS Studio

STEP 5: *Select a SAS program:* Note the list of files at the left of the **SAS Studio** window. See Figure E.3. This shows `SECOND.SAS` listed as well as some other files. Doubleclick on `SECOND.SAS` and code appears in the main **SAS Studio** window.

Figure E.3 List of Uploaded files in SAS Studio

STEP 6: *Run the program:* Click on the running man icon to see the results of this SAS program. See Table E.1.

TABLE E.1 Results of SECOND.SAS, run in SAS ODA

	The MEANS Procedure			
	Analysis Variable: AGE			
N	Mean	Std Dev	Minimum	Maximum
50	10.4600000	2.4261332	4.0000000	15.0000000

HANDS ON EXAMPLE E.1

1. Using the technique shown above, upload the SAS file named SAMPLEPROGRAM.SAS. We'll use this to provide a second illustration for using ODA. This program illustrates how to access a saved dataset named SURVEY. For this program to run properly, this file must be accessible.
2. To make the SAS dataset accessible, upload the file SURVEY.SAS7BDAT.
3. Create a SAS Library. Click on **Libraries** at the bottom left of the **Studio** screen. Click on the left icon under **Libraries** (**New Library**) A dialog box similar to Figure E.4 appears. Name this library MYSASLIB. Click on **Browse for Path** and select **Home**. The name that appears on your screen will be a little different than what is shown here. Also select **Recreate this library** to make the library permanent. Click **OK**. This makes your home directory a library named MYSASLIB, which allows programs from this book to run. (If you've uploaded appropriate datasets.)
4. Under Files(home) double click on SAMPLEPROGRAM.SAS. This will place SAMPLEPROGRAM.SAS in the **Code (editor) Window**. Click the running man icon. Examine the **Log Window** to make sure there are no errors in the program (there should not be.)
5. *A series of result tables will appear, showing the results of the program. Part of the first table is shown as Table E.2.*

(*Continued*)

(*Continued*)

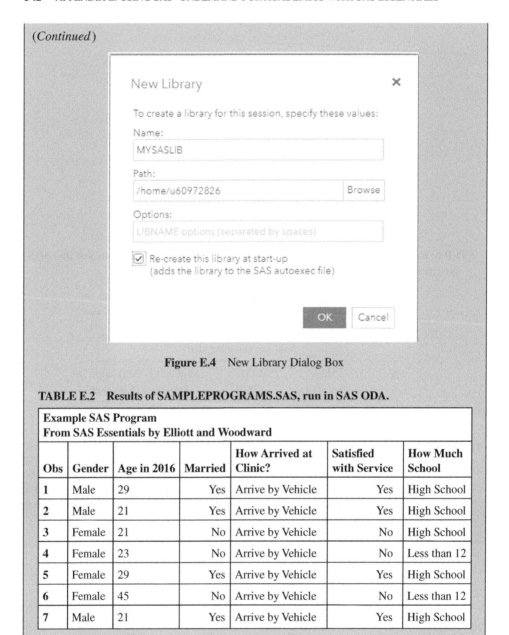

Figure E.4 New Library Dialog Box

TABLE E.2 Results of SAMPLEPROGRAMS.SAS, run in SAS ODA.

Example SAS Program From SAS Essentials by Elliott and Woodward							
Obs	Gender	Age in 2016	Married	How Arrived at Clinic?	Satisfied with Service	How Much School
1	Male	29	Yes	Arrive by Vehicle	Yes	High School
2	Male	21	Yes	Arrive by Vehicle	Yes	High School
3	Female	21	No	Arrive by Vehicle	No	High School
4	Female	23	No	Arrive by Vehicle	No	Less than 12
5	Female	29	Yes	Arrive by Vehicle	Yes	High School
6	Female	45	No	Arrive by Vehicle	No	Less than 12
7	Male	21	Yes	Arrive by Vehicle	Yes	High School

HANDS ON EXAMPLE E.2

1. To enter new code, click on **Code** to enter the **Code (editor) Window**, and clear any old code from the **Code Window** by clicking on the double-X "Clear Code" icon.

2. Enter the following code:

```
PROC SGPLOT DATA=MYSASLIB.SURVEY;
    HISTOGRAM STAYMINUTES;
TITLE "SGPLOT Histogram";
RUN;
```

3. Click on the running man icon. Check the Log Window for errors. The following history should appear, as shown in Figure E.5.

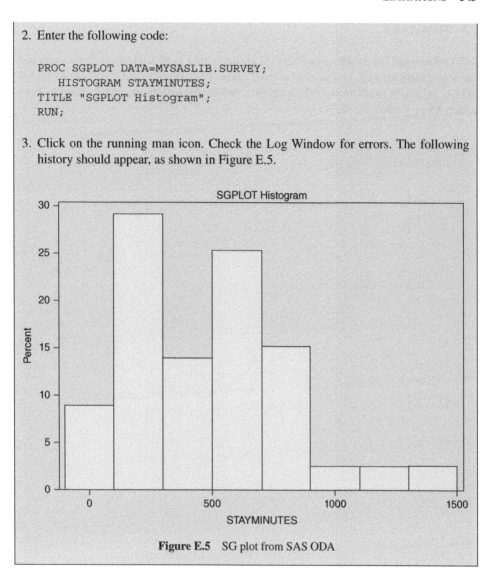

Figure E.5 SG plot from SAS ODA

There are many other aspects of SAS **OnDemand for Academics** and **SAS Studio** that are not covered in this very brief introduction. We've introduced it here in case you do not have access to an installed version of SAS. Using ODA, you should be able to run most of the examples in this book and learn SAS programming.

E.2   LIMITATIONS

A few SAS PROCS are not included in ODA, so some examples in this book that use these PROCS may not run. SAS changes over time, so you'll have to experiment to see which PROCS do not work.

E.3 SUMMARY

SAS **OnDemand for Academics** (ODA) is available free to all students and educators and can be run from the web. You need an active Internet connection to run this version of SAS. ODA is sufficient to do most of the examples in *SAS Essentials* and can be used to learn basic SAS programming skills.

APPENDIX F

SAS® Example BASE Exam Questions

The following exam questions are modelled after a typical SAS Basic Certification exam. These are not actual exam questions but are similar. Knowing how to answer these questions correctly will help you prepare for the actual exam. The answers to the question may be found on the book's www.alanelliott.com/sas webpage.

Be aware that no text can teach all of the possible scenarios that are posed in the SAS exam questions. Therefore, the exam expects you to either have sufficient experience in using SAS or be able to logically determine the best answer from your knowledge of SAS programming even if you haven't encountered the exact scenario presented in a question.

1. The following program is submitted.
    ```
    DATA SAMPLE;
       INPUT NAME $ SCORE;
    DATALINES;
    MARY -22;
    FRED +23;
    RUN;
    ```

    Which values are stored in the output data set?
    ```
    A.
    NAME              SCORE
    --------------------
    MARY              -22
    FRED              +23
    ```

SAS® Essentials: Mastering SAS for Data Analytics, Third Edition. Alan C. Elliott and Wayne A. Woodward.
© 2023 John Wiley & Sons, Inc. Published 2023 by John Wiley & Sons, Inc.
Companion website: https://www.alanelliott.com/SASED3/

B.
```
NAME              SCORE
---------------------
MARY              (missing value)
JOHN              (missing value)
```

C.
```
Name              Age
---------------------
(missing value)   (missing value)
(missing value)   (missing value)
```

D. The DATA step fails execution due to data errors.

2. In the following SAS program, the data files TEMP.SALES and WORK.RECEIPT are sorted by the NAMES variable:

```
LIBNAME TEMP 'C:\SASDATA';
DATA TEMP.SALES;
MERGE TEMP.SALES WORK.RECEIPT;
BY NAMES;
RUN;
```

Which one of the following results occurs when this program is submitted?

A. The program executes successfully and a temporary SAS data set is created.
B. The program executes successfully and a permanent SAS data set is created.
C. The program fails execution because the same SAS data set is referenced for both read and write operations.
D. The program fails execution because the SAS data sets on the MERGE statement are in two different libraries.

3. Given the SAS data set WORK.ONE

```
Id    Char1
---   -----
182   A
190   B
250   C
720   D
```

and the SAS data set WORK.TWO:

```
Id    Char2
---   -----
182   E
623   F
720   G
```

The following program is submitted:

```
DATA WORK.BOTH;
   MERGE WORK.ONE WORK.TWO;
   BY ID;
RUN;
```

What is the first observation in the SAS data set WORK.BOTH?

A.
```
Id   Char1   Char2
---  -----   -----
182   A
```

B.
```
Id   Char1   Char2
---  -----   -----
182            E
```

C.
```
Id   Char1   Char2
---  -----   -----
182   A        E
```

D.
```
Id   Char1   Char2
---  -----   -----
720   D        G
```

4. When the following SAS program is submitted, the data set EX4 contains 500 observations:

```
OPTIONS OBS = 50;
PROC PRINT DATA= EX4 (FIRSTOBS = 10);
RUN;

OPTIONS OBS=MAX;
PROC MEANS DATA= EX4 (FIRSTOBS = 50);
RUN;
```

How many observations are processed by each procedure?

```
A. 40 for PROC PRINT, 450 for PROC MEANS
B. 41 for PROC PRINT, 451 for PROC MEANS
C. 41 for PROC PRINT, 450 for PROC MEANS
D. 50. for PROC PRINT, 500 for PROC MEANS
```

5. The following SAS program is submitted:

```
DATA NEW;
LENGTH WORD $7;
COUNT= 3;
IF COUNT = 5 THEN WORD = '.LIZARD';
ELSE IF COUNT= 3 THEN WORD= 'BEAR';
ELSE WORD= 'NONE!!!';
COUNT= 5;
RUN;
```

Which one of the following represents the values of the COUNT and WORD variables?

```
A. COUNT=5  AND WORD = BEAR
B. COUNT =5 AND WORD = LIZARD
C. COUNT = 3 AND WORD=BEAR
D. COUNT =3 AND WORD = ' ' (MISSING CHARACTER VALUE)
```

6. Which one of the following is true of the SUM statement in a SAS DATA step program?
   A. It is only valid in conjunction with SUM function.
   B. It is not valid with the SET, MERGE and UPDATE statements.
   C. Returns a result that is the sum of all nonmissing values in the argument list.
   D. It does not retain the accumulator, variable value from one iteration of the SAS DATA step to the next.

7. Given the text file COLORS.TXT:

   ```
   ----+----1----+----2----+----
   RED     ORANGE  YELLOW  GREEN
   BLUE    INDIGO  PURPLE  VIOLET
   CYAN    WHITE   FUCSIA  BLACK
   GRAY    BROWN   PINK    MAGENTA
   ```

   The following SAS program is submitted:

   ```
   data WORK.COLORS;
      infile 'COLORS.TXT';
      input @1 Var1 $ @8 Var2 $ @;
      input @1 Var3 $ @8 Var4 $ @;
   run;
   ```

   What will the data set WORK.COLORS contain?

   A.
   ```
   Var1    Var2    Var3    Var4
   ------  ------  ------  ------
   RED     ORANGE  RED     ORANGE
   BLUE    INDIGO  BLUE    INDIGO
   CYAN    WHITE   CYAN    WHITE
   GRAY    BROWN   GRAY    BROWN
   ```

   B.
   ```
   Var1    Var2    Var3    Var4
   ------  ------  ------  ------
   RED     ORANGE  BLUE    INDIGO
   CYAN    WHITE   GRAY    BROWN
   ```

   C.
   ```
   Var1    Var2    Var3    Var4
   ------  ------  ------  ------
   RED     ORANGE  YELLOW  GREEN
   BLUE    INDIGO  PURPLE  VIOLET
   ```

   D.
   ```
   Var1    Var2    Var3    Var4
   ------  ------  ------  ------
   RED     ORANGE  YELLOW  GREEN
   BLUE    INDIGO  PURPLE  VIOLET
   CYAN    WHITE   FUCSIA  BLACK
   GRAY    BROWN   PINK    MAGENTA
   ```

8. The following SAS program is submitted:

   ```
   DATA WORK.SETS;
   DO UNTIL (PROD GT 7);
   PROD + 2;
   END;
   RUN;
   ```

   Which one of the following is the value of the variable PROD in the output data set?

   A. 6
   B. 7
   C. 8
   D. 9

9. Given the SAS data set WORK.INPUT:

VAR1	VAR2
One	orange
Two	green
Three	red
Four	green
Five	orange

   And the code:

   ```
   data WORK.ONE WORK.TWO;
     set WORK.INPUT;
     if Var2='green' then output WORK.ONE;
     output;
   run;
   ```

   How many records are in WORK.ONE?

   A. 6
   B. 7
   C. 8
   D. 9

10. The following SAS program is submitted:

    ```
    ODS PDF;
    PROC PRINT DATA= MYSASLIB.FAMILY;
    RUN;
    <INSERT OPTIONS STATEMENT HERE>;
    PROC MEANS DATA= MYSASLIB.BOXSTORE;
    RUN;
    ODS PDF CLOSE;
    ```

    Which one of the following OPTIONS statements resets the page number to 1 for the second report? (Assume your output is to a destination that has page numbers such as RTF or PDF.)

    A. options pageno = 1;
    B. options pagenum = 1;
    C. options reset pageno = 1;
    D. options reset pagenum = 1;

11. The following SAS program is submitted:

    ```
    data WORK.LOOP;
      VAL = 0;
      do ICNT = 1 to 10  by  2;
        VAL = ICNT;
      end;
    run;
    ```

    After running this program, what are the values of the variables VAL and ICNT in WORK.LOOP?

    A. VAL = 9, ICNT = 9
    B. VAL = 5, ICNT = 9
    C. VAL = 9, ICNT = 10
    D. VAL = 9, ICNT = 11

12. The following SAS program is submitted:

    ```
    DATA SOME;
    INPUT MASK  5. @10 VOLUME;
    DATALINES;
    12345$56.89
    ;
    RUN;
    ```

    Which one of the following is the value of the VOLUME variable?

    A. 56.89
    B. $56.89
    C. 89
    D. . (missing numeric value)
    E. No value is stored as the program fails to execute due to errors.

13. The following SAS program is submitted:

    ```
    proc format;
      value score  1  - 50  = 'Fail'
                   51 - 100 = 'Pass';
    run;
    ```

    Which one of the following PRINT procedure steps correctly applies the format?

    A.
    ```
    proc print data = MYSASLIB.SURVEY;
      var SATISFACTION;
      format SATISFACTION SCORE;
    run;
    ```

    B.
    ```
    proc print data = MYSASLIB.SURVEY;
            var SATISFACTION;
            format SATISFACTION SCORE.;
    run;
    ```

C.
```
proc print data = MYSASLIB.SURVEY format = SCORE;
   var SATISFACTION;
run;
```
D.
```
proc print data = MYSASLIB.SURVEY format = SCORE.;
   var SATISFACTION;
run;
```

14. The contents of the raw data file TYPECOLOR are listed below:
    ```
    ----|----10---|----20---|----30
    MISTYPURPLE
    ```

    The following SAS program is submitted:
    ```
    DATA FLOWERS;
    INFILE 'TYPECOLOR';
    INPUT TYPE $ 1-5 +1 COLOR $;
    RUN;
    ```

    Which one of the following represents the values of TYPE and COLOR?
    A. MISTY PURPLE
    B. MISTYPURPLE
    C. MISTY URPLE
    D. No values are stored as the program fails to execute due to syntax errors,

15. This question asks you to provide a line of missing code. Given the following data set WORK.INCOME:
    ```
    SalesID  SalesJan  FebSales  MarchAmt
    -------  --------  --------  --------
    W6790          50       400       350
    W7693          25       100       125
    W1387           .       300       250
    ```

    The following SAS program is submitted:
    ```
    data WORK.QTR1;
       set WORK.SALES;
       array month{3} SalesJan FebSales MarchAmt;
       <insert code here>
    run;
    ```

    Which statement should be inserted to produce the following output?

Obs	Sales ID	Jan	Feb	March	Qtr1
1	W1234	50	40	35	125
2	W5678	25	10	12	47
3	W2468		30	25	55

    A. `Qtr1 = sum(of month{_ALL_});`
    B. `Qtr1 = month{1} + month{2} + month{3};`
    C. `Qtr1 = sum(of month{*});`
    D. `Qtr1 = sum(of month{3});`

16. The following SAS program is submitted and reads 100 records from a raw data file:
```
DATA MYDATA;
INFILE 'C:\SASDATA\EXAMPLE.TXT' END=EOF;
INPUT ID $ 1-3 GP $ 5 AGE 6-9
      TIME1 10-14 TIME2 15-19 TIME3 20-24;
<INSERT STATEMENT HERE>;
RUN;
PROC PRINT DATA=MYDATA;
RUN;
```

Which one of the following IF statements writes the last observation to the output data set?

A. if end = 0;
B. if eof= 0;
C. if end= 1;
D. if eof= 1;

17. A raw data record has the following contents:
```
----|----10---|----20---|----30
son,Adams,
```

The following output is desired:
```
relation firstname
son      Adams
```

Which one of the following SAS programs reads the data correctly?

A
```
DATA FAMILY / DLM=',';
INFILE datalines;
INPUT RELATION $ FIRSTNAME $;
DATALINES;
son,Adams,
;
RUN;
```

B
```
DATA FAMILY;
INFILE datalines delimiter=',';
INPUT RELATION $ FIRSTNAME $;
DATALINES;
son,Adams,
;
RUN;
```

C.
```
DATA FAMILY;INFILE datalines DLM=  ',';
INPUT RELATION $ FIRSTNAME $;
DATALINES;
son,Adams,
;
RUN;
```

```
D.
DATA FAMILY;
INFILE datalines;
INPUT RELATION $ FIRSTNAME $/ DLM = ',';
DATALINES;
son,Adams,
RUN;
```

18. The following SAS program is submitted:

    ```
    LIBNAME RAWDATA1 'C:\SASDATA';
    FILENAME RAWDATA2 'C:\SASDATA\EXAMPLE.TXT';
    DATA WORK.TESTDATA;
    INFILE < FILL-IN CODE >;
      INPUT ID $ 1-3 GP $ 5 AGE 6-9
            TIME1 10-14 TIME2 15-19 TIME3 20-24;
    RUN;
    ```

    Which one of the following <FILL-IN CODE> is needed to complete the program correctly?

    A. RAWDATA1
    B. RAWDATA2
    C. 'RAWDATA1'
    D. 'RAWDATA2'

19. The contents of two SAS data sets named EMPLOYEE and SALARY are listed below:

WORKER	WAGES
Variables NAME AGE	Variables NAME SALARY
Dan 35	Dan 37000
Dwaine 30	Dan .
	Dwaine 40000
	Dwaine 35000

    The following SAS program is submitted:

    ```
    DATA WORK.EMPWAGES;
    MERGE WORK.WORKER (IN= INWORK)
    WORK.WAGES (IN = INWAGE);
    BY NAME;
    IF INWORK AND INWAGE;
    RUN;
    ```

    How many observations will the data set WORK.EMPSALARY contain?

    A. 2
    B. 4
    C. 5
    D. 6

20. The SAS data sets WORK.EMPLOYEE and WORK.SALARY are listed below:

WORKER	WAGES
Variables NAME AGE	Variables NAME SALARY
Dan 35	Dan 37000
Dwaine 30	Dan .
	Dwaine 40000
	Dwaine 35000

The following SAS program is submitted:
```
DATA WORKER;
MERGE WORKER WAGES;
BY NAME;
TOTSAL + WAGES;
RUN;
```

How many variables are output to the WORKER data set?

A. 3
B. 4
C. 5
D. No variables are output to the data set as the program fails to execute due to errors.

21. The following SAS SORT procedure step generates an output data set:
```
PROC SORT DATA = MYSASLIB.SURVEY OUT = REPORT;
BY AGE;
RUN;
```

In which library is the output data set saved?

A. WORK
B. REPORT
C. HOUSES
D. SASUSER

22. The following SAS DATA step is submitted:
```
LIBNAME TEMP 'C:\SASDATA';
DATA TEMP.REPORT;
    SET SASUSER.HOUSES;
    NEWVAR=PRICE* 1.04;
RUN;
```

Which one of the following statements is true regarding the program above?

A. The program is reading from a temporary data set and writing to a temporary data set.
B. The program is reading from a temporary data set and writing to a permanent data set.
C. The program is reading from a permanent data set and writing to a temporary data set.
D. The program is reading from a permanent data set and writing to a permanent data set.

23. Which of the following SAS DATA steps saves the temporary data set named MYDATA as a permanent data set?

    A.
    ```
    LIBNAME SASDATA 'SAS-DATA-LIBRARY';
    DATA SASDATA.MYDATA;
    COPY MYDATA;
    RUN;
    ```

    B.
    ```
    LIBNAME SASDATA 'SAS-DATA-LIBRARY';
    DATA SASDATA.MYDATA;
    KEEP MYDATA;
    RUN;
    ```

    C.
    ```
    LIBNAME SASDATA 'SAS"DATA•LIBRARY';
    DATA SASDATA.MYDATA;
    SAVE MYDATA;
    RUN;
    ```

    D.
    ```
    LIBNAME SASDATA 'SAS-DATA-LIBRARY';
    DATA SASDATA.MYDATA;
    SET MYDATA; RUN;
    ```

24. The following SAS DATA step is submitted:
    ```
    DATA SASDATA.ATLANTA SASDATA.BOSTON
    WORK.PORTLAND WORK.PHOENIX;
         SET COMPANY.PRDSALES;
              IF REGION='NE' THEN OUTPUT BOSTON;
              IF REGION='SE' THEN OUTPUT ATLANTA;
              IF REGION='SW' THEN OUTPUT PHOENIX;
              IF REGION='NW' THEN OUTPUT PORTLAND;
    RUN;
    ```

    Which one of the following is true regarding the output data sets?

    A. No library references are required.
    B. The data sets listed on all the IF statements require a library reference.
    C. The data sets listed in the last two IF statements require a library reference.
    D. The data sets listed in the first two IF statements require a library reference.

25. The following SAS program is submitted:
    ```
    data WORK.TEMP;
      Char1='0123456789';
      Char2=substr(Char1,3,4);
    run;
    ```

What is the value of Char2?

A. 23
B. 34
C. 345
D. 2345

Answers to BASE Exam Sample Test

Also, see the SAS file named BASE_EXAM_ANSWERS.SAS located on the www.alanelliott.com/sas webpage for code that explains each answer.

1. D
2. B
3. C
4. B
5. A
6. C
7. A
8. C
9. B
10. A
11. D
12. C
13. B
14. C
15. C
16. D
17. C
18. B
19. B
20. B
21. A
22. D
23. D
24. D
25. D

INDEX

Symbols
#n indicator, 40
%DO Loop (macro), 197
%INCLUDE statement, 197, 394
* (comment), 121
/* (comment), 122
@, 42, 91
@@, 29, 42
\SASDATA folder, 4
ERROR, 74
_N_, 74

A
ABS function, 142
ACCESS and ACCESSCS files, 69
ACROSS option (REPORT), 447, 481
ADJUST= option (multiple comparisons), 351, 379
AIC (Akaike Information Criteria), 379
ALPHA= option, 254
analysis of variance, one-way. *See* One-Way ANOVA
analysis of covariance (ANCOVA), 363ff, 529
analysis of variance (ANOVA/AOV), 347ff
analysis of variance, two factor. *See* Two Factor ANOVA

ANALYSIS option (REPORT), 449
ANGLE (A=) option, 437
ANNOTATE= option, 430
annotation data sets, 465
Answers, SAS Sample Exam, 556
appending data sets, 99
AR(1). *See* Autoregressive
ARCOS function, 511
area under the curve (AUC). *See* ROC Curve
arithmetic/mathematical functions, 142, 508
arithmetic operators, 83
ARRAY statement, 171ff
ARSIN function, 511
ASCII file/format, 15, 38, 63
assigning missing values. *See* Missing Values
asterisk (*) for comments, 121
ATAN function, 511
AUC (area under the curve). *See* ROC Curve
automated model selection, 337–339
autoregressive AR(1) covariance structure, 379
AXIS statement, 436

B
BACKGROUND= (color) option, 505
BACKWARD procedure, 337–339
bar charts, 441ff, 446ff

bar charts (using GCHART), 446–447
bar fill, 465
bar patterns, 500
binary data, 500
binary logistic regression. *See* Logistic Regression
Bonferroni test, 350
box-and-whiskers plot, 353, 390, 451, 453
box plot. *See* Box-and-Whiskers Plot
BOXSTYLE= option, 451
BUBBLE statement (SGPLOT), 430, 453–454
BY CLASS, 263
BY DESCENDING statement, 96
BY GROUP, 96, 126
BY statement, 96, 126

C

CALCULATED , SQL, 241
calculations in SQL, 240
CALL SYMPUT, 199
CALL SYMPUT, CALL SYMPUTN, 521
CALL SYMPUT Function, 521
CASE Clause, SQL, 231
case functions (upper, lower), 516
CAT, CATS, CATT, CATX functions, 515
categorical data, 492–493
CAXIS= option, 264, 433, 441
CBOXES= option (BOXPLOT), 451
CBOXFILL= option (BOXPLOT), 451
CEIL function, 519
center output. *See* NOCENTER
CFILL statement, 263–264
CFRAME option, 264, 433, 441
character functions, 480
Character (Text) Functions, 515
character variables, 25
chi-square test, 529
chi-square test (PROC FREQ), 290–297
choosing a SAS procedure, 528ff
CITRIX, 4, 22, 55
CLASS statement, 256, 275, 379, 400
CLDIFF multiple comparisons, 350
Clear All command, 9
CLI statement, 334
CLM= option (GCHART), 442
COCHRAN option, 312
Cochran Q, 529
Cochran Q test, 529
code file, 15
color choices, 499
COLOR= option (C=), 434, 435, 437
Colors, Specifying, 499
column input, 30

column name, SQL, 230
COLUMN statement (REPORT), 481
compact data entry method, 28
comparing K dependent samples, 392
comparing K independent samples, 387
comparing three or more means using one-way ANOVA, 348ff
comparing three or more repeated measures, 529
comparing two dependent samples, 529
comparing two independent samples, 529
comparison statistics, choosing, 529
comparison tests, choosing, 529
COMPBL function, 515
compile SAS code, 73
compound symmetry (CS) covariance structure, 379
COMPRESS function, 515
COMPUTE option(REPORT), 480, 481, 487
concatenation functions, 515
conditionally read a data set line, 91
conditional operators, 86
confidence limits, plotting, 440, 449
CONTENTS option, 67
contingency tables (PROC FREQ), 290–297
continuity adj. chi-square, 295
CONTRAST statement, 359
Cook's D plot (regression), 331
Cook's D statistic, 343
C= option. *See* Color Option
correlation analysis, 319ff
correlation analysis using WITH option, 325
correlation and regression, 530
correlation coefficient (Pearson), 319ff
correlations, factor analysis, 423
COS function, 511
COSH function, 511
COUTLINE option, 441
covariance structure, 379
CREATE TABLE, SQL, 234
creating a SAS library, 56ff
creating new variables, 82ff
creating SAS data set, 21, 25, 52ff
Creating SAS Datasets in SQL, 234
crosshatch pattern, 500
cross-tabulation, 528
CS covariance structure. *See* Compound Symmetry
CSV (comma separated values), 43–45, 64–69
CTABLE option, 401, 409
CTEXT option, 433, 441
Ctrl-Break, 12
cut off value, logistic regression, 409

INDEX 559

D
DATALINES statement, 26
data sets. *See* SAS Data Sets
DATA statement. *See* DATA step
data step, 10, 20ff
data types. *See* Variable types
DATDIF function, 512
date and time functions, 511, 511ff
DATE function, 512
DATE option. *See* NODATE
DATEPART function, 512
dates in SAS, 511ff
DATETIME function, 512
date variables, 25
DAY function, 513
day of week format. *See* DOWNAME
day of week, naming, 523
dBase files, 69
DBMS sources, 69
DEFINE statement (REPORT), 481
density plots, 263, 457
dependent variables, 327, 333, 399
DEQUOTE function, 515
DESCENDING option, 399, 400
descriptive procedures, choosing, 528
descriptive statistics, 528
descriptive statistics, choosing procedure, 528
DHMS function, 163, 513
DIF function, 521
DISCRETE option (GCHART), 441, 454
DISPLAY statement (REPORT), 481
DLM option, 42
DLMSTR, 42
DOCX (Microsoft Word), 207, 209, 210
DO LOOP, 175ff
DONUT option(GCHART), 441
dot (.). *See* Missing Values
DO UNTIL loop, 175, 176
DO WHILE loop, 175–177
DOWNAME format, 506
download sample files, 4
drinking analysis, 290–292
DROP statement, 92
DSD option, 42
Duncan multiple comparison test, 350
DUNNETT multiple comparisons, 350
Dunn's test (multiple comparisons), 394

E
editor. *See* Enhanced editor
editor window, 5
eigenvalues (factor analysis), 418, 425

END AS, SQL, 231
enhanced editor, 5, 12
equality of variances test, 311
EQUAMAX option, 415
ERRORBAR= option (GCHART), 442
EVENT= option, 399–400
EXACT statement, 281, 385
excel, 15, 23, 63, 69, 82, 479
EXCEL output destination. *See* TAGSETS
EXCEL, saving table to (TABULATE), 479
EXCEPT, Unions, SQL, 243
execute SAS code, 73
EXP function, 508
Export. *See* PROC EXPORT

F
FACT function, 508
factor analysis, 414ff
FILENAME command, 553
FILE PRINT statement, 492
files handling, uploading in SAS Studio, 540
FILLATTRS= option (SGPLOT), 457
financial functions, 519
FIND function, 162, 515
finding FIRST and LAST values, 112
FIRSTOBS= option, 43, 134, 137
FIRST values, find, 112
FISHER option, 282
Fisher's exact test, 281–282, 295ff
fixed factors, 370
FLOOR function, 519
FLOWOVER option, 43
FLYOVER= option, 505
FONT_FACE= option, 505
FONT= option, 505
FONT_SIZE= option, 505
FONTS, SAS, 498
fonts, specifying, 498
FONT_WEIGHT= option, 505
FOOTNOTE statement, 118, 532
FOREGROUND= (color) option, 505
FORMAT= option, 231, 266, 468
formats, 35–36
formats, creating. *See* PROC FORMAT
formats, input, 505
formats library, 108–111
formats, output, 505–506
formats, permanent. *See* PROC FORMAT
FORMATS, Using, 505
formatted input, 33
FORWARD selection, 334, 337–338
freeform list input, 25

560 INDEX

FREQ statement, 349
frequency tables, 283, 286
Friedman ANOVA, 529
Friedman's test, 392
FROM statement, SQL, 233
FULL OUTER JOIN, SQL, 243
function keys. *See* SAS Function Keys
functions reference, 526–527
functions, SAS, 141–146, Appendix B
FUZZ function, 519

G

GAXIS option, 436
GLM (General Linear Model). *See* PROC GLM
GMAPS, 220ff
goodness-of-fit analysis, 283–289, 529
GOPTIONS statement, 430, 439, 444
graphs, 429ff
Graph Template Language (GTL), 465
GRID option, 433
GROUP Clause, SQL, 236
grouping factor, 258, 347, 363, 377ff
GROUP option (GCHART), 441
GROUP option (REPORT), 484
GTL. *See* Graph Template Language

H

HAVING clause, SQL, 237
HAXIS option, 433, 436
HBAR3D option (GCHART), 441
HBAR option (GCHART), 441, 457
HBOUND (arrays), 173
help, 15
HISTOGRAM(SGPLOT), 441ff
HISTOGRAM (PROC UNIVARIATE), 272
HMS function, 513
homogeneity hypothesis test, 529
homogeneity, test for, 529
horizontal bar chart (HBAR), 441
Hosmer and Lemeshow test, 401, 408ff
HOUR function, 513
HREF option (GCHART), 433, 455
HTML (HyperText Markup Language), 15, 206, 281, 441, 454ff, 532
HTML= option (GCHART), 442, 454ff
hyperlinks. *See* HTML
HyperText Markup Language (HTML), 206

I

ID statement, 127
IF-THEN-ELSE statements, 86ff
IF-THEN for missing values, 88

IF-THEN to subset data sets, 89
IF to conditionally read a data set, 91
importing data, 63ff
importing data use code. *See* PROC IMPORT
importing data using import wizard, 63ff
INCLUDE=k statement, 334, 401
independent variables, 97, 327, 333, 399
INDEX function, 516
INFILE statement, 38, 42, 532
informats, 33
INFORMATS, Using, 505
INNER JOIN, SQL, 243
INPUT function, 162, 521
INSET statement, 262, 275, 451
INSIDE= option (GCHART), 442
INTCK function, 143–144, 513
interaction effects, 369ff
INTERPOLATION= option (I=), 434, 439
interquartile range, 268, 453
inter-rater reliability. *See* Kappa
INTERSECT, Unions, SQL, 243
INT function, 519
INTNX function, 143–144, 513
I= option. *See* Interpolation Option
IRR function, 519
IRR (Internal Rate of Return) Function, 519

J

JOBSCORE example, 335ff
JOINS in SQL, 241
JUSTIFICATION= option, 437
JUST= option, 505

K

kappa, 262–298ff, 282, 529
KEEP statement, 92
KEYLEGEND option (SGPLOTS), 457
Kruskal-Wallis, 529
Kruskal-Wallis multiple comparisons, 393ff
Kruskal-Wallis (KW) test, 387ff, 529
kurtosis, 255, 267

L

labeling variables, 80–82
LABEL= option, 437
LABEL statement, 71, 80–82, 129, 532
labels, variables, SQL, 230
LACKFIT option, 401, 408
LAG function, 521, 523
landscape orientation. *See* ORIENTATION
LAST value, finding, 112
launching SAS, 5

INDEX   561

LBOUND (Arrays), 173
LEFT function, 516
LEFT OUTER JOIN, SQL, 243
LEGEND = option, 433
LENGTH function, 516
libraries. *See* SAS Library
library, create in SAS Studio, 541
library nicknames, 6, 22, 56
linear regression, 529
linear regression, multiple, 297ff
linear regression, simple, 289ff
LINE= option, 434, 440
line plots, 429, 440
LINESIZE option, 137
LINE style. *See* LINE= option
line styles, specifying, 503
LISTING output destination, 206
log file, 15
LOG function, 508
logistic analysis, multiple, 406ff
logistic equation, 398ff
logistic regression, 397, 530
logistic regression, graphs, 404ff
logistic analysis, simple, 402ff
log window, 5, 9, 12–14
Lotus 1-2-3 files, 69
LOWCASE function, 156, 516
Lower Case function, 516
LRECL option, 43
LSD multiple comparisons, 350
LSMEANS statement, 350, 359, 382

M

macro routine, %Include, 196ff
macro routines, 192ff
macros, SAS. *See* SAS Macros
macro variable, creating with CALL SYMPUT, 199
macro variables, 188ff
main effects, 369
MAJOR= option, 437
Mann-Whitney, 529
Mann-Whitney test. *See* Wilcoxon-Mann-Whitney
mathematical functions, 508
matrix of histograms, 263, 273–276
matrix of scatterplots, 322–324
MAXDEC=n option, 190
MAX function, 143, 508
maximum likelihood estimates, 398, 403–405
MAXIS option, 436
McNemar test, 529

MDY function, 144, 188, 513
mean bars (GCHART), 448ff
MEAN function, 508
MEANS statement (multiple comparisons), 350
MEDIAN function, 142–143, 509
median test, 385, 387
menus. *See* SAS Menus
merge, few-to-many, 103ff
merging data sets, 99ff
messy data, cleaning, 153ff
Microsoft Access, 24, 63, 69
Microsoft Excel. *See* Excel
Microsoft Word, 207, 209, 210
MIDPOINTS option, 263
MINEIGEN=n option, 415
MIN function, 142–143, 509
MINOR= option, 437
MINUTE function, 70
missing values, assign, 88ff
MISSOVER option, 42, 46
MIXED model. *See* PROC MIXED
mode, 255
MODEL I ANOVA, 370
MODEL II ANOVA, 370
MODEL III ANOVA, 370, 374ff
MODEL statement, 328, 333, 348–349, 365, 399
MOD function, 509
moments, 267
MONNAME format, 506
MONTH function, 513
month name format. *See* MONNAME
month, naming, 523
MORT function, 519
MORT (Mortgage) Function, 519
MS Access. *See* Access
MS Excel. *See* Excel
multiple comparisons, 350ff
multiple linear regression, 332ff, 530
MUO= option (Univariate), 262

N

naming conventions, 22–24
new library, create in SAS Studio, 541
NFACTORS option, 415
N function, 143, 509
NMISS function, 143, 509
NMISS statistic, 255, 473
NOCENTER option, 137
NODATE option, 137
NOFRAME option, 266, 397, 433, 437
NOLEGEND option, 433, 442
nonparametric analysis, 384ff

nonparametric ANOVA, 529
nonparametric multiple comparisons, 393
NONUMBER option, 137
NOPRINT option, 262, 281, 283, 327, 334, 349, 385, 400, 415
NOPROB option, 320
N option, 132
normal curve on graph, 263, 457
normality, test for, 271
normal probability plot, 262, 271
NOSIMPLE option, 320
NOTCHES option (BOXPLOT), 451
NPV (Net Present Value) function, 519
numeric variables, 25

O

OBS= option, 43, 132
odds ratios (OR), 297, 398
ODDSRATIO Statement, 400
ODS (Output Delivery System), 205ff, 532
ODS CLOSE, 207
ODS destinations, 205–206
ODS EXCLUDE statement, 211–213
ODS GRAPHICS, 216–218
ODS interactive plot, 454
ODS OUTPUT statement, 205–208
ODS preferences, 208
ODS, reset defaults. *See* ODS Preferences
ODS SELECT statement, 211–213
ODS STYLE attributes, 504–505
ODS styles, 208–211
ODS TAGSETS, 207, 479
OFFSET= option, 437
olympic data analysis, 421ff
OnDemand, 539
one-sample t-test, 306ff
one-way ANOVA, 348ff, 529
open VMS, 4
OPTIONS. *See* System options
OPTIONS FMTSEARCH, 110
OPTIONS NOFMTERR, 111
OR (odds ratios). *See* Odds Ratio
ORDER BY, SQL, 228, 238
ORDER= option, 281, 349, 437, 468, 482
ORDER option (REPORT), 481, 483
ordinal data, 528–530
ORDINAL function, 509
ORIENTATION= option, 137
OUTER UNION, SQL, 243
Output Delivery System (ODS). *See* ODS
OUTPUT OUT= statement, 256, 400
output window. *See* Results viewer

OUTROC= option, 401
OUTSIDE= option (GCHART), 442

P

PAD option, 43, 46
PAGESIZE option, 137
PAGESIZE= option, 137
paired (dependent) samples, 529
paired t-test, 315ff, 529
pairwise comparisons, 350, 356, 360, 366, 389, 393
patterns for GPLOT/PROC UNIVARIATE, 443ff
patterns, specifying, 500
PATTERN statement, 430, 441, 443–445, 500–502
PCL (Print Command Language), 206
PDF (Portable Document Format), 206
PDIFF multiple comparisons, 351
PDV. *See* Program Data Vector
Pearson correlation coefficient, 320ff, 529
Pearson's correlation, 528
permanent SAS data sets, 22, 55–58, 61
PFILL statement, 265
pie charts, 441ff
PIE3D option (GCHART), 441
PIE option (GCHART), 441
PLOTS option, 349, 400
PLOT statement, 430ff
plotting symbols, 504
pointer control commands, 42
Portable Document Format (PDF), 206
portrait orientation. *See* ORIENTATION
postscript. *See* PS (Postscript Output Destination)
PowerPoint, 206, 207, 221
PPTX (PowerPoint), 206, 207, 221
predicted values, linear regression, 330, 339ff
predicted values, logistic regression, 409ff
prediction bands, linear regression, 330
predictions, linear regression, 339ff
principal component analysis, 415
Print Command Language (PCL), 206
PRIORS=option, 415
PROBSIG= option, 137
PROC ANOVA, 348ff, 534
PROC APPEND, 105
PROC BOXPLOT, 315, 450ff, 536
PROC CATALOG, 111
PROC CONTENTS, 72
PROC CORR, 319ff, 528, 530, 534
PROC DATASETS, 71
PROC DOCUMENT, 222

PROC EXPORT, 70
PROC FACTOR, 414ff, 535
PROC FORMAT, 106ff, 274, 533
PROC FREQ, 280ff, 393, 534
PROC GCHART, 441ff, 454, 535
PROC GLM, 348–351, 354–361, 363–374, 534
PROC GPLOT, 380, 424ff, 535
PROC GSLIDE, 196
PROC IMPORT, 68, 532
PROC LOGISTIC, 399ff, 535
PROC MEANS, 10, 39, 253ff, 344, 532
PROC MIXED, 374ff, 534
PROC NPAR1WAY, 385ff, 535
PROC PRINT, 131ff, 533
PROC PRINT traffic lighting, 218–220
PROC REG, 325ff, 535
PROC REPORT, 480ff, 536
PROC SGPLOT, 457ff
PROC SORT, 95ff, 380, 533
PROC SQL, 225ff
PROC statement syntax, 123ff
PROC TABULATE, 467ff, 536
PROC TEMPLATE, 208
PROC TRANSPOSE, 146ff
PROC TTEST, 307ff, 534
PROC UNIVARIATE, 261ff, 391, 500, 534
PROC UNIVARIATE inset, 265–266
program code. *See* code files
program data vector, 72–75
PROMAX option, 415
PS (Postscript Output Destination), 206
PUT function, 521
PUT statement option, 492
PUT statement reports, 492, 536
PUT Statements for Creating Reports, 536
p-value one-tailed, 310
p-values in multiple comparisons, 349, 366
p-value two-tailed, 310

Q
Q-Q plot, 310, 314
QTR function, 513
quantiles, 268
quantitative data, 253ff, 319–320, 348, 363, 370, 398, 528–530
QUARTIMAX option, 415
QUIT statement, 123, 332
QUOTE function, 516

R
random factors, 370ff
random numbers, creating, 509

range, 268
RANNOR function, 509
RANUNI function, 509
RAXIS option, 436
reading data into SAS, 25ff, 54ff
reading/writing data, 54ff
Receiver Operating Characteristic (ROC). *See* ROC Curve
REG option (SGPLOT), 457
regression, linear, 530
regression, logistic, 530
relational analyses, choosing, 530
relational tests, choosing, 530
relative risk (PROC FREQ), 282, 297ff
RELRISK risk measure, 282, 297
repeated measures, 529
repeated measures ANOVA, 355ff
repeated measures ANOVA with Grouping Factor, 377ff
REPEATED statement, 349, 379
REPEAT function, 516
residual analysis, regression, 342ff
residual by predicted value plot, 330
residual by quartile plot, 331
residuals by percent plot, 331
response profile table, 403
results viewer, 6
RETAIN statement, 181ff
REVERSE function, 481
Rich Text Format (RTF), 126, 131, 194
RIGHT function, 516
RIGHT OUTER JOIN, SQL, 243
RISKLIMITS option, 401, 407, 409, 535
ROC curve, 401, 411
ROTATE= option, 415, 419
ROUND function, 142, 519
r (Pearson correlation coefficient), 320ff, 530
R-Square, 329
RStudent by leverage plot, 331
RStudent by predicted value plot, 330
RTF (Rich Text Format), 132, 137, 206
running man icon, 7, 123
running programs in SAS Studio, 539
RUN statement, 123

S
SAS7BDAT file, 15, 22
SAS Code Examples, 531
SAS colors, 499–500
SAS data sets, 21
SAS data sets in SQL, 234
SAS fonts, 498

SAS function keys, 13
SAS graphs, 429ff
SAS help, 15
SAS library, 53ff
SAS macros, 187ff, 394
SAS menus, 14
SAS OnDemand for Academics, 539
SAS Profile, creating, 540
SAS program, complete example, 537
SAS Sample Exam, 545
SAS SQL Reference, 239–240
SAS SQL unions and joins reference, 245–246
SAS Studio, 539
SAS Studio, limitations, 543
SAS University Edition (also see SAS Studio), 22, 539ff
SAS Windows Interface, 5
Save As command, 8–9, 17
SCAN function, 516
SCANOVER option, 43
SCATTER option (SGPLOT), 457, 462
scatterplots, 393, 396, 399
Scheffe multiple comparisons, 350, 359, 367, 369
SCHEMATIC boxplot option, 416, 417
SCREE plot and option, 320, 322, 330, 431–432, 435, 438
SECOND function, 513
selecting variables (regression), 337–338, 401, 406
SELECTION= statement, 334, 401
SELECT statement, 151ff
SELECT statement, SQL, 228, 230ff
sensitivity, 409
SET statement, 93ff
signed rank test, 262, 268, 307, 390ff, 529
SIGN function, 509
sign test, 262, 268, 390, 529
simple linear regression, 325ff, 530
SIN function, 511
SINH function, 511
skewness, 255
SLENTRY= option, 334, 401
slope, test, 329
SLSTAY= option, 334, 338, 401
SMC (factor analysis), 416ff
SNK multiple comparisons, 350, 351
sorting data. *See* PROC SORT
SOUNDEX function, 516
SPACE= option (GCHART), 442
Spearman correlations, 320ff, 528, 530

special use functions, 142, 519
specificity, 409
SPLIT= statement, 134
splitting titles, 132, 134ff
SQL definition, history, 226
SQL JOINS, 243
SQL nomenclature, 227
SQL statement order, 229
SQL UNIONS, 241
SQRT function, 507ff
squared multiple correlations. *See* SMC
square root function (SQRT). *See* SQRT function
stacked bar charts, 445ff
standard deviation (STDDEV), 215, 255, 473
STAR option (GCHART), 380, 434, 441, 504
STATISTICNAME option (REPORT), 482
stem-and-leaf plot, 262, 269
STEPWISE procedure, 334, 401
STOPOVER Option, 43
Student's t-test, 262, 268, 306ff, 348, 355, 385, 529
STYLE= option, 451
STYLE option (ODS), 208
SUBGROUP option (GCHART), 442
subsetting data sets, 97ff
SUBSTR function, 162, 516ff
SUM function, 142–143, 174, 507
summarized frequencies (PROC FREQ), 286ff
summarized two-way frequencies, 293
SUM Statement, 81
SUMVAR option (GCHART), 442
SYMBOLn = option, 503
symbols, specifying, 504
SYMPUT, 199
SYMPUT (CALL) function, 521
SYMPUT, SYMPUTN, 521
syntax for PROC statements, 123–124
system options, 137

T

TABLES statement, 218, 281ff, 290, 297
TAGSETS, 207, 479
TAGSETS.EXCELXP output destination, 207
TAN function, 511
TANH function, 511
temporary SAS data sets, 22, 54
test for location, 308
text (character) functions, 515
TIME function, 512ff
TIMEPART function, 513
tips and tricks, 11

TITLE statement, 118ff
TODAY function, 143, 513
traffic lighting (PROC PRINT), 218ff
TRANSLATE function, 516
TRANSPARANCY= option (SGPLOT), 457
transposing data sets. *See* PROC TRANSPOSE
TRANWRD function, 516
trigonometric functions, 511
TRIM, TRIMN functions, 516
truncation functions, 518ff
TRUNCOVER option, 43–46
t-test, one-sample, 306ff
t-test, paired, 315ff
t-tests. *See* Student's t-test
t-test, two-sample, 310ff, 529
TUKEY multiple comparisons, 350ff, 394
two-factor ANOVA, 369ff
two independent samples, comparing, 385ff
two-sample t-test, 310ff, 529
two-way ANOVA using, 369ff
Two-Way Tables (PROC FREQ), 290ff
TYPE= option (GCHART), 442

U

uncorrected sum of squares (USS), 267
UN covariance structure. *See* Unstructured
UNIFORM option, 142, 430
UNION ALL, SQL, 243
UNIONS in SQL, 241
University Edition (also see SAS Studio), 22, 501
UNIX, 4
unstructured (UN) covariance structure, 379, 380
UPCASE function, 156ff, 170, 235, 507
uploading files in SAS Studio, 540
upper case function, 516

V

VALUE= option (V=), 433, 435
Van der Waerden test, 385
Variable Labels, SQL, 230
variable name rules, 24
variable selection (regression), 225, 230ff, 401
variable types, 25
variance components (VC) covariance structure, 374, 379
variance selection in SQL, 230
VARIMAX option, 415ff
VAR statement, 107, 125ff
VAXIS option, 433, 436, 438, 440

VBAR3D option (GCHART), 441, 445
VBAR option (GCHART), 441ff, 457, 460
VC covariance structure. *See* Variance Components
VERIFY function, 516
viewing data, 14, 60
view, SQL, 227
viewtable, 60
vocabulary data analysis, 416ff
VREF option, 433
VSCALE option, 263

W

Washington DC crime analysis, 344–345
WAXIS option, 263
WBARLINE option, 263, 265
WEEKDAY function, 513
weighted kappa, 301–305
WEIGHT statement, 267, 286, 349
WHEN statement, SQL, 231
WHERE, SQL, 228, 235
WHERE statement, 130, 131, 204, 243, 252
WIDTH= option, 44, 436–438, 443, 446, 447, 449, 502
Wilcoxon-Mann-Whitney, 385–390, 393, 529
Windows interface. *See* SAS Windows Interface
Wireless Markup Language (WML), 207
WITH statement, 325
WML (Wireless Markup Language), 207
Word (Microsoft), 207, 209, 210
WOUTLINE option, 207, 209, 210
WPGM. *See* Enhanced editor

X

XLS and XLSX files, 15, 69, 70, 206, 207, 479
XML (Extensible Markup Language Output Destination), 207, 480

Y

Yate's chi-square, 295
YEARCUTOFF= option, 137, 512
YEAR function, 513, 523
YRDIF function, 146, 513
YYQ function, 513

Z

ZIPCITYDISTANCE function, 522
ZIPCITY function, 522
zipcode functions, 522
ZIPSTATE function, 522
z/OS, 4